Lecture Notes in Computer Science   9807

Commenced Publication in 1973
Founding and Former Series Editors:
Gerhard Goos, Juris Hartmanis, and Jan van Leeuwen

More information about this series at http://www.springer.com/series/7407

Jasmin Christian Blanchette · Stephan Merz (Eds.)

Interactive Theorem Proving

7th International Conference, ITP 2016
Nancy, France, August 22–25, 2016
Proceedings

Editors
Jasmin Christian Blanchette
Inria Nancy – Grand Est
Villers-lès-Nancy
France

Stephan Merz
Inria Nancy – Grand Est
Villers-lès-Nancy
France

ISSN 0302-9743   ISSN 1611-3349 (electronic)
Lecture Notes in Computer Science
ISBN 978-3-319-43143-7   ISBN 978-3-319-43144-4 (eBook)
DOI 10.1007/978-3-319-43144-4

Library of Congress Control Number: 2016945777

LNCS Sublibrary: SL1 – Theoretical Computer Science and General Issues

This Springer imprint is published by Springer Nature
The registered company is Springer International Publishing AG Switzerland

Preface

The International Conference on Interactive Theorem Proving (ITP) is the premier venue for publishing research in the area of logical frameworks and interactive proof assistants, ranging from theoretical foundations, technology, and implementation aspects to their applications in areas such as verifying algorithms and programs, ensuring their safety and security, or formalizing significant mathematical theories. ITP grew out of the TPHOLs conferences and ACL2 workshops organized since the early 1990s.

Previous editions of ITP took place in Edinburgh, Nijmegen, Princeton, Rennes, Vienna, and Nanjing. The seventh edition (ITP 2016) was organized by the Inria research center Nancy – Grand Est in Nancy, France, during August 22–25, 2016. In all, 55 submissions were received for ITP 2016. Each submitted paper was reviewed by at least three members of the Program Committee or external reviewers, and the Program Committee decided to accept 27 regular contributions and five rough diamonds. Viktor Kuncak, Grant Olney Passmore, and Nikhil Swamy were invited to present keynote talks at the conference. The main conference was followed by workshops dedicated to the Coq and Isabelle systems, as well as to the Mathematical Components library.

The present volume collects the scientific contributions accepted for publication at ITP 2016. It also contains abstracts of the keynote presentations.

We are very grateful to the members of the ITP Steering Committee for their guidance and advice. Our colleagues in the Program Committee and the external reviewers did an excellent job in preparing timely and helpful reviews as a basis for selecting the accepted contributions. We extend our thanks to the authors of all submitted papers and the ITP community at large, without which the conference would not exist.

The Inria research center Nancy – Grand Est, and in particular the delegate for colloquia Anne-Lise Charbonnier, provided professional support for the organization of ITP 2016. We gratefully acknowledge financial support by Aesthetic Integration, Communauté Urbaine du Grand Nancy, Microsoft Research, Région Alsace Champagne-Ardenne Lorraine, and Springer. As in previous years, Springer accepted to publish the proceedings of ITP 2016 as a volume in the LNCS series, and we would like to thank the editorial team for the very smooth interaction.

June 2016

Jasmin Christian Blanchette
Stephan Merz

Preface

Organization

Steering Committee

David Basin	ETH Zürich, Switzerland
Jasmin Christian Blanchette	Inria Nancy – Grand Est, France
Amy Felty	University of Ottawa, Canada
Panagiotis Manolios	Northeastern University, USA
César Muñoz	NASA Langley Research Center, USA
Christine Paulin-Mohring	Université Paris Sud, France
Lawrence Paulson	University of Cambridge, UK
Michael Norrish	NICTA, Australia
Tobias Nipkow	Technische Universität München, Germany
Sofiène Tahar	Concordia University, Canada
Christian Urban	King's College London, UK

Program Committee

Andrea Asperti	University of Bologna, Italy
Jeremy Avigad	Carnegie Mellon University, USA
Yves Bertot	Inria Sophia Antipolis – Méditerranée, France
Lars Birkedal	Aarhus University, Denmark
Jasmin Christian Blanchette	Inria Nancy – Grand Est, France
Adam Chlipala	MIT, USA
Nils Anders Danielsson	University of Gothenburg and Chalmers University of Technology, Sweden
Amy Felty	University of Ottawa, Canada
Herman Geuvers	Radboud University Nijmegen, The Netherlands
Georges Gonthier	Microsoft Research, UK
John Harrison	Intel, USA
Hugo Herbelin	Inria Paris, France
Cătălin Hriţcu	Inria Paris, France
Cezary Kaliszyk	University of Innsbruck, Austria
Matt Kaufmann	University of Texas at Austin, USA
Gerwin Klein	NICTA and UNSW, Australia
Xavier Leroy	Inria Paris, France
Andreas Lochbihler	ETH Zürich, Switzerland
Frédéric Loulergue	Université d'Orleans and LIFO, France
Assia Mahboubi	Inria Saclay – Île-de-France, France
Panagiotis Manolios	Northeastern University, USA
Stephan Merz	Inria Nancy – Grand Est, France

Magnus O. Myreen	Chalmers University of Technology, Sweden
Adam Naumowicz	University of Białystok, Poland
Tobias Nipkow	Technische Universität München, Germany
Michael Norrish	NICTA, Australia
Sam Owre	SRI International, USA
Christine Paulin-Mohring	Université Paris-Sud and LRI, France
Lawrence C. Paulson	University of Cambridge, UK
Andrei Popescu	University of Middlesex, UK
Gert Smolka	Saarland University, Germany
Matthieu Sozeau	Inria Paris, France
René Thiemann	University of Innsbruck, Austria
Laurent Théry	Inria Sophia Antipolis – Méditerranée, France
Andrew Tolmach	Portland State University, USA
Christian Urban	King's College London, UK
Viktor Vafeiadis	MPI-SWS, Germany

Additional Reviewers

Bahr, Patrick	Chau, Cuong Kim
Cohen, Cyril	Contejean, Evelyne
Doczkal, Christian	Felgenhauer, Bertram
Färber, Michael	Giero, Mariusz
Goel, Shilpi	Heule, Marijn J.H.
Hölzl, Johannes	Imine, Abdessamad
Immler, Fabian	Jimenez, Raul Pardo
Joosten, Sebastiaan	Krebbers, Robbert
Lammich, Peter	Li, Wenda
Mahmoud, Mohamed Yousri	Maric, Ognjen
Mehta, Mihir	Narboux, Julien
Oster, Gérald	Schäfer, Steven
Selfridge, Benjamin	Sibut-Pinote, Thomas
Spitters, Bas	Sternagel, Christian
Struth, Georg	Tan, Yong Kiam
Tuerk, Thomas	Wenzel, Makarius
Wiedijk, Freek	Yamada, Akihisa

Sponsors

Aesthetic Integration
Communauté Urbaine du Grand Nancy
Microsoft Research
Région Alsace Champagne-Ardenne Lorraine
Springer International Publishing AG

Abstracts of Keynote Presentations

Propositions as Programs, Proofs as Programs

Viktor Kuncak

École Polytechnique Fédérale de Lausanne (EPFL)

Leon is a system that (among other features) enables writing verified programs and their properties in a purely functional subset of Scala. The key specification statement in Leon is that a function satisfies its contract for all inputs. Leon proves properties and finds counterexamples using SMT solvers and an unfolding strategy for recursive functions. A newly developed link with Isabelle provides an additional safety net for soundness of the approach.

Due to Leon's unfolding mechanism, it is possible to write additional, semantically redundant, expressions that help Leon prove a theorem. We attempt to formalize this "accidental" feature of Leon. In our view, propositions, as well as proofs, are just terminating programs. This makes Leon statements and proofs (syntactically) accessible to the half a million of Scala developers. We explain some limitations of this approach in writing proof tactics and controlling the space of assumptions, suggesting that a form of reflection would provide benefits of Turing-complete tactic language without ever leaving the paradise of purely functional Scala programs.

Formal Verification of Financial Algorithms, Progress and Prospects

Grant Olney Passmore

Aesthetic Integration and University of Cambridge

Many deep issues plaguing today's financial markets are symptoms of a fundamental problem: The complexity of algorithms underlying modern finance has significantly outpaced the power of traditional tools used to design and regulate them. At Aesthetic Integration, we've pioneered the use of formal verification for analysing the safety and fairness of financial algorithms. With a focus on financial infrastructure (e.g., the matching logics of exchanges and dark pools), we'll describe the landscape, and illustrate our Imandra formal verification system on a number of real-world examples. We'll sketch many open problems and future directions along the way.

Dijkstra Monads for Free: A Framework for Deriving and Extending F*'s Effectful Semantics

Nikhil Swamy

Microsoft Research

F* is a higher-order effectful language with dependent types. It aims to provide equal support for general purpose programming (as in the ML family of languages) as well as for developing formal proofs (like other type-theory based proof assistants, e.g., Coq, Agda or Lean). By making use of an SMT solver while type-checking, F* provides automation for many routine proofs.

At the heart of F* is the manner in which effects and dependent types are combined: this presents several well-known difficulties. Our basic approach to solving these difficulties is not surprising: effectful computations are encapsulated within monad-like structures. More specifically, F* interprets effectful computations using monads of predicate transformers, so called "Dijkstra monads" that compute weakest pre-conditions for arbitrary post-conditions. These Dijkstra monads are arranged in a lattice of effect inclusions, e.g., pure computations are included within stateful ones.

In this talk, I will describe a new technique for deriving F*'s Dijkstra monad lattice by CPS'ing (with result type Prop) purely functional definitions of monads corresponding to F*'s effects. Several benefits ensue:

1. For starters, programmers are able to customize F*'s effect lattice using familiar Haskell-style monadic definitions, while gaining for each such monad a weakest pre-condition calculus suitable for Hoare-style verification of programs.
2. Next, several useful properties, e.g., monotonicity of predicate transformers, are guaranteed by the derivation, reducing the proof obligations for adding an effect to F*.
3. Third, our technique supports a mechanism to break the abstraction of a monadic effect in a controlled manner, reifying an effectful computation as its pure counterpart, and reflecting pure reasoning on the reified program back on to the effectful code.

I will also provide a general introduction to F* and its applications, notably its use in Everest, a new project to build and deploy a verified, secure implementation of HTTPS, including Transport Layer Security, TLS-1.3.

F* is open source and developed on github by researchers at Microsoft Research and Inria. For more information, visit https://fstar-lang.org.

Contents

Rough Diamonds

Regular Contributions

An Isabelle/HOL Formalisation
of Green's Theorem

Mohammad Abdulaziz[1,2(✉)] and Lawrence C. Paulson[3]

[1] Canberra Research Laboratory, NICTA, Canberra, Australia
mohammad.abdulaziz@nicta.com.au
[2] Australian National University, Canberra, Australia
[3] Computer Laboratory, University of Cambridge, Cambridge, England

Abstract. We formalise a statement of Green's theorem in Isabelle/HOL, which is its first formalisation to our knowledge. The theorem statement that we formalise is enough for most applications, especially in physics and engineering. An interesting aspect of our formalisation is that we neither formalise orientations nor region boundaries explicitly, with respect to the outwards-pointing normal vector. Instead we refer to equivalences between paths.

1 Introduction

The *Fundamental Theorem of Calculus* (FTC) is a theorem of immense importance in differential calculus and its applications, relating a function's differential to its integral. Having been conceived in the seventeenth century in parallel to the development of infinitesimal calculus, more general forms of the FTC have been developed, the most general of which is referred to as the *General Stokes' Theorem*.

A generalisation of the FTC or a special case of the General Stokes' Theorem in $\mathbb{R}^2$ was published in 1828 by George Green [2], with applications to electromagnetism in mind. This generalisation is referred to as Green's Theorem, and it is the main topic of this work. In modern terms the theorem can be stated as follows:

Theorem 1. *Given a domain D with an "appropriate" positively oriented boundary ∂D, and a field F, with components F_x and F_y "appropriately" defined on D, the following identity holds:*

$$\int_D \frac{\partial F_y}{\partial x} - \frac{\partial F_x}{\partial y} \; dxdy = \oint_{\partial D} F_x dx + F_y dy,$$

NICTA is funded by the Australian Government through the Department of Communications and the Australian Research Council through the ICT Centre of Excellence Program.

J.C. Blanchette and S. Merz (Eds.): ITP 2016, LNCS 9807, pp. 3–19, 2016.
DOI: 10.1007/978-3-319-43144-4_1

where the left hand side is a double integral and the right hand side is a line integral[1] in $\mathbb{R}^2$.

Many statements of Green's theorem define with varying degrees of generality what is an appropriate boundary (i.e. the geometrical assumptions), and what is an appropriate field (i.e. the analytic assumptions). This mainly depends on how geometrically sophisticated the proof is, and the underlying integral. The prevalent text book form of Green's theorem asserts that, geometrically, the region can be divided into *elementary regions* and that, analytically, the field is continuous and has continuous partial derivatives throughout the region. Also, usually the underlying integral is a Riemann integral.

Despite this being enough for most applications, more general forms of the theorem have been proved in the analysis literature. Although a full literature review is not appropriate here, we present some examples of very general formulations of Green's theorem. For example, Michael [7] proves a statement of the theorem that generalises the geometrical assumptions, where it only assumes that the region has a rectifiable cycle as its boundary. Jurkat et al. [6] prove a statement of the theorem with very general analytic assumptions on the field. They only assume that the field is continuous in the region, and that the *total* derivative of the field exists in the region except for a σ_1-finite set of points in the region. Then, they also derive a very general form of Cauchy's integral theorem.

Having Green's theorem formalised is significant because of its wide range of applications, too many to list in full. There are applications in physics (electrodynamics, mechanics, etc.) and engineering (deriving moments of inertia, hydrodynamics, the basis of the planimeter, etc.). In mathematics, Green's theorem is a fundamental result: it can be used to derive Cauchy's integral theorem.

We formalise a statement of Green's theorem for Henstock-Kurtzweil gauge integrals in the interactive theorem prover Isabelle/HOL [8]. Our work builds on the work of Hölzl et al., where we use the Isabelle/HOL multivariate analysis library [5] and the probability library [4]. We also build on the second author's porting of John Harrison's complex analysis library [3]. Our formalisation does not strictly follow a certain proof, but it was inspired by Zorich and Cooke [11], Spivak [10] and Protter [9].

2 Basic Concepts and Lemmas

In this section we discuss the basic lemmas we need to prove Green's theorem. However, we need to firstly discuss two basic definitions needed to state the theorem statement: *line integrals* and *partial derivatives*. Definitions of both of those two concepts are ubiquitous in the literature, nonetheless, we had to adapt

---

[1] This line integral can be physically interpreted as the work done by F on the ∂D, making this statement a special case of the 3-dimensional Kelvin-Stokes' theorem. If the line integral is replaced with $\oint_{\partial D} F_x dx - F_y dy$, it can be interpreted as the flux of F through ∂D and the theorem would be the 2-dimensional special case of the divergence theorem.

them to be defined on the Euclidean spaces type class in the Isabelle multivariate analysis library.

We define the line integral of a function F on the parameterised path γ as follows:

Definition 1. *Line Integral*

$$\int_\gamma F\lfloor_B = \int_0^1 \Sigma_{b\epsilon B}((F(\gamma(t))\cdot b)(\gamma'(t)\cdot b))dt$$

A difference in our definition is that we add the argument B, a set of vectors, to which F and γ, and accordingly the line integral are projected. The reasons for adding the B argument will become evident later.. Above, $\cdot$ denotes the inner product of two vectors. Formally, the definition is:

```
definition line_integral::
  "('a::euclidean_space ⇒ 'a ) ⇒ 'a set ⇒ (real ⇒ 'a) ⇒ real"
  where "line_integral F B γ =
    integral {0 .. 1}
      (λt. ∑b∈B. (F(γ t) · b) *
                   (vector_derivative γ (at t within {0..1}) · b))"
definition line_integral_exists where
  "line_integral_exists F B γ =
    ((λt. ∑b∈B. F (γ t) · b *
                  (vector_derivative γ (at t within {0..1}) · b))
      integrable_on {0..1})"
```

Note that `integral` refers to the Henstock-Kurzweil gauge integral implementation in Isabelle/HOL library. As one would expect, the line integral distributes over unions of sets of vectors and path joins.

```
lemma line_integral_sum_gen:
    assumes "finite B"
            "line_integral_exists F B1 γ"
            "line_integral_exists F B2 γ"
            "B1 ∪ B2 = B" "B1 ∩ B2 = {}"
      shows "line_integral F B γ =
              (line_integral F  B1 γ) + (line_integral F B2 γ)"

lemma line_integral_distrib:
    assumes "line_integral_exists f B γ1"
            "line_integral_exists f B γ2"
            "valid_path γ1" "valid_path γ2"
      shows "line_integral f B (γ1 +++ γ2) =
              line_integral f B γ1 + line_integral f B γ2"
```

The line integral also admits a transformation equivalent to integration by substitution. This lemma applies to paths where all components are defined as a function in terms of one component orthogonal to all of them. It is a critical lemma for proving Green's theorem for "elementary" regions (to be defined later).

lemma *line_integral_on_pair_path:*
 fixes *F::"('a::euclidean_space) ⇒ ('a)"*
       g::"real ⇒ 'a"
       γ::"(real ⇒ 'a)"
       b::'a
 assumes *"b ∈ Basis"*
           "∀x. g(x) · b = 0"
           *"γ = (λt. f(t) *ᵣ b + g(f(t)))"*
           "γ C1_differentiable_on {0 .. 1}"
           "continuous_on (f ` {0..1}) g"
           "(pathstart γ) · b ≤ (pathfinish γ) · b"

           "continuous_on (path_image γ) (λx. F x · b)"
 shows *"(line_integral F {b} γ)*
           = integral (cbox ((pathstart γ) · b)
                       ((pathfinish γ) · b))
                   *(λf_var. (F (f_var *ᵣ b + g(f_var)) · b))"*

Partial derivatives are defined on the Euclidean space type class implemented in Isabelle/HOL. For a function F defined on a Euclidean space, we define its partial derivative to be w.r.t. the change in the magnitude of a component vector b of its input. At a point a, the partial derivative is defined as:

Definition 2. *Partial Derivative*

$$\left.\frac{\partial F(v)}{\partial b}\right|_{v=a} = \left.\frac{dF(a + (x - a \cdot b)b)}{dx}\right|_{x=a \cdot b}$$

Again, this definition is different from the classical definition in that the partial derivative is w.r.t. the change of magnitude of a vector rather than the change in one of the variables on which F is defined. However, our definition is similar to the classical definition of a partial derivative, when b is a base vector. Formally we define it as:

definition *has_partial_vector_derivative::*
"('a::euclidean_space ⇒ 'b::euclidean_space) ⇒ 'a ⇒ 'b ⇒ 'a
⇒ bool" **where**
"has_partial_vector_derivative F b F' a
  *= ((λx. F((a - ((a · b) *ᵣ b)) + x *ᵣ b))*
     has_vector_derivative F') (at (a · b))"

definition *partially_vector_differentiable* **where**
"partially_vector_differentiable F b p
   = (∃F'. has_partial_vector_derivative F b F' p)"

definition *partial_vector_derivative::*
"('a::euclidean_space ⇒ 'b::euclidean_space) ⇒ 'a ⇒ 'a ⇒ 'b"
where *"partial_vector_derivative F b a*
       = (vector_derivative
               *(λx. F((a - ((a · b) *ᵣ b)) + x *ᵣ b)) (at (a · b)))"*

The following FTC for the partial derivative follows from the FTC for the vector derivative that is proven in Isabelle/HOL analysis library.

lemma *fundamental_theorem_of_calculus_partial_vector_gen:*
  fixes *k1 k2::"real"*
        F::"('a::euclidean_space ⇒ 'b::euclidean_space)"
        b::"'a"
        F_b::"('a::euclidean_space ⇒ 'b)"
  assumes *"k1 ≤ k2"*
      "b · b = 1"
      "c · b = 0 "
      "∀ p ∈ D. has_partial_vector_derivative F b (F_b p) p"
      "{v. ∃x. k1 ≤ x ∧ x ≤ k2 ∧ v = x *R b + c} ⊆ D"
  shows *"((λx. F_b(x *R b + c)) has_integral*
                F(k2 *R b + c) - F(k1 *R b + c)) (cbox k1 k2)"

Given these definitions and basic lemmas, we can now start elaborating on our formalisation of Green's theorem. The first issue is how to formalise $\mathbb{R}^2$. We use pairs of *real* to refer to members of $\mathbb{R}^2$, where we define the following *locale* that fixes the base vector names:

locale *R2 =*
    fixes *i j*
    assumes *i_is_x_axis: "i = (1::real,0::real)"* **and**
          j_is_y_axis: "j = (0::real, 1::real)"

Proofs of Green's theorem usually start by proving "half" of the theorem statement for every type of "elementary regions" in $\mathbb{R}^2$. These regions are referred to as Type I, Type II or Type III regions, defined below.[2]

Definition 3. *Elementary Regions*
A region D (modelled as a set of real pairs) is Type I iff there are C^1 smooth functions g_1 and g_2 such that for two constants a and b:

$$D = \{(x,y) \mid a \le x \le b \wedge g_2(x) \le y \le g_1(x)\}.$$

Similarly D would be called type II iff for g_1, g_2, a and b

$$D = \{(x,y) \mid a \le y \le b \wedge g_2(y) \le x \le g_1(y)\}.$$

Finally, a region is of type III if it is both of type I and type II.

To prove Green's theorem a common approach is to prove the following two "half" Green's theorems, for any regions D_x and D_y that are type I and type II, respectively, and their positively oriented boundaries:

$$\int_{D_x} -\frac{\partial(F_i)}{\partial j} \, dx dy = \int_{\partial D_x} F|_{\{i\}},$$

---

[2] Using elementary regions that are bounded by C^1 smooth functions is as general as using piece-wise smooth functions because it can be shown that the latter can be divided into regions of the former type (see [9]).

and

$$\int\limits_{D_y} \frac{\partial(F_j)}{\partial i}\, dxdy = \int\limits_{\partial D_y} F|_{\{j\}}.$$

Here i and j are the base vectors while F_i and F_j are the x-axis and y-axis components, respectively, of F. However, the difference in the expressions for the type I and type II regions is because of the asymmetry of the x-axis and the y-axis w.r.t. the orientation. We refer to the top expression as the x-axis Green's theorem, and the bottom one as the y-axis Green's theorem. Below is the statement of the x-axis Green's theorem for type I regions as we have formalised it in Isabelle/HOL. For the boundary, we model its paths explicitly as functions of type `real ⇒ (real * real)`, where $\gamma 1, \gamma 2, \gamma 3$ and $\gamma 4$ are the bottom, right, top and left sides, respectively.

lemma `Greens_thm_type_I`:
fixes `F::"((real*real) ⇒ (real * real))"`
          `γ1 γ2 γ3 γ4 ::"(real ⇒ (real * real))"`
          `a::"real" b::"real"`
          `g1::"(real ⇒ real)" g2::"(real ⇒ real)"`
assumes `"Dx = {(x,y) . x ∈ cbox a b ∧ y ∈ cbox (g2 x) (g1 x)}"`
          `"γ1 = (λt. (a + (b - a) * t, g2(a + (b - a) * t)))"`
          `"γ1 C1_differentiable_on {0..1}"`
          `"γ2 = (λt. (b, g2(b) + t  *R (g1(b) - g2(b))))"`
          `"γ3 = (λt. (a + (b - a) * t, g1(a + (b - a) * t)))"`
          `"γ3 C1_differentiable_on {0..1}"`
          `"γ4 = (λt. (a,  g2(a) + t *R (g1(a) - g2(a))))"`
          `"analytically_valid Dx (λp. F(p) · i) j"`
          `"∀x ∈ cbox a b. (g2 x) ≤ (g1 x)"`
          `"a < b"`
shows `"(line_integral F {i} γ1) + (line_integral F {i} γ2) -`
          `(line_integral F {i} γ3) - (line_integral F {i} γ4) =`
          `(integral Dx`
          `   (λa. - (partial_vector_derivative (λp. F(p) · i) j a)))"`

Proving the lemma above depends on the observation that for a path γ (e.g. $\gamma 1$ above) that is straight along a vector x (e.g. i), $\int_\gamma F|_{\{x\}} = 0$, for an F continuous on γ. (Formally, this observation follows immediately from theorem `work_on_pair_path`.) The rest of the proof boils down to an application of Fubini's theorem and the FTC to the double integral, the integral by substitution to the line integrals and some algebraic manipulation ([11, p. 238]). Nonetheless, this algebraic manipulation proved to be quite tedious when done formally in Isabelle/HOL.

However, we did not discuss the predicate `analytically_valid`, which represents the analytic assumptions of our statement of Green's theorem, to which an "appropriate" field has to conform. Firstly let 1_s be the indicator function for a set s. Then, for the x-axis Green's theorem, our analytic assumptions are that

(i) F_i is continuous on D_x
(ii) $\frac{\partial (F_i)}{\partial j}$ exists everywhere in D_x
(iii) the product $1_{D_x}(x, y)\frac{\partial (F_i)}{\partial j}(x, y)$ is Lebesgue integrable
(iv) the product $1_{[a,b]}(x)\int_{g_1(x)}^{g_2(x)} F(x, y)dy$, where the integral is a Henstock-Kurzweil gauge integral, is a borel measurable function

These assumptions vary symmetrically for the y-axis Green's theorem, so to avoid having two symmetrical definitions, we defined the predicate `analytically_valid` to take the axis (as a base vector) as an argument.

definition `analytically_valid::`
  `"'a::euclidean_space set ⇒`
   `('a ⇒ 'b::{euclidean_space, times, one, zero}) ⇒ 'a ⇒ bool"`
  where
  `"analytically_valid s F b =`
   `((∀ a ∈ s. partially_vector_differentiable F b a) ∧`
`continuous_on s F ∧`
`integrable lborel`
   `(λp. (partial_vector_derivative F i) p * indicator s p) ∧`
`(λx. integral UNIV`
       `(λy. (partial_vector_derivative F i`
                 `(y *ᴿ b + x *ᴿ (∑ b ∈(Basis - {i}). b))) *`
              `(indicator s`
                 `(y *ᴿ b + x *ᴿ (∑ b ∈(Basis - {i}). b)) )))`
 `∈ borel_measurable lborel)"`

These conditions refer to Lebesgue integrability and to measurability because we use Fubini's theorem for the Lebesgue integral in Isabelle/HOL's probability library to derive a Fubini like result for the Henstock-Kurzweil integral. Proving Fubini's theorem for the gauge integral would allow for more general analytic assumptions. However, the rest of our approach would still be valid.

We prove the y-axis Green's theorem for type II regions similarly, where this is its conclusion.

shows `"(line_integral F {j} γ1) + (line_integral F {j} γ2) -`
        `(line_integral F {j} γ3) - (line_integral F {j} γ4) =`
        `(integral Dy`
           `(partial_vector_derivative (λp. F(p) · j) i))"`

3   More General Structures

Now that we described some of the basic definitions and how to derive Green's theorem for elementary regions, the remaining question is how to prove the theorem for more general regions. As we stated earlier in the introduction, a lot of the text book proofs of Green's theorem are given for regions that can be divided into elementary regions. It can be shown that any *regular* region can be divided into elementary regions [9,11]. Regular regions (see their definition in [9, p. 235]), are enough for a lot of applications, especially practical applications in physics and engineering.

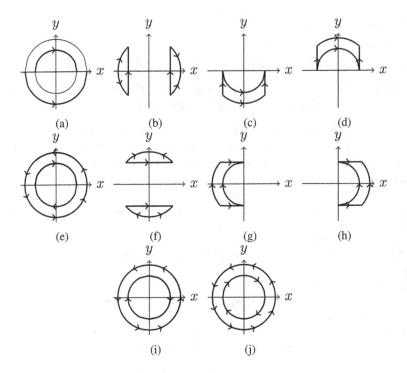

Fig. 1. An annulus and its partitioning in type I and type II regions. In this figure every 1-cube (i.e. path) is represented with an arrow whose direction is the same as the orientation of the 1-cube. (a) The positively oriented boundary of the annulus. (b), (c) and (d) The members of a type I partitioning of the annulus. (e) A 1-chain that includes all the horizontal boundaries in the type I partition. (f), (g) and (h) The members of a type II partitioning of the annulus. (i) A 1-chain that includes all the vertical boundaries in the type II partition. (j) A common subdivision of the chains in (e) and (i).

In this section we describe how we prove Green's theorem for regions that can be divided both into type I regions and type II regions *only* using vertical and horizontal edges, respectively. We believe that for most practical purposes, the additional assumption that the division is done only by vertical and horizontal edges is equivalent to assuming just the existence of type I and type II divisions. Indeed, we conjecture that the additional constraints do not lead to any loss of generality, however, we would not pursue the proof of this claim in the current paper.

Figure 1 shows an example of a region and its type I and type II partitions. In this example, some of the elementary regions appear to have a missing edge. This is because the type I or the type II partitioning induced a *one-point path*: a function mapping the interval $[0, 1]$ to a single point in $\mathbb{R}^2$. For instance, the left edge in the left 1-chain in 1b is a point on the x-axis.

3.1   Chains and Cubes

For tackling more general regions and their boundaries we use the concepts of cubes and chains [10, Chap. 8]. One use of cubes is to represent parameterisable (sub)surfaces (regions in $\mathbb{R}^2$ and paths in our case). A k-dimensional such surface embedded in $\mathbb{R}^n$ is represented by a function whose domain is a space homeomorphic to $\mathbb{R}^k$ and whose codomain is $\mathbb{R}^n$. Roughly speaking, we model cubes as functions and chains as sets of cubes. We use the existing Isabelle/HOL formalisation of paths, where we model 1-cubes as functions defined on the interval *{0..1}*. We model a 1-chain as a set of pairs of *int* (coefficients) and 1-cubes. For example, the following definition shows the lifting of the line integrals to 1-chains.

definition *one_chain_line_integral* *::*
  *"((real*real) => (real *real)) => ((real*real) set)* $\Rightarrow$
    *(int * (real* $\Rightarrow$ *real*real)) set* $\Rightarrow$ *real"* **where**
"one_chain_line_integral F b C =
      *($\sum$ (k,γ)$\in$C. k * (line_integral F b γ))"*

We extend the way we model 1-cubes to model 2-cubes, which we model as functions of type *(real * real* $\Rightarrow$ *real * real)* defined on the interval *{ (0, 0) .. (1, 1) }*.

definition *cubeImage::*
  *"(real*real* $\Rightarrow$ *real*real)* $\Rightarrow$ *((real*real) set)"* **where**
"cubeImage twoC = (twoC ' (cbox (0,0) (1,1)))"

The orientation of the boundary of a 2-cube (a 1-chain) is taken to be counterclockwise. A 1-cube is given the coefficient -1 if the path's direction is against the counter-clockwise boundary traversal order, otherwise it is given the coefficient 1. Formally this is defined as follows:

definition *horizontal_boundary* *::*
  *"(real*real* $\Rightarrow$ *real*real)* $\Rightarrow$ *(real* * *(real* $\Rightarrow$ *real*real)) set"*
  where *"horizontal_boundary twoC =*
        { (1, (λx. twoC(x,0))), (-1, (λx. twoC(x,1)))}"

definition *vertical_boundary* *::*
  *"(real*real* $\Rightarrow$ *real*real)* $\Rightarrow$ *(real* * *(real* $\Rightarrow$ *real*real)) set"*
  where *"vertical_boundary twoC =*
        { (-1, (λy. twoC(0,y))), (1, (λy. twoC(1,y)))}"

definition *boundary* **where**
"boundary twoC =
      horizontal_boundary twoC $\cup$ *vertical_boundary twoC"*

We follow the convention and define the 2-cubes in such a way that the top and left edges are against the counter-clockwise orientation (e.g. see the 2-cube in Fig. 1c). Accordingly both the left and top edges take a -1 coefficient in the 1-cube representation. Defining 2-cubes in that way makes it easier to define predicates identifying type I and type II 2-cubes, as follows.

definition `typeI_twoCube ::`
`"((real * real) ⇒ (real * real)) ⇒ bool"` **where**
`"typeI_twoCube twoC ⟷`
`   (∃ (a::real) (b::real) (g1::real⇒real) (g2::real ⇒ real).`
`    a < b ∧ (∀x ∈ {a..b}. g2 x ≤ g1 x) ∧`
`    twoC =`
`       (λ(t1,t2). ((1-t1)*a + t1*b,`
`                   (1 - t2) * (g2 ((1-t1)*a + x*b)) +`
`                   t2 * (g1 ((1-t1)*a + t1*b))))∧`
`    g1 C1_differentiable_on {a .. b} ∧`
`    g2 C1_differentiable_on {a .. b})"`

Although we do not render it here, an analogous predicate is defined for type II 2-cubes. We also require that all 2-cubes conform to the following predicate:

definition `valid_two_cube` **where**
`"valid_two_cube twoC = (card (boundary twoC) = 4)"`

This predicate filters out cases where 2-cubes have either: (i) right and top edges that are both one point paths, or (ii) left and bottom edges that are both one point paths. Although this assumption potentially leaves our theorems less general regarding some corner cases, it makes our computations much smoother. After defining these concepts on 2-cubes, we derive the following statement of x-axis Green's theorem (and its equivalent y-axis Green's theorem) in terms of 2-cubes.

`GreenThm_typeI_twoCube:`
 assumes `"typeI_twoCube twoC"`
`    "valid_two_cube twoC"`
`    "analytically_valid (cubeImage twoC) (λx. (F x) · i) j"`
 shows `"integral (cubeImage twoC)`
`            (λa. - partial_vector_derivative (λp. F p · i) j  a)`
`         = one_chain_line_integral F {i} (boundary twoC)"`

Although we anticipated that proving this theorem would be a straightforward unfolding of definitions and usage of `GreenThm_typeI`, it was a surprisingly long and tedious proof that took a few hundred lines.

For 2-chains, we model them as sets of 2-cubes, which suits our needs in working in the context of Green's theorem. We define the boundary of a 2-chain as follows:

definition `two_chain_boundary ::`
`  "(((real*real) ⇒ (real*real))) set) ⇒`
`  ((real* (real ⇒ (real*real))) set)"` **where**
`"two_chain_boundary twoChain  = ⋃ (boundary ` twoChain)"`

We similarly defined the functions `two_chain_horizontal_boundary` and `two_chain_vertical_boundary`. We also lift the double integral to 2-chains as follows.

definition `two_chain_integral::`
`"(((((real*real) ⇒ (real*real)))) set) ⇒`
`((real*real)⇒(real)) ⇒ real" `**where**
`"two_chain_integral twoChain F =`
`        (∑ (C)∈twoChain. (integral (cubeImage C) F))"`

Lastly, to smoothe our computations on integrals over 2-chains and their boundaries, we require that a 2-chain: (i) only has valid 2-cubes members, (ii) edges of different 2-cubes only coincide if they have opposite orientations, and (iii) different 2-cubes have different images. These requirements are formally defined in the following predicate:

definition `valid_two_chain` **where**
`"valid_two_chain twoChain =`
`  ((∀ twoCube ∈ twoChain. valid_two_cube twoCube) ∧`
`   (∀ twoCube1 ∈ twoChain.`
`     ∀ twoCube2 ∈ twoChain.`
`       twoCube1 ≠ twoCube2 ⟶`
`             boundary twoCube1 ∩ boundary twoCube2 = {}) ∧`
`   inj_on cubeImage twoChain)"`

Given these definitions on 2-chains, we lift our x-axis Green's theorem from 2-cubes to 2-chains, as shown in the following statement.

lemma `GreenThm_typeI_twoChain:`
  assumes `"∀ twoCube ∈ twoChain. typeI_twoCube twoCube"`
          `"valid_two_chain twoChain"`
          `"∀ twoC ∈ twoChain.`
            `analytically_valid (cubeImage twoC) (λx. (F x) · i) j"`
  shows `"two_chain_integral twoChain`
            `(λa. - partial_vector_derivative`
                        `(λx. (F x) · i) j a) =`
        `one_chain_line_integral F {i}`
                    `(two_chain_boundary twoChain)"`

A similar lemma can be derived for a 2-chain whose members are all of Type II (with the obvious consequences of orientation asymmetry).

After proving the x-axis and y-axis Green's theorems, the next step is, after algebraic and analytic manipulation, to add the line integral sides of the x-axis Green's theorem to its counterpart in the y-axis theorem and similarly add the double integrals of both theorems. Given `GreenThm_typeI_twoChain` and its type II equivalent, we can sum up both sides of the equalities in the conclusion and get Green's theorem in terms of 2-chains and their boundaries. However, the main goal of the paper is to obtain the theorem directly for a region and its boundary, given that the region can be *vertically* sliced intro regions of type I and *horizontally* sliced into regions of type II.

The first (and easier) part in proving this is to prove the equivalence of the double integral on a region and the integral on a 2-chain that divides that region. Before deriving such a theorem we generalised the notion of `division_of`, defined in Isabelle/HOL's multivariate analysis library, to work when the division is not constituted of rectangles.

definition `gen_division` **where**
```
"gen_division s S =
        (finite S ∧ (⋃S = s) ∧
         (∀u ∈ S. ∀t ∈ S. u ≠ t ⟶ negligible (u ∩ t)))"
```

Then we show the following equivalence:

lemma `two_chain_integral_eq_integral_divisable:`
assumes `"∀ twoCube ∈ twoChain. F integrable_on cubeImage twoCube"`
        `"gen_division s (cubeImage ` twoChain)"`
        `"valid_two_chain twoChain"`
shows `"integral s F = two_chain_integral twoChain F"`

The other part concerning the line integrals, proved to be trickier, and we will explain it in the next section.

3.2   Dealing with Boundaries

What remains now is to prove an equivalence between the line integral on the 1-chain boundary of the region under consideration and the line integral on the 1-chain boundary of the region's elementary division (i.e. the 2-chain division of the region). The classical approach depends on that the line integrals on the introduced boundaries will cancel each other, leaving out the line integral on the region's original boundary. For example, the vertical-straight-line paths in Figs. 1b, c and d, are the introduced boundaries to obtain the type I division of the annulus. In this example, the line integrals on the introduced vertical-straight-line paths will cancel each other because of their opposite orientations, thus, leaving out the integral on the original boundary.

To prove this formally, the classical approach needs to define what is a positively oriented boundary, which requires an explicit definition of the boundary of a region, and also defining the exterior normal of the region (examples of this approach can be found in [1,9]). However, we use a different approach that does not depend on these definitions and avoids a lot of the resulting geometrical and analytic complications. Our approach depends on two observations:

O1: if a path γ is straight along a vector x, then $\int_{\gamma} F \lfloor_{\{x\}} = 0$, for an F continuous on γ.

O2: partitioning the region in type I/type II regions was done by introducing only vertical/horizontal boundaries.

For a type I 2-chain division of a region, consider the 1-chain γ_x, that: (i) includes *all* the horizontal boundaries of the dividing 2-chain, and (ii) includes *some* subpaths of the vertical boundaries of the dividing 2-chain (call this condition Cx). Based on O1, the line integral on the vertical edges in the 1-chain boundary of the type I division and accordingly γ_x, projected on i, will be zero. Accordingly we can prove the x-axis Green's theorem for γ_x. Formal statements of Cx (formally: `only_vertical_division`), and the consequence of a 1-chain conforming to it are:

definition `only_vertical_division` **where**
`"only_vertical_division one_chain two_chain =`
`(∃V. finite V ∧`
   `(∀ (k, γ) ∈ V.`
      `(∃ (k', γ') ∈ two_chain_vertical_boundary two_chain.`
        `(∃ a ∈ {0 .. 1}.`
         `∃ b ∈ {0..1}.`
          `a ≤ b ∧ subpath a b γ' = γ))) ∧`
    `one_chain = V ∪ (two_chain_horizontal_boundary two_chain))"`

lemma `GreenThm_typeI_divisible_region_boundary_gen`:
  assumes `"valid_typeI_division s twoChain"`
       `"∀twoC ∈ twoChain.`
         `analytically_valid (cubeImage twoC) (λa. F(a) · i) j"`
       `"only_vertical_division γ_x  twoChain"`
  shows `"integral s`
              `(λx. - partial_vector_derivative (λa. F(a) · i) j x)`
          `= one_chain_line_integral F {i} γ_x "`

In the previous lemma, `valid_typeI_division` is an abbreviation that, for a region and a 2-chain, means that the 2-chain constitutes only valid type I cubes and that this 2-chain is a division of the given region.

An analogous condition, Cy, for a type II partitioning asserts that the 1-chain includes *all* the vertical boundaries of the dividing 2-chain, and includes *some* subpaths of the horizontal boundaries of the dividing 2-chain. For Cy, we formally define the predicate (`only_horizontal_division`) and prove an analogous theorem for type II partitions, with the obvious changes in the conclusion.

From the second observation, O2, we can conclude that there will always be

- a 1-chain, γ_x, whose image is the boundary of the region under consideration and that satisfies Cx for the type I division.
- a 1-chain, γ_y, whose image is the boundary of the region under consideration and that satisfies Cy for the type II division, where it is *not* necessary that $\gamma_x = \gamma_y$.

Figure 1e and i show two 1-chains that satisfy Cx and Cy for the type I and type II divisions of the annulus. Notice that in this example, those two 1-chains are *not equal* even though they have the same *orientation* and *image*.

Now, if we can state and formalise the equivalence between γ_x and γ_y, and that this equivalence lifts to equal line integrals, we can obtain Green's theorem in terms of the region, which is our goal. One way to formalise path equivalence is to explicitly define the notion of orientation. Then the equivalence between γ_x and γ_y can be characterised by their having similar orientations and images. An advantage of this approach is that it can capture equivalence in path orientations regardless of the path image.

However, we do not need this generality in the context of proving the equivalence of 1-chains that have the same image and orientation, especially that this generality will cost a lot of analytic and geometric complexities to be formalised. Instead we choose to foramlise the notion of equivalence in terms of having a *common subdivision*. For example the 1-chain shown in Fig. 1j is a subdivision of each of the 1-chains in Fig. 1e and Fig. 1i as well as the original boundary 1-chain in Fig. 1a, i.e. a common subdivision between the three 1-chains. We now formally define the concept of a common subdivision between 1-chains, where we mainly focus on "boundary" 1-chains, defined as follows.

definition `boundary_chain` **where**
`"boundary_chain s = (∀ (k, γ) ∈ s. k = 1 ∨ k = -1)"`

First, we lift the `path_join` operator defined in the Isabelle/HOL multivariate analysis library, to act on 1-chains ordered into lists as follows.

fun `rec_join` `::`
`"(int *(real ⇒ (real*real))) list ⇒ (real ⇒ (real * real))"`
where
`  "rec_join [] = (λx.0)" |`
`  "rec_join ((k, γ) # xs) =`
`      (if (k = 1) then γ else reversepath γ)   +++ (rec_join xs)"`

To use the theory for paths developed in the multivariate analysis library, we need the joined chains to be piece-wise C^1 smooth in the sense that is defined in that library (`valid_path`). A necessary condition for a path to be valid, is that the ending point of every piece of that path to be the starting point of the next. Accordingly we define the following predicate for the validity 1-chains ordered into lists.

fun `valid_chain_list` **where**
`  "valid_chain_list [] = True" |`
`  "valid_chain_list ((k, γ)#l) =`
`   (( if (k = 1) then (pathfinish γ = pathstart (rec_join l))`
`      else (pathfinish (reversepath γ) = pathstart (rec_join l)))`
`   ∧ valid_chain_list l)"`

Based on those concepts we now define what it means for a 1-chain to be a subdivision of a path, which is a straightforward definition.

definition `chain_subdiv_path::`
`  "(real ⇒ real * real) ⇒ ((int * (real ⇒ real * real)) set)`
`  ⇒ bool"` **where**
`  "chain_subdiv_path γ one_chain`
`        = (∃l. set l = one_chain ∧ distinct l ∧ rec_join l = γ ∧`
`                valid_chain_list l)"`

We call a 1-chain γ, a subdivision of another 1-chain η, if one can map every cube in η to a sub-chain of γ that is a subdivision of it. Formally this is defined as follows:

definition `chain_subdiv_chain` **where**
```
"chain_subdiv_chain one_chain1 subdiv =
  (∃ f. ((⋃ (f ' one_chain1)) = subdiv) ∧
        (∀ (k,γ)∈one_chain1.
          if k = 1 then
            chain_subdiv_path γ (f(k, γ))
          else
            chain_subdiv_path (reversepath γ) (f(k,γ))) ∧
        (∀ p∈one_chain1.
          ∀p'∈one_chain1. p ≠ p' ⟶ f p ∩ f p' = {}) ∧
        (∀ x ∈ one_chain1. finite (f x)))"
```

After proving that each of the previous notions of equivalence implies equality of line integrals, we define equivalence of 1-chains in terms of having a common subdivision, and prove that it implies equal line integrals. We define it as having a boundary 1-chain that is a subdivision for each of the 1-chains under consideration. Formally this definition and the equality of line integrals that it implies are as follows:

definition `common_boundary_sudivision_exists` **where**
```
"common_boundary_sudivision_exists one_chain1 one_chain2 =
   (∃ subdiv. chain_subdiv_chain one_chain1 subdiv ∧
              chain_subdiv_chain one_chain2 subdiv ∧
              (∀ (k, γ) ∈ subdiv. valid_path γ)∧
              (boundary_chain subdiv))"
```

lemma `common_subdivision_imp_eq_line_integral`:
  assumes `"(common_boundary_sudivision_exists`
                             `one_chain1 one_chain2)"`
        `"boundary_chain one_chain1"`
        `"boundary_chain one_chain2"`
        `"∀ (k, γ)∈one_chain1. line_integral_exists F B γ"`
        `"finite one_chain1"`
        `"finite one_chain2"`
  shows `"one_chain_line_integral F B one_chain1 =`
             `one_chain_line_integral F B one_chain2"`
        `"∀ (k, γ)∈one_chain2. line_integral_exists F B γ"`

Based on this lemma, finally, we prove the following statement of Green's theorem.

lemma `GreenThm_typeI_typeII_divisible_region:`
assumes `"valid_typeI_division s twoChain_typeI"`
`      "valid_typeII_division s twoChain_typeII"`
`      "∀ twoC ∈ twoChain_typeI.`
`         analytically_valid (cubeImage twoC) (λx. (F x) · i) j"`
`      "∀ twoC ∈ twoChain_typeII.`
`         analytically_valid (cubeImage twoC) (λx. (F x) · j) i"`
`      "only_vertical_division γ_x twoChain_typeI"`
`      "boundary_chain γ_x"`
`      "only_horizontal_division γ_y twoChain_typeII"`
`      "boundary_chain γ_y"`
`      "common_boundary_sudivision_exists γ_x γ_y"`
shows `"integral s`
`           (λx. partial_vector_derivative (λx. (F x) · j) i x -`
`               partial_vector_derivative (λx. (F x) · i) j x)`
`           = one_chain_line_integral F {i, j} γ_x"`
`      "integral s`
`           (λx. partial_vector_derivative (λx. (F x) · j) i x -`
`               partial_vector_derivative (λx. (F x) · i) j x)`
`           = one_chain_line_integral F {i, j} γ_y"`

This theorem does not require the 1-chains γ_x and γ_y to have as their image exactly the boundary of the region. However, of course it applies to the 1-chains if their image is the boundary of the region. Accordingly it fits as Green's theorem for a region that can be divided into elementary regions just by vertical and horizontal slicing.

It is worth noting that although this statement seems to have a lot of assumptions, its analytic assumptions regarding the field are strictly more general than the those in [9,11], where they retuire the field and both of its partial derivatives to be continuous in the region. For the geometric assumptions, on the other hand, we have two extra requirements: the type I and type II divisions should be obtained using only vertical slicing and only horizontal slicing, respectively.

4 Conclusion and Future Work

We formalised a statement of Green's theorem that is enough for a lot of practical applications. Theory and concepts we developed here can be used in proving more general statements of Green's theorem [6,7]. Most such proofs depend on approximating both line and double integrals on a region by corresponding integrals on a region that can be divided into elementary regions. An interesting aspect of our work is that we avoided defining the region's boundary and its orientation explicitly. We did so by assuming that the division was done by inserting only vertical edges for the type I division, and only horizontal edges for the type II division. We claim that this added condition on the division represents no loss of generality, and intend to prove this claim in the future.

Isabelle Notation and Availability. All blocks starting with isabelle keywords: **lemma, definition, fun** have been generated automatically using

Isabelle/HOL's LATEX pretty-printing utility. Sometimes we have edited them slightly to improve readability, but the full sources are available online.[3]

Acknowledgement. This research was supported in part by an Australian National University - International Alliance of Research Universities Travel Grant and by an Australian National University, College of Engineering and Computer Science Dean's Travel Grant Award. Also, the first author thanks Katlyn Quenzer for helpful discussions.

References

1. Federer, H.: Geometric Measure Theory. Springer, Heidelberg (2014)
2. Green, G.: An essay on the application of mathematical analysis to the theories of electricity and magnetism (1828)
3. Harrison, J.: Formalizing basic complex analysis. In: From Insight to Proof: Festschrift in Honour of Andrzej Trybulec. Studies in Logic, Grammar and Rhetoric, vol. 10(23), pp. 151–165 (2007)
4. Hölzl, J., Heller, A.: Three chapters of measure theory in Isabelle/HOL. In: van Eekelen, M., Geuvers, H., Schmaltz, J., Wiedijk, F. (eds.) ITP 2011. LNCS, vol. 6898, pp. 135–151. Springer, Heidelberg (2011)
5. Hölzl, J., Immler, F., Huffman, B.: Type classes and filters for mathematical analysis in Isabelle/HOL. In: Blazy, S., Paulin-Mohring, C., Pichardie, D. (eds.) ITP 2013. LNCS, vol. 7998, pp. 279–294. Springer, Heidelberg (2013)
6. Jurkat, W., Nonnenmacher, D.: The general form of Green's theorem. Proc. Am. Math. Soc. **109**(4), 1003–1009 (1990)
7. Michael, J.: An approximation to a rectifiable plane curve. J. Lond. Math. Soc. **1**(1), 1–11 (1955)
8. Nipkow, T., Paulson, L.C., Wenzel, M.: Isabelle/HOL: a proof assistant for higher-order logic, vol. 2283. Springer, Heidelberg (2002)
9. Protter, M.H.: Basic Elements of Real Analysis. Springer Science & Business Media, New York (2006)
10. Spivak, M.: A Comprehensive Introduction to Differential Geometry. Publish or Perish, Inc., University of Tokyo Press (1981)
11. Zorich, V.A., Cooke, R.: Mathematical Analysis II. Springer, Heidelberg (2004)

---

[3] bitbucket.org/MohammadAbdulaziz/isabellegeometry/.

HOL Zero's Solutions for Pollack-Inconsistency

Mark Adams[1,2(✉)]

[1] Proof Technologies Ltd, Worcester, UK
mark@proof-technologies.com
[2] Radboud University, Nijmegen, The Netherlands

Abstract. HOL Zero is a basic theorem prover that aims to achieve the highest levels of reliability and trustworthiness through careful design and implementation of its core components. In this paper, we concentrate on its treatment of concrete syntax, explaining how it manages to avoid problems suffered in other HOL systems related to the parsing and pretty printing of HOL types, terms and theorems, with the goal of achieving well-behaved parsing/printing and Pollack-consistency. Included are an explanation of how Hindley-Milner type inference is adapted to cater for variable-variable overloading, and how terms are minimally annotated with types for unambiguous printing.

1 Introduction

1.1 Overview

HOL Zero [16] is a basic theorem prover for the HOL logic [8]. It differs from other systems in the HOL family [7] primarily due to its emphasis on reliability and trustworthiness. One innovative area of its design is its concrete syntax for HOL types, term and theorems, and the associated parsers and pretty printers. These are not only important for usability, to enable the user to input and read expressions without fuss or confusion during interactive proof, but also for high assurance work, when it is important to know that what has been proved is what is supposed to have been proved.

In this paper we cover aspects of HOL Zero's treatment of concrete syntax, explaining how it manages to avoid classic pitfalls suffered in other HOL systems and achieve, or so we claim, two desirable qualities outlined by Wiedijk [15], namely well-behaved parsing/printing and Pollack-consistency.

In Sect. 2, we provide motivation for a thorough treatment of concrete syntax. In Sect. 3, we cover the concrete syntax problems suffered in other HOL systems. In Sect. 4, we explain the solutions provided in HOL Zero by its lexical syntax. In Sect. 5, we describe HOL Zero's solution for interpreting terms involving variable-variable overloading, through its variant of the Hindley-Milner type inference algorithm. In Sect. 6, we describe HOL Zero's solution for printing terms and theorems that would be ambiguous without type annotation, through its algorithm for minimal type annotation. In Sect. 7, we present our conclusions.

© Springer International Publishing Switzerland 2016
J.C. Blanchette and S. Merz (Eds.): ITP 2016, LNCS 9807, pp. 20–35, 2016.
DOI: 10.1007/978-3-319-43144-4_2

1.2   Concepts, Terminology and Notation

In [15], a theorem prover's term parser and printer are *well-behaved* if the result of parsing the output from printing any well-formed term is always the same as the original term. This can be thought of as $\forall$ tm • parse(print(tm)) = tm. A term parser is *input-complete* if every well-formed term can be parsed from concrete syntax. A term or theorem printer is *Pollack-consistent* if provably different terms or theorems can never be printed the same. For any printer, by *unambiguous printing* we mean that different internal representation can never be printed the same. Note that well-behaved parsing/printing implies input-completeness, unambiguous printing and Pollack-consistency for terms. Also note that none of these notions fully address the issue of *faithfulness*, where internal representation and concrete syntax correctly correspond. A printer that printed false as true and true as false might be Pollack-consistent but would not be faithful.

By *entity* we mean a HOL constant, variable, type constant or type variable. Note that the same entity can occur more than once in a given type or term. Two different entities are *overloaded* if they occur in the same scope and have the same name. A *syntax function* is an ML function dedicated to a particular syntactic category of HOL type or term, for constructing a type or term from components, destructing into components, or testing for the syntactic category. A *quotation* is concrete syntax for a HOL type or term that can be parser input or printer output. An *antiquotation* is an embedding of ML code within HOL concrete syntax (this is only supported by some HOL systems). A *symbolic* name is a non-empty string of symbol characters such as !, +, |, #, etc. An *alphanumeric* name is a non-empty string of letter, numeric, underscore or single-quote characters. An *irregular* name is a name that is neither symbolic nor alphanumeric. Note that the precise definitions of *symbolic* and *alphanumeric* vary between systems.

In Table 1 we summarise the concrete syntax constructs used in this paper.

Table 1. Syntactic constructs used in our examples. Constructs are listed in increasing order of binding power, so for example x + y <= z is the same as (x + y) <= z, but note that equality binds less tightly than implication in HOL4.

Construct	HOL4, HOL Light and HOL Zero	Isabelle/HOL	ProofPower HOL
logical entailment	P |- Q	Q [P]	P ⊢ Q
universal quantification	!v. P	ALL v. P	∀ v• P
existential quantification	?v. P	EX v. P	∃ v• P
lambda abstraction	\v. E	%v. P	λ v• E
implication	P ==> Q	P --> Q	P ⇒ Q
disjunction	P \/ Q	P | Q	P ∨ Q
conjunction	P /\ Q	P & Q	P ∧ Q
equality	x = y	x = y	x = y
non-equality	x <> y	x ~= y	x ≠ y
less than or equals	x <= y	x <= y	x ⩽ y
addition	x + y	x + y	x + y

2   The Need for Good Parsers and Printers

2.1   Proof Auditing

In recent years, HOL systems have been employed in large-scale high-assurance projects such as the verification of the seL4 operating system kernel [10], the verification of safety-critical avionics in the EuroFighter Typhoon [2], and the Flyspeck project formalising the Kepler Conjecture proof [6], a major result in mathematics. Projects of such importance should be independently audited to reduce the risk that they contain fundamental errors. As argued in [13], not only is it vital that the inference steps performed in a formal proof are correct, but also that the formal proof is proving what is intended to be proved. In a software verification project, there may be a large specification of required program behaviour and/or various formal statements of properties that the program must satisfy, and these should be reviewed. In a mathematics formalisation, the statement of the ultimate theorem being proved should be reviewed. The definitions of the constants used in any of these statements also need to be reviewed.

To be reviewed, these expressions need to be written in human-readable form. But how can the auditor be sure that the expressions seen in concrete syntax correctly correspond to their internal representation that gets manipulated in the formal proof? They must rely on the parsers and/or pretty printers to faithfully and unambiguously convert between the two representations. However, as with most theorem provers that support concrete syntax, the parsers and printers of HOL systems (other than HOL Zero) are known to suffer from problems such as input-incompleteness, ambiguous printing and Pollack-inconsistency, as pointed out in [15] and detailed further in Sect. 3.

As discussed in [1], the auditor can avoid the need to trust the system's parsers, as well as the need to review the project's proof scripts, by examining the system's state after the formal proof has been processed. This would normally require the full concrete syntax printers to be trusted.

Arguably, the auditor could also avoid the need to trust the full concrete syntax printers, either by using relatively-simple primitive syntax printers, or by using syntax functions to decompose expressions, or by directly viewing internal representation in ML. However, any of these measures would greatly reduce readability (for example, see how the small expression in Fig. 1 balloons up), and for non-trivial content, the process of review would itself become error-prone. It would be far better if the auditor could trust the correctness of the printers for full concrete syntax, because this would greatly simplify the review process.

Many of the known problems with printing are obscure cases. However, we do not consider it satisfactory to dismiss them as too unlikely to occur in practice. For two reasons, they are more likely than might be imagined. Firstly, because industrial-sized formal proof projects inevitably involve extending the theorem prover with bespoke automated proof routines, problematic obscure syntax that would not be used in interactive proof could quite conceivably be generated by poorly written code, for example in generating a variable name. Secondly, because formal proofs can be outsourced, as was done in Flyspeck, we

```
'!x y. x > 1 /\ y > 1 ==> x * y > 1'
```

```
'((!):(num->bool)->bool)
   (\(x:num).
      ((!):(num->bool)->bool)
        (\(y:num).
           ((==>):bool->bool->bool)
             (((/\):bool->bool->bool)
               (((>):num->num->bool) (x:num) (NUMERAL (BIT1 _0)))
               (((>):num->num->bool) (y:num) (NUMERAL (BIT1 _0))))
             (((>):num->num->bool)
               (((*):num->num->num) (x:num) (y:num))
               (NUMERAL (BIT1 _0)))))'
```

Fig. 1. Concrete and primitive syntax for the same expression in HOL Light.

cannot discount the possibility of users maliciously exploiting flaws in a pretty printer in order to cheat the system and get paid for theorems they did not really prove.

Neither do we consider it realistic that the auditor can properly rule out the possibility of subtle printer exploits by reviewing the project's proof scripts. These can run to tens or even hundreds of thousands of lines of code, and unless they adhere to a language subset that rules out exploits, they are virtually impossible to review properly in reasonable time (e.g. see [1]).

Finally, the auditor not only needs to trust the pretty printers, but also needs to know *why* they can be trusted. Thus it is preferable if their implementation is simple or has a simple architectural argument for why it is fail-safe, or ideally a formal proof of correctness. Complex code implementing a complex architecture of interaction does not help in this regard.

2.2 Usability

Another concern, quite separate from the need to review results, is usability during interactive proof. Users naturally expect not to be distracted with using convoluted mechanisms to ensure their input gets parsed to the intended internal representation, or with worries about whether a printed expression gets wrongly printed, or whether it gets ambiguously printed and wrongly interpreted by the user. In addition, users reasonably expect to always be able to parse back in printed terms to result in the same internal term. However, there is no HOL system (other than HOL Zero) that meets these expectations in basic usage.

3   Classic Problems with HOL Parsers and Printers

In this section, we attempt to enumerate all the common problems that exist in the treatment of concrete syntax, and comment on how well they are addressed in each of the four main HOL systems: HOL4 [14], Isabelle/HOL [12], HOL Light [9] and ProofPower HOL [3]. The problems relate to basic interaction with the system, in monochrome and using the system's standard character set (UTF-8 for HOL4, ASCII for Isabelle/HOL and HOL Light, extended ASCII for Proof-Power HOL). We illustrate some problems with examples, providing code for reproducing simple but seemingly absurd theorem results (thus showing Pollack-inconsistency). The reader need not understand the code, but just the concrete syntax (see Table 1) of the resulting theorem, printed at the bottom.

Some systems have facilities that can be employed to help. Both HOL4 and Isabelle/HOL support coloured syntax highlighting in printed output when used with certain GUIs, distinguishing between free variables, bound variables, constants and keywords. However, this restricts choice of GUI, complicates the trusted code base and does not help the parser parse the output. HOL4, Isabelle/HOL and ProofPower HOL support antiquotation in parsed type and term quotations, allowing entities to be constructed outside of concrete syntax. However, such expressions are quite difficult to read, and the user must know how to use syntax functions for constructing entities. Also, HOL4 and Isabelle/HOL do not print using antiquotation, and so this facility does not help the auditor.

HOL4, Isabelle/HOL and ProofPower HOL each have a type annotation printing mode for their term and theorem printers, which can provide solutions but are unset by default. HOL4's (set by the show_types flag) annotates one occurrence of each variable entity, and each instance of a polymorphic constant unless it is a function with at least one argument supplied. Isabelle/HOL's (set by the show_types flag) annotates one occurrence of each variable entity but no constants. ProofPower HOL's (set by the pp_show_HOL_types flag) annotates every occurrence of each variable but no constants.

3.1 Entities with Irregular Names

In HOL abstract syntax, there are no restrictions on the form of the name that can be given to an entity.[1] In HOL4 and Isabelle/HOL, irregular names can only be parsed by using antiquotation, and are printed incorrectly and without delimitation. In HOL Light, irregular names cannot be parsed, and are printed incorrectly without delimitation. ProofPower HOL does cater for irregular entity names, by allowing names to be enclosed within $"" delimiters, except that type variable names that do not begin with the single-quote character cannot be parsed, and entity names that begin with a double-quote character can only be parsed using antiquotation and are printed with antiquotation.

Example 1. The following theorem about natural numbers gets misleadingly printed in HOL Light due to the irregular variable name "!y. x".

```
# let x = mk_var ("x",':num') and y = mk_var ("y",':num') in
  let v = mk_var ("!y. x",':num') in
  let tm1 = list_mk_comb ('(+)',[v;y]) in
  EXISTS (mk_exists (x, mk_eq (tm1,x)), tm1) (REFL tm1);;
  val it : thm = |- ?x. !y. x + y = x
```

---

[1] Although in hol90, type variable names must start with a single-quote character.

3.2   Entities with Keyword Names

In HOL abstract syntax, there is nothing preventing an entity from having the same name as a keyword from a given system's concrete syntax. In HOL4, such entities are prefixed with $ in their identifier for both parsing and printing, and can also be distinguished from the keyword by syntax highlighting. In Isabelle/HOL, such entities can only be parsed by using antiquotation, and in printed output can only be distinguished from keywords by using coloured syntax highlighting. In HOL Light, such entities cannot be parsed, and are printed incorrectly as though they were keywords. ProofPower HOL uses $"" delimiters for entities overloaded with keywords, and does not suffer from any problems like those for irregular names because no keywords involve the single or double quote characters.

3.3   Variable-Constant Overloading

In HOL abstract syntax, constants cannot be overloaded with other constants, and type constants cannot be overloaded with other type constants.[2] However, variables may be overloaded with constants, and type variables overloaded with type constants. In HOL4 and Isabelle/HOL, such variables and type variables can only be parsed by using antiquotation, and in printed output can only be distinguished from constants and type constants by using coloured syntax highlighting. In HOL Light, such variables and type variables cannot be parsed, and are printed incorrectly as though they are the corresponding constants or type constants respectively. In ProofPower HOL, such variables and type variables can only be parsed by using antiquotation, and are printed using antiquotation.

3.4   Variable-Variable Overloading

HOL abstract syntax allows different variables with the same name but different types to occur in the same scope. This causes problems for the term quotation parser in any system that uses the basic algorithm for Hindley-Milner type inference, because this algorithm does not cater for variable-variable overloading (as explained in Sect. 5.1). HOL4, Isabelle/HOL, HOL Light and ProofPower HOL all use the Hindley-Milner algorithm, and so cannot parse terms with overloaded variables, even if antiquotation is used. Neither can these systems show distinction between overloaded variables in default printed output, although the type annotation printing modes of HOL4, Isabelle/HOL and ProofPower HOL do.

Example 2. The bound variable for the inner universal quantification in the following theorem proved in ProofPower HOL is a boolean x, rather than the natural number x used elsewhere in the theorem.

---

[2] We refer here to the "vanilla" version of the HOL language, implemented by all HOL systems except Isabelle/HOL and HOL Omega.

```
:) let val x1 = mk_var ("x",⌜:ℕ⌝) and x2 = mk_var ("x",⌜:BOOL⌝)
       and v = mk_var ("a",⌜:ℕ⌝)
       val tm = mk_eq (x1,v)
   in
     ∀_intro x1
       (∃_intro (mk_∃ (v, mk_⇒ (tm, mk_∀ (x2,tm))))
         (⇒_intro tm (∀_intro x2 (asm_rule tm))))
   end;
 val it = ⊢ ∀ x• ∃ a• x = a ⇒ (∀ x• x = a): THM
```

3.5   Type Ambiguity in Printed Terms

Typically, the default mode for printing in HOL systems, including in HOL4, Isabelle/HOL, HOL Light and ProofPower HOL, is to not provide any type annotation when printing terms and theorems. This is problematic because an expression involving variables or polymorphic constants then gets printed ambiguously when there is more than one type-correct way of assigning types to the atoms in the expression. However, HOL4's type annotation printing mode will perform sufficient annotation to remove ambiguity, except potentially in the obscure circumstance of a polymorphic function constant application with types not fully resolved by the arguments supplied. Isabelle/HOL's and ProofPower's will also suffice, so long as no constant annotation is required.

Example 3. The following theorem in HOL4 is proved for the one-valued type unit, but by default is not printed with type annotation and so looks like it has been proved for any type.

```
> let val x = ''x:unit''and y = ''y:unit''
  in
    GENL [x,y]
      (DISCH ''(x:unit)<>y''
        (TRANS (SPEC x oneTheory.one)
               (SYM (SPEC y oneTheory.one)) ))
  end;
# val it = |- !x y. x <> y ==> (x = y): thm
```

3.6   Juxtaposition of Lexical Tokens in Printed Expressions

Concrete syntax is generally printed with spacing characters separating lexical tokens, to ensure that the boundaries between tokens are obvious. In some circumstances, however, it tends to be more readable to be more concise and print without spacing, for example between parentheses and the expression they enclose. This is safe providing that boundaries between tokens are still clear, so that juxtaposed tokens without intervening space cannot be confused for a single token, or vice versa. HOL Light, however, does not ensure boundaries still exist when it prints without spacing in the printing of bindings, pairs and lists. We do not know of any problem cases in HOL4, Isabelle/HOL or ProofPower HOL.

Example 4. In the following theorem in HOL Light, no space is printed in the second conjunct between the binder !, binding variable # and "." keyword, making it indistinguishable from the variable "!#." in the first conjunct.

```
# let x = mk_var ("x",':num') and v1 = mk_var ("#",':num')
  and v2 = mk_var ("!#.",':num->num') in
  let tm1 = mk_comb (v2,v1) in
  let tm2 = mk_exists (x, list_mk_comb ('(<=)',[tm1;x])) in
  CONJ (EXISTS (tm2,tm1) (SPEC tm1 LE_REFL))
       (MESON [ARITH_RULE '!x. ~ (SUC x <= x)']
              '~ ?x. ! # . # <= x');;
  val it : thm = |- (?x. !#. # <= x) /\ ~(?x. !#. # <= x)
```

4   Lexical Solutions

Some of the problems highlighted in Sect. 3 are tackled in HOL Zero purely through lexical considerations. This section covers those solutions.

4.1   Identifier Delimiters

For the problems of irregular entity names (see Sect. 3.1) and entities with the same name as keywords (see Sect. 3.2), HOL Zero allows an entity's identifier to wrap double-quote delimiters ("") around the entity name. Any entity name can be wrapped with these delimiters, but an irregular or keyword name must be wrapped in order to get properly parsed. An entity identifier gets printed with the delimiters if and only if the entity name is irregular or a keyword.

This is essentially the same solution as used in ProofPower HOL, but there are no corner cases that cause problems. The backslash character (\) functions as an escape in the identifier, where double-quote and backslash are preceded by a backslash, and unprintable ASCII characters (such as tab and line feed) and back-quote (used in HOL Zero to delimit HOL type and term quotations) are denoted by their 3-digit decimal ASCII code preceded by a backslash.

Example 5. The irregular variable name in Example 1 is parsed and printed in HOL Zero using double-quote delimiters.

```
# '?x. "!y. x" + y = x';;
- : term = '?x. "!y. x" + y = x'
```

4.2   Variable Marking

For the problem of variables being overloaded with constants (see Sect. 3.3), HOL Zero allows a variable identifier to indicate that it denotes a variable, by preceding the variable's name with a percent character (%). This marking must immediately precede the variable name, without intervening space. Similarly, a type variable identifier can be marked as such by preceding the name with a single-quote character ('). If a variable is overloaded with a constant, or a type variable with a type constant, then it necessarily gets marked as such in printed output.

These marking characters must occur outside any double-quote delimiters used in the entity identifier (see Sect. 4.1). Note that a single-quote mark at the start of a type variable identifier cannot get confused with a single-quote at the start of a type variable name, because alphanumeric names in HOL Zero cannot start with a single-quote and so the name would be classed as irregular and get enclosed within double-quote delimiters. Similarly, a percent mark at the start of a variable identifier cannot get confused with a percent at the start of a variable name, because in HOL Zero percent is not classed as a symbolic character.

Example 6. The variable `true` is overloaded with the constant, and so needs a variable mark when parsed and printed.

```
# '%true + 1 < 5';;
- : term = '%true + 1 < 5'
```

Example 7. The type variable `nat` is overloaded with the type constant for natural numbers, and so needs a type-variable mark when parsed and printed. Note that, by default, all type variables are printed with the type variable mark.

```
# '!(x:'nat) (y:A). ?z. z = (x,y)';;
- : term = '!(x:'nat) (y:'A). ?z. z = (x,y)'
```

4.3   Spacing Between Tokens

For the problem of juxtaposed lexical tokens (see Sect. 3.6), HOL Zero ensures that in any situation where tokens get printed without any intervening space, a check is performed to make sure that the two tokens are compatible, i.e. that there is a lexical boundary between them when printed together, and if this check fails then spacing is inserted.

Example 8. In HOL Zero, the juxtaposed symbolic tokens in the head of the universal quantification in the second conjunct in Example 4 are printed with spacing to make clear they are separate tokens.

```
# '(?x. !#. # <= x) /\ ~ (?x. ! # . # <= x)';;
- : term = '(?x. !#. # <= x) /\ ~ (?x. ! # . # <= x)'
```

5   Type Inference for Variable-Variable Overloading

In order to cater for variable-variable overloading in the interpretation of HOL concrete syntax (see Sect. 3.4), it is not sufficient to use the classic Hindley-Milner type inference algorithm [11] that is used in other HOL systems. This algorithm was originally designed for parsing the programming language ML, which supports parametric polymorphism, needed for HOL's polymorphic constants, but not ad-hoc polymorphism, needed for HOL's variable-variable overloading. In this section, we explain how the Hindley-Milner algorithm is adapted in HOL Zero to cater for both.

Note that the discussion takes place at the level of HOL primitive syntax, because this is the level of representation used in the abstract syntax tree (AST) upon which type inference is carried out, and thus *binding* refers to lambda abstraction.

5.1 Hindley-Milner Type Inference

The Hindley-Milner type inference algorithm first involves assigning provisional types to all constant and variable atoms in the AST of the expression being parsed. Each instance of a non-polymorphic constant is simply assigned the constant's type, each instance of a polymorphic constant is assigned the constant's generic type but with unique meta type variables replacing any placeholder types, and each variable is given a unique meta type variable which is assigned to each occurrence of the variable (according to the scoping rules explained below). These meta type variables are then resolved bottom-up when function applications are encountered in ascending the AST, where the domain type of a function is unified with the type of its argument.

The notion of a variable identifier's syntactic scope in Hindley-Milner is straightforward. All variable atoms with a given name refer to the same entity throughout the AST of the expression up to the first common binding for a variable with that name (if such a binding exists), or otherwise the top level of the expression (when the variable is free). The types of all these variable atoms must unify, or otherwise the expression is ill-typed. Above the binding in the AST, the variable ceases to be in scope and any variable atoms with the given name refer to a different entity. This notion of scope does not allow for variable-variable overloading.

Example 9. Consider the following expression in HOL Zero concrete syntax:

  x \/ (!(x:nat). x = 5 \/ true = x)

Using purely primitive syntax, this is written as follows (where $ is used to strip an operator of its fixity):[3]

  $\/ x ($! (\(x:nat). $\/ ($= x 5) ($= true x)))

We uniquely number variables that are different according to the Hindley-Milner scoping rules, as well as each instance of the same polymorphic constant, so that they can be treated distinctly.

  $\/ x_1 ($! (\(x_2:nat). $\/ ($=_1$ x_2 5) ($=_2$ true x_2)))

---

[3] Note that, in HOL Zero, numerals are not atoms of primitive syntax, but for illustrative purposes it suffices to treat 5 as an atom in this example.

Provisional types get assigned as follows, with τ_n denoting the nth generated meta type variable:

\\/	bool $\rightarrow$ bool $\rightarrow$ bool	x_1	τ_4
!	($\tau_1 \rightarrow$ bool) $\rightarrow$ bool	x_2	τ_5
$=_1$	$\tau_2 \rightarrow \tau_2 \rightarrow$ bool		
$=_2$	$\tau_3 \rightarrow \tau_3 \rightarrow$ bool		
5	nat		
true	bool		

Type inference on the AST then resolves meta type variables bottom-up, starting as follows on the branches of the inner \\/:

$$\tau_2 = \tau_5 = \texttt{nat}$$
$$\tau_3 = \texttt{bool} = \tau_5$$

At this point, there is a conflict: τ_5 cannot be both nat and bool. Thus the expression is ill-typed according to traditional Hindley-Milner rules.

5.2 A New Notion of Syntactic Scope for Variables

Clearly we need a new notion of a variable's syntactic scope in order to cater for variable-variable overloading. As in Hindley-Milner type inference, we want this notion to be simple and intuitive for users, as well as relatively light in the amount of type annotation required. We do not want this notion, for example, to require every occurrence of every variable to be type-annotated, because this would be too inconvenient for the user.

In HOL Zero, the scoping rules are exactly the same as for Hindley-Milner for the basic case when there is no variable overloading – all variable atoms with a given name denote the same entity up to the binding for the variable name, or the top level of the expression if there is no such binding, so long as the types of these variable atoms all unify. The difference lies in the case when the variable atoms for a given variable name don't all unify. Unlike Hindley-Milner, HOL Zero allows for an extra case, for when the variable atoms each have fully-resolved types (i.e. not involving meta type variables) purely from local type resolution that can take place within the binding alone, i.e. not considering contextual type information from outside the binding in the AST. In this special case, according to the HOL Zero scoping rules there is a variable in scope for each of the fully resolved types, and only the variable corresponding to the binding variable ceases to be in scope above the binding in the AST.

Thus, in a nutshell, the HOL Zero scoping rules say that at a binding, all variables with the same name as the binding variable must unify or otherwise have types that can be fully resolved locally. Although more complicated than in Hindley-Milner, this new notion of variable scope is still relatively straightforward to understand and to use. All expressions that parse under Hindley-Milner type inference will still parse under HOL Zero type inference, to result in the same internal term. However, any expression involving variable-variable over-loading can also be parsed, so long as there is sufficient type-annotation. Note that one consequence of the rules is that the scope of a variable can only be determined at the type inference stage of parsing, rather than earlier, but we do not see this as a problem.

5.3 Adjustments to the Hindley-Milner Algorithm

To support the HOL Zero notion of variable scope, the Hindley-Milner algorithm needs a few adjustments. The first is that, in the initial assignment of provisional types throughout the AST, each variable atom, rather than variable entity, is given its own unique meta type variable. This is because the variable referred to by a given variable atom is only determined as type inference unfolds, and at the start of type inference each variable atom could potentially refer to a different variable.

The second adjustment is to the variable environment that is inevitably passed around in the algorithm's implementation, that provides the partially resolved type for a given variable name. This environment needs to be adjusted to carry a list of types, rather than a single type, with one type for each potentially distinct variable.

The third adjustment is another restriction on how the algorithm is implemented. In the Hindley-Milner algorithm, the variable environment for the components of a compound expression may build on the variable environment for a sibling component. However, this is not allowed in general in the HOL Zero type inference implementation, because the scoping rules must consider whether variables' types are fully resolved purely locally.

The fourth adjustment is to type inference at a binding, to cater for the extra case, of having variables overloaded with the binding variable, instead of simply raising an error.

Example 10. In the expression from Example 9, we uniquely number each variable atom, because at this stage they are potentially all different according to the HOL Zero scoping rules.

$$\$\backslash/\ x_1\ (\$!\ (\backslash(x_2:\texttt{nat}).\ \$\backslash/\ (\$=_1\ x_3\ 5)\ (\$=_2\ \texttt{true}\ x_4)))$$

Provisional types get assigned as follows:

$\backslash/$	$\texttt{bool} \rightarrow \texttt{bool} \rightarrow \texttt{bool}$	x_1	τ_4
!	$(\tau_1 \rightarrow \texttt{bool}) \rightarrow \texttt{bool}$	x_2	τ_5
$=_1$	$\tau_2 \rightarrow \tau_2 \rightarrow \texttt{bool}$	x_3	τ_6
$=_2$	$\tau_3 \rightarrow \tau_3 \rightarrow \texttt{bool}$	x_4	τ_7
5	$\texttt{nat}$		
true	$\texttt{bool}$		

Meta type variables are resolved bottom-up. On reaching the binding, the following type information has been inferred:

$$\tau_2 = \tau_6 = \text{nat}$$
$$\tau_3 = \text{bool} = \tau_7$$
$$\tau_5 = \text{nat}$$

This means that types for the variables local to the binding are thus:

x_2	nat
x_3	nat
x_4	bool

According to HOL Zero scoping rules, given that x is the name of the binding variable and the types of the different x atoms do not all unify, all of the local x atoms at this point must have fully resolved types, which they do. The atom for the binding variable is x_2 and has the same resolved type as x_3, and so these refer to the same variable, and are both removed from scope, leaving just x_4 in the local variable environment. Type inference can now proceed upwards from the binding.

$$\tau_1 = \tau_5$$
$$\tau_4 = \text{bool}$$

Having reached the top level of the expression, there are no conflicts, so the expression has passed type inference. The atoms are thus resolved as follows:

$\backslash/$	bool $\rightarrow$ bool $\rightarrow$ bool		x_1	bool
!	(nat $\rightarrow$ bool) $\rightarrow$ bool		x_2	nat
$=_1$	nat $\rightarrow$ nat $\rightarrow$ bool		x_3	nat
$=_2$	bool $\rightarrow$ bool $\rightarrow$ bool		x_4	bool
5	nat			
true	bool			

Thus there are two variables with name x: one with type nat, which is a bound variable, and one with type bool, which is free.

6   Minimal Unambiguous Type Annotation

Unless the types of its variables and polymorphic constants can be fully resolved from context, a printed term or theorem needs to be type-annotated in order to avoid type ambiguity (see Sect. 3.5). However, type annotating every variable and constant atom significantly reduces readability of output. HOL4's type annotation printing mode offers a better solution by annotating each variable entity only once in a given term, and only annotating instances of polymorphic constants. However, this level of annotation can still be excessive for large expressions.

In this section, we describe HOL Zero's algorithm for minimal type annotation when printing, so that there is just enough to avoid ambiguity and the output stays readable. This algorithm works for HOL Zero's notion of variable scoping that caters for variable-variable overloading (see Sect. 5.2).

Example 11. No type annotation is necessary when types can be inferred, as in the example in Fig. 1.

`'!x y. x > 1 /\ y > 1 ==> x * y > 1'`

Example 12. It is not necessary to type-annotate variable a from Example 2, because its type can be inferred:

`'!x. ?a. x = a ==> (!(x:bool). (x:nat) = a)'`

The basic outline of the algorithm is first to make a copy of the term to be printed but with each atom replaced with a provisional type, in exactly the same way as is done for HOL Zero type inference, with unique meta type variables for each variable atom and for each placeholder type in a polymorphic constant's generic type. The mapping from meta type variables to actual types is then worked out by matching this term copy with the original term. A subset of the atoms in the term copy is then picked for annotation, in such a way as to be minimal but sufficient to remove ambiguity from the printed term. The chosen atoms are then annotated with their types according to the meta type variable mapping to actual types. Before being printed, the resulting annotated term is passed to the type analyser to check that type resolution results in an internal term that is identical to the original term. This final check means that the relatively-complicated minimal type annotation algorithm does not need to be trusted, at the expense of having to trust the relatively-simple type analyser.

The only part of this that needs any further description is the process for picking atoms for annotation from the term copy. This first involves calculating four lists of atoms from the term copy, corresponding to the overloaded variables, the free variables, the bound variables and the polymorphic constants of the expression. The atoms in these four lists are the candidates for type annotation. Type inference is then performed on the term copy (with its meta type variables in place of actual types), to result in a partial instantiation list for the meta type variables based purely on what can be deduced from the term copy without any type annotations added.

This partial instantiation list is then used as the basis for removing atoms from the four atom lists, to remove those atoms that have types that can be fully resolved without any type annotation. The remaining atoms are selected down into a list to be type annotated. This is done by first ordering the atoms according to priority, with overloaded variables having highest priority, then bound variables, then free variables and polymorphic constants having lowest priority. The list is then scanned to remove atoms coming later in the list that will not need annotating because their types are inferred by annotating atoms earlier on in the list. The remaining atoms are then returned for annotation.

7 Conclusions

In this paper, we have enumerated the classic problems that have plagued default usage of HOL system parsers and pretty printers over the years. Some HOL systems fare better than others, and offer a patchwork of solutions such as antiquotation and coloured syntax highlighting which can be employed to solve some problems. However, these complicate the trusted code base, complicate output and restrict how the auditor can work, and in any case do not completely solve all problems. Our view is that simpler and more comprehensive solutions are possible, that work in monochrome ASCII and are easier to trust. We find it surprising that such solutions are not already prevalent for a logic that is so established as HOL.

We have shown how HOL Zero manages to overcome all the classic problems we list and fit our criteria for simplicity, by employing three main solutions. The first is a better lexical syntax regime that supports delimiting and marking for otherwise problematic entity names, and that ensures sufficient spacing between lexical tokens. The second is an adapted form of Hindley-Milner type inference that enables term quotations with variable-variable overloading to be represented in concrete syntax. The third is an algorithm for performing minimal type annotation on the atoms of a term, so that it can be printed in a way that is both unambiguous and about as readable as can be expected.

Together, these solutions enable parsing and printing of concrete syntax to be more complete and less ambiguous than in the other HOL systems. This boosts HOL Zero's credentials as a trustworthy system for auditing formal proofs. We believe HOL Zero does not suffer from any incompleteness or ambiguity in its parsers or printers, and printed output can always be parsed back in to give the same internal representation. This would make HOL Zero's parsers and printers well-behaved and Pollack-consistent. As far as we know, this would be a first amongst not only HOL systems, but also various other theorem proving systems that support concrete syntax, such as Coq and Mizar.

It would be interesting to establish for certain whether our claims of correctness in HOL Zero's parsing and printing are true. HOL Zero has a bounty [16] that, as well as for logical unsoundness, gets paid out for "printer unsoundness", i.e. if the pretty printer can be made to produce output that is ambiguous or not faithful to internal representation. Six printer-related problems have been reported, but none since August 2011. Trustworthiness could be further bolstered by adding a check for well-behaved parsing/printing to the printers.

Ideally there would be a formal proof about the correctness of HOL Zero's parsers and printers, which would cover questions of faithfulness as well as well-behaved parsing/printing and Pollack-consistency. The challenges in achieving this would presumably be quite different from those in formally verifying [5] the parser and printer of the Milawa theorem prover [4], which has no concrete syntax as such and so effectively boiled down to verifying the underlying Lisp implementation's parser and pretty printer for s-expressions.

Our solutions could conceivably be implemented in other HOL systems, to improve their usability as well as trustworthiness. It should be possible to incorporate type inference for overloaded variables and minimal type annotation without any major backwards compatibility issues, because the former just expands the set of term quotations that can be parsed, and the latter could be an optional printing mode. Incorporating delimiters for problematic entity names and marking for variables would require adjustment to the lexical syntax as well as the parser, and may have knock-on effects, but these are unlikely to be severe.

References

1. Adams, M.: Proof auditing formalised mathematics. J. Formalized Reasoning **9**(1), 3–32 (2016)
2. Adams, M., Clayton, P.B.: ClawZ: cost-effective formal verification for control systems. In: Lau, K.-K., Banach, R. (eds.) ICFEM 2005. LNCS, vol. 3785, pp. 465–479. Springer, Heidelberg (2005)
3. Arthan, R., Jones, R.: Z in HOL in ProofPower. In: Issue 2005–1 of the British Computer Society Specialist Group Newsletter on Formal Aspects of Computing Science, pp. 39–54 (2005)
4. Davis, J.: A Self-Verifying Theorem Prover. PhD Thesis, University of Texas at Austin (2009)
5. Davis, J., Myreen, M.: The reflective Milawa theorem prover is sound. J. Autom. Reasoning **55**(2), 117–183 (2015). Springer
6. Hales, T., et al.: A Formal Proof of the Kepler Conjecture. arXiv:1501.02155v1 [math.MG]. arXiv.org (2015)
7. Gordon, M.: From LCF to HOL: a short history. In: Proof, Language and Interaction, pp. 169–186. MIT Press (2000)
8. Gordon, M., Melham, T.: Introduction to HOL: A Theorem Proving Environment for Higher Order Logic. Cambridge University Press, Cambridge (1993)
9. Harrison, J.: HOL Light: an overview. In: Berghofer, S., Nipkow, T., Urban, C., Wenzel, M. (eds.) TPHOLs 2009. LNCS, vol. 5674, pp. 60–66. Springer, Heidelberg (2009)
10. Klein, G., et al.: seL4: formal verification of an OS Kernel. In: Proceedings of the ACM SIGOPS 22nd Symposium on Operating Systems Principles, pp. 207–220. ACM (2009)
11. Milner, R.: A theory of type polymorphism in programming. J. Comput. Syst. Sci. **17**, 348–375 (1978). Elsevier
12. Nipkow, T., Paulson, L.C., Wenzel, M. (eds.): Isabelle/HOL: A Proof Assistant for Higher-Order Logic. LNCS, vol. 2283. Springer, Heidelberg (2002)
13. Pollack, R.: How to believe a machine-checked proof. In: Twenty-Five Years of Constructive Type Theory, chap. 11. Oxford University Press (1998)
14. Slind, K., Norrish, M.: A brief overview of HOL4. In: Mohamed, O.A., Muñoz, C., Tahar, S. (eds.) TPHOLs 2008. LNCS, vol. 5170, pp. 28–32. Springer, Heidelberg (2008)
15. Wiedijk, F.: Pollack-inconsistency. Electron. Not. Theoret. Comput. Sci. **285**, 85–100 (2012). Elsevier Science
16. HOL Zero homepage. http://www.proof-technologies.com/holzero/

Infeasible Paths Elimination by Symbolic Execution Techniques
Proof of Correctness and Preservation of Paths

Romain Aissat[✉], Frédéric Voisin, and Burkhart Wolff

LRI, Univ Paris-Sud, CNRS, CentraleSupélec,
Université Paris-Saclay, Orsay, France
romainaissat@gmail.com, wolff@lri.fr

Abstract. TRACER [8] is a tool for verifying safety properties of sequential C programs. TRACER attempts at building a finite symbolic execution graph which over-approximates the set of all concrete reachable states and the set of feasible paths.

We present an abstract framework for TRACER and similar CEGAR-like systems [2,3,5,6,9]. The framework provides (1) a graph-transformation based method for reducing the feasible paths in control-flow graphs, (2) a model for symbolic execution, subsumption, predicate abstraction and invariant generation. In this framework we formally prove two key properties: correct construction of the symbolic states and preservation of feasible paths. The framework focuses on core operations, leaving to concrete prototypes to "fit in" heuristics for combining them.

Keywords: TRACER · CEGAR · Symbolic execution · Feasible paths · Control-flow graphs · Graph transformation

1 Introduction

TRACER [8] is a symbolic execution-based tool for verifying safety properties of imperative programs. TRACER tries to build from a program control-flow graph (CFG) a finite symbolic execution tree which over-approximates the set of reachable states. To this end, TRACER avoids the full enumeration of symbolic paths by learning from infeasible paths, i.e. from paths for which no input state exists allowing their execution. This learning phase uses *interpolants* for each program point. An interpolant is a formula characterizing a set of program states. If an interpolant allows to establish that a symbolic state is *subsumed* by a previous state in its path, TRACER annotates the symbolic execution tree by so-called subsumption links turning the tree into a graph. Thus, this annotated tree can represent infinite sets over-approximating the feasible paths.

Finding accurate approximations of the feasible paths of a program is of wide-spread interest for static analysis techniques, worst-time analyzers, code optimization and code-slicing techniques and structural test-case generation.

© Springer International Publishing Switzerland 2016
J.C. Blanchette and S. Merz (Eds.): ITP 2016, LNCS 9807, pp. 36–51, 2016.
DOI: 10.1007/978-3-319-43144-4_3

Since in many programs the ratio of infeasible paths to feasible ones may be very high, a lot of computing power in a static analyser could be addressed at more rewarding targets, while dramatically improving the quality of results by discarding information stemming from infeasible paths. Our motivation is in random-testing of imperative programs: there exist efficient algorithms that draw in a statistically uniform way long paths from very large graphs [4]. If the probability to find a feasible path from a (transformed) CFG is high, one could use these methods to randomly draw long paths, compute their path predicate, and test the program along an instance of the path predicate against a user-defined post-predicate (note that this method does not depend on user-defined loop-invariants). Thus, the method could be extended to an effective statistical structural (white-box) testing method.

When adapting TRACER mechanisms to our own purposes, we found that the presented proof sketches in the accompanying literature revealed a sensible gap to a formal proof development. We therefore built a formal theory in Isabelle/HOL of an abstract version of the TRACER algorithm, called ATRACER. ATRACER is a highly non-deterministic model of TRACER, consisting on five graph transformations of a so-called *red-black graph*, where the red part roughly corresponds to the analyzed symbolic execution tree gained by partial unfolding of the CFG and the black part is the initial CFG of the program. For ATRACER, two major theoretic results were established:

1. *correctness*: for every path in the new graph, there exists a path with the same trace in the original one, and
2. *preservation of feasible paths*: each transformation preserves the set of feasible paths. This very important property is often claimed in papers without a complete proof.

These results extend to an entire family of TRACER-like algorithms, which add to ATRACER specific heuristics in their goal to provide approximations of feasible paths of a given program. These heuristics fill in the non-deterministic "gaps" of ATRACER: which node to select, which interpolant to choose, which learned invariant to inject, etc. Note that our goal is not to provide a formal proof of TRACER implementation: heuristics aspects are not modeled, and ATRACER uses completely different data-structures. ATRACER is a rational reconstruction of TRACER identifying the core operations performed on symbolic execution graphs (SEGs) in order to prove the two above properties. In this paper, TRACER is essentially used as an instance of such systems.

This paper proceeds as follows: After providing a short introduction into Isabelle/HOL and the notations we need, we present in Sect. 3 ATRACER by a small example. Section 4.1 is devoted to the introduction of the formal machinery of red-black graphs and their symbolic execution. We present in Sect. 4.2 the formalization of graph-transformations. In Sect. 4.3 we state formally the correctness and preservation properties and outline the proofs. The entire formalization and proof effort in Isabelle/HOL consists of about one hundred definitions or abbreviations and two hundred lemmas, representing about 8k lines of code. All proofs were written using the Isar proof language in a structured manner.

No fancy theorem proving technologies were needed: the most expensive tactic used is force. The sources are available under https://www.lri.fr/~wolff/atracer.zip.

2   Background: Isabelle and Isabelle/HOL

Isabelle/HOL [10] is an interactive theorem proving environment for Church's higher-order logic (HOL), a classical logic based on a simply typed λ-calculus extended by parametric polymorphism. HOL provides the usual logical connectives like $\_ \wedge \_$, $\_ \rightarrow \_$, $\neg\_$ as well as the object-logical quantifiers $\forall x.\ P\ x$ and $\exists x.\ P\ x$; in contrast to first-order logic, quantifiers may range over arbitrary types, including total functions $f :: \alpha \Rightarrow \beta$. HOL is centered around extensional equality $\_ = \_ :: \alpha \Rightarrow \alpha \Rightarrow$ bool.

Isabelle/HOL comes with rich libraries for lists, typed sets, total and partial functions, etc. We introduce some library notations used throughout this paper: wrt. to sets, we use the usual {x. P(x)} for set comprehensions, x ∈ S for inclusion, A ∪ B, A ∩ B for union and intersection, etc. Lists were built by the constructors Nil and _ # _. Of particular importance for this paper is the use of record notation; records are basically cartesian products where the components have a *tag-name*. As example, we declare a record by the specification construct:

record ('α, 'β) point = x ::"'α" y :: "'β"

This specification construct introduces a number of operations on record types (such as ('α, 'β)point). For example, P = ⦇ x = 4, y = True ⦈ is a constructor of a record (of type (int, bool) point), where as P' = P(x := 3) is an update function of the record at the component x. The tag-names implicitly define selector functions on records; thus y P' is equivalent to True.

3   A Guided Run of ATRACER

TRACER avoids the full enumeration of symbolic paths by learning from infeasible paths and computing *interpolants* for program points. In this context, an interpolant is a logical formula associated with a program point that constraints a set of program states: whenever symbolic execution reaches that program point in a state satisfying the interpolant, it is ensured that the *final* program point can be reached from that point (without going through a given error statement). Once an interpolant has been synthesized for a program point, any symbolic execution path that reaches that point in a symbolic state satisfying the interpolant needs not to be extended further: it is ensured that it will reach the final program point. The new path is said to be *subsumed* by the previous one.

To avoid unrolling loops infinitely, when reaching a loop header TRACER checks if that program point can be subsumed by one of its prior occurrences on the path. Detection of subsumptions at loop headers is performed by computing abstraction between symbolic states, that is weakening constraints on the symbolic states for that point. Abstraction can be performed, for example, by removing or weakening (e.g. turning equalities into inequalities) constraints from the path predicates. Abstraction can be seen as a synthesis of a loop invariant. If the synthesized invariant is not strong enough, this can result in "false negatives" where paths that are infeasible in the original program are not ruled out by the abstraction of the program states. Such abstractions have to be refined: Whenever a symbolic path leading to an error statement is produced from a point where an abstraction occurred, TRACER checks if that path exists without the abstraction. If this is the case, the error statement is truly reachable. Otherwise, information about the unfeasability is collected and an interpolant characterizing the unreachability of the error statement is attached to the node where the faulty abstraction was made. The analysis is restarted from that point, with its new interpolant serving as a safeguard for abstractions: abstractions at that point must now satisfy the interpolant. When it is not possible to find an abstraction between different occurrences of a loop header, the loop is unrolled one more time in the hope of later subsumptions. Absence of valid subsumptions leads TRACER to unroll loops infinitely. Otherwise, abstraction and subsumption result in a SEG that includes all the behaviors of the original program with respect to the reachability of error statements.

3.1   Presenting ATRACER by an Example

Our abstraction of the original TRACER is conceived as a set of graph transformations of an annotated CFG we call *red-black graph*. Its transitions are annotated with basic blocks of assignments, the skip-statement, or a guard that has to hold when executing this transition. A red-black graph represents the over-approximated set of feasible paths and is made of two parts: the *black part* is the initial CFG and remains unchanged throughout the transformations; the *red part* consists initially of a single vertex and is extended by unfolding the initial CFG using our graph-transformation operators, i.e. by adding transitions that are symbolically executed, subsumption links, etc. The red and black parts are the *known* and *unknown* parts, respectively.

We illustrate ATRACER with the example drawn from Jaffar et al. [7]. The program in Fig. 1a implements a lock acquisition algorithm. The goal of the authors is to check that statement at line 8 is not reachable on any feasible path, ensuring that the lock is held when the execution exits the loop. The condition of statement if (*) at line 4 abstracts a call to an external condition (like a function or a system call) that returns true if the lock is held by another process. Hence at each traversal either branch of such a conditional can be taken independently of the current state of the execution. Doing so is equivalent to executing a true-guard. In Fig. 1b we give the CFG for the *lock* program.

Since we are interested in illustrating the graph transformation operators, not in finding how to combine them in an actual system, in ATRACER we proceed as if we always guess correctly the next step. Thus, our sequence of elementary transformations differs from the one described by the authors in [7], whose order is controlled by several heuristics. However, we end up with the same final SEG as the original algorithm.

Notation: to distinguish the different *occurrences of program points* in the red part, we decorate the original location label (the line number) with a superscript. Superscripts start at 0 and increase with every further visit. Vertices labels without superscript refer to the black part, those with a superscript to the red part. In Fig. 2a and latter, some red vertices are linked to their black counterpart by dotted edges. These links represent the continuation of the computation in the original program, i.e. parts that has not been symbolically executed yet.

```
1    lock = 0; new = old + 1;
2    while (new != old){
3        lock = 1; old = new;
4        if (*){
5            lock = 0; new = new + 1;}
6    }
7    if (lock == 0)
8        error ();
```

(a)

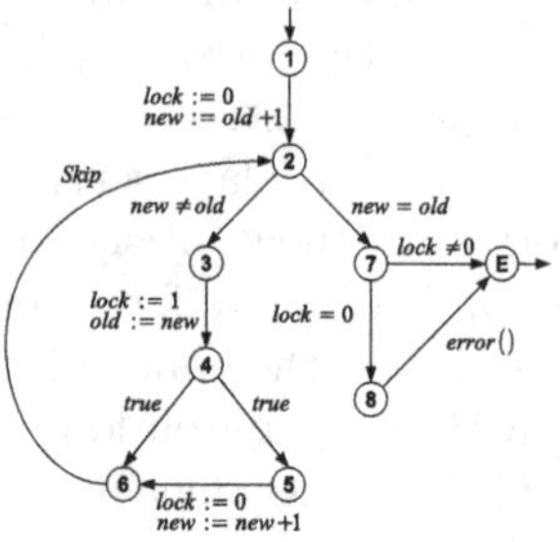

(b)

Fig. 1. The 'lock' example and its CFG.

In the red part (depicted with square vertices), vertices are implicitly annotated with *configurations*. Configurations are tuples: the first component, called the symbolic state, is a function associating symbolic variables with program variables; the second component is the path predicate, i.e. the conjunction of constraints over symbolic variables that are accumulated during symbolic execution of the current path up to that point. Path predicates are written under static single assignment form, introducing new symbolic variables for each assignment.

Initialization: We start the symbolic execution of the program in Fig. 1a with the configuration $(\{lock \mapsto lock_0, new \mapsto new_0, old \mapsto old_0\}, true)$ The red part consists of the single red vertex 1^0, corresponding to the entry program point in the black part and is linked to the latter (Fig. 2a).

Symbolic execution of assignments: from 1^0, we perform symbolic execution over the black transition leading from 1 to 2. This results in the addition of a red transition from 1^0 to 2^0. The symbolic state at 2^0 is obtained from the one at 1^0 by associating fresh symbolic variables with variables *lock* and *new* and adding two constraints to the path predicate. The configuration for 2^0 is: $(\{lock \mapsto lock_1, new \mapsto new_1, old \mapsto old_0\}, lock_1 = 0 \wedge new_1 = old_0 + 1)$.

Symbolic execution of guards: assuming the first symbolic path enters the loop, symbolic execution is performed from 2^0 over the transition from 2 to 3. The path predicate at 3^0 is the conjunct of path predicate at 2^0 with the constraint $new_1 \neq old_0$, obtained by substituting occurrences of program variables in the guard by the symbolic variables they are associated

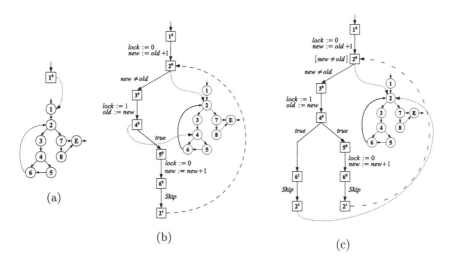

Fig. 2. Partial unfoldings of the lock acquisition CFG.

with in the symbolic state. Assuming we follow first the *then* branch of the inner conditional, i.e. statements at line 5, symbolic execution is performed until the statement at line 2 (the loop header) is reached for the second time, completing one loop iteration. The configuration for 2^1 is ($\{lock \mapsto lock_3, new \mapsto new_2, old \mapsto old_1\}, lock_1 = 0 \wedge new_1 = old_0 + 1 \wedge lock_2 = 1 \wedge old_1 = new_1 \wedge lock_3 = 0 \wedge new_2 = new_1 + 1$).

Subsumption between loop headers: 2^0 and 2^1 are two occurrences on the current path of the same loop header. Given two occurrences v and v' of a same program point, v' can be subsumed by v if it is a particular case of v. When candidates for subsumption are discovered, in most cases the subsumption cannot occur directly. Subsuming a vertex often requires the configuration of the subsumer to be *abstracted*, that is relaxing some constraints of its path predicate, to force the subsumee to imply its subsumer. This is not needed here, since 2^1 and 2^0 can be shown to be logically equivalent. Vertex 2^1 is subsumed by 2^0 and the small dotted edge linking 2^1 to the black part is replaced by a subsumption link from 2^1 to 2^0 in Fig. 2b (depicted by the big dotted edge). After that subsumption, symbolic execution resumes at 4^0 and extends up to 2^2, a new target for a subsumption.

Limiting abstractions with interpolants: Before processing 2^2, the interpolant $new \neq old$ (written between brackets in Fig. 2c) is added at 2^0 to prevent the subsumption of 2^2 by 2^0. Labeling a vertex with an interpolant needs to show that the configuration entail the interpolant, which is the case here. The symbolic state at 2^2 is not a particular case of the one at 2^0 and abstraction is forbidden by the interpolant: subsumption cannot occur and the loop is unrolled, performing symbolic execution from 2^2 to 3^1.

Marking nodes as unsatisfiable: The path predicate at 3^1 is unsatisfiable, as it requires *new* and *old* to be both equal and different. We *mark* (with a

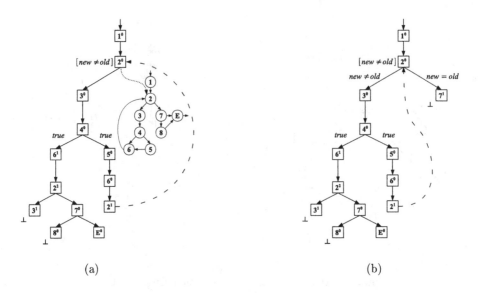

(a)   (b)

Fig. 3. A partial and a complete unfolding of the lock acquisition CFG.

⊥ symbol, in Fig. 3a) nodes known as unsatisfiable. In practice, this would result from a call to a constraint solver: in ATRACER only configurations proved as unsatisfiable can be marked. As we do not expect the user to call a solver at every node, for performances reasons, we let it be an explicit action. We do not disallow to pursue from a marked configuration, but symbolic execution will carry the mark to its successors.

Symbolic execution resumes at node 2^2, and follows the exit branch of the loop. The exit point and the error location are reached, respectively at E^0 and 8^0, and the latter is marked since its path predicate is unsatisfiable. Symbolic execution now resumes at node 2^0, the last pending point, and exits the loop, reaching 7^1. As the configuration at 2^0 does not satisfy the exit condition, 7^1 is marked as unsatisfiable (Fig. 3b).

In Fig. 3b, every leaf of the red part is either marked as unsatisfiable, subsumed or an occurrence of the final black location. Every feasible path of the black part is contained in the red part! The error statement at line 8 is no longer reachable from a feasible symbolic path (from the red entry) either in the red part or in the black part (red vertices linked to the black part are all marked as unsatisfiable) which can therefore be pruned.

4   The Formalization of ATRACER

4.1   The Theory of Red-Black Graphs

In this section, we introduce the main concepts needed in order to formalize red-black graphs and to state and prove the main theorems we are interested in.

We first introduce the definitions we use to model graphs and CFGs, then present main definitions and facts about symbolic execution.

Basic Definitions About Graphs. The notion of graph and associated concepts like sub-paths are central to our formalization because we need an abstraction of sharing in the abstract syntax as well as arbitrary cycles in the CFG. Moreover, we need to consider paths going through subsumption links, a notion that is specific to our approach. We start with a conventional definition of a graph over some type of vertices 'a as a set of arcs linking the nodes:

```
record'a arc = src :: "'a"        record 'a rgraph = root :: "'a"
                tgt :: "'a"                         arcs :: "'a arc set"
```

Our notion of graph assumes that they have one single `root` (which comes in handy when modeling CFG's). So far the definitions do neither imply that the graph is connected and that `root` has any connection to the arcs. These kind of side-conditions are captured by additional predicates, and sometimes managed by the Isabelle concept of a *locale*, covered by Ballarin in [1].

On this basis, a rich theory of auxiliary concepts must be developed. For instance, we need the concept of a "consistent arc sequence":

```
fun cas :: "'a ⇒ 'a arc list ⇒  'a ⇒ bool"
where
  "cas v1 [] v2 = (v1 = v2)"
| "cas v1 (a#seq) v2 = (src a = v1 ∧ cas (tgt a) seq v2)"
```

which paves the way to the concept of a `subpath`:

```
"subpath g v1 seq v2 ≡ cas v1 seq v2 ∧ v1 ∈ verts g ∧ set seq ⊆ arcs g"
```

and **path** (a sub-path from the root). Both concepts were borrowed from Nochinski's Graph Library for Isabelle [11]. Here, we define as **vert** the nodes which are either root, source or target of an arc, and add the usual notions of in- or outgoing arcs. We add abstract operations on graphs (like adding arcs) and establish a number of properties wrt. vertices, paths, inarcs, and outarcs.

We then define graphs equipped with a subsumption relation. In the following, subsumptions only involve vertices of the red part that represent different occurrences of a same vertex of the black part. We represent subsumption relations by sets of pairs of indexed vertices:

```
type_synonym 'a sub_t = "(('a × nat) × ('a × nat))"
type_synonym 'a sub_rel_t = "'a sub_t set"
```

Again, paths and sub-paths in a graph equipped with such a relation are defined using the notion of consistency of an arc sequence. An arc sequence is consistent in presence of a subsumption relation if it is made of a number of consecutive consistent (without the subsumption relation) sequences whose extremities are linked throughout elements of the subsumption relation.

If we add an arc labeling function to graphs, we speak of *labeled transition systems* (lts). Their type is defined by the record-extension:

```
record ('a,'b,'c) lts = "'a rgraph" + labf :: "'a arc ⇒('b,'c) label"
```

where 'c is the type of values taken by program variables. It enriches the 'a rgraph with a labeling function; these labels turning an arbitrary graph into a CFG have a richer structure that we describe in the following.

Main Definitions and Facts About Symbolic Execution. First, we define corresponding to program variables 'a the symbolic variables with their super-scripts (cf. Sect. 3.1) by a type synonym (pairing that program variable with an integer) and define the concept of a store for bookkeeping the current association of a program variable with its symbolic variable represented by its superscript.

```
type_synonym 'b symvar = "'b × nat"
type_synonym 'b store  = "'b ⇒ nat"
```

States are used to give values to variables. Arithmetic and boolean expressions are modeled in shallow embedding style, by total functions from variables to their domain and to boolean values, respectively.

```
type_synonym ('b,'c) state = "'b ⇒'c"
type_synonym ('b,'c) aexp  = "('b,'c) state ⇒'c"
type_synonym ('b,'c) bexp  = "('b,'c) state ⇒ bool"
```

This way of modeling expressions has the advantage that there is no need to for-malize the different operators on expressions, which would have been necessary using a syntactic approach. Moreover, shallow embedding allows the use of the existing Isabelle notations and theorems about functions.

On the other hand, it makes it a bit harder to describe the set of variables of such expressions, which is needed when reasoning about the freshness of some symbolic variable for a configuration. We define the set of variables of an arith-metic (resp. boolean) expression as the set of variables that can actually have an influence over the value of this expression.

```
definition vars :: "('b,'c) aexp ⇒'b set" where
  "vars e = {v. ∃σ val. e (σ(v := val)) ≠ e σ}"
```

Since configurations and subsumption between configurations have been introduced in Sect. 3.1, we skip their formal definitions here and go directly to symbolic execution. We note $c \sqsubseteq c'$ the fact that configuration c is subsumed by configuration c'.

Symbolic execution is defined as an inductive predicate se that takes two configurations c and c' and a label l and evaluates to *true* if c' is a *result* of the symbolic execution of l from c. Results are defined up to the way fresh indexes are chosen in the case of Assign labels. We prove that fresh indexes exist when needed, assuming expressions in labels and configurations from which symbolic execution is performed have finite sets of variables.

Labels are either *Skip*, *Assume* ϕ, where ϕ is a boolean expression, or *Assign* v e where v is a program variable and e an arithmetic expression.

```
datatype ('b,'c) label =
  Skip | Assume "('b,'c) bexp" | Assign 'b "('b,'c) aexp"
```

```
inductive se :: "('b,'c) conf ⇒('b,'c) label ⇒('b,'c) conf ⇒bool"
where
  "se c Skip c"
| "se c (Assume e) (| store c, pred = pred c ∪ {adapt_bexp e (store c)} |)"
| "fresh_symvar (v,i) c ⟹
   se c (Assign v e)
   (| store = (store c)(v := i),
      pred  = pred c ∪ {(λ σ. σ (v,i) = (adapt_aexp e (store c)) σ)} |)"
```

Here, `adapt_aexp e s` (resp. `adapt_bexp`) represent the expression obtained
from the arithmetic (resp. boolean) expression e by substituting every occurrence of program variables by their symbolic counterpart given by s. It would
have been possible to define `se` as a function, but the assumption about freshness
in the case of an assignment would require a special treatment. This could be
done in a number of ways. For example, `se` could be a partial function defined
only in those cases where the new symbolic variable is indeed fresh.[1] In the end,
we found that using a predicate was the simplest way to model `se`, and also
yields simpler proofs in the rest of the formalization.

We extend symbolic execution to sequences of labels, and model it by an
inductive predicate `se_star` that takes two configurations and a sequence of
labels, and evaluates to *true* if the second configuration is a possible result of
symbolic execution of the given sequence from the first configuration.

To prove the key properties of our approach, one first proves that symbolic
execution is monotonic with respect to the previous definition of subsumption.
We only state the theorem for `se`, a similar one holds for `se_star`.

```
theorem se_mono_for_sub :
assumes "se c1 l c1'"
assumes "se c2 l c2'"
assumes "c2 ⊑ c1"
shows   "c2' ⊑ c1'"
```

The proof is obtained by case distinction on l, expressing the states of $c1'$ and
$c2'$ as functions of the states of $c1$ and $c2$, respectively. In the case of sequence
of labels ls, the proof is obtained by induction on ls, using `se_mono_for_sub`.

4.2   Graph-Transformations on Red-Black Graphs

We are ready to give the structure of `('a,'b,'c) pre_RedBlack` before defining
what means to be a `('a,'b,'c) RedBlack` graph.

```
record ('a,'b,'c) pre_RedBlack =
  red          :: "('a ×nat) rgraph"
  black        :: "('a,'b,'c) lts"
  subs         :: "(('a ×nat) ×('a × nat)) set"
  init_conf    :: "'b conf"
```

---

[1] Given an arbitrary configuration, there is no guarantee that there exists a fresh
symbolic variable for a given program variable, since expressions are defined as total
functions.

```
confs          :: "('a ×nat) ⇒ ('b,'c) conf"
marked         :: "('a ×nat) ⇒ bool"
strengthenings :: "('a ×nat) ⇒('b,'c) bexp"
```

The fields **red** and **black** represent the red and black parts, respectively. **init_conf** is the configuration initially chosen to start the analysis. **subs** is the subsumption relation which contains the subsumption links between the vertices of **red**. Finally, **confs**, **marked** and **strengthenings** are functions associating to the vertices of **red** their current configuration, the fact that they are marked as unsatisfiable or not, and their current interpolant, respectively.

We now specify what we call the GT-calculus, i.e. the five graph transformations and the set of reachable red-black graphs from an initial configuration containing just a black part and an empty well-formed red part.[2] The construction proceeds per inductive definition as follows:

```
inductive RedBlack :: "('a,'b,'c) pre_RedBlack ⇒bool" where
  init :
    "fst (root (red rb)) = init (black rb)         ⟹
    arcs (red rb) = {}                              ⟹
    subs rb = {}                                    ⟹
    (confs rb) (root (red rb)) = init_conf rb       ⟹
    marked rb = (λ rv. False)                        ⟹
    strengthenings rb = (λ rv. (λ σ. True))          ⟹ RedBlack rb"
| se_step :
    "RedBlack rb ⟹ se_extends rb ra c' rb'          ⟹ RedBlack rb'"
| mark_step :
    "RedBlack rb ⟹ mark_extends rb rv rb'           ⟹ RedBlack rb'"
| subsum_step :
    "RedBlack rb ⟹ subsum_extends rb sub rb'        ⟹ RedBlack rb'"
| abstract_step :
    "RedBlack rb ⟹ abstract_extends rb rv e rb'     ⟹ RedBlack rb'"
| strengthen_step :
    "RedBlack rb ⟹ strengthen_extends rb rv e rb'   ⟹RedBlack rb'"
```

where operations **se_extends**, **mark_extends**, **subsum_extends**, **abstract_-extends** and **strengthen_extends** are abbreviations (macros) for a number of constraints necessary to describe, one by one, the graph transformations informally introduced in Sect. 3.1. We pick the graph transformation **se_extends** as example:

```
abbreviation se_extends ::
  "('a,'b,'c) pre_RedBlack ⇒('a × nat) arc ⇒ ('b,'c) conf ⇒
  ('a,'b,'c) pre_RedBlack ⇒bool"
where
  "se_extends rb ra c' rb' ≡
    ui_arc ra ∈ arcs (black rb)                    (* 1 *)
    ∧ ArcExt.extends (red rb) ra (red rb')         (* 2 *)
    ∧ src ra ∉ subsumees (subs rb)                 (* 3 *)
```

---

[2] This is ensured by a number of constraints on the free variable **rb** forcing the root of the red part to be the initial location of the black part, etc.

```
∧ se (confs rb (src ra)) (labf (black rb)(ui_arc ra)) c'          (* 4 *)
∧ rb' = (|   red        = red rb',
             black      = black rb,
             subs       = subs rb,
             init_conf = init_conf rb,
             confs      = (confs rb) (tgt ra := c'),
             marked     = (marked rb)(tgt ra := marked rb (src ra)),
             strengthenings = strengthenings rb |)          (* 5 *)"
```

The constraints describe formally the following side-conditions (we follow the labels in comments above):

1. `ui_arc ra`, the (unindexed) black counterpart of red arc ra must exist in the black graph,
2. `ArcExt.extends` is an abbreviation that states that the source of ra must be an existing vertex of the red graph, but not its target, and that the new red graph is obtained by adding ra to the arcs of the old one,
3. the source of ra is not already subsumed,[3]
4. c' is the new configuration obtained by symbolic execution of ra
5. the new red-black graph rb' is constructed from the old one by the following updates:
   - ra is added to the red graph
   - the new configuration is added at the target of ra
   - the satisfiability-flag of the target of ra is set to the one of its source. Recall that we want registration of unsatisfiablity to be an explicit action.

The amount of detail that must be added when reasoning precisely over the correctness issues of these type of graph-based static analysis algorithms is quite substantial and makes it a valuable target for a machine-checked analysis.

From now on, we call red-black graphs the set of **pre_RedBlack** reachable by the predicate **RedBlack**.

4.3   Main Theorems of ATRACER

Relation Between Red Vertices. In the case of a classical symbolic execution tree, one would prove that, given one sub-path in the tree, the symbolic state at its end has been obtained by symbolic execution of its trace from the symbolic state at its beginning. This property is too strong for red graphs obtained by the GT-calculus. We must handle sub-paths that go through subsumption links and configurations along these sub-paths may have been abstracted, both inducing a loss of information about program states.[4] In ATRACER, the configuration at the end of a sub-path merely subsumes the one obtained by classical symbolic execution. This is expressed by the following theorem.

---

[3] The conjunction of 2 and 3 is equivalent to say that the source of ra is a pending point in the analysis.

[4] Note that in the second assumption of **gt_calc_se_rel**, unlike in Sect. 4.1, **subpath** has a fifth parameter: the subsumption relation of rb.

```
theorem gt_calc_se_rel :
assumes "RedBlack rb"
assumes "subpath (red rb) r1 s r2 (subs rb)"
assumes "se_star (confs rb r1) (trace (ui_as s) (labelling(black rb))) c"
shows   "c ⊑ (confs rb r2)"
```

The proof is obtained by rule induction on RedBlack, i.e. the five transformation operators maintain the property. All cases are quite straightforward, except for adding a subsumption link. The details of its proof are quite tedious and numerous, so we skip them here and just give the main idea. The problem is that we do not know how many times the considered sub-path goes through the new subsumption, if it does. But as we consider finite sub-paths only, this number is finite: the proof is obtained by a backward induction on s, using the fact that subsumption between configurations is a partial ordering for which symbolic execution is monotonic.

Red-Black Sub-paths and Paths. Before stating our two main theorems, we formalize the notion of sub-path of a red-black graph and its set of paths. Given a vertex rv of the red graph, we first define the set of red-black sub-paths starting from rv as the union of the two following sets:

- the sets of black sub-paths entirely represented in the red graph by sub-paths starting at rv and ending in a non-marked red vertex,
- the sets of black sub-paths that have a prefix represented in the red graph leading to a non-marked red vertex rv', which is not subsumed and from which there exist black arcs that have not been symbolically executed yet. Moreover, the remaining black suffix must have no (non-empty) prefixes represented in the red graph (starting at rv').

As in Sect. 4.1, we define the set of red-black paths as the set of red-black sub-paths starting at the root of the red graph. This complex definition is needed to ensure that what we call the set of red-black paths is not simply the set of paths of the black graph.

Correctness of the GT-Calculus. Our first main theorem states that red-black paths all come from paths of the black part. More precisely, every red-black sub-path starting at some red vertex rv is also a sub-path starting at the black vertex represented by rv in the black graph. Thus, our approach is correct in the sense that it does not introduce new paths in the red-black graph and preserve program behavior.

```
theorem gt_calc_correct :
assumes "RedBlack rb"
shows   "RedBlack_subpaths_from rb rv
         ⊆ Graph.subpaths_from (black rb) (fst rv)"
```

The theorem relies on the fact that arcs added to the red part are simply indexed versions of black arcs, and that subsumption links only involve different occurrences of the same black vertices.

Preservation of Feasible Paths. Finally, we prove that the original set of feasible paths is contained in the red-black graph. Our main theorem is the following: given a red vertex *rv*, every feasible black sub-path *bs* starting at the black vertex represented by *rv* from the configuration associated to *rv* is a red-black sub-path starting at *rv*.

```
theorem gt_calc_preserves :
assumes "RedBlack rb"
assumes "rv ∈ red_vertices rb"
assumes "bs ∈ feasible_subpaths_from (black rb) (confs rb rv) (fst rv)"
shows    "bs ∈ RedBlack_subpaths_from rb rv"
```

As for correctness, the proof (which is 2.3k loc, and can be found in file RB.thy) is obtained by rule induction on RedBlack. For each operator, we use the induction hypothesis to get that *bs* is also a red-black sub-path of the old red-graph, before proving that it is not ruled out by the current transformation. The initial case is trivial, as well as those of abstracting a configuration and adding an interpolant, since the former only makes the set of red-black sub-paths larger and the latter does not modify the graph structure but only prevents future abstractions. The case of adding a red arc is simple but tedious as one needs to treat the numerous sub-cases. Marking a red vertex as unsatisfiable is proved using the fact that the vertices that *bs* goes through cannot be marked, otherwise *bs* would not be feasible. The case of adding a subsumption link is the difficult one, for the same reasons as previously. Again, the proof is obtained by a backward induction on the considered sub-path before proceeding by case distinction.

We rephrase our main theorem in more readable way:

```
theorem gt_calc_preserves2 :
assumes "RedBlack rb"
shows    "feasible_paths (black rb) (init_conf rb) ⊆ RedBlack_paths rb"
```

It is proved using the fact that the initial configuration of a red-black graph is subsumed by the one associated with the root of its red part, hence the set of feasible paths starting from the former is a subset of the set of the latter.

4.4   Summary

The formalization of ATRACER presented a number of challenges. We first attempted to formalize the whole TRACER's algorithm, heuristics aspects included. At this time, the SEG was modeled as a tree, whose nodes and leaves could have different types: *simple, unsatisfiable, subsumed, subsumer,* and were decorated with much information, like configurations, the identity of the subsumer, etc. We then faced major difficulties. First, this structure is not suitable to describe inductively how its set of paths evolves after adding a new node, a subsumption, etc. Our current modeling of graphs equipped with subsumption relations makes this task far more easy. Second, it is very difficult to model in details the heuristics aspects, like graph traversals, how subsumptions are detected, or how abstractions are refined in practice, for example. We finally chose to "break" TRACER's algorithm into pieces in order to identify and formalize the core operations it performs on SEG, and to give up the heuristics aspects, since they have no influence on the preservation of feasible paths. Finally, due to the nature of the problem - symbolic execution in presence of unbounded loops, TRACER-like algorithms might not terminate. In practice, this is handled using some kind of time-out condition. When such conditions are triggered, the only way to preserve all feasible paths is to "plug" the actual SEGs into the original graph. In ATRACER, this is represented by the black part and the complex definition of red-black paths. This is also what motivates identifying the core operations, since the problem of preservation is reduced to showing that each operator never rules out feasible paths.

5   Conclusion

Related Work. Our work is inspired by Tracer [8] and the more wider class of Cegar-like systems [2,3,5,6,9] based on predicate abstraction. However, we did not attempt any code-verification of these systems and rather opted for their rational reconstruction allowing for a clean separation of heuristics and fundamental parts. Moreover, our treatment of Assume and Assign-labels is based on shallow encodings for reasons of flexibility and model simplification, which these systems lack. There is a substantial amount of formal developments of graph-theories in HOL, most closest is perhaps by Lars Noschinski [11] in the Isabelle AFP. However, we do not use any deep graph-theory in our work; graphs were just used as a kind of abstract syntax allowing sharing and arbitrary cycles in the control-flow. And there are a large number of works representing programming languages, be it by shallow or deep embedding; on the Isabelle system alone, there is most notably the works on Ninja, NanoJava, IMP, etc. However, these works represent the underlying abstract syntax by a free data-type and are not concerned with the introduction of sharing in the program presentation; to our knowledge, our work is the first approach that describes optimizations by a series of graph transformations on CFG's in HOL.

Summary. We formally proved the correctness of a set of graph transformations used by systems that compute approximations of sets of (feasible) paths by building symbolic evaluation graphs. Formalizing all the details needed for a machine-checked proof was a substantial work. To our knowledge, such formalization was not done before.

The ATRACER model separates the fundamental aspects and the heuristic parts of the algorithm. Additional graph transformations for restricting abstractions or for computing interpolants or invariants can be added to the current framework, reusing the existing machinery for graphs, paths, configurations, etc.

Future Work. Currently, we are implementing a prototype "by hand" that must not only preserve feasible paths but heuristically generate abstractions and subsumptions. It would be possible to generate the core operations on red-black graphs by the Isabelle code-generator, by introducing un-interpreted function symbols for concrete heuristic functions that were mapped to implementations written by hand. This represents a substantial, albeit rewarding effort that has not yet been undertaken.

Acknowledgement. We thank Marie-Claude Gaudel for her support while doing this work and for her remarks on this article.

References

1. Ballarin, C.: Locales: a module system for mathematical theories. J. Autom. Reasoning **52**(2), 123–153 (2014)
2. Beyer, D., Henzinger, T.A., Jhala, R., Majumdar, R.: The software model checker blast. STTT **9**(5–6), 505–525 (2007)
3. Clarke, E.M., Kroening, D., Sharygina, N., Yorav, K.: SATABS: SAT-based predicate abstraction for ANSI-C. In: Proceedings of TACAS 2005, pp. 570–574 (2005)
4. Denise, A., Gaudel, M.-C., Gouraud, S.-D., Lassaigne, R., Oudinet, J., Peyronnet, S.: Coverage-biased random exploration of large models and application to testing. Int. J. Softw. Tools Technol. Transfer **14**(1), 73–93 (2011). ISSN 1433-2787
5. Grebenshchikov, S., Lopes, N.P., Popeea, C., Rybalchenko, A.: Synthesizing software verifiers from proof rules. In: Proceedings of PLDI 2012, pp. 405–416 (2012)
6. Ivančić, F., Yang, Z., Ganai, M.K., Gupta, A., Shlyakhter, I., Ashar, P.: F-SOFT: software verification platform. In: Etessami, K., Rajamani, S.K. (eds.) CAV 2005. LNCS, vol. 3576, pp. 301–306. Springer, Heidelberg (2005)
7. Jaffar, J., Navas, J.A., Santosa, A.E.: Unbounded symbolic execution for program verification. In: Proceedings of RV 2011 (2011)
8. Jaffar, J., Murali, V., Navas, J.A., Santosa, A.E.: TRACER: a symbolic execution tool for verification. In: Madhusudan, P., Seshia, S.A. (eds.) CAV 2012. LNCS, vol. 7358, pp. 758–766. Springer, Heidelberg (2012)
9. McMillan, K.L.: Lazy abstraction with interpolants. In: Ball, T., Jones, R.B. (eds.) CAV 2006. LNCS, vol. 4144, pp. 123–136. Springer, Heidelberg (2006)
10. Nipkow, T., Paulson, L.C., Wenzel, M. (eds.): Isabelle/HOL–A Proof Assistant for Higher-Order Logic. LNCS, vol. 2283. Springer, Heidelberg (2002)
11. Noschinski, L.: A Graph Library for Isabelle. Math. Comput. Sci. **9**(1), 23–39 (2015). doi:10.1007/s11786-014-0183-z. ISSN 1661–8289. http://dx.doi.org/10.1007/s11786-014-0183-z

Proof of OS Scheduling Behavior in the Presence of Interrupt-Induced Concurrency

June Andronick[1,2]([⊠]), Corey Lewis[1], Daniel Matichuk[1,2], Carroll Morgan[1,2], and Christine Rizkallah[1,2]

[1] Data61, CSIRO (formerly NICTA), Sydney, Australia
{june.andronick,corey.lewis,daniel.matichuk,
carroll.morgan,christine.rizkallah}@data61.csiro.au
[2] UNSW, Sydney, Australia
{june.andronick,daniel.matichuk,carroll.morgan,
christine.rizkallah}@unsw.edu.au

Abstract. We present a simple yet scalable framework for formal reasoning and machine-assisted proof of interrupt-driven concurrency in operating-system code, and use it to prove the principal scheduling property of the embedded, real-time *eChronos OS*: that the running task is always the highest-priority runnable task. The key differentiator of this verification is that the *OS* code itself runs with interrupts *on*, even within the scheduler, to minimise latency. Our reasoning includes context switching, interleaving with interrupt handlers and nested interrupts; and it is formalised in Isabelle/HOL, building on the Owicki-Gries method for fine-grained concurrency. We add support for explicit concurrency control and the composition of multiple independently-proven invariants. Finally, we discuss how scalability issues are addressed with proof engineering techniques, in order to handle thousands of proof obligations.

1 Introduction

We address the problem of providing strong machine-checked guarantees for (uniprocessor) operating-system code with a high-degree of interrupt-driven concurrency, but without hardware-enforced memory protection. Our contribution is a technique to reason feasibly about preemptive real-time kernels; we demonstrate its effectiveness on the commercially-used embedded *eChronos OS* [3].

Rather than inventing our own, new concurrency formalism from scratch, we chose to "go for simplicity". The Owicki-Gries method [19], more than 40 years old, was invented when mechanised theorem proving was scarce, when the principal concern was compact, elegant formalisms applied to small intricate problems.[1] Proving an elegant property of a small, intricate operating system thus seemed to be an ideal experiment; an added attraction was that the system is in real-world use. We further motivate the choice of Owicki-Gries in Sect. 4.1.

There were, however, two immediate concerns: even a small *operating system kernel* is nowhere near small enough to be reasoned about by hand, as the

---

[1] Broadly speaking, this was the "Hoare/Dijkstra/Gries School" of Formal Methods.

© Springer International Publishing Switzerland 2016
J.C. Blanchette and S. Merz (Eds.): ITP 2016, LNCS 9807, pp. 52–68, 2016.
DOI: 10.1007/978-3-319-43144-4_4

OG (Owicki-Gries) pioneers typically did [8]; and the *OG* concurrency model did not at first seem right for reasoning about the coarse-grained concurrency of switching between tasks. In *OG*, atomic actions are arbitrarily interleaved, and *OG*'s "await statements" (Sect. 2 below) are not designed for reasoning about interrupt-driven scheduling *including* the scheduler and context-switching code itself.

The former concern would, we hoped, be taken care of by the increased power and sophistication of theorem provers in the decades since *OG* was introduced: we use Isabelle/HOL [18]. The latter concern is handled by our novel style of *OG* reasoning, presented in a previous paper [4], that adapts *OG* to allow reasoning about interrupt-induced and scheduler-controlled concurrency. Although conceptually simple, this style introduces significant extra text that we hide through modern techniques: we call it "await painting". Await-painting introduces an active-task variable that tracks which task is allowed to execute, and wraps every atomic statement with an *AWAIT* using that variable to restrict the allowed interleavings. It is what allows the program code itself to control the interleaving between tasks, something not normally done in *OG*.

The top-level theorem we prove is a scheduling property, not directly expressible in *OG* without the await-painting step: that the currently executing task is the highest-priority runnable task. We prove it on a model of the interleaving between the *OS* and the (possibly nested) interrupt handlers. In future work, we aim to prove that this model is a correct abstraction of the existing *eChronos OS* implementation. Our proof assumes that all application-provided code is well behaved; that is, it does not change any *OS* variables, and application-provided interrupt handlers only call a specific API function. Although the *eChronos OS* does not explicitly export any functions modifying *OS* variables, the *OS* is not able to enforce these constraints as the system runs on hardware with no memory protection. These assumptions can, however, be statically checked.

Our specific contributions are the following ones. We extend the model presented previously [4] (Contribution 2), while the proofs themselves are new (Contributions 1, 3, 4, 5). Most of our model and proof framework is generic and should apply to systems that support interruptible OS-es, preemptible applications and nested interrupts. All our proofs and model are available online [1].

1. We provide a proof framework, using a formalisation of *OG* in Isabelle/HOL [20], to reason about interrupt-driven and scheduler-controlled concurrency. Our framework is driven by the aim to handle complex parallel composition which requires that invariants can be proved compositionally. (Sect. 4.1)
2. We give an updated model of our interleaving framework and instantiation to the *eChronos OS*. It extends the one presented previously [4] by including system calls that can influence the scheduling decisions, introduces nondeterminism to properly represent under-specified operations, and properly separates generic interleaving and *eChronos* instantiation. (Sect. 3)
3. We show that proving the scheduling property for the *eChronos* system is within the capabilities of modern theorem provers, at least for an application

of this size. This includes handling the *OG*-characteristic of quadratically many "interference-freedom" verification conditions. (Sect. 4.2)

4. We develop a number of proof engineering techniques to address some observed problems that occur in a proof of this scale. (Sect. 4.3)

5. We contribute to the real-world utility of an existing preemptive kernel that is in widespread use, in particular in medical devices.

2   Background and Big Picture

The goal of our work is to provide a verification framework for *OS* code involving interrupt-induced concurrency, in particular real-time embedded *OS*-es.

A real-time *OS* (RTOS), like the *eChronos OS*, is typically used in tightly constrained embedded devices, running on micro-controllers with limited memory and no memory-protection support. The role of the *OS* is closer to that of a library than of a fully-fledged operating environment, allowing the application running on top to be organised in multiple independent tasks and providing a set of API functions that the application tasks can call to synchronise (signals, semaphores, mutexes). The *OS* also provides the underlying mechanism for switching from one task to another, and is responsible for sharing the available time between tasks, by scheduling them according to some given *OS*-specific policy. For instance, tasks can cooperatively yield control to each other (*cooperative scheduling*); or tasks can be scheduled according to their assigned *priority*, and their execution must then be *preempted* if a higher priority is made available (*preemptive scheduling*). The system typically also reacts to external events via interrupts. An *interrupt handler* needs to be defined for each interrupt by the application. When an interrupt occurs, the hardware ensures that the corresponding interrupt handler is executed (unless the interrupt is disabled/masked).

The job of the scheduler is to ensure that at any given point the running task is the correct one, as defined by the scheduling policy of the system.

For instance, in a priority-based preemptive system, when a task in unblocked (e.g. by an interrupt handler sending the signal it was waiting for) a context switch should occur if this task is at a higher priority than the currently running one. This defines the *correctness of the scheduling behavior* and is the target of our proof about the *eChronos OS*.

To reason about such an RTOS, and prove such a scheduling property, we present a *verification framework supporting the concurrency reasoning required by preemption and interrupt handling* (on uniprocessor hardware).

In previous work [4] we provided a model of interleaving that faithfully represents the interaction between application code, OS code, interrupt handler and scheduler, in such an RTOS. Roughly, the system is modelled as a parallel composition $A_1||...||A_n||Sched||H_1||...||H_m$, where the code for each application A_i is parameterised (including calls to *OS* API functions), as well as the code for each interrupt handler H_j, and the code for the scheduler. The key feature of the framework is that the interleaving in the parallel composition is *controlled* using a small formalised API of the hardware mechanisms for taking interrupts, returning from interrupts, masking interrupts, etc. We have formalised our logic in Isabelle/HOL, based on [20].

In this paper we present (a) a logic to prove invariants about such parallel composition, with support for handling complex proofs and (b) instantiation of the model to the *eChronos OS* and proof of its scheduling behavior. Namely, the property we prove is

$$\|-_b \ \{\!| scheduler\text{-}invariant |\!\} \ \{\!| True |\!\} \ eChronos\text{-}sys \ \{\!| False |\!\} \tag{1}$$

where $\|-_b$ is the derivability of a "bare" program (i.e. with no annotations), and is defined in terms of $\|-_i$ at the end of Sect. 4.1. The notation $\|-_i \ I \ p \ c \ q$ means that if the (annotated) parallel program c starts in a state satisfying the precondition p, then the invariant I holds at all reachable execution steps of c, and the postcondition q holds if c terminates. The definition, explained in detail in Sect. 4.1, adds an invariant to the original Owicki-Gries statement $\|- \ p \ c \ q$, which in turn is an extension of traditional Hoare-logic statement $\vdash \ p \ c \ q$. Owicki-Gries extends the sequential programs of Hoare-logic with two constructs: the parallel composition $c_1 \| c_2$ and the *AWAIT*-statement *AWAIT b DO c OD*. The execution of $c_1 \| c_2$ is the execution of the current instruction of *either* c_1 *or* c_2. The statement *AWAIT b DO c OD* can only execute if condition b is satisfied, in which case c is executed *atomically* (meaning that b is still true as c begins, and reasoning within c is purely sequential).

eChronos_sys is our model of an *eChronos* system. The *eChronos OS*[2] provides a priority-based preemptive scheduler with static priorities. It comprises about 500 lines of C code and runs on ARM uniprocessor hardware.[3] Our model is an instantiation of the generic model of interleaving [4], with definitions for the scheduler, and for the API system calls (the ones that may influence the scheduling decisions) which are called from application or handler code. This model is given in Sect. 3.

Coming back to the property (1), it says that the *eChronos* system, starting in *any* initial state and never terminating, will satisfy the *scheduler_invariant* at every point of execution. The precondition *True* is always trivially satisfied and the postcondition *False* is valid because the system is an infinite loop of execution. The invariant property for the *eChronos OS* states that *the running task is always the highest priority runnable task*. We describe its formal definition and proof in Sect. 4.2.

Owicki-Gries reasoning introduces quadratically many proof obligations due to parallelism: indeed our proof for the *eChronos* scheduling behavior initially generates thousands. However, using a combination of (a) compositional proofs

---

[2] The *eChronos OS* [3] comes in many variants, varying in the hardware they run on, the scheduling policy they enforce and the synchronisation primitives they offer. In this paper we simply refer to the *eChronos OS* for the specific variant that we are targeting, called *Kochab*, which supports the features that create interesting reasoning challenges (preemption, nested-interrupts, etc.).

[3] We specifically target an ARM Cortex-M4 platform, simply referred to as ARM here.

for proving invariants; (b) controlled interleaving to eliminate unfeasible executions; and (c) proof engineering techniques to automate discharging a large number of conditions, we show the feasibility of this approach for a preemptive and interruptible real-time OS running on a uniprocessor (Sect. 4.3).

3   The Model

In recent work [4] we presented a model of interleaving between application code, interrupt handler code, and scheduler code, for an ARM platform that supports both direct and delayed calls to the scheduler. The model was designed to be generic and we then instantiated it to the *eChronos OS*.

Here we present this model, with several improvements. We explicitly separate the generic portion to clarify how one could use the framework to formalise a different system. We extended the formalisation of the *eChronos OS* to include system calls that can influence the scheduling decisions. Finally, we introduce non-determinism to properly represent under-specified operations.

3.1   A Generic Model of Interrupt-Driven Interleaving

In the generic part of the model we focus both on formalising the hardware mechanisms that control interleaving and on faithfully representing the concurrency induced by interrupts. The system is modelled as the parallel composition $svc_aTake||svc_a||svc_s||H_1||...||H_m||A_1||...||A_n$, where the scheduler (*Sched* in Sect. 2) is taking into account here both direct/synchronous calls to the scheduler (svc_s) and delayed/asynchronous ones (svc_a). ARM provides a direct (synchronous) *supervisor call (SVC)* mechanism that can be thought of as a program-initiated interrupt. It is triggered by the execution of the SVC instruction (SVC_now), which results in the execution switching to an SVC handler (svc_s). ARM also provides a delayed (asynchronous) supervisor call, also behaving like an interrupt, with instructions allowing programs to enable and disable it. It is triggered by raising a flag (svc_aReq), whose status is constantly checked by the hardware (modelled by svc_aTake). If the flag is raised and the asynchronous SVC is enabled, the execution will switch to a specific handler (svc_a). In the case of the *eChronos OS*, both SVC handlers will execute the scheduler.

The formal model of interleaving is presented in Fig. 1.[4] The code for the application initialisation, application tasks, interrupt handlers and SVC handlers are parameters as they are system-specific.

The code for each part of the parallel composition is in fact wrapped in an infinite *WHILE* loop, reflecting the reactive nature of the system. Moreover, to faithfully represent the controlled interleaving allowed by the hardware

---

[4] The model is written in a simple formalised imperative language with parallel composition and await statements, which has the following syntax:
$c \equiv x := v \mid c;; c \mid IF\ b\ \ THEN\ c\ \ ELSE\ c\ FI \mid WHILE\ b\ DO\ c\ OD \mid$
$\quad\quad AWAIT\ b\ \ THEN\ c\ END \mid (COBEGIN\ c\ \|\ c\ COEND)$
The *SCHEME* constructor models a parametric number of parallel programs, as seen in [20]. Here this is the number of handlers plus the number of application tasks.

definition *interleaving app-init svc_a-code svc_s-code handler-code app-code* $\equiv$
    hardware-init;;
    app-init;;
  (*COBEGIN*
    WHILE True DO svc_a Take OD
    ‖ *WHILE True DO control svc_a svc_a-code OD*
    ‖ *WHILE True DO control svc_s svc_s-code OD*
    ‖ *SCHEME [user0 $\leq$ i $<$ user0 + nbRoutines]*
    IF i $\in$ I THEN
      WHILE True DO ITake i;; control i (handler-code i) OD
    ELSE
      WHILE True DO control i (app-code i) OD
    FI
  COEND)

Fig. 1. Definition of generic interrupt-driven interleaving in Isabelle/HOL

we *await-paint* most of the code. This means that we introduce an active-task variable, AT, and associate each task, including the interrupt handlers, with a unique identifier. Every atomic statement c in Task t is then converted into a statement *AWAIT AT=t THEN c END*. As described in Sect. 2 this prevents the execution of c until the await-condition holds. Only when the command $AT := t$ is performed will Task t be able to execute. In particular, this means that no other Task t' with $t' \neq t$ will be able to interfere. In the model this await-painting is performed by the *control* function, which recursively wraps every command with an *AWAIT*.

To be precise, we await-paint all of the code except for where concurrency can actually occur: during the background hardware routine $svc_a Take$, and when an interrupt is taken, *ITake*. These represent our model of the hardware mechanisms that control the interleaving and context switching. We define them as below, with an *AWAIT* with the condition that the interrupt is allowed to be taken. We have previously described these functions in detail [4], but abstractly they save the AT variable on a stack (the notation $x\#xs$ adds x to the list xs) and switch to the interrupt or SVC handler. While these are the only places where concurrency is not controlled, they are still guarded by the condition that the interrupt is enabled (is in the set EIT of enabled interrupts), is not already running (or itself interrupted), and is allowed to interrupt the active task.

can-interrupt i $\equiv$ i $\in$ EIT $-$ ATStack $\wedge$ i $\in$ interrupt-policy AT

ITake i $\equiv$ AWAIT can-interrupt i
THEN ATStack := AT # ATStack;; AT := i END

svc_a Take $\equiv$ AWAIT svc_a Req $\wedge$ can-interrupt svc_a
THEN ATStack := AT # ATStack;; AT := svc_a;; svc_a Req := False END

One of the central features of the *eChronos OS* is that while *OS* code is interruptible, it is not preemptible. In practice, this means that although standard interrupts are handled immediately, the call to the scheduler via the SVC$_a$

interrupt is delayed until the OS code is completed. To achieve this, SVC_a is temporarily removed from EIT, ensuring that $svc_a Take$ cannot execute.

We also provide a model of the SVC_now and $IRet$ hardware instructions that are called by OS functions.

$SVC\text{-}now \equiv ATStack := AT \mathbin{\#} ATStack;;\ AT := svc_s$

$IRet \equiv IF\ svc_a Req \wedge can\text{-}interrupt'\ svc_a$
$THEN\ AT := svc_a;;\ svc_a Req := False$
$ELSE\ AT := hd\ ATStack;;\ ATStack := tl\ ATStack\ FI$

$can\text{-}interrupt'\ i \equiv i \in EIT - ATStack \wedge i \in interrupt\text{-}policy\ (hd\ ATStack)$

SVC_now is used to directly switch to the SVC interrupt handler. $IRet$ returns control from an interrupt handler: it either switches control to svc_a (if svc_a has both been requested and is allowed to interrupt the *head of ATStack*) or returns control to the head of $ATStack$, which was saved as part of $ITake$. Although not explicitly part of *interleaving*, we require that the last command of svc_a_code, svc_s_code, and $handler\_code$ is $IRet$. This can be checked when instantiating the interleaving model to a specific system.

3.2 Instantiation to the *eChronos OS*

To model the *eChronos OS* we now just need to instantiate the above framework with the OS specific code. We give an overview of this instantiation below[5] while the full details can be found online [1] or in our previous paper [4].

$eChronos\text{-}sys \equiv interleaving\ eChronos\text{-}init\ eChronos\text{-}svc_a\text{-}code\ eChronos\text{-}svc_s\text{-}code$
$\qquad\qquad\qquad\qquad eChronos\text{-}handler\text{-}code\ eChronos\text{-}app\text{-}code$

$eChronos\text{-}svc_a\text{-}code \equiv schedule;;\ context\text{-}switch\ True;;\ IRet$

$eChronos\text{-}svc_s\text{-}code \equiv schedule;;\ context\text{-}switch\ False;;\ IRet$

$eChronos\text{-}handler\text{-}code\ i \equiv E :\in \{E'\mid E \subseteq E'\};;\ svc_a Request;;\ IRet$

$eChronos\text{-}app\text{-}code\ i \equiv userSyscall :\in \{SignalSend,\ Block\};;$
$IF\ userSyscall = SignalSend$
$THEN\ svc_a Disable;;\ R :\in \{R' \mid \forall i.\ R\ i = Some\ True \longrightarrow R'\ i = Some\ True\};;$
$\quad svc_a Request;;\ svc_a Enable;;\ WHILE\ svc_a Req\ DO\ SKIP\ OD$
$ELSE\ IF\ userSyscall = Block$
$\quad THEN\ svc_a Disable;;\ R := R(i \mapsto False);;\ SVC\text{-}now;;\ svc_a Enable;;$
$\quad\quad WHILE\ svc_a Req\ DO\ SKIP\ OD$
$\quad FI$
FI

---

[5] For presentation purposes, we omit ghost variables added to the program for verification purposes. The notation $x :\in S$ stands for non-deterministically updating x to be any element of S.

In this work we focus on the scheduling behaviour, modelling only the parts that might affect scheduling decisions. These decisions depend on two variables, R for runnable tasks and E for the events signalled by interrupt handlers.

The parameters $eChronos\_svc_a\_code$ and $eChronos\_svc_s\_code$ are almost identical and are used by the OS to call the scheduler. First, *schedule*, defined below, picks the next task to run by first updating R through handling the unprocessed events E before using whichever scheduling policy is in place. After choosing the task a context switch is performed, with the old task being saved and the new task being placed on the stack.

$schedule \equiv nextT := None;;$
$WHILE\ nextT = None$
$DO\ E\text{-}tmp := E;;\ R := handle\text{-}events\ E\text{-}tmp\ R;;\ E := E - E\text{-}tmp;;$
$nextT := sched\text{-}policy\ R\ OD$

$context\text{-}switch\ preempt\text{-}enabled \equiv$
$contexts := contexts(curUser \mapsto (preempt\text{-}enabled,\ ATStack));;$
$curUser := the\ nextT;;\ ATStack := snd\ (the\ (contexts\ curUser));;$
$IF\ fst\ (the\ (contexts\ curUser))$
$THEN\ svc_a Enable$
$ELSE\ svc_a Disable\ FI$

Next, $eChronos\_handler\_code$ is mostly application-provided and is only allowed to affect the behaviour of the OS by expanding the set of events E. A flag is then raised saying that the scheduler should be run as soon as enabled. To finish, the handler, by calling $IRet$, either returns control to the previously executing context or, if allowed, switches control to the scheduler.

Finally, the only way the application code can affect the interleaving behaviour is via system calls. We model two representative syscalls, *signal_send* and *block*. In the *eChronos OS*, syscalls run with interrupts enabled, but preemption disabled; that is, they are surrounded by disabling and enabling the svc_a interrupt. This is to delay a call to the scheduler requested by an interrupt handler until *after* the OS syscall is finished. Each syscall ends with a loop that ensures that, if required, the scheduler executes before the OS returns control to the application. The syscall *signal_send* increases the set of runnable tasks and sets a flag indicating that the scheduler needs to be run, while *block* modifies R so that the specific application task is no longer runnable and then directly calls svc_s via SVC_now.

4   Proof Framework and Scheduler Proof

4.1   Framework and Compositionality Lemma

In this section we explain our definition of *derivability of a (bare) parallel program c with respect to an invariant I, precondition p and postcondition q*, denoted $\|-_b\ I\ p\ c\ q$. We present the framework that we build to ease the proof of such a statement, by assuming helper invariants, and decomposing the proof into composable subproofs. In Sect. 4.2, we use this framework to state and prove the scheduler correctness $\|-_b \{ scheduler\text{-}invariant \} \{ True \}$ $eChronos\text{-}sys \{ False \}$.

We use the Owicki-Gries (OG) treatment of concurrency, captured in Isabelle/HOL by Prensa [20]. Reasoning about high-performance shared-variable system code requires a very low level of abstraction [4], and motivated our choosing OG over alternatives such as the more structured Rely-Guarantee method. Futhermore, our goal was to verify existing code rather than to synthesise new code, i.e. a bottom-up proof rather than a top-down correctness-by-construction exercise. An attractive possibility, however, is to now use the invariants and assertions that OG and the code helped us to synthesise, and to explore whether with that "head start" a Rely-Guarantee approach would be possible: probably it would suggest proof-motivated modification to the code.

The OG method, introduced 40 years ago, extends the Hoare-style assertional-proof technique to reason about a number of individually sequential processes that are executed collectively in parallel. Namely, OG provides (1) a definition of *validity* of a Hoare triple over a (*fully annotated*) parallel composition of programs, denoted $\|= p\ c\ q$; (2) a set of proof rules for efficient verification of such a statement, with an associated *derivability* statement, denoted $\|- p\ c\ q$; (3) a *soundness* theorem of the rules w.r.t validity, namely $\|- p\ c\ q \longrightarrow \|= p\ c\ q$; and finally (4) an automated verification condition generator (VCG), i.e. a tactic *oghoare* in Isabelle/HOL to decompose a derivability statement into subgoals.

We explain these standard OG definitions before going into our extensions, which are proved sound with respect to the concurrency semantics. In the following, c and ac are mutually recursive datatypes; c is sequential code, which can contain a parallel composition of annotated code, ac. The parallel composition consists of a list of annotated programs with their postconditions. An annotated program can contain an AWAIT statement, whose body is a sequential program.

$$c \equiv \quad x := v \mid c;; c \mid IF\ b\ THEN\ c\ ELSE\ c\ FI \mid WHILE\ b\ DO\ c\ OD \mid$$
$$COBEGIN\ ts\ COEND$$
$$ts \equiv [\ ] \mid (aco, \{\!|a|\!\}) \# ts$$
$$aco \equiv None \mid Some\ ac$$
$$ac \equiv \{\!|a|\!\}\ x := v \mid ac;; ac \mid \{\!|a|\!\}\ IF\ b\ THEN\ ac\ ELSE\ ac\ FI \mid$$
$$\{\!|a|\!\}\ WHILE\ b\ INV\ \{\!|a|\!\}\ DO\ c\ OD \mid \{\!|a|\!\}\ AWAIT\ b\ THEN\ c\ END$$

In the above b is a boolean expression, and a is an assertion. Validity $\|= p\ c\ q$ is defined in terms of the execution semantics of the program, as in Hoare logic (all states reachable via multiple steps of execution from initial states satisfying the precondition will satisfy the postcondition). The execution of the standard language constructs is also defined as in Hoare logic. For parallel composition, one of the programs is at each step non-deterministically chosen to make progress. For the *AWAIT* statement, the body is executed, under the condition that the guard is satisfied (and that the body does not contain any parallel composition). The derivability rules ($\|- p\ c\ q$) are also the same as for Hoare logic. The key feature of OG is providing a proof rule for parallel composition, which consists in showing *local correctness* and *interference-freedom* for a list

[(*Some* ac_1, q_1), ..., (*Some* ac_n, q_n)] of annotated programs. Each program ac_i and postcondition q_i is first proved correct in isolation using standard sequential Hoare logic rules. Then, each assertion a in ac_i is proved to not be interfered with by any (annotated) statement $\{a'\}$ st' in another program ac_j (shown using standard Hoare logic as well: $\{a \wedge a'\}$ st' $\{a\}$). This interference-freedom requirement makes the *OG* technique non-compositional and quadratic. However, in systems with limited concurrency like ours, the complexity is reduced and we apply proof engineering techniques to make it scale to verify real *OS* scheduling behavior.

Our first, small, extension to the original definition of derivability is to explicitly talk about the *invariant* of the program. The programs we target are infinite loops, where the postcondition is not reached. Therefore, their correctness can be expressed better in terms of an invariant over their execution. An invariant for an annotated program is merely a property repeated in all annotations. However, manually inserting it everywhere is tedious, error-prone and results in bad readability. Instead, we define the derivability of invariants as follows:

$$\|-_i I\ p\ c\ q \equiv \|- p\ (add\text{-}inv\text{-}com\ I\ c)\ q$$

where *add_inv_com I c* simply inducts over the structure of program c and adds a conjunction with I to all annotations.

Our second extension is to be able to *assume* a helper invariant, while proving a main invariant. This feature is necessary in larger proofs where the property of interest relies on a number of other invariants. These invariants might need different sets of annotations; proving them all together quickly becomes unreadable, and even infeasible due to the explosion of complexity. It also makes it hard for multiple people to work on a single proof. We modify the original set of *OG* derivability rules to allow assuming an invariant, denoted $I\ \|- p\ c\ q$, as follows: preconditions get an extra conjunction with I (i.e. I can be assumed true initially) and postconditions get an extra implication from I (i.e. the postcondition itself only need to be proven if I holds). Then $\|- p\ c\ q$ simply stands for *UNIV* $\|- p\ c\ q$ (*UNIV* is the universal set) and $I'\ \|-_i I\ p\ c\ q$ stands for $I'\ \|- p\ (add\text{-}inv\text{-}com\ I\ c)\ q$. Putting things together, we provide a *compositionality lemma* to decompose the proof along the invariants.

$$\frac{I'\ \|-_i I\ p\ c\ q \qquad \|-_i I'\ p'\ c'\ q' \qquad merge\text{-}prog\text{-}com\ c\ c' = Some\ c''}{\|-_i (I' \cap I)\ (p \cap p')\ c''\ (q \cap q')} \tag{2}$$

where the merge of two programs requires the programs to only differ on annotations (i.e. have identical program text), and if so, returns the same program text with merged annotations (by conjunction). Our proof of the *eChronos* scheduler uses this lemma extensively, and would have not been tractable without it.

Finally we define the derivability of an invariant I over a *bare* program (i.e. not annotated) as the existence of an appropriate annotation sufficient to prove I, as follows:

$$\|-_b \; I \; p \quad c \; q \equiv \exists \, c'. \; extract\text{-}prg \; c' = c \wedge \|-_i \; I \; p \; c' \; q \qquad (3)$$

Since invariants are merely annotations, we can prove an introduction rule for derivability, which allows us to directly introduce helper invariants:

$$\frac{\exists \, c'. \; extract\text{-}prg \; c' = c \wedge \|-_i \; (I \cap I') \; p \; c' \; q}{\|-_b \; I \; p \quad c \; q} \qquad (4)$$

4.2   The Statement and Its Proof

Now that we have defined our framework, we present the statement of *eChronos'* scheduler correctness:

$$\|-_b \; \{\!|scheduler\text{-}invariant|\!\} \; \{\!|True|\!\} \; eChronos\text{-}sys \; \{\!|False|\!\} \qquad (1)$$

The definition of *eChronos_sys* is described in Sect. 3. Here we define *scheduler_invariant* and explain its proof.

As previously mentioned, the key property enforced by the *eChronos OS* is that the running application task is always the highest priority runnable task. We express this property as an invariant *scheduler_invariant*, defined as follows:

scheduler-invariant $x \equiv$
$AT \; x \in U \wedge svc_a \in EIT \; x \wedge \neg \; svc_a Req \; x \longrightarrow$
sched-policy $(handle\text{-}events \; (E \; x) \; (R \; x)) = Some \; (AT \; x)$

where x is the current state. The statement says that whenever the currently active task is a user (i.e. not an interrupt handler and not the scheduler), and we are not inside a system call (we will come back to that), then that user is indeed the one supposed to be running, according to the scheduling policy. The latter is expressed by the fact that the scheduling policy would choose the running user if re-run with the current values of events E and of the runnable set R.

The condition of not being in a system call is because, as explained in Sect. 3, preemption is turned off during system calls, meaning that any asynchronous request for the scheduler is delayed until the system call finishes running. Therefore, when the currently active task *is* a user, but is *inside* a system call, it might not be of highest priority. However, as soon as the system call is finished, the execution must *not* go back to that user but must instead immediately call the scheduler. The invariant should, therefore, only be checked outside of system calls. Being outside a system call is defined by the asynchronous scheduler being enabled.[6] The third premise represents the specific situation where preemption is turned back on, but the request for asynchronous scheduling is still on, waiting for the hardware to do the switch (as explained in Sect. 3). The execution only goes back to the user when this asynchronous scheduling request has been handled.

---

[6] Disabling the scheduler is one of the functions that the *eChronos OS* does *not* export, to keep control of latency, as mentioned in Sect. 1.

Now we describe how we prove (1). We use lemma (4), and for this we create a suitable complete annotation of *eChronos_sys* sufficient to prove the invariant *scheduler_invariant*. The details of the annotations are not particularly insightful, but the process of identifying them and incrementally building them is discussed at the end of this section. The main theorem we prove is:

$$\|-_i \ (\{scheduler\text{-}invariant\} \cap helper\text{-}invs) \ \{True\} \ eChronos\text{-}sys\text{-}ann \ \{False\} \quad (5)$$

where *eChronos_sys_ann* is the fully annotated program, whose extracted program text is *eChronos_sys*, and where *helper_invs* are a set of nine invariants about *eChronos* state variables and data structures, required to prove *scheduler_invariant*. We prove lemma (5) by applying the compositionality lemma. We first prove the scheduler invariant *assuming* all the helper invariants:

$$helper\text{-}invs \ \|-_i \ \{scheduler\text{-}invariant\} \ \{True\} \ eChronos\text{-}sys\text{-}ann \ \{False\} \quad (6)$$

We then prove each helper invariant independently (and this can be done by different people, increasing efficiency). These invariants reveal much about the data structures but do not represent a high level correctness property of the *eChronos OS*. We omit their definitions for space reasons (they are available online [1]), and just give two representative examples:

$$last\text{-}stack\text{-}inv \ x \equiv last \ (AT \ x \ \# \ ATStack \ x) \in U$$
$$ghostP\text{-}inv \ x \equiv ghostP \ x \longrightarrow AT \ x \in I \cup \{svc_a, \ svc_s\}$$

The first invariant describes the allowed shape of the stack, namely that its last element is always a user task. It is representative of the invariants about the data structures. The second invariant is representative of the need for ghost variables to express where certain programs are in their execution. Here *ghostP* is a ghost flag that represents the fact that the asynchronous scheduler is running. The invariant *ghostP_inv* ensures that *ghostP* cannot be set if the active task is a user application. It is needed in the proofs of interference-freedom of user applications' assertions: it tells us that the asynchronous scheduler instructions cannot violate them as they cannot be running.

For each of the nine helper invariants, we prove that it is preserved by *eChronos_sys_ann*. Some of them rely on others so we reuse the compositionality lemma for these.

We proved all of these helper lemmas along with lemma (6) in an iterative process to discover the required annotations. Roughly, we start with minimal annotations, and run the *oghoare* tactic to generate the proof obligation for local correctness and interference-freedom. We apply the techniques discussed in Sect. 4.3 to reduce the number of subgoals by removing duplication and automatically discharging as many as possible. We are then left with a manageable set of subgoals, where we can identify which assertion in the program is being proved, and can start augmenting assertions as required to prove these subgoals.

4.3 Proof-Engineering Considerations

The *oghoare* proof tactic, offered in the Isabelle distribution and derived from [20], is the VCG used for decomposing an annotated program into subgoals. Each of these goals is ultimately either a judgement that the precondition for each program step is sufficient to demonstrate its postcondition or that a given annotation is not interfered with by anything else running in parallel.

The tactic is defined as a mutually recursive function that decomposes program sequencing, program (user-defined) annotations, and parallel composition. The provided implementation of this tactic results in a quadratic explosion of proof obligations: a ~200 line parallel program takes *oghoare* ~90 s to generate ~3,000 subgoals.

Rather than solve each of these goals by hand, we chose to write a single custom tactic which was powerful enough to solve all of them. Here we leveraged Isabelle's existing proof automation and parallelisation infrastructure [17]. With some instrumentation and custom lemmas, Isabelle's *simplifier* [18, Sect. 3] can discharge almost all of the subgoals produced from *oghoare*. Isabelle's provided PARALLEL_GOALS tactical allows us to apply our custom tactic to all subgoals simultaneously in parallel, resulting in a significant reduction in overall proof processing time. Despite this infrastructure, however, these ~3,000 subgoals can still take over an hour to prove. This is impractical from a proof engineering perspective, as this proof needs to be re-run every time the tactic is adjusted or the program annotations are changed. This prompted the development of several proof engineering methodologies that, although generally applicable, were instrumental in the completion of this proof.

Subgoal Deduplication and Memoization. An initial investigation revealed that many of the proof obligations produced by *oghoare* were identical. Isabelle's provided *distinct_subgoals* tactic can remove duplicate subgoals, but takes over 30 s to complete on 3,000 subgoals. We found that we could instead store proof obligations as *goal hypotheses* as they are produced, which are efficiently de-duplicated by Isabelle's proof kernel. This adds negligible overhead, and results in approximately a 3-fold reduction in the total number of proof obligations.

This large number of duplicate subgoals is a consequence of having many identical program annotations. The *oghoare* tactic recurses on the syntax of the annotated program, generating non-interference verification conditions for each annotation. Rather than complicate the implementation of *oghoare*, we chose to simply de-duplicate these proof obligations as they are produced.

Although this de-duplication reduces the total time required to finish the proof, it still indicates that the *oghoare* tactic is doing redundant computation and that the observed ~90 s overhead could be reduced. To address this, we developed a new tactical for memoization, SUBGOAL_CACHE, which caches the result of applying a given tactic to the current subgoal. When the tactic is subsequently invoked again, the cache is consulted to determine if it contains a previously-computed result for the current subgoal. On a cache hit, the stored result is simply applied rather than having the tactic re-compute it. Isabelle's

LCF-style proof kernel guarantees that such a cache is sound, as each cached result is a previously-checked subgoal that was produced by the kernel.

We applied SUBGOAL_CACHE to each of the mutually recursive tactics that *oghoare* comprises. Including subgoal de-duplication, this change reduces the running time of *oghoare* from $\sim$90 s to $\sim$5 s (on a $\sim$200 line program), without requiring any change to the underlying algorithm.

Subgoal Proof Skipping. Even once these $\sim$3,000 subgoals have been de-duplicated down to $\sim$1,000 distinct subgoals, discharging them all can take between 5 and 30 min, depending on the particular annotations. The development strategy was to run the simplifier on all the subgoals and then analyse those that remained unsolved. Each iteration required adding additional program annotations or providing the simplifier with additional lemmas in order to discharge more subgoals. This would then require waiting for up to 30 min again to see if the change was successful.

To save time, we added another tactical, PARALLEL_GOALS_SKIP, which builds on Isabelle's *skip_proofs* mode and PARALLEL_GOALS tactical. This tactical is equivalent to the existing PARALLEL_GOALS, but records which subgoals were successfully discharged as global state data. This global record can then be accessed if the tactic is re-executed in Isabelle/jEdit (after, for example, going back and adding another annotation to the function). When re-executing the tactic, subgoals that were previously discharged are instead simply *skipped* and assumed solved. In practice, this reduces the effective iteration time from minutes to seconds, depending on the significance of the change. When the proof is complete, PARALLEL_GOALS_SKIP is then replaced with PARALLEL_GOALS in order to avoid skipping proofs and guarantee soundness.

Together these methodologies make this approach far more tractable and scalable than was previously thought possible.

5   Related Work

We discuss models of interrupts, verification of operating systems, models of real-world systems with concurrency, and automation and mechanisation of *OG*.

The closest work to ours formalises interrupts explicitly [9,12,13], using "ownership" to reason about resource sharing. That provides verification modularity, but the run-time discipline it induces limits the effective concurrency unacceptably for a real-time system where low latency is paramount. Indeed, they assume that interrupts are disabled when data is shared and during scheduling and context switching; we do not. They also do not support nested interrupts, although some [9] do suggest how they could. However, one [12] does support multicore.

Other works in *OS* verification, less closely related, either do not model interrupts, or target systems where *OS* code runs with interrupts disabled. Close to the *eChronos OS* is FreeRTOS [2], a real-time *OS* for embedded microcontrollers. Its verification has been the target of several projects: in [6,7,10], the focus of the verification is on the scheduling policy itself (picking the next task), or on the correct handling of the data-structure lists and tasks by the scheduler. While FreeRTOS runs with interrupts mostly enabled, interrupt handling is not modelled in these works, nor is context-switching. That work is complementary to ours, where we leave the policy generic. In [5], the authors target progress properties (absence of data-race and deadlock) of their proposed multicore version of FreeRTOS. Again, this is orthogonal to our focus on correctness. Another embedded real-time *OS* that has been verified [14] is used in the OSEK/VDX automotive standard. It has been model-checked in CSP, and the interleaving model has some similarities with ours, where tasks are in parallel composition with the scheduler. But again, interrupts are out of scope.

In *OS* verification generally, existing, larger *OS*-es that are formally verified [16,21], run with interrupts disabled throughout the executions of system calls from applications, making those calls' executions sequential.

Finally, a notable verification effort outside the pure *OS* world is on-the-fly garbage collection (GC) in a relaxed memory model [11]. They too chose the rigour of Isabelle/HOL, and used a system-wide invariant. Concurrency control is via message passing, while the *eChronos OS* uses shared variables.

A GC is also the target of the main existing use of formalised and mechanised Owicki-Gries. Prensa, the author of the *OG* formalisation [20] in Isabelle/HOL, on which our work is based, used her framework to verify a simple GC algorithm. In terms of scale, Prensa's model contains only two threads in parallel, one of which contains only 2–3 instructions: this generates only ~100 verification conditions. Our proof effort generates ~3,000 *VC*'s, and so requires significant proof engineering [15] to be feasible. Also, Prensa's work does not extract the correctness property in a separate, well-identified invariant annotation. Nor does it allow control of concurrency and interleaving between the parallel processes, that is, the inclusion of a task-scheduler which is itself subject to verification.

6   Conclusion

Our contribution has been the intersection of three ideas: that modern proof-automation now makes Owicki-Gries reasoning about concurrency feasible for much larger programs than before; that *OG* can be used in a style that allows reasoning about programs that control and limit their own concurrency; and that an ideal target to test these ideas is a small, highly interleaved preemptive real-time operating system. To our knowledge this is the first proof of an *OS* system running with interrupts enabled even during scheduling, and allowing nested interrupts. The proof does make assumptions about application code conventions, as remarked in Sect. 1, precisely because the *OS* is not hardware protected. But these are statically checkable and reasonable for applications running on a real-time *OS*. Our experience in doing this proof should be useful to the wider ITP community: we can contribute proof-engineering insights for dealing with a huge amount of goals. Furthermore, our proof of scheduler correctness for a real-time *OS* already in commercial use in medical devices has real, practical value.

The work we have done so far sits roughly in the middle of a complete verification of an application running on the *eChronos OS* platform. Above, further work could provide a verified *OS* API specification that application programmers could use to prove their programs' correct behavior. Below, we have yet to prove refinement between the large-grained atomic steps and the low-level primitives for concurrency, and that the *OG* model on which our proof is based accurately captures the behaviour of our target processor. Those last two will be our next step, as well as continuing to develop proof-engineering techniques crucially needed for efficient and scalable concurrency software verification.

Acknowledgements. The authors would like to thanks Gerwin Klein and Stefan Götz for their feedback on drafts of this paper. NICTA is funded by the Australian Government through the Department of Communications and by the Australian Research Council through the ICT Centre-of-Excellence Program.

References

1. eChronos model and proofs. https://github.com/echronos/echronos-proofs
2. FreeRTOS. http://www.freertos.org/
3. The eChronos OS. http://echronos.systems
4. Andronick, J., Lewis, C., Morgan, C.: Controlled Owicki-Gries concurrency: reasoning about the preemptible eChronos embedded operating system. In: Workshop on Models for Formal Analysis of Real Systems (MARS) (2015)
5. Chandrasekaran, P., Kumar, K.B.S., Minz, R.L., D'Souza, D., Meshram, L.: A multi-core version of FreeRTOS verified for datarace and deadlock freedom. In: MEMOCODE, pp. 62–71. IEEE (2014)
6. Cheng, S., Woodcock, J., D'Souza, D.: Using formal reasoning on a model of tasks for FreeRTOS. Formal Aspects Comput. **27**(1), 167–192 (2015)
7. Divakaran, S., D'Souza, D., Kushwah, A., Sampath, P., Sridhar, N., Woodcock, J.: Refinement-based verification of the FreeRTOS scheduler in VCC. In: Butler, M., Conchon, S., Zaïdi, F. (eds.) Formal Methods and Software Engineering. LNCS, vol. 9047, pp. 170–186. Springer, Heidelberg (2015)
8. Feijen, W.H.J., van Gasteren, A.J.M.: On a Method of Multiprogramming. Monographs in Computer Science. Springer, New York (1999)
9. Feng, X., Shao, Z., Guo, Y., Dong, Y.: Certifying low-level programs with hardware interrupts and preemptive threads. J. Autom. Reasoning **42**(2–4), 301–347 (2009)
10. Ferreira, J.F., Gherghina, C., He, G., Qin, S., Chin, W.N.: Automated verification of the FreeRTOS scheduler in HIP/SLEEK. Int. J. Softw. Tools Technol. Transf. **16**(4), 381–397 (2014)
11. Gammie, P., Hosking, T.A., Engelhardt, K.: Relaxing safely: verified on-the-fly garbage collection for x86-TSO. In: Blackburn, S. (ed.) PLDI 2015: The 36th annual ACM SIGPLAN conference on Programming Language Design and Implementation, p. 11. ACM, New York (2015)
12. Gotsman, A., Yang, H.: Modular verification of preemptive OS kernels. J. Funct. Program. **23**(4), 452–514 (2013)
13. Guo, Y., Zhang, H.: Verifying preemptive kernel code with preemption control support. In: 2014 Theoretical Aspects of Software Engineering Conference, TASE 2014, Changsha, China, 1–3 September 2014, pp. 26–33. IEEE (2014)

14. Huang, Y., Zhao, Y., Zhu, L., Li, Q., Zhu, H., Shi, J.: Modeling and verifying the code-level OSEK/VDX operating system with CSP. In: Theoretical Aspects of Software Engineering (TASE), pp. 142–149. IEEE (2011)
15. Klein, G.: Proof engineering considered essential. In: Jones, C., Pihlajasaari, P., Sun, J. (eds.) FM 2014. LNCS, vol. 8442, pp. 16–21. Springer, Heidelberg (2014)
16. Klein, G., Andronick, J., Elphinstone, K., Murray, T., Sewell, T., Kolanski, R., Heiser, G.: Comprehensive formal verification of an OS microkernel. Trans. Comput. Syst. **32**(1), 2:1–2:70 (2014)
17. Matthews, D.C., Wenzel, M.: Efficient parallel programming in Poly/ML and Isabelle/ML. In: Petersen, L., Pontelli, E. (eds.) POPL 2010 WS Declarative Aspects of Multicore Programming, pp. 53–62. ACM, New York (2010)
18. Nipkow, T., Paulson, L., Wenzel, M.: Isabelle/HOL – A Proof Assistant for Higher-Order Logic. LNCS, vol. 2283. Springer, Heidelberg (2002)
19. Owicki, S., Gries, D.: An axiomatic proof technique for parallel programs. Acta Informatica **6**, 319–340 (1976)
20. Prensa Nieto, L.: Verification of parallel programs with the Owicki-Gries and rely-guarantee methods in Isabelle/HOL. Ph.D. thesis, T.U. München (2002)
21. Yang, J., Hawblitzel, C.: Safe to the last instruction: automated verification of a type-safe operating system. In: 2010 PLDI, pp. 99–110. ACM (2010)

POSIX Lexing with Derivatives of Regular Expressions (Proof Pearl)

Fahad Ausaf[1], Roy Dyckhoff[2], and Christian Urban[1(✉)]

[1] King's College London, London, UK
fahad.ausaf@icloud.com, christian.urban@kcl.ac.uk
[2] University of St Andrews, St Andrews, UK
roy.dyckhoff@st-andrews.ac.uk

Abstract. Brzozowski introduced the notion of derivatives for regular expressions. They can be used for a very simple regular expression matching algorithm. Sulzmann and Lu cleverly extended this algorithm in order to deal with POSIX matching, which is the underlying disambiguation strategy for regular expressions needed in lexers. Sulzmann and Lu have made available on-line what they call a "rigorous proof" of the correctness of their algorithm w.r.t. their specification; regrettably, it appears to us to have unfillable gaps. In the first part of this paper we give our inductive definition of what a POSIX value is and show (*i*) that such a value is unique (for given regular expression and string being matched) and (*ii*) that Sulzmann and Lu's algorithm always generates such a value (provided that the regular expression matches the string). We also prove the correctness of an optimised version of the POSIX matching algorithm. Our definitions and proof are much simpler than those by Sulzmann and Lu and can be easily formalised in Isabelle/HOL. In the second part we analyse the correctness argument by Sulzmann and Lu and explain why the gaps in this argument cannot be filled easily.

Keywords: POSIX matching · Derivatives of regular expressions · Isabelle/HOL

1 Introduction

Brzozowski [2] introduced the notion of the *derivative* $r\backslash c$ of a regular expression r w.r.t. a character c, and showed that it gave a simple solution to the problem of matching a string s with a regular expression r: if the derivative of r w.r.t. (in succession) all the characters of the string matches the empty string, then r matches s (and *vice versa*). The derivative has the property (which may almost be regarded as its specification) that, for every string s and regular expression r and character c, one has $cs \in L(r)$ if and only if $s \in L(r\backslash c)$. The beauty of Brzozowski's derivatives is that they are neatly expressible in any functional language, and easily definable and reasoned about in theorem provers—the definitions just consist of inductive datatypes and simple recursive functions.

© Springer International Publishing Switzerland 2016
J.C. Blanchette and S. Merz (Eds.): ITP 2016, LNCS 9807, pp. 69–86, 2016.
DOI: 10.1007/978-3-319-43144-4_5

A mechanised correctness proof of Brzozowski's matcher in for example HOL4 has been mentioned by Owens and Slind [9]. Another one in Isabelle/HOL is part of the work by Krauss and Nipkow [6]. And another one in Coq is given by Coquand and Siles [3].

If a regular expression matches a string, then in general there is more than one way of how the string is matched. There are two commonly used disambiguation strategies to generate a unique answer: one is called GREEDY matching [4] and the other is POSIX matching [7,11,13]. For example consider the string xy and the regular expression $(x + y + xy)^\star$. Either the string can be matched in two 'iterations' by the single letter-regular expressions x and y, or directly in one iteration by xy. The first case corresponds to GREEDY matching, which first matches with the left-most symbol and only matches the next symbol in case of a mismatch (this is greedy in the sense of preferring instant gratification to delayed repletion). The second case is POSIX matching, which prefers the longest match.

In the context of lexing, where an input string needs to be split up into a sequence of tokens, POSIX is the more natural disambiguation strategy for what programmers consider basic syntactic building blocks in their programs. These building blocks are often specified by some regular expressions, say r_{key} and r_{id} for recognising keywords and identifiers, respectively. There are two underlying (informal) rules behind tokenising a string in a POSIX fashion according to a collection of regular expressions:

- *The Longest Match Rule* (or *"maximal munch rule"*): The longest initial substring matched by any regular expression is taken as next token.
- *Priority Rule:* For a particular longest initial substring, the first regular expression that can match determines the token.

Consider for example a regular expression r_{key} for recognising keywords such as *if, then* and so on; and r_{id} recognising identifiers (say, a single character followed by characters or numbers). Then we can form the regular expression $(r_{key} + r_{id})^\star$ and use POSIX matching to tokenise strings, say *iffoo* and *if*. For *iffoo* we obtain by the Longest Match Rule a single identifier token, not a keyword followed by an identifier. For *if* we obtain by the Priority Rule a keyword token, not an identifier token—even if r_{id} matches also.

One limitation of Brzozowski's matcher is that it only generates a YES/NO answer for whether a string is being matched by a regular expression. Sulzmann and Lu [11] extended this matcher to allow generation not just of a YES/NO answer but of an actual matching, called a [lexical] *value.* They give a simple algorithm to calculate a value that appears to be the value associated with POSIX matching. The challenge then is to specify that value, in an algorithm-independent fashion, and to show that Sulzmann and Lu's derivative-based algorithm does indeed calculate a value that is correct according to the specification.

The answer given by Sulzmann and Lu [11] is to define a relation (called an "order relation") on the set of values of r, and to show that (once a string to be matched is chosen) there is a maximum element and that it is computed by their derivative-based algorithm. This proof idea is inspired by work of Frisch and Cardelli [4] on a GREEDY regular expression matching algorithm. However, we were not able to establish transitivity and totality for the "order relation" by Sulzmann and Lu. In Sect. 5 we identify some inherent problems with their approach (of which some of the proofs are not published in [11]); perhaps more importantly, we give a simple inductive (and algorithm-independent) definition of what we call being a *POSIX value* for a regular expression r and a string s; we show that the algorithm computes such a value and that such a value is unique. Our proofs are both done by hand and checked in Isabelle/HOL. The experience of doing our proofs has been that this mechanical checking was absolutely essential: this subject area has hidden snares. This was also noted by Kuklewicz [7] who found that nearly all POSIX matching implementations are "buggy" [11, Page 203] and by Grathwohl et al. [5, Page 36] who wrote:

> *"The POSIX strategy is more complicated than the greedy because of the dependence on information about the length of matched strings in the various subexpressions."*

Contributions: We have implemented in Isabelle/HOL the derivative-based regular expression matching algorithm of Sulzmann and Lu [11]. We have proved the correctness of this algorithm according to our specification of what a POSIX value is (inspired by work of Vansummeren [13]). Sulzmann and Lu sketch in [11] an informal correctness proof: but to us it contains unfillable gaps.[1] Our specification of a POSIX value consists of a simple inductive definition that given a string and a regular expression uniquely determines this value. Derivatives as calculated by Brzozowski's method are usually more complex regular expressions than the initial one; various optimisations are possible. We prove the correctness when simplifications of $\mathbf{0} + r$, $r + \mathbf{0}$, $\mathbf{1} \cdot r$ and $r \cdot \mathbf{1}$ to r are applied.

2 Preliminaries

Strings in Isabelle/HOL are lists of characters with the empty string being represented by the empty list, written [], and list-cons being written as $\_ :: \_$. Often we use the usual bracket notation for lists also for strings; for example a string consisting of just a single character c is written $[c]$. By using the type *char* for characters we have a supply of finitely many characters roughly corresponding to the ASCII character set. Regular expressions are defined as usual as the elements of the following inductive datatype:

$$r := \mathbf{0} \mid \mathbf{1} \mid c \mid r_1 + r_2 \mid r_1 \cdot r_2 \mid r^\star$$

where $\mathbf{0}$ stands for the regular expression that does not match any string, $\mathbf{1}$ for the regular expression that matches only the empty string and c for matching a character literal. The language of a regular expression is also defined as usual by the recursive function L with the six clauses:

---

[1] An extended version of [11] is available at the website of its first author; this extended version already includes remarks in the appendix that their informal proof contains gaps, and possible fixes are not fully worked out.

(1) $L(\mathbf{0}) \stackrel{def}{=} \varnothing$

(2) $L(\mathbf{1}) \stackrel{def}{=} \{[]\}$

(3) $L(c) \stackrel{def}{=} \{[c]\}$

(4) $L(r_1 \cdot r_2) \stackrel{def}{=} L(r_1) @ L(r_2)$

(5) $L(r_1 + r_2) \stackrel{def}{=} L(r_1) \cup L(r_2)$

(6) $L(r^\star) \stackrel{def}{=} (L(r))\star$

In clause (4) we use the operation $\_ @ \_$ for the concatenation of two languages (it is also list-append for strings). We use the star-notation for regular expressions and for languages (in the last clause above). The star for languages is defined inductively by two clauses: (i) the empty string being in the star of a language and (ii) if s_1 is in a language and s_2 in the star of this language, then also $s_1 @ s_2$ is in the star of this language. It will also be convenient to use the following notion of a *semantic derivative* (or *left quotient*) of a language defined as

$$Der\ c\ A \stackrel{def}{=} \{s \mid c :: s \in A\}\ .$$

For semantic derivatives we have the following equations (for example mechanically proved in [6]):

$$\begin{aligned}
Der\ c\ \varnothing &\stackrel{def}{=} \varnothing \\
Der\ c\ \{[]\} &\stackrel{def}{=} \varnothing \\
Der\ c\ \{[d]\} &\stackrel{def}{=} if\ c = d\ then\ \{[]\}\ else\ \varnothing \\
Der\ c\ (A \cup B) &\stackrel{def}{=} Der\ c\ A \cup Der\ c\ B \\
Der\ c\ (A @ B) &\stackrel{def}{=} (Der\ c\ A @ B) \cup (if\ [] \in A\ then\ Der\ c\ B\ else\ \varnothing) \\
Der\ c\ (A\star) &\stackrel{def}{=} Der\ c\ A @ A\star
\end{aligned} \tag{1}$$

Brzozowski's derivatives of regular expressions [2] can be easily defined by two recursive functions: the first is from regular expressions to booleans (implementing a test when a regular expression can match the empty string), and the second takes a regular expression and a character to a (derivative) regular expression:

$$\begin{aligned}
nullable\ (\mathbf{0}) &\stackrel{def}{=} False \\
nullable\ (\mathbf{1}) &\stackrel{def}{=} True \\
nullable\ (c) &\stackrel{def}{=} False \\
nullable\ (r_1 + r_2) &\stackrel{def}{=} nullable\ r_1 \vee nullable\ r_2 \\
nullable\ (r_1 \cdot r_2) &\stackrel{def}{=} nullable\ r_1 \wedge nullable\ r_2 \\
nullable\ (r^\star) &\stackrel{def}{=} True
\end{aligned}$$

$$\begin{aligned}
(\mathbf{0})\backslash c &\stackrel{def}{=} \mathbf{0} \\
(\mathbf{1})\backslash c &\stackrel{def}{=} \mathbf{0} \\
d\backslash c &\stackrel{def}{=} if\ c = d\ then\ \mathbf{1}\ else\ \mathbf{0} \\
(r_1 + r_2)\backslash c &\stackrel{def}{=} (r_1\backslash c) + (r_2\backslash c) \\
(r_1 \cdot r_2)\backslash c &\stackrel{def}{=} if\ nullable\ r_1\ then\ (r_1\backslash c) \cdot r_2 + (r_2\backslash c)\ else\ (r_1\backslash c) \cdot r_2 \\
(r^*)\backslash c &\stackrel{def}{=} (r\backslash c) \cdot r^\star
\end{aligned}$$

We may extend this definition to give derivatives w.r.t. strings:

$$r\backslash[] \overset{def}{=} r$$
$$r\backslash(c :: s) \overset{def}{=} (r\backslash c)\backslash s$$

Given the equations in (1), it is a relatively easy exercise in mechanical reasoning to establish that

Proposition 1
(1) nullable r if and only if $[] \in L(r)$, and
(2) $L(r\backslash c) = Der\ c\ (L(r))$.

With this in place it is also very routine to prove that the regular expression matcher defined as

$$match\ r\ s \overset{def}{=} nullable\ (r\backslash s)$$

gives a positive answer if and only if $s \in L(r)$. Consequently, this regular expression matching algorithm satisfies the usual specification for regular expression matching. While the matcher above calculates a provably correct YES/NO answer for whether a regular expression matches a string or not, the novel idea of Sulzmann and Lu [11] is to append another phase to this algorithm in order to calculate a [lexical] value. We will explain the details next.

3   POSIX Regular Expression Matching

The clever idea by Sulzmann and Lu [11] is to define values for encoding *how* a regular expression matches a string and then define a function on values that mirrors (but inverts) the construction of the derivative on regular expressions. *Values* are defined as the inductive datatype

$$v := () \mid Char\ c \mid Left\ v \mid Right\ v \mid Seq\ v_1\ v_2 \mid Stars\ vs$$

where we use *vs* to stand for a list of values. (This is similar to the approach taken by Frisch and Cardelli for GREEDY matching [4], and Sulzmann and Lu for POSIX matching [11]). The string underlying a value can be calculated by the *flat* function, written $|\_|$ and defined as:

$$|()| \overset{def}{=} []$$
$$|Char\ c| \overset{def}{=} [c]$$
$$|Left\ v| \overset{def}{=} |v|$$
$$|Right\ v| \overset{def}{=} |v|$$

$$|Seq\ v_1\ v_2| \overset{def}{=} |v_1|\ @\ |v_2|$$
$$|Stars\ []| \overset{def}{=} []$$
$$|Stars\ (v :: vs)| \overset{def}{=} |v|\ @\ |Stars\ vs|$$

Sulzmann and Lu also define inductively an inhabitation relation that associates values to regular expressions:

$$() : \mathbf{1} \qquad Char \; c : c$$

$$\frac{v_1 : r_1}{Left \; v_1 : r_1 + r_2} \qquad \frac{v_2 : r_1}{Right \; v_2 : r_2 + r_1}$$

$$\frac{v_1 : r_1 \qquad v_2 : r_2}{Seq \; v_1 \; v_2 : r_1 \cdot r_2}$$

$$\frac{}{Stars \; [] : r^\star} \qquad \frac{v : r \qquad Stars \; vs : r^\star}{Stars \; (v :: vs) : r^\star}$$

Note that no values are associated with the regular expression $\mathbf{0}$, and that the only value associated with the regular expression $\mathbf{1}$ is $()$, pronounced (if one must) as *Void*. It is routine to establish how values "inhabiting" a regular expression correspond to the language of a regular expression, namely

Proposition 2. $L(r) = \{|v| \mid v : r\}$

In general there is more than one value associated with a regular expression. In case of POSIX matching the problem is to calculate the unique value that satisfies the (informal) POSIX rules from the Introduction. Graphically the POSIX value calculation algorithm by Sulzmann and Lu can be illustrated by the picture in Fig. 1 where the path from the left to the right involving *derivatives/nullable* is the first phase of the algorithm (calculating successive Brzozowski's derivatives) and *mkeps/inj*, the path from right to left, the second phase. This picture shows the steps required when a regular expression, say r_1, matches the string $[a, \, b, \, c]$. We first build the three derivatives (according to a, b and c). We then use *nullable* to find out whether the resulting derivative regular expression r_4 can match the empty string. If yes, we call the function *mkeps* that produces a value v_4 for how r_4 can match the empty string (taking into account the POSIX constraints in case there are several ways). This function is defined by the clauses:

$$mkeps \; (\mathbf{1}) \quad \overset{def}{=} \; ()$$
$$mkeps \; (r_1 \cdot r_2) \quad \overset{def}{=} \; Seq \; (mkeps \; r_1) \; (mkeps \; r_2)$$
$$mkeps \; (r_1 + r_2) \quad \overset{def}{=} \; if \; nullable \; r_1 \; then \; Left \; (mkeps \; r_1) \; else \; Right \; (mkeps \; r_2)$$
$$mkeps \; (r^\star) \quad \overset{def}{=} \; Stars \; []$$

Note that this function needs only to be partially defined, namely only for regular expressions that are nullable. In case *nullable* fails, the string $[a, \, b, \, c]$ cannot be matched by r_1 and the null value *None* is returned. Note also how this function makes some subtle choices leading to a POSIX value: for example if an alternative regular expression, say $r_1 + r_2$, can match the empty string and furthermore r_1 can match the empty string, then we return a *Left*-value. The *Right*-value will only be returned if r_1 cannot match the empty string.

The most interesting idea from Sulzmann and Lu [11] is the construction of a value for how r_1 can match the string $[a, \, b, \, c]$ from the value how the last derivative, r_4 in Fig. 1, can match the empty string. Sulzmann and Lu achieve this by stepwise "injecting back" the characters into the values thus inverting the

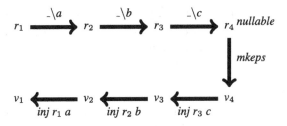

Fig. 1. The two phases of the algorithm by Sulzmann and Lu [11], matching the string $[a, b, c]$. The first phase (the arrows from left to right) is Brzozowski's matcher building successive derivatives. If the last regular expression is *nullable*, then the functions of the second phase are called (the top-down and right-to-left arrows): first *mkeps* calculates a value v_4 witnessing how the empty string has been recognised by r_4. After that the function *inj* "injects back" the characters of the string into the values.

operation of building derivatives, but on the level of values. The corresponding function, called *inj*, takes three arguments, a regular expression, a character and a value. For example in the first (or right-most) *inj*-step in Fig. 1 the regular expression r_3, the character c from the last derivative step and v_4, which is the value corresponding to the derivative regular expression r_4. The result is the new value v_3. The final result of the algorithm is the value v_1. The *inj* function is defined by recursion on regular expressions and by analysing the shape of values (corresponding to the derivative regular expressions).

$$
\begin{aligned}
&(1) &&inj \ d \ c \ () &&\overset{\text{def}}{=} \ Char \ d \\
&(2) &&inj \ (r_1 + r_2) \ c \ (Left \ v_1) &&\overset{\text{def}}{=} \ Left \ (inj \ r_1 \ c \ v_1) \\
&(3) &&inj \ (r_1 + r_2) \ c \ (Right \ v_2) &&\overset{\text{def}}{=} \ Right \ (inj \ r_2 \ c \ v_2) \\
&(4) &&inj \ (r_1 \cdot r_2) \ c \ (Seq \ v_1 \ v_2) &&\overset{\text{def}}{=} \ Seq \ (inj \ r_1 \ c \ v_1) \ v_2 \\
&(5) &&inj \ (r_1 \cdot r_2) \ c \ (Left \ (Seq \ v_1 \ v_2)) &&\overset{\text{def}}{=} \ Seq \ (inj \ r_1 \ c \ v_1) \ v_2 \\
&(6) &&inj \ (r_1 \cdot r_2) \ c \ (Right \ v_2) &&\overset{\text{def}}{=} \ Seq \ (mkeps \ r_1) \ (inj \ r_2 \ c \ v_2) \\
&(7) &&inj \ (r^\star) \ c \ (Seq \ v \ (Stars \ vs)) &&\overset{\text{def}}{=} \ Stars \ (inj \ r \ c \ v :: vs)
\end{aligned}
$$

To better understand what is going on in this definition it might be instructive to look first at the three sequence cases (clauses (4)–(6)). In each case we need to construct an "injected value" for $r_1 \cdot r_2$. This must be a value of the form $Seq \_ \_$. Recall the clause of the *derivative*-function for sequence regular expressions:

$$(r_1 \cdot r_2) \backslash c \ \overset{\text{def}}{=} \ if \ nullable \ r_1 \ then \ (r_1 \backslash c) \cdot r_2 + (r_2 \backslash c) \ else \ (r_1 \backslash c) \cdot r_2$$

Consider first the *else*-branch where the derivative is $(r_1 \backslash c) \cdot r_2$. The corresponding value must therefore be of the form $Seq \ v_1 \ v_2$, which matches the left-hand side in clause (4) of *inj*. In the *if*-branch the derivative is an alternative, namely $(r_1 \backslash c) \cdot r_2 + (r_2 \backslash c)$. This means we either have to consider a *Left*- or *Right*-value. In case of the *Left*-value we know further it must be a value for a sequence regular expression. Therefore the pattern we match in the clause

(5) is *Left* (*Seq* v_1 v_2), while in (6) it is just *Right* v_2. One more interesting point is in the right-hand side of clause (6): since in this case the regular expression r_1 does not "contribute" to matching the string, that means it only matches the empty string, we need to call *mkeps* in order to construct a value for how r_1 can match this empty string. A similar argument applies for why we can expect in the left-hand side of clause (7) that the value is of the form *Seq* v (*Stars* vs)—the derivative of a star is $(r \backslash c) \cdot r^*$. Finally, the reason for why we can ignore the second argument in clause (1) of *inj* is that it will only ever be called in cases where $c = d$, but the usual linearity restrictions in patterns do not allow us to build this constraint explicitly into our function definition.[2]

The idea of the *inj*-function to "inject" a character, say c, into a value can be made precise by the first part of the following lemma, which shows that the underlying string of an injected value has a prepended character c; the second part shows that the underlying string of an *mkeps*-value is always the empty string (given the regular expression is nullable since otherwise *mkeps* might not be defined).

Lemma 1
(1) If $v : r \backslash c$ then $|inj\ r\ c\ v| = c :: |v|$.
(2) If nullable r then $|mkeps\ r| = []$.

Proof. Both properties are by routine inductions: the first one can, for example, be proved by induction over the definition of *derivatives*; the second by an induction on r. There are no interesting cases.   □

Having defined the *mkeps* and *inj* function we can extend Brzozowski's matcher so that a [lexical] value is constructed (assuming the regular expression matches the string). The clauses of the Sulzmann and Lu lexer are

$$lexer\ r\ [] \quad \overset{def}{=}\ if\ nullable\ r\ then\ Some\ (mkeps\ r)\ else\ None$$
$$lexer\ r\ (c :: s) \overset{def}{=}\ case\ lexer\ (r \backslash c)\ s\ of$$
$$None \Rightarrow None$$
$$|\ Some\ v \Rightarrow Some\ (inj\ r\ c\ v)$$

If the regular expression does not match the string, *None* is returned. If the regular expression *does* match the string, then *Some* value is returned. One important virtue of this algorithm is that it can be implemented with ease in any functional programming language and also in Isabelle/HOL. In the remaining part of this section we prove that this algorithm is correct.

The well-known idea of POSIX matching is informally defined by the longest match and priority rule (see Introduction); as correctly argued in [11], this needs formal specification. Sulzmann and Lu define an "ordering relation" between values and argue that there is a maximum value, as given by the derivative-based algorithm. In contrast, we shall introduce a simple inductive definition

---

[2] Sulzmann and Lu state this clause as *inj c c* () $\overset{def}{=}$ *Char c*, but our deviation is harmless.

that specifies directly what a *POSIX value* is, incorporating the POSIX-specific choices into the side-conditions of our rules. Our definition is inspired by the matching relation given by Vansummeren [13]. The relation we define is ternary and written as $(s, r) \to v$, relating strings, regular expressions and values.

$$\frac{}{([], 1) \to ()}P1 \qquad \frac{}{([c], c) \to Char\ c}Pc$$

$$\frac{(s, r_1) \to v}{(s, r_1 + r_2) \to Left\ v}P+L \qquad \frac{(s, r_2) \to v \qquad s \notin L(r_1)}{(s, r_1 + r_2) \to Right\ v}P+R$$

$$\frac{\begin{array}{c}(s_1, r_1) \to v_1 \qquad (s_2, r_2) \to v_2 \\ \nexists s_3\ s_4.\ s_3 \neq [] \land s_3\ @\ s_4 = s_2 \land s_1\ @\ s_3 \in L(r_1) \land s_4 \in L(r_2)\end{array}}{(s_1\ @\ s_2, r_1 \cdot r_2) \to Seq\ v_1\ v_2}PS$$

$$\frac{}{([], r^\star) \to Stars\ []}P[]$$

$$\frac{\begin{array}{c}(s_1, r) \to v \qquad (s_2, r^\star) \to Stars\ vs \qquad |v| \neq [] \\ \nexists s_3\ s_4.\ s_3 \neq [] \land s_3\ @\ s_4 = s_2 \land s_1\ @\ s_3 \in L(r) \land s_4 \in L(r^\star)\end{array}}{(s_1\ @\ s_2, r^\star) \to Stars\ (v :: vs)}P\star$$

We can prove that given a string s and regular expression r, the POSIX value v is uniquely determined by $(s, r) \to v$.

Theorem 1
(1) If $(s, r) \to v$ then $s \in L(r)$ and $|v| = s$
(2) If $(s, r) \to v$ and $(s, r) \to v'$ then $v = v'$.

Proof. Both by induction on the definition of $(s, r) \to v$. The second parts follows by a case analysis of $(s, r) \to v'$ and the first part.   □

We claim that our $(s, r) \to v$ relation captures the idea behind the two informal POSIX rules shown in the Introduction: Consider for example the rules $P+L$ and $P+R$ where the POSIX value for a string and an alternative regular expression, that is $(s, r_1 + r_2)$, is specified—it is always a *Left*-value, *except* when the string to be matched is not in the language of r_1; only then it is a *Right*-value (see the side-condition in $P+R$). Interesting is also the rule for sequence regular expressions (PS). The first two premises state that v_1 and v_2 are the POSIX values for (s_1, r_1) and (s_2, r_2) respectively. Consider now the third premise and note that the POSIX value of this rule should match the string $s_1\ @\ s_2$. According to the longest match rule, we want that the s_1 is the longest initial split of $s_1\ @\ s_2$ such that s_2 is still recognised by r_2. Let us assume, contrary to the third premise, that there *exist* an s_3 and s_4 such that s_2 can be split up into a non-empty string s_3 and a possibly empty string s_4. Moreover the longer string $s_1\ @\ s_3$ can be matched by r_1 and the shorter s_4 can still be matched by r_2. In this case s_1 would *not* be the longest initial split of $s_1\ @\ s_2$ and therefore $Seq\ v_1\ v_2$ cannot be a POSIX value for $(s_1\ @\ s_2, r_1 \cdot r_2)$. The main point is that our side-condition ensures the longest match rule is satisfied.

A similar condition is imposed on the POSIX value in the $P\star$-rule. Also there we want that s_1 is the longest initial split of $s_1 @ s_2$ and furthermore the corresponding value v cannot be flattened to the empty string. In effect, we require that in each "iteration" of the star, some non-empty substring needs to be "chipped" away; only in case of the empty string we accept *Stars* [] as the POSIX value.

Next is the lemma that shows the function *mkeps* calculates the POSIX value for the empty string and a nullable regular expression.

Lemma 2. *If nullable r then* $([], r) \rightarrow$ *mkeps r.*

Proof. By routine induction on r.   □

The central lemma for our POSIX relation is that the *inj*-function preserves POSIX values.

Lemma 3. *If* $(s, r\backslash c) \rightarrow v$ *then* $(c :: s, r) \rightarrow$ *inj r c v.*

Proof. By induction on r. We explain two cases.

- Case $r = r_1 + r_2$. There are two subcases, namely (a) $v = Left\ v'$ and $(s, r_1\backslash c) \rightarrow v'$; and (b) $v = Right\ v', s \notin L(r_1\backslash c)$ and $(s, r_2\backslash c) \rightarrow v'$. In (a) we know $(s, r_1\backslash c) \rightarrow v'$, from which we can infer $(c :: s, r_1) \rightarrow inj\ r_1\ c\ v'$ by induction hypothesis and hence $(c :: s, r_1 + r_2) \rightarrow inj\ (r_1 + r_2)\ c\ (Left\ v')$ as needed. Similarly in subcase (b) where, however, in addition we have to use Proposition 1(2) in order to infer $c :: s \notin L(r_1)$ from $s \notin L(r_1\backslash c)$.
- Case $r = r_1 \cdot r_2$. There are three subcases:
  - (a) $v = Left\ (Seq\ v_1\ v_2)$ and *nullable* r_1
  - (b) $v = Right\ v_1$ and *nullable* r_1
  - (c) $v = Seq\ v_1\ v_2$ and $\neg$ *nullable* r_1

  For (a) we know $(s_1, r_1\backslash c) \rightarrow v_1$ and $(s_2, r_2) \rightarrow v_2$ as well as

$$\nexists s_3\ s_4.\ s_3 \neq []\ \wedge\ s_3 @ s_4 = s_2\ \wedge\ s_1 @ s_3 \in L(r_1\backslash c)\ \wedge\ s_4 \in L(r_2)$$

From the latter we can infer by Proposition 1(2):

$$\nexists s_3\ s_4.\ s_3 \neq []\ \wedge\ s_3 @ s_4 = s_2\ \wedge\ c :: s_1 @ s_3 \in L(r_1)\ \wedge\ s_4 \in L(r_2)$$

We can use the induction hypothesis for r_1 to obtain $(c :: s_1, r_1) \rightarrow inj\ r_1\ c\ v_1$. Putting this all together allows us to infer $(c :: s_1 @ s_2, r_1 \cdot r_2) \rightarrow Seq\ (inj\ r_1\ c\ v_1)\ v_2$. The case (c) is similar.
For (b) we know $(s, r_2\backslash c) \rightarrow v_1$ and $s_1 @ s_2 \notin L((r_1\backslash c) \cdot r_2)$. From the former we have $(c :: s, r_2) \rightarrow inj\ r_2\ c\ v_1$ by induction hypothesis for r_2. From the latter we can infer

$$\nexists s_3\ s_4.\ s_3 \neq []\ \wedge\ s_3 @ s_4 = c :: s\ \wedge\ s_3 \in L(r_1)\ \wedge\ s_4 \in L(r_2)$$

By Lemma 2 we know $([], r_1) \rightarrow$ *mkeps* r_1 holds. Putting this all together, we can conclude with $(c :: s, r_1 \cdot r_2) \rightarrow Seq\ (mkeps\ r_1)\ (inj\ r_2\ c\ v_1)$, as required.

Finally suppose $r = r_1^*$. This case is very similar to the sequence case, except that we need to also ensure that $|inj\ r_1\ c\ v_1| \neq []$. This follows from $(c :: s_1, r_1) \rightarrow inj\ r_1\ c\ v_1$ (which in turn follows from $(s_1, r_1\backslash c) \rightarrow v_1$ and the induction hypothesis). $\qquad\square$

With Lemma 3 in place, it is completely routine to establish that the Sulzmann and Lu lexer satisfies our specification (returning the null value *None* iff the string is not in the language of the regular expression, and returning a unique POSIX value iff the string *is* in the language):

Theorem 2
(1) $s \notin L(r)$ if and only if lexer r s = None
(2) $s \in L(r)$ if and only if $\exists v.\ lexer\ r\ s$ = Some $v \wedge (s,\ r) \rightarrow v$

Proof. By induction on s using Lemmas 2 and 3. $\qquad\square$

In (2) we further know by Theorem 1 that the value returned by the lexer must be unique. A simple corollary of our two theorems is:

Corollary 1
(1) lexer r s = None if and only if $\nexists v.\ (s,\ r) \rightarrow v$
(2) lexer r s = Some v if and only if $(s,\ r) \rightarrow v$

This concludes our correctness proof. Note that we have not changed the algorithm of Sulzmann and Lu,[3] but introduced our own specification for what a correct result—a POSIX value—should be. A strong point in favour of Sulzmann and Lu's algorithm is that it can be extended in various ways.

4   Extensions and Optimisations

If we are interested in tokenising a string, then we need to not just split up the string into tokens, but also "classify" the tokens (for example whether it is a keyword or an identifier). This can be done with only minor modifications to the algorithm by introducing *record regular expressions* and *record values* (for example [12]):

$$r := \ldots \mid (l : r) \qquad\qquad v := \ldots \mid (l : v)$$

where l is a label, say a string, r a regular expression and v a value. All functions can be smoothly extended to these regular expressions and values. For example $(l : r)$ is nullable iff r is, and so on. The purpose of the record regular expression is to mark certain parts of a regular expression and then record in the calculated value which parts of the string were matched by this part. The label can then serve as classification for the tokens. For this recall the regular expression $(r_{key} + r_{id})^*$ for keywords and identifiers from the Introduction. With the

---

[3] All deviations we introduced are harmless.

record regular expression we can form $((key : r_{key}) + (id : r_{id}))^*$ and then traverse the calculated value and only collect the underlying strings in record values. With this we obtain finite sequences of pairs of labels and strings, for example

$$(l_1 : s_1), ..., (l_n : s_n)$$

from which tokens with classifications (keyword-token, identifier-token and so on) can be extracted.

Derivatives as calculated by Brzozowski's method are usually more complex regular expressions than the initial one; the result is that the derivative-based matching and lexing algorithms are often abysmally slow. However, various optimisations are possible, such as the simplifications of $0 + r, r + 0, 1 \cdot r$ and $r \cdot 1$ to r. These simplifications can speed up the algorithms considerably, as noted in [11]. One of the advantages of having a simple specification and correctness proof is that the latter can be refined to prove the correctness of such simplification steps. While the simplification of regular expressions according to rules like

$$0 + r \Rightarrow r \qquad r + 0 \Rightarrow r \qquad 1 \cdot r \Rightarrow r \qquad r \cdot 1 \Rightarrow r \qquad (2)$$

is well understood, there is an obstacle with the POSIX value calculation algorithm by Sulzmann and Lu: if we build a derivative regular expression and then simplify it, we will calculate a POSIX value for this simplified derivative regular expression, *not* for the original (unsimplified) derivative regular expression. Sulzmann and Lu [11] overcome this obstacle by not just calculating a simplified regular expression, but also calculating a *rectification function* that "repairs" the incorrect value.

The rectification functions can be (slightly clumsily) implemented in Isabelle/HOL as follows using some auxiliary functions:

$$F_{Right}\ f\ v \stackrel{\text{def}}{=} Right\ (f\ v)$$
$$F_{Left}\ f\ v \stackrel{\text{def}}{=} Left\ (f\ v)$$
$$F_{Alt}\ f_1\ f_2\ (Right\ v) \stackrel{\text{def}}{=} Right\ (f_2\ v)$$
$$F_{Alt}\ f_1\ f_2\ (Left\ v) \stackrel{\text{def}}{=} Left\ (f_1\ v)$$
$$F_{Seq1}\ f_1\ f_2\ v \stackrel{\text{def}}{=} Seq\ (f_1\ ())\ (f_2\ v)$$
$$F_{Seq2}\ f_1\ f_2\ v \stackrel{\text{def}}{=} Seq\ (f_1\ v)\ (f_2\ ())$$
$$F_{Seq}\ f_1\ f_2\ (Seq\ v_1\ v_2) \stackrel{\text{def}}{=} Seq\ (f_1\ v_1)\ (f_2\ v_2)$$

$$simp_{Alt}\ (\mathbf{0},\ \_)\ (r_2, f_2) \stackrel{\text{def}}{=} (r_2, F_{Right}\ f_2)$$
$$simp_{Alt}\ (r_1, f_1)\ (\mathbf{0},\ \_) \stackrel{\text{def}}{=} (r_1, F_{Left}\ f_1)$$
$$simp_{Alt}\ (r_1, f_1)\ (r_2, f_2) \stackrel{\text{def}}{=} (r_1 + r_2, F_{Alt}\ f_1\ f_2)$$
$$simp_{Seq}\ (\mathbf{1}, f_1)\ (r_2, f_2) \stackrel{\text{def}}{=} (r_2, F_{Seq1}\ f_1\ f_2)$$
$$simp_{Seq}\ (r_1, f_1)\ (\mathbf{1}, f_2) \stackrel{\text{def}}{=} (r_1, F_{Seq2}\ f_1\ f_2)$$
$$simp_{Seq}\ (r_1, f_1)\ (r_2, f_2) \stackrel{\text{def}}{=} (r_1 \cdot r_2, F_{Seq}\ f_1\ f_2)$$

The functions $simp_{Alt}$ and $simp_{Seq}$ encode the simplification rules in (2) and compose the rectification functions (simplifications can occur deep inside the regular expression). The main simplification function is then

$$simp\ (r_1 + r_2) \stackrel{\text{def}}{=} simp_{Alt}\ (simp\ r_1)\ (simp\ r_2)$$
$$simp\ (r_1 \cdot r_2) \stackrel{\text{def}}{=} simp_{Seq}\ (simp\ r_1)\ (simp\ r_2)$$
$$simp\ r \stackrel{\text{def}}{=} (r,\ id)$$

where id stands for the identity function. The function $simp$ returns a simplified regular expression and a corresponding rectification function. Note that we do not simplify under stars: this seems to slow down the algorithm, rather than speed it up. The optimised lexer is then given by the clauses:

$$lexer^+\ r\ [] \stackrel{\text{def}}{=} \textit{if nullable r then Some (mkeps r) else None}$$
$$lexer^+\ r\ (c :: s) \stackrel{\text{def}}{=} \textit{let } (r_s,\ f_r) = simp\ (r \backslash c)\ \textit{in}$$
$$\textit{case } lexer^+\ r_s\ s\ \textit{of}$$
$$\textit{None} \Rightarrow \textit{None}$$
$$|\ \textit{Some } v \Rightarrow \textit{Some } (inj\ r\ c\ (f_r\ v))$$

In the second clause we first calculate the derivative $r \backslash c$ and then simplify the result. This gives us a simplified derivative r_s and a rectification function f_r. The lexer is then recursively called with the simplified derivative, but before we inject the character c into the value v, we need to rectify v (that is construct $f_r\ v$). Before we can establish the correctness of $lexer^+$, we need to show that simplification preserves the language and simplification preserves our POSIX relation once the value is rectified (recall $simp$ generates a (regular expression, rectification function) pair):

Lemma 4
(1) $L(fst\ (simp\ r)) = L(r)$
(2) If $(s, fst\ (simp\ r)) \rightarrow v$ then $(s, r) \rightarrow snd\ (simp\ r)\ v$.

Proof. Both are by induction on r. There is no interesting case for the first statement. For the second statement, of interest are the $r = r_1 + r_2$ and $r = r_1 \cdot r_2$ cases. In each case we have to analyse four subcases whether $fst\ (simp\ r_1)$ and $fst\ (simp\ r_2)$ equals $\mathbf{0}$ (respectively $\mathbf{1}$). For example for $r = r_1 + r_2$, consider the subcase $fst\ (simp\ r_1) = \mathbf{0}$ and $fst\ (simp\ r_2) \neq \mathbf{0}$. By assumption we know $(s, fst\ (simp\ (r_1 + r_2))) \rightarrow v$. From this we can infer $(s, fst\ (simp\ r_2)) \rightarrow v$ and by IH also (*) $(s, r_2) \rightarrow snd\ (simp\ r_2)\ v$. Given $fst\ (simp\ r_1) = \mathbf{0}$ we know $L(fst\ (simp\ r_1)) = \varnothing$. By the first statement $L(r_1)$ is the empty set, meaning (**) $s \notin L(r_1)$. Taking (*) and (**) together gives by the $P+R$-rule $(s, r_1 + r_2) \rightarrow Right\ (snd\ (simp\ r_2)\ v)$. In turn this gives $(s, r_1 + r_2) \rightarrow snd\ (simp\ (r_1 + r_2))\ v$ as we need to show. The other cases are similar. □

We can now prove relatively straightforwardly that the optimised lexer produces the expected result:

Theorem 3. $lexer^+ \ r \ s = lexer \ r \ s$

Proof. By induction on s generalising over r. The case $[]$ is trivial. For the cons-case suppose the string is of the form $c :: s$. By induction hypothesis we know $lexer^+ \ r \ s = lexer \ r \ s$ holds for all r (in particular for r being the derivative $r \backslash c$). Let r_s be the simplified derivative regular expression, that is $fst \ (simp \ (r \backslash c))$, and f_r be the rectification function, that is $snd \ (simp \ (r \backslash c))$. We distinguish the cases whether (*) $s \in L(r \backslash c)$ or not. In the first case we have by Theorem 2(2) a value v so that $lexer \ (r \backslash c) \ s = Some \ v$ and $(s, r \backslash c) \to v$ hold. By Lemma 4(1) we can also infer from (*) that $s \in L(r_s)$ holds. Hence we know by Theorem 2(2) that there exists a v' with $lexer \ r_s \ s = Some \ v'$ and $(s, r_s) \to v'$. From the latter we know by Lemma 4(2) that $(s, r \backslash c) \to f_r \ v'$ holds. By the uniqueness of the POSIX relation (Theorem 1) we can infer that v is equal to $f_r \ v'$—that is the rectification function applied to v' produces the original v. Now the case follows by the definitions of $lexer$ and $lexer^+$.

In the second case where $s \notin L(r \backslash c)$ we have that $lexer \ (r \backslash c) \ s = None$ by Theorem 2(1). We also know by Lemma 4(1) that $s \notin L(r_s)$. Hence $lexer \ r_s \ s = None$ by Theorem 2(1) and by IH then also $lexer^+ \ r_s \ s = None$. With this we can conclude in this case too.   □

5   The Correctness Argument by Sulzmann and Lu

An extended version of [11] is available at the website of its first author; this includes some "proofs", claimed in [11] to be "rigorous". Since these are evidently not in final form, we make no comment thereon, preferring to give general reasons for our belief that the approach of [11] is problematic. Their central definition is an "ordering relation" defined by the rules (slightly adapted to fit our notation):

$$\frac{v_1 >_{r_1} v_1'}{Seq \ v_1 \ v_2 >_{r_1 \cdot r_2} Seq \ v_1' \ v_2'} \text{(C2)} \qquad \frac{v_2 >_{r_2} v_2'}{Seq \ v_1 \ v_2 >_{r_1 \cdot r_2} Seq \ v_1 \ v_2'} \text{(C1)}$$

$$\frac{len \ |v_1| < len \ |v_2|}{Right \ v_2 >_{r_1 + r_2} Left \ v_1} \text{(A1)} \qquad \frac{len \ |v_2| \leq len \ |v_1|}{Left \ v_1 >_{r_1 + r_2} Right \ v_2} \text{(A2)}$$

$$\frac{v_1 >_{r_2} v_2}{Right \ v_1 >_{r_1 + r_2} Right \ v_2} \text{(A3)} \qquad \frac{v_1 >_{r_1} v_2}{Left \ v_1 >_{r_1 + r_2} Left \ v_2} \text{(A4)}$$

$$\frac{|Stars \ (v :: vs)| = []}{Stars \ [] >_{r^\star} Stars \ (v :: vs)} \text{(K1)} \qquad \frac{|Stars \ (v :: vs)| \neq []}{Stars \ (v :: vs) >_{r^\star} Stars \ []} \text{(K2)}$$

$$\frac{v_1 >_r v_2}{Stars \ (v_1 :: vs_1) >_{r^\star} Stars \ (v_2 :: vs_2)} \text{(K3)} \qquad \frac{Stars \ vs_1 >_{r^\star} Stars \ vs_2}{Stars \ (v :: vs_1) >_{r^\star} Stars \ (v :: vs_2)} \text{(K4)}$$

The idea behind the rules (A1) and (A2), for example, is that a *Left*-value is bigger than a *Right*-value, if the underlying string of the *Left*-value is longer or of equal length to the underlying string of the *Right*-value. The order is reversed, however, if the *Right*-value can match a longer string than a *Left*-value. In this way the POSIX value is supposed to be the biggest value for a given string and regular expression.

Sulzmann and Lu explicitly refer to the paper [4] by Frisch and Cardelli from where they have taken the idea for their correctness proof. Frisch and Cardelli introduced a similar ordering for GREEDY matching and they showed that their GREEDY matching algorithm always produces a maximal element according to this ordering (from all possible solutions). The only difference between their GREEDY ordering and the "ordering" by Sulzmann and Lu is that GREEDY always prefers a *Left*-value over a *Right*-value, no matter what the underlying string is. This seems to be only a very minor difference, but it has drastic consequences in terms of what properties both orderings enjoy. What is interesting for our purposes is that the properties reflexivity, totality and transitivity for this GREEDY ordering can be proved relatively easily by induction.

These properties of GREEDY, however, do not transfer to the POSIX "ordering" by Sulzmann and Lu, which they define as $v_1 \geq_r v_2$. To start with, $v_1 \geq_r v_2$ is not defined inductively, but as $(v_1 = v_2) \vee (v_1 >_r v_2 \wedge |v_1| = |v_2|)$. This means that $v_1 >_r v_2$ does not necessarily imply $v_1 \geq_r v_2$. Moreover, transitivity does not hold in the "usual" formulation, for example:

Falsehood 1. *Suppose $v_1 : r$, $v_2 : r$ and $v_3 : r$. If $v_1 >_r v_2$ and $v_2 >_r v_3$ then $v_1 >_r v_3$.*

If formulated in this way, then there are various counter examples: For example let r be $a + ((a + a)\cdot(a + \mathbf{0}))$ then the v_1, v_2 and v_3 below are values of r:

$$v_1 = Left\ (Char\ a)$$
$$v_2 = Right\ (Seq\ (Left\ (Char\ a))\ (Right\ ()))$$
$$v_3 = Right\ (Seq\ (Right\ (Char\ a))\ (Left\ (Char\ a)))$$

Moreover $v_1 >_r v_2$ and $v_2 >_r v_3$, but *not* $v_1 >_r v_3$! The reason is that although v_3 is a *Right*-value, it can match a longer string, namely $|v_3| = [a, a]$, while $|v_1|$ (and $|v_2|$) matches only $[a]$. So transitivity in this formulation does not hold—in this example actually $v_3 >_r v_1$!

Sulzmann and Lu "fix" this problem by weakening the transitivity property. They require in addition that the underlying strings are of the same length. This excludes the counter example above and any counter-example we were able to find (by hand and by machine). Thus the transitivity lemma should be formulated as:

Conjecture 1. *Suppose $v_1 : r$, $v_2 : r$ and $v_3 : r$, and also $|v_1| = |v_2| = |v_3|$. If $v_1 >_r v_2$ and $v_2 >_r v_3$ then $v_1 >_r v_3$.*

While we agree with Sulzmann and Lu that this property probably (!) holds, proving it seems not so straightforward: although one begins with the assumption that the values have the same flattening, this cannot be maintained as one descends into the induction. This is a problem that occurs in a number of places in the proofs by Sulzmann and Lu.

Although they do not give an explicit proof of the transitivity property, they give a closely related property about the existence of maximal elements.

They state that this can be verified by an induction on r. We disagree with this as we shall show next in case of transitivity. The case where the reasoning breaks down is the sequence case, say $r_1 \cdot r_2$. The induction hypotheses in this case are

IH r_1:

$\forall \; v_1, v_2, v_3.$
$$v_1 : r_1 \wedge v_2 : r_1 \wedge v_3 : r_1$$
$$\wedge \; |v_1| = |v_2| = |v_3|$$
$$\wedge \; v_1 >_{r_1} v_2 \wedge v_2 >_{r_1} v_3$$
$$\Rightarrow v_1 >_{r_1} v_3$$

IH r_2:

$\forall \; v_1, v_2, v_3.$
$$v_1 : r_2 \wedge v_2 : r_2 \wedge v_3 : r_2$$
$$\wedge \; |v_1| = |v_2| = |v_3|$$
$$\wedge \; v_1 >_{r_2} v_2 \wedge v_2 >_{r_2} v_3$$
$$\Rightarrow v_1 >_{r_2} v_3$$

We can assume that

$$Seq \; v_{1l} \; v_{1r} >_{r_1 \cdot r_2} Seq \; v_{2l} \; v_{2r} \quad \text{and} \quad Seq \; v_{2l} \; v_{2r} >_{r_1 \cdot r_2} Seq \; v_{3l} \; v_{3r} \quad (3)$$

hold, and furthermore that the values have equal length, namely:

$$|Seq \; v_{1l} \; v_{1r}| = |Seq \; v_{2l} \; v_{2r}| \quad \text{and} \quad |Seq \; v_{2l} \; v_{2r}| = |Seq \; v_{3l} \; v_{3r}| \quad (4)$$

We need to show that $Seq \; v_{1l} \; v_{1r} >_{r_1 \cdot r_2} Seq \; v_{3l} \; v_{3r}$ holds. We can proceed by analysing how the assumptions in (3) have arisen. There are four cases. Let us assume we are in the case where we know

$$v_{1l} >_{r_1} v_{2l} \quad \text{and} \quad v_{2l} >_{r_1} v_{3l}$$

and also know the corresponding inhabitation judgements. This is exactly a case where we would like to apply the induction hypothesis IH r_1. But we cannot! We still need to show that $|v_{1l}| = |v_{2l}|$ and $|v_{2l}| = |v_{3l}|$. We know from (4) that the lengths of the sequence values are equal, but from this we cannot infer anything about the lengths of the component values. Indeed in general they will be unequal, that is

$$|v_{1l}| \neq |v_{2l}| \quad \text{and} \quad |v_{1r}| \neq |v_{2r}|$$

but still (4) will hold. Now we are stuck, since the IH does not apply. As said, this problem where the induction hypothesis does not apply arises in several places in the proof of Sulzmann and Lu, not just for proving transitivity.

6   Conclusion

We have implemented the POSIX value calculation algorithm introduced by Sulzmann and Lu [11]. Our implementation is nearly identical to the original and all modifications we introduced are harmless (like our char-clause for *inj*). We have proved this algorithm to be correct, but correct according to our own specification of what POSIX values are. Our specification (inspired from work by Vansummeren [13]) appears to be much simpler than in [11] and our proofs are nearly always straightforward. We have attempted to formalise the original proof by Sulzmann and Lu [11], but we believe it contains unfillable gaps. In the online version of [11], the authors already acknowledge some small problems, but our experience suggests that there are more serious problems.

Having proved the correctness of the POSIX lexing algorithm in [11], which lessons have we learned? Well, this is a perfect example for the importance of the *right* definitions. We have (on and off) explored mechanisations as soon as first versions of [11] appeared, but have made little progress with turning the relatively detailed proof sketch in [11] into a formalisable proof. Having seen [13] and adapted the POSIX definition given there for the algorithm by Sulzmann and Lu made all the difference: the proofs, as said, are nearly straightforward. The question remains whether the original proof idea of [11], potentially using our result as a stepping stone, can be made to work? Alas, we really do not know despite considerable effort.

Closely related to our work is an automata-based lexer formalised by Nipkow [8]. This lexer also splits up strings into longest initial substrings, but Nipkow's algorithm is not completely computational. The algorithm by Sulzmann and Lu, in contrast, can be implemented with ease in any functional language. A bespoke lexer for the Imp-language is formalised in Coq as part of the Software Foundations book by Pierce et al. [10]. The disadvantage of such bespoke lexers is that they do not generalise easily to more advanced features. Our formalisation is available from the Archive of Formal Proofs [1] under http://www.isa-afp.org/entries/Posix-Lexing.shtml.

Acknowledgements. We are very grateful to Martin Sulzmann for his comments on our work and moreover for patiently explaining to us the details in [11]. We also received very helpful comments from James Cheney and anonymous referees.

References

1. Ausaf, F., Dyckhoff, R., Urban, C.: POSIX Lexing with Derivatives of Regular Expressions. Archive of Formal Proofs (2016). http://www.isa-afp.org/entries/Posix-Lexing.shtml, Formal proof development
2. Brzozowski, J.A.: Derivatives of regular expressions. J. ACM **11**(4), 481–494 (1964)
3. Coquand, T., Siles, V.: A decision procedure for regular expression equivalence in type theory. In: Jouannaud, J.-P., Shao, Z. (eds.) CPP 2011. LNCS, vol. 7086, pp. 119–134. Springer, Heidelberg (2011)
4. Frisch, A., Cardelli, L.: Greedy regular expression matching. In: Díaz, J., Karhumäki, J., Lepistö, A., Sannella, D. (eds.) ICALP 2004. LNCS, vol. 3142, pp. 618–629. Springer, Heidelberg (2004)
5. Grathwohl, N.B.B., Henglein, F., Rasmussen, U.T.: A Crash-Course in Regular Expression Parsing and Regular Expressions as Types. Technical report, University of Copenhagen (2014)
6. Krauss, A., Nipkow, T.: Proof pearl: regular expression equivalence and relation algebra. J. Autom. Reasoning **49**, 95–106 (2012)
7. Kuklewicz, C.: Regex Posix. https://wiki.haskell.org/Regex_Posix

8. Nipkow, T.: Verified lexical analysis. In: Grundy, J., Newey, M. (eds.) TPHOLs 1998. LNCS, vol. 1479, pp. 1–15. Springer, Heidelberg (1998)
9. Owens, S., Slind, K.: Adapting functional programs to higher order logic. High. Order Symbolic Comput. **21**(4), 377–409 (2008)
10. Pierce, B.C., Casinghino, C., Gaboardi, M., Greenberg, M., Hriţcu, C., Sjöberg, V., Yorgey, B.: Software Foundations. Electronic Textbook (2015). http://www.cis.upenn.edu/~bcpierce/sf
11. Sulzmann, M., Lu, K.Z.M.: POSIX regular expression parsing with derivatives. In: Codish, M., Sumii, E. (eds.) FLOPS 2014. LNCS, vol. 8475, pp. 203–220. Springer, Heidelberg (2014)
12. Sulzmann, M., van Steenhoven, P.: A flexible and efficient ML lexer tool based on extended regular expression submatching. In: Cohen, A. (ed.) CC 2014 (ETAPS). LNCS, vol. 8409, pp. 174–191. Springer, Heidelberg (2014)
13. Vansummeren, S.: Type inference for unique pattern matching. ACM Trans. Program. Lang. Syst. **28**(3), 389–428 (2006)

CoSMed: A Confidentiality-Verified Social Media Platform

Thomas Bauereiß[1](✉), Armando Pesenti Gritti[2,3], Andrei Popescu[2,4], and Franco Raimondi[2]

[1] German Research Center for Artificial Intelligence (DFKI), Bremen, Germany
thomas@bauereiss.name
[2] School of Science and Technology, Middlesex University, London, UK
[3] Global NoticeBoard, London, UK
[4] Institute of Mathematics Simion Stoilow of the Romanian Academy, Bucharest, Romania

Abstract. This paper describes progress with our agenda of formal verification of information-flow security for realistic systems. We present CoSMed, a social media platform with verified document confidentiality. The system's kernel is implemented and verified in the proof assistant Isabelle/HOL. For verification, we employ the framework of *Bounded-Deducibility (BD) Security*, previously introduced for the conference system CoCon. CoSMed is a second major case study in this framework. For CoSMed, the static topology of declassification bounds and triggers that characterized previous instances of BD security has to give way to a dynamic integration of the triggers as part of the bounds.

1 Introduction

Web-based systems are pervasive in our daily activities. Examples include enterprise systems, social networks, e-commerce sites and cloud services. Such systems pose notable challenges regarding confidentiality [1].

Recently, we have started a line of work aimed at addressing information flow security problems of realistic web-based systems by interactive theorem proving—using our favorite proof assistant, Isabelle/HOL [26,27]. We have introduced a security notion that allows a very fine-grained specification of what an attacker can observe about the system, and what information is to be kept confidential and in which situations. In our case studies, we assume the observers to be users of the system, and our goal is to verify that, by interacting with the system, the observers cannot learn more about confidential information than what we have specified. As a first case study, we have developed CoCon [18], a conference system (*à la* EasyChair) verified for confidentiality. We have verified a comprehensive list of confidentiality properties, systematically covering the relevant sources of information from CoCon's application logic [18, Sect. 4.5]. For example, besides authors, only PC members are allowed to learn about the content of submitted papers, and nothing beyond the last submitted version before the deadline.

© Springer International Publishing Switzerland 2016
J.C. Blanchette and S. Merz (Eds.): ITP 2016, LNCS 9807, pp. 87–106, 2016.
DOI: 10.1007/978-3-319-43144-4_6

This paper introduces a second major end product of this line of work:
CoSMed, a confidentiality-verified social media platform. CoSMed allows users
to register and post information, and to restrict access to this information based
on friendship relationships established between users. Architecturally, CoSMed
is an I/O automaton formalized in Isabelle, exported as Scala code, and wrapped
in a web application (Sect. 2).

For CoCon, we had proved that information only flows from the stored documents to the users in a suitably *role-triggered* and *bounded* fashion. In CoSMed's
case, the "documents" of interest are friendship requests, friendship statuses,
and posts by the users. The latter consist of title, text, and an optional image.
The roles in CoSMed include admin, owner and friend. Modeling the restrictions
on CoSMed's information flow poses additional challenges (Sect. 3), since here
the roles vary dynamically. For example, assume we prove a property analogous to those for CoCon: A user U1 learns nothing about the friend-only posts
posted by a user U2 *unless* U1 becomes a friend of U2. Although this property
makes sense, it is too weak—given that U1 may be "friended" and "unfriended"
by U2 multiple times. A stronger confidentiality property would be: U1 learns
nothing about U2's friend-only posts *beyond* the updates performed *while* U1
and U2 were friends. For the verification of both CoCon and CoSMed, we have
employed Bounded-Deducibility (BD) Security (Sect. 3.2), a general framework
for the verification of rich information flow properties of input/output automata.
BD security is parameterized by declassification *bounds* and *triggers*. While for
CoCon a fixed topology of bounds and triggers was sufficient, CoSMed requires
a more dynamic approach, where the bounds incorporate trigger information
(Sect. 3.3). The verification proceeds by providing suitable unwinding relations,
closely matching the bounds (Sect. 4).

CoSMed has been developed to fulfill the functionality and security needs of a
charity organization [4]. The current version is a prototype, not yet deployed for
the charity usage. Both the formalization and the running website are publicly
available [5].

Notation. Given $f : A \to B$, $a : A$ and $b : B$, we write $f(a := b)$ for the
function that returns b for a and otherwise acts like f. [] denotes the empty list
and @ denotes list concatenation. Given a list xs, we write last xs for its last
element. Given a predicate P, we write filter P xs for the sublist of xs consisting
of those elements satisfying P. Given a function f, we write map f xs for the
list resulting from applying the function f to each element of xs. Given a record
σ, field labels $l_1, \ldots, l_n$ and values $v_1, \ldots, v_n$ respecting the types of the labels,
we write $\sigma(l_1 := v_1, \ldots, l_n := v_n)$ for σ with the values of the fields l_i updated
to v_i. We let $l_i \, \sigma$ be the value of field l_i stored in σ.

2   System Description

In this section we describe the system functionality as formalized in Isabelle
(Sect. 2.1)—we provide enough detail so that the reader can have a good grasp

of the formal confidentiality properties discussed later. Then we sketch CoSMed's overall architecture (Sect. 2.2).

2.1   Isabelle Specification

Abstractly, the system can be viewed as an I/O automaton, having a state and offering some actions through which the user can affect the state and retrieve outputs. The state stores information about users, posts and the relationships between them, namely:

- user information: pending new-user requests, the current user IDs and the associated user info, the system's administrator, the user passwords;
- post information: the current post IDs and the posts associated to them, including content and visibility information;
- post-user relationships: the post owners;
- user-user relationships: the pending friend requests and the friend relationships.

All in all, the **state** is represented as an Isabelle record:

```
RECORD state =
(* User info: *)
    pendingUReqs : userID list    userReq : userID → req    userIDs : userID list
    user : userID → user    pass : userID → password    admin : userID
(* Friend info: *)
    pendingFReqs : userID → userID list    friendReq : userID → userID → req
    friendIDs : userID → userID list
(* Post info: *)
    postIDs : postID list    post : postID → post    owner : postID → userID
```

Above, the types userID, postID, password, and req are essentially strings (more precisely, datatypes with one single constructor embedding strings). Each pending request (be it for user or for friend relationship) stores a request info (of type req), which contains a message of the requester for the recipient (the system admin or a given user). The type user contains user names and information. The type post of posts contains tuples (*title, text, img, vis*), where the title and the text are essentially strings, *img* is an (optional) image file, and *vis* ∈ {FriendV, PublicV} is a visibility status that can be assigned to posts: FriendV means visibility to friends only, whereas PublicV means visibility to all users.

The **initial state** of the system is completely empty: there are empty lists of registered users, posts, etc. Users can interact with the system via six categories of **actions**: start-up, creation, deletion, update, reading and listing.

The actions take varying numbers of parameters, indicating the user involved and optionally some data to be loaded into the system. Each action's behavior is specified by two functions:

– An effect function, actually performing the action, possibly changing the state and returning an output
– An enabledness predicate (marked by the prefix "e"), checking the conditions under which the action should be allowed

When a user issues an action, the system first checks if it is enabled, in which case its effect function is applied and the output is returned to the user. If it is not enabled, then an error message is returned and the state remains unchanged.

The **start-up action**, startSys : state $\rightarrow$ userID $\rightarrow$ password $\rightarrow$ state, initializes the system with a first user, who becomes the admin:

$$\begin{aligned}
&\text{startSys } \sigma \ uid \ p \ \equiv \\
&\quad \sigma(\text{admin} := uid, \ \text{userIDs} := [uid], \ \text{user} := (\text{user } \sigma)(uid := \text{emptyUser}), \\
&\quad \ \ \text{pass} := (\text{pass } \sigma)(uid := p))
\end{aligned}$$

The start-up action is enabled only if the system has no users:

$$\text{e\_startSys } \sigma \ uid \ p \ \equiv \ \text{userIDs } \sigma = [\,]$$

Creation actions perform registration of new items in the system. They include: placing a new user registration request; the admin approving such a request, leading to registration of a new user; a user creating a post; a user placing a friendship request for another user; a user accepting a pending friendship request, thus creating a friendship connection.

The three main kinds of items that can be created/registered in the system are users, friends and posts. Post creation can be immediately performed by any user. By contrast, user and friend registration proceed in two stages: first a request is created by the interested party, which can later be approved by the authorized party. For example, a friendship request from uid to uid' is first placed in the pending friendship request queue for uid'. Then, upon approval by uid', the request turns into a friendship relationship. Since friendship is symmetric, both the list of uid''s friends and that of uid's friends are updated, with uid and uid' respectively.

There is only one **deletion action** in the system, namely friendship deletion ("unfriending" an existing friend).

Update actions allow users with proper permissions to modify content in the system: user info, post content, visibility status, etc. For example, the following action is updating, on behalf of the user uid, the text of a post with ID pid to the value $text$.

$$\begin{aligned}
&\text{updateTextPost } \sigma \ uid \ p \ pid \ text \ \equiv \\
&\quad \sigma \, (\text{post} := (\text{post } \sigma)(pid := \text{setTextPost } (\text{post } \sigma \ pid) \ text))
\end{aligned}$$

It is enabled if both the user ID and the post ID are registered, the given password matches the one stored in the state and the user is the post's owner. Besides the text, one can also update the title and the image of a post.

Reading actions allow users to retrieve content from the system. One can read user and post info, friendship requests and status, etc. Finally, the **listing actions** allow organizing and listing content by IDs. These include the listing of: all the pending user registration requests (for the admin); all users of the system; all posts; one's friendship requests, one's own friends, and the friends of them.

Action Syntax and Dispatch. So far we have discussed the action behavior, consisting of effect and enabledness. In order to keep the interface homogeneous, we distinguish between an action's behavior and its *syntax*. The latter is simply the input expected by the I/O automaton. The different kinds of actions (start-up, creation, deletion, update, reading and listing) are wrapped in a single datatype through specific constructors:

DATATYPE act = Sact sAct | Cact cAct | Dact dAct | Uact uAct | Ract rAct | Lact lAct

In turn, each kind of action forms a datatype with constructors having varying numbers of parameters, mirroring those of the action behavior functions. For example, the following datatypes gather (the syntax of) all the update and reading actions:

DATATYPE uAct =
 uUser userID password password name info
| uTitlePost userID password postID title
| uTextPost userID password postID text
| uImgPost userID password postID img
| uVisPost userID password postID vis

DATATYPE rAct =
 rUser userID password userID
| rNUReq userID password userID
| rNAReq userID password appID
| rAmIAdmin userID password
| rTitlePost userID password postID
| rTextPost userID password postID
| rImgPost userID password postID
| rVisPost userID password postID
| rOwnerPost userID password postID
| rFriendReqToMe userID password userID
| rFriendReqFromMe userID password userID

We have more reading actions than update actions. Some items, such as new-user and new-friend request info, are readable but not updatable.

The naming convention we follow is that a constructor representing the syntax of an action is named by abbreviating the name of that action. For example, the constructor uTextPost corresponds to the effect function updateTextPost.

The overall **step function**, step : state $\rightarrow$ act $\rightarrow$ out $\times$ state, proceeds as follows. When given a state σ and an action a, it first pattern-matches on a to discover what kind of action it is. For example, for the update action Uact (uTextPost *uid p pid text*), the corresponding enabledness predicate is called on the current state (say, σ) with the given parameters, e_updateTextPost σ *uid p pid text*. If this returns False, the result is (outErr, σ), meaning that the state has not changed and an error output is produced. If it returns True, the effect function is called, updateTextPost σ *uid p pid text*, yielding a new state σ'. The result is then $(\mathsf{outOK}, \sigma')$, containing the new state along with an output indicating that the update was successful.

Note that start, creation, deletion and update actions change the state but do not output non-trivial data (besides outErr or outOK). By contrast, reading actions do not change the state, but they output data such as user info, post content and friendship status. Likewise, listing actions output lists of IDs and other data. The datatype out, of the overall system outputs, wraps together all these possible outputs, including outErr and outOK.

In summary, all the heterogeneous parametrized actions and outputs are wrapped in the datatypes act and out, and the step function dispatches any request to the corresponding enabledness check and effect. The end product is a single I/O automaton.

2.2 Implementation

For CoSMed's implementation, we follow the same approach as for CoCon [18, Sect. 2]. The I/O automaton formalized by the initial state istate : state and the step function step : state $\rightarrow$ act $\rightarrow$ out $\times$ state represents CoSMed's kernel—it is this kernel that we formally verify. The kernel is automatically translated to isomorphic Scala code using Isabelle's code generator [15].

Around the exported code, there is a thin layer of trusted (unverified) code. It consists of an API written with the Scalatra framework and a web application that communicates with the API. Although this architecture involves trusted code, there are reasons to believe that the confidentiality guarantees of the kernel also apply to the overall system. Indeed, the Scalatra API is a thin layer: it essentially forwards requests back and forth between the kernel and the outside world. Moreover, the web application operates by calling combinations of primitive API operations, without storing any data itself. User authentication, however, is also part of this unverified code. Of course, complementing our secure kernel with a verification that "nothing goes wrong" in the outer layer (by some language-based tools) would give us stronger guarantees.

3   Stating Confidentiality

Web-based systems for managing online resources and workflows for multiple users, such as CoCon and CoSMed, are typically programmed by distinguishing between various roles (e.g., author, PC member, reviewer for CoCon, and admin, owner, friend for CoSMed). Under specified circumstances, members with specified roles are given access to (controlled parts of) the documents.

Access control is understood and enforced *locally*, as a property of the system's *reachable states*: that a given action is only allowed if the agent has a certain role and certain circumstances hold. However, the question whether access control achieves its purpose, i.e., really restricts undesired information flow, is a *global* question whose formalization simultaneously involves *all the system's execution traces*. We wish to restrict not only what an agent can access, but also what an agent can infer, or learn.

3.1   From CoCon to CoSMed

For CoCon, we verified properties with the pattern: A user can learn nothing about a document *beyond* a certain amount of information *unless* a certain event occurs. E.g.:

- A user can learn nothing about the uploads of a paper *beyond* the last uploaded version in the submission phase *unless* that user becomes an author.
- A user can learn nothing about the updates to a paper's review *beyond* the last updated version before notification *unless* that user is a non-conflicted PC member.

The "beyond" part expresses a *bound* on the amount of disclosed information. The "unless" part indicates a *trigger* in the presence of which the bound is not guaranteed to hold. This bound-trigger tandem has inspired our notion of BD security—applicable to I/O automata and instantiatable to CoCon. But let us now analyze the desired confidentiality properties for CoSMed. For a post, we may wish to prove:

> (P1) A user can learn nothing about the updates to a post content *unless that user is the post's owner, or he becomes friends with the owner, or the post is marked as public.*

And indeed, the system can be proved to satisfy this property. But is this strong enough? Note that the trigger, emphasized in (P1) above, expresses a condition in whose presence our property stops guaranteeing anything. Therefore, since both friendship and public visibility can be freely switched on and off by the owner at any time, relying on such a strong trigger simply means giving up too easily. We should aim to prove a stronger property, describing confidentiality along several iterations of issuing and disabling the trigger. A better candidate property is the following:[1]

> (P2) A user can learn nothing about the updates to a post content *beyond* those updates that are performed *while* one of the following holds: either that user is the post's owner, or he is a friend of the owner, or the post is marked as public.

In summary, the "beyond"-"unless" bound-trigger combination we employed for CoCon will need to give way to a "beyond"-"while" scheme, where "while" refers to the periods in a system run during which observers are allowed to learn about confidential information. We will call these periods "access windows." To formalize them, we will incorporate (and iterate) the trigger inside the bound. As we show below, this is possible with the price of enriching the notion of secret to record changes to the "openness" of the access window. In turn, this leads to more complex bounds having more subtle definitions. But first let us recall BD security formally.

---

[1] As it will turn out, this property needs to be refined in order to hold. We'll do this in Sect. 3.3.

3.2   BD Security Recalled

We focus on the security of systems specified as I/O automata. In such an automaton, we call the inputs "actions." We write state, act, and out for the types of states, actions, and outputs, respectively, istate : state for the initial state, and step : state $\to$ act $\to$ out $\times$ state for the one-step transition function. Transitions are tuples describing an application of step:

$$\text{DATATYPE trans} = \text{Trans state act out state}$$

A transition $trn = $ Trans $\sigma\ a\ o\ \sigma'$ is called valid if it corresponds to an application of the step function, namely step $\sigma\ a = (o, \sigma')$. Traces are lists of transitions:

$$\text{TYPE\_SYNONYM trace} = \text{trans list}$$

A trace $tr = [trn_0, \ldots, trn_{n-1}]$ is called valid if it starts in the initial state istate and all its transitions are valid and compose well, in that, for each $i < n - 1$, the target state of trn_i coincides with the source state of trn_{i+1}. Valid traces model the runs of the system: at each moment in the lifetime of the system, a certain trace has been executed. All our formalized security definitions and properties quantify over valid traces and transitions—to ease readability, we shall omit the validity assumption, and pretend that the types trans and trace contain only valid transitions and traces.

   We want to verify that there are no unintended flows of information to attackers who can observe and influence certain aspects of the system execution. Hence, we specify

1. what the capabilities of the attacker are,
2. which information is (potentially) confidential, and
3. which flows are allowed.

The first point is captured by a function O taking a trace and returning the observable part of that trace. Similarly, the second point is captured by a function S taking a trace and returning the sequence of (potential) secrets occurring in that trace. For the third point, we add a parameter B, which is a binary relation on sequences of secrets. It specifies a lower *bound* on the uncertainty of the observer about the secrets, in other words, an upper bound on these secrets' *declassification*. In this context, BD security states that O *cannot learn anything about* S *beyond* B. Formally:

> For all valid system traces tr and sequence of secrets sl' such that B (S tr) sl' holds, there exists a valid system trace tr' such that S $tr' = sl'$ and O $tr' = $ O tr.

Thus, BD security requires that, if B $sl\ sl'$ holds, then observers cannot distinguish the sequence of secrets sl from sl'—if sl is consistent with a given observation, then so must be sl'. Classical nondeducibility [29] corresponds to B being the total relation—the observer can then deduce *nothing* about the secrets.

Smaller relations B mean that an observer may deduce some information about the secrets, but nothing beyond B—for example, if B is an equivalence relation, then the observer may deduce the equivalence class, but not the concrete secret within the equivalence class.

The original formulation of BD security in [18] includes an additional parameter T, a *declassification trigger*: The above condition is only required to hold for traces *tr* where T does not occur. Hence, as soon as the trigger occurs, the security property no longer offers any guarantees. This was convenient for CoCon, but for CoSMed this is too coarse-grained, as discussed in Sect. 3.1. Since, in general, an instance of BD security *with* T can be transformed into one *without*,[2] in this paper we decide to drop T and use the above trigger-free formulation of BD security.

Regarding the parameters O and S, we assume that they are defined in terms of functions on individual transitions:

- isSec : trans $\rightarrow$ bool, filtering the transitions that produce secrets
- getSec : trans $\rightarrow$ secret, producing a secret out of a transition
- isObs : trans $\rightarrow$ bool, filtering the transitions that produce observations
- getObs : trans $\rightarrow$ obs, producing an observation out of a transition

We then define O = map getObs $\circ$ filter isObs and S = map getSec $\circ$ filter isSec. Thus, O uses filter to select the transitions in a trace that are (partially) observable according to isObs, and then maps this sequence of transitions to the sequence of their induced observations, via getObs. Similarly, S produces sequences of secrets by filtering via isSec and mapping via getSec.

All in all, BD security is parameterized by the following data:

- an I/O automaton (state, act, out, istate, step)
- a security model, consisting of:
  - a secrecy infrastructure (secret, isSec, getSec)
  - an observation infrastructure (obs, isObs, getObs)
  - a declassification bound B

3.3 CoSMed Confidentiality as BD Security

Next we show how to capture CoSMed's properties as BD security. We first look in depth at one property, post confidentiality, expressed informally by (P2) from Sect. 3.1.

Let us attempt to choose appropriate parameters in order to formally capture a confidentiality property in the style of (P2). The I/O automaton will of course be the one described by the state, actions and outputs from Sect. 2.1.

For the security model, we first instantiate the observation infrastructure (obs, isObs, getObs). The observers are users. Moreover, instead of assuming a single user observer, we wish to allow coalitions of an arbitrary number of users—this will provide us with stronger security guarantees. Finally, from a transition

---

[2] By modifying S to produce a dedicated value as soon as T occurs, and modifying B to only consider sequences without that value.

Trans σ a o σ' issued by a user, it is natural to allow that user to observe both their own action a and the output o.

Formally, we take the type obs of observations to be act $\times$ out and the observation-producing function getObs : trans $\rightarrow$ obs to be getObs (Trans _ a o _) $\equiv$ (a, o). We fix a set UIDs of user IDs and define the observation filter isObs : trans $\rightarrow$ obs by

$$\text{isObs} (\text{Trans } \sigma \ a \ o \ \sigma') \equiv \text{userOf } a \in \text{UIDs}$$

where userOf a returns the user who performs the action. In summary, the observations are all actions issued by members of a fixed set UIDs of users together with the outputs that these actions are producing.

Let us now instantiate the secrecy infrastructure (secret, isSec, getSec). Since the property (P2) talks about the text of a post, say, identified by PID : postID, a first natural choice for secrets would be the text updates stored in PID via updateTextPost actions. That is, we could have the filter isSec a hold just in case a is such a (successfully performed) action, say, updateTextPost σ uid p pid $text$, and have the secret-producing function getSec a return the updated secret, here $text$. But later, when we state the bound, how would we distinguish updates that should not be learned from updates that are OK to be learned because they happen while the access is legitimate for the observers—e.g., while a user in UIDs is the owner's friend? We shall refer to the portions of the trace when the observer access is legitimate as *open access windows*, and refer to the others as *closed access windows*. The bound clearly needs to distinguish these. Indeed, it states that nothing should be learned beyond the updates that occurred during open access windows.

To enable this distinction, we enrich the notion of secret to include not only the post text updates, but also marks for the shift between closed and open access windows. To this end, we define the state predicate open to express that PID is registered and one of the users in UIDs is entitled to access the text of PID—namely, is the owner or a friend of the owner, or the post is public.

$$\begin{aligned}
\text{open } \sigma \equiv \ & \text{PID} \in \text{postIDs } \sigma \ \wedge \\
& \exists uid \in \text{UIDs.} \ uid \in \text{userIDs } \sigma \ \wedge \\
& \quad (uid = \text{owner } \sigma \ pid \ \vee \ uid \in \text{friendIDs } \sigma \ (\text{owner } \sigma \ pid) \ \vee \\
& \quad \ \text{visPost } (\text{post } \sigma \ \text{PID}) = \text{PublicV})
\end{aligned}$$

Now, the secret selector isSec : trans $\rightarrow$ bool will record both successful post-text updates and the changes in the truth value of open for the state of the transition:

$$\begin{aligned}
\text{isSec } (\text{Trans } \_ (\text{Uact } (\text{uTextPost } pid \ \_ \ \_ \ text)) \ o \ \_) &\equiv pid = \text{PID} \ \wedge \ o = \text{outOK} \\
\text{isSec } (\text{Trans } \sigma \ \_ \ \_ \ \sigma') &\equiv \text{open } \sigma \neq \text{open } \sigma'
\end{aligned}$$

In consonance with the filter, the type of secrets will have two constructors

$$\text{DATATYPE secret} = \text{TSec text} \mid \text{OSec bool}$$

$$\frac{textl \neq [] \rightarrow textl' \neq []}{\text{B (map TSec } textl) \text{ (map TSec } textl')} \text{ (1)} \qquad \text{BO (map TSec } textl) \text{ (map TSec } textl) \text{ (2)}$$

$$\frac{\text{BO } sl \ sl' \qquad textl \neq [] \leftrightarrow textl' \neq [] \qquad textl \neq [] \rightarrow \text{last } textl = \text{last } textl'}{\text{B (map TSec } textl \text{ @ OSec True @ } sl) \text{ (map TSec } textl' \text{ @ OSec True @ } sl')} \text{ (3)}$$

$$\frac{\text{B } sl \ sl'}{\text{BO (map TSec } textl \text{ @ OSec False @ } sl) \text{ (map TSec } textl \text{ @ OSec False @ } sl')} \text{ (4)}$$

Fig. 1. The bound for post text confidentiality

and the secret-producing function getSec : trans $\rightarrow$ secret will retrieve either the updated text or the updated openness status:

$$\text{getSec (Trans \_ (Uact (uTitlePost \_ \_ \_ } text)) \text{ \_ \_)} \equiv \text{TSec } text$$
$$\text{getSec (Trans \_ \_ \_ } \sigma') \equiv \text{OSec (open } \sigma')$$

In order to formalize the desired bound B, we first note that all sequences of secrets produced from system traces consist of:

- a (possibly empty) block of text updates TSec $text^1_1, \ldots,$ TSec $text^1_{n_1}$
- possibly followed by a shift to an open access window, OSec True
- possibly followed by another block of text updates TSec $text^2_1, \ldots,$ TSec $text^2_{n_2}$
- possibly followed by a shift to a closed access window, OSec False
- ... and so on ...

We wish to state that, given any such sequence of secrets sl (say, produced from a system trace tr), any other sequence sl' that coincides with sl on the open access windows (while being allowed to be *arbitrary* on the closed access windows) is equally possible as far as the observer is concerned—in that there exists a trace tr' yielding the same observations as tr and producing the secrets sl'.

The purpose of B is to capture this relationship between sl and sl', of coincidence on open access windows. But which part of a sequence of secrets sl represents such a window? It should of course include all the text updates that take place during the time when one of the observers has legitimate access to the post—namely, all blocks of sl that are immediately preceded by an OSec True secret.

But there are other secrets in the sequence that properly belong to this window: the last updated text before the access window is open, that is, the secret TSec $text^k_{n_k}$ occurring *immediately before* each occurrence of OSec True. For example, when the post becomes public, a user can see not only upcoming updates to its text, but also the current text, i.e., the last update before the visibility upgrade.

The definition of B reflects the above discussion, using an auxiliary predicate BO to cover the case when the window is open. The predicates are defined mutually inductively as in Fig. 1.

Clause (1), the base case for B, describes the situation where the original system trace has made no shift from the original closed access window. Here, the produced sequence of secrets sl consists of text updates only, i.e., $sl =$ map TSec $textl$. It is indistinguishable from any alternative sequence of updates $sl' =$ map TSec $textl'$, save for the corner case where an observer can learn that sl is empty by inferring that the post does not exist, e.g. because the system has not been started yet, or because no users other than the observers exist who could have created the post. Such harmless knowledge is factored in by asking that sl' (i.e., $textl'$) be empty whenever sl (i.e., $textl$) is.

Clause (2), the base case for BO, handles sequences of secrets produced during open access windows. Since here information is entirely exposed, the corresponding sequence of secrets from the alternative trace has to be identical to the original.

Clause (3), the inductive case for B, handles sequences of secrets map TSec $textl$ produced during closed access windows. The difference from clause (1) is that here we know that there will eventually be a shift to a closed access window—this is marked by the occurrences of OSec True in the conclusion, followed by a remaining sequence sl. As previously discussed, the only constraint on the sequence of secrets produced by the alternative trace, map TSec $textl'$, is that it ends in the same secret—hence the condition that the sequences be empty at the same time and have the same last element. Finally, clause (4), the inductive case for BO, handles the secrets produced during open access window on a trace known to eventually move to an open access window.

With all the parameters in place, we have a formalization of post text confidentiality: the BD-security instance for these parameters. However, we saw that the legitimate exposure of the posts is wider than initially suggested, hence (P2) is bogus as currently formulated. Namely, we need to factor in the last updates *before* open access windows in addition to the updates performed *during* open access windows. If we also factor in the generalization from a single user to a coalition of users, we obtain:

(P3) A coalition of users can learn nothing about the updates to a post content beyond those updates that are performed while one of the following holds *or the last update before one of the following starts to hold*:

- a user in the coalition is the post's owner or a friend of the post's owner, or
- there is at least one user in the coalition and the post is marked as public.

3.4 More Confidentiality Properties

So far, we have discussed confidentiality for post content (i.e., text). However, a post also has a title and an image. For these, we want to verify the same confidentiality properties as in Sect. 3.3, only substituting text content by titles and images, respectively. In addition to posts, another type of information with confidentiality ramifications is that about friendship between users: who is friends with whom, and who has requested friendship with whom. We consider the confidentiality of the friendship information of two arbitrary but fixed users UID1 and UID2 who are *not* in the coalition of observers:

(P4) A coalition of users UIDs can learn nothing about the updates to the friendship status between two users UID1 and UID2 beyond those updates that are performed while a member of the coalition is friends with UID1 or UID2, or the last update before there is a member of the coalition who becomes friends with UID1 or UID2.

(P5) A coalition of users UIDs can learn nothing about the friendship requests between two users UID1 and UID2 beyond the existence of a request before each successful friendship establishment.

Formally, we declare open access window to friendship information when either an observer is friends with UID1 or UID2 (since the listing of friends of friends is allowed), or the two users have not been created yet (since observers know statically that there is no friendship if the users do not exist yet).

$$\text{open}_F\ \sigma \equiv (\exists uid \in \text{UIDs}.\ uid \in \text{friendIDs}\ \sigma\ \text{UID1} \vee uid \in \text{friendIDs}\ \sigma\ \text{UID2})$$
$$\vee\ \text{UID1} \notin \text{userIDs}\ \sigma\ \vee\ \text{UID2} \notin \text{userIDs}\ \sigma$$

The relevant transitions for the secrecy infrastructure are the creation of users and the creation and deletion of friends or friend requests. The creation and deletion of friendship between UID1 and UID2 produces an FSec True or FSec False secret, respectively. In the case of openness changes, OSec is produced just as for the post confidentiality. Moreover, for (P5), we let the creation of a friendship request between UID1 and UID2 produce FRSec *uid text* secrets, where *uid* indicates the user that has placed the request, and *text* is the request message.

The main inductive definition of the two phases of the declassification bounds for friendship (P4) is given in Fig. 2, where *fs* ranges over friendship statuses, i.e., Booleans. Note that it follows the same "while"-"last update before" scheme as Fig. 1 for the post confidentiality, but with FSec instead of TSec. The overall bound is then defined as $\text{BO}_F\ sl\ sl'$ (since we start in the open phase where

$$\text{BO}_F\ (\text{map FSec}\ fs)\ (\text{map FSec}\ fs)\ (1) \quad \text{BC}_F\ (\text{map FSec}\ fs)\ (\text{map FSec}\ fs')\ (2)$$

$$\frac{\text{BO}_F\ sl\ sl' \qquad fs \neq [] \longleftrightarrow fs' \neq [] \qquad fs \neq [] \to \text{last}\ fs = \text{last}\ fs'}{\text{BC}_F\ (\text{map FSec}\ fs\ @\ \text{OSec True}\ @\ sl)\ (\text{map FSec}\ fs'\ @\ \text{OSec True}\ @\ sl')}\ (3)$$

$$\frac{\text{BC}_F\ sl\ sl'}{\text{BO}_F\ (\text{map FSec}\ fs\ @\ \text{OSec False}\ @\ sl)\ (\text{map FSec}\ fs\ @\ \text{OSec False}\ @\ sl')}\ (4)$$

Fig. 2. The bound on friendship status secrets

UID1 and UID2 do not exist yet) plus a predicate on the values that captures the static knowledge of the observers: that the FSec's form an *alternating* sequence of "friending" and "unfriending."

For (P5), we additionally require that at least one FRSec and at most two FRSec secrets from different users have to occur before each FSec True secret. Beyond that, we require nothing about the request values. Hence, the bound for friendship requests states that observers learn nothing about the requests between UID1 and UID2 beyond the existence of a request before each successful friendship establishment. In particular, they learn nothing about the "orientation" of the requests (i.e., which of the two involved users has placed a given request) and the contents of the request messages.

4   Verifying Confidentiality

Next we recall the unwinding proof technique for BD security (Sect. 4.1) and show how we have employed it for CoSMed (Sect. 4.2).

4.1   BD Unwinding Recalled

In [18], we have presented a verification technique for BD security inspired by Goguen and Meseguer's *unwinding* technique for noninterference [13]. Classical noninterference requires that it must be possible to purge all secret transitions from a trace, without affecting the outputs of observable actions. The unwinding technique uses an equivalence relation on states, relating states with each other that are supposed to be indistinguishable for the observer. The proof obligations are that 1. equivalent states produce equal outputs for observable actions, 2. performing an observable action in two equivalent states again results in two equivalent states, and 3. the successor state of a secret transition is equivalent to the source state. This guarantees that purging secret transitions preserves observations. The proof proceeds via an induction on the original trace.

For BD security, the situation is different. Instead of purging all secret transitions, we have to *construct* a *different* trace tr' that produces the same observations as the original trace tr, but produces precisely a given sequence of secrets sl' for which B (S tr) sl' holds.

The idea is to construct tr' incrementally, in synchronization with tr, but "keeping an eye" on sl' as well. The unwinding relation [18, Sect. 5.1] is therefore not a relation on states, but a relation on (state × secret list), or equivalently, a set of tuples $(\sigma, sl, \sigma', sl')$. Each of these tuples represents a possible configuration of the unwinding "synchronization game": σ and sl represent the current state reached by a potential original trace and the secrets that are still to be produced by it; and similarly for σ' and sl' w.r.t. the alternative trace.

To keep proof size manageable, the framework supports the decomposition of Δ into smaller unwinding relations $\Delta_0, \ldots, \Delta_n$ focusing on different phases of the synchronization game. The unwinding conditions require that, from any such configuration for which one of the relations hold, say, $\Delta_i \ \sigma \ sl \ \sigma' \ sl'$, the alternative trace can "stay in the game" by choosing to (1) either act independently or (2) wait for the original trace to act and then choose how to react to it: (1.a) either ignore that transition or (1.b) match it with an own transition. For the resulting configuration, one of the unwinding relations has to hold again. More precisely, the allowed steps in the synchronization game are the following:

INDEPENDENT ACTION: *There exists* a transition $trn' = $ Trans σ' _ _ σ'_1 that is unobservable (i.e., $\neg$ isObs trn'), produces the first secret in sl', and leads to a configuration that is again in one of the relations, Δ_j σ sl σ'_1 sl'_1 for $j \in \{1, \ldots, n\}$

REACTION: *For all* possible transitions $trn = $ Trans σ _ _ σ_1 one of the following holds:

    IGNORE: trn is unobservable and again leads to a related configuration Δ_k σ_1 sl_1 σ' sl' for $k \in \{1, \ldots, n\}$

    MATCH: There exists an observationally equivalent transition $trn' = $ Trans σ' _ _ σ'_1 (i.e., isObs trn $\leftrightarrow$ isObs trn' and isObs trn $\rightarrow$ getObs $trn = $ getObs trn') that together with trn leads to a related configuration Δ_l σ_1 sl_1 σ'_1 sl'_1 for $l \in \{1, \ldots, n\}$

If one of these conditions is satisfied for any configuration, then the unwinding relations can be seen as forming a graph: For each i, Δ_i is connected to all the relations into which it "unwinds," i.e., the relations Δ_j, Δ_k or Δ_l appearing in the above conditions. We use these conditions in the inductive step of the proof of the soundness theorem below.

Finally, we require that the initial relation Δ_0 is a proper generalization of the bound for the initial state, $\forall sl$ sl'. B sl sl' $\rightarrow$ Δ_0 istate sl istate sl'. This corresponds to initializing the game with a configuration that loads any two sequences of secrets satisfying the bound.

Theorem 1. [18] *If* $\Delta_0, \ldots, \Delta_n$ *form a graph of unwinding relations, and* B sl sl' *implies* Δ_0 istate sl istate sl' *for all* sl *and* sl', *then (the given instance of) BD security holds.*

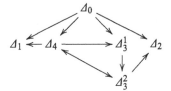

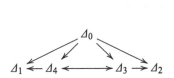

Fig. 3. Graph of unwinding relations   **Fig. 4.** Refined graph

4.2   Unwinding Relations for CoSMed

In a graph $\Delta_0, \ldots, \Delta_n$ of unwinding relations, Δ_0 generalizes the bound B. In turn, Δ_0 may unwind into other relations, and in general any relation in the graph may unwind into its successors. Hence, we can think of Δ_0 as "taking over the bound," and of all the relations as "maintaining the bound" together with state information. It is therefore natural to design the graph to reflect the definition of B.

We have applied this strategy to all our unwinding proofs. The graph in Fig. 3 shows the unwindings of the post-text confidentiality property (P3). In addition to the initial relation Δ_0, there are 4 relations Δ_1–Δ_4 with Δ_i corresponding to clause (i) for the definition of B from Fig. 1. The edges correspond to the possible causalities between the clauses. For example, if B sl sl' has been obtained applying clause (3), then, due to the occurrence of BO in the assumptions, we know the previous clauses must have been either (2) or (4)—hence the edges from Δ_3 to Δ_2 and Δ_4. Each Δ_i also provides a relationship between the states σ and σ' that fits the situation. Since we deal with repeated opening and closing of the access window, we naturally require:

- that $\sigma = \sigma'$ when the window is open
- that $\sigma =_{\mathsf{PID}} \sigma'$, i.e., σ and σ' are equal everywhere save for the value of PID's text, when the window is closed

Indeed, only when the window is open the observer would have the power to distinguish different values for PID's text; hence, when the window is closed the secrets are allowed to diverge. Open windows are maintained by the clauses for BO, (2) and (4), and hence by Δ_2 and Δ_4. Closed windows are maintained by the clauses for B, (1) and (3), with the following exception for clause (3): When the open-window marker OSec True is reached, the PID text updates would have synchronized (last $textl$ = last $textl'$), and therefore the relaxed equality $=_{\mathsf{PID}}$ between states would have shrunk to plain equality—this is crucial for the switch between open and closed windows.

To address this exception, we refine our graph as in Fig. 4, distinguishing between clause (3) applied to nonempty update prefixes where we only need $\sigma =_{\mathsf{PID}} \sigma'$, covered by Δ_3^1, and clause (3) with empty update prefixes where we need $\sigma = \sigma'$, covered by Δ_3^2. Figure 5 gives the formal definitions of the relations. Δ_0 covers the prehistory of PID—from before it was created. In Δ_1–Δ_4, the conditions on sl and sl' essentially incorporate the inversion rules corresponding to clauses (1)-(4) in B's definition, while the conditions on σ and σ' reflect the access conditions, as discussed.

Proposition 2. The relations in Fig. 5 form a graph of unwinding relations, and therefore (by Theorem 1) the post-text confidentiality property (P3) holds.

For unwinding the friendship confidentiality properties, we proceed analogously. We define unwinding relations, corresponding to the different clauses in Fig. 2, and prove that they unwind into each other and that B sl sl' implies Δ_0 istate sl istate sl'. In the open phase, we require that the two states are equal up to pending friendship requests between UID1 and UID2. In the closed phase, the two states may additionally differ on the friendship *status* of UID1 and UID2. Again, we need to converge back to the same friendship status when changing

$\Delta_0\ \sigma\ sl\ \sigma'\ sl' \equiv \neg\ \mathsf{PID} \in \mathsf{postIDs}\ \sigma \wedge \sigma = \sigma'$

$\Delta_1\ \sigma\ sl\ \sigma'\ sl' \equiv \mathsf{PID} \in \mathsf{postIDs}\ \sigma \wedge \sigma =_{\mathsf{PID}} \sigma' \wedge \neg\ \mathsf{open}\ \sigma \wedge$
$\qquad \exists textl\ textl'.\ sl = \mathsf{map\ TSec}\ textl \wedge sl' = \mathsf{map\ TSec}\ textl' \wedge$
$\qquad textl = [] \rightarrow textl' = []$

$\Delta_2\ \sigma\ sl\ \sigma'\ sl' \equiv \mathsf{PID} \in \mathsf{postIDs}\ \sigma \wedge \sigma = \sigma' \wedge \mathsf{open}\ \sigma \wedge$
$\qquad \exists textl.\ sl = \mathsf{map\ TSec}\ textl \wedge sl' = \mathsf{map\ TSec}\ textl$

$\Delta_3^1\ \sigma\ sl\ \sigma'\ sl' \equiv \mathsf{PID} \in \mathsf{postIDs}\ \sigma \wedge \sigma =_{\mathsf{PID}} \sigma' \wedge \neg\ \mathsf{open}\ \sigma \wedge$
$\qquad \exists textl\ textl'\ sl_1\ sl'_1.\ sl = \mathsf{map\ TSec}\ textl\ @\ \mathsf{OSec\ True}\ \#\ sl_1 \wedge$
$\qquad sl' = \mathsf{map\ TSec}\ textl'\ @\ \mathsf{OSec\ True}\ \#\ sl'_1 \wedge$
$\qquad \mathsf{BO}\ sl_1\ sl'_1 \wedge textl \neq [] \wedge textl' \neq [] \wedge \mathsf{last}\ textl = \mathsf{last}\ textl'$

$\Delta_3^2\ \sigma\ sl\ \sigma'\ sl' \equiv \mathsf{PID} \in \mathsf{postIDs}\ \sigma \wedge \sigma = \sigma' \wedge \neg\ \mathsf{open}\ \sigma \wedge$
$\qquad \exists sl_1\ sl'_1.\ sl = \mathsf{OSec\ True}\ \#\ sl_1 \wedge sl' = \mathsf{OSec\ True}\ \#\ sl'_1 \wedge \mathsf{BO}\ sl_1\ sl'_1$

$\Delta_4\ \sigma\ sl\ \sigma'\ sl' \equiv \mathsf{PID} \in \mathsf{postIDs}\ \sigma \wedge \sigma = \sigma' \wedge \mathsf{open}\ \sigma \wedge$
$\qquad \exists textl\ sl_1\ sl'_1.\ sl = \mathsf{map\ TSec}\ textl\ @\ \mathsf{OSec\ False}\ \#\ sl_1 \wedge$
$\qquad sl' = \mathsf{map\ TSec}\ textl'\ @\ \mathsf{OSec\ False}\ \#\ sl'_1 \wedge \mathsf{B}\ sl_1\ sl'_1$

Fig. 5. The unwinding relations for post-text confidentiality

from the closed into the open phase. Hence, we maintain the invariant in the closed phase that if an OSec True secret follows later in the sequence of secrets, then the last updates before OSec True must be equal, analogous to Δ_3^1 for post texts, and the friendship status must be equal in the two states immediately before an OSec True secret, analogous to Δ_3^2 for post texts.

5   Verification Summary

The whole formalization consists of around 9700 Isabelle lines of code (LOC). The (reusable) BD security framework takes 1800 LOC. CoSMeD's kernel implementation represents 700 LOC. Specifying and verifying the confidentiality properties for CoSMeD represents the bulk, 6500 LOC. Some additional 200 LOC are dedicated to various safety properties to support the confidentiality proofs—e.g., that two users cannot be friends if there are pending friendship requests between them. Unlike the confidentiality proofs, which required explicit construction of unwindings, safety proofs were performed automatically (by reachable-state induction).

Yet another kind of properties were formulated in response to the following question: We have shown that a user can only learn about updates to posts that were performed during times of public visibility or friendship, and about the last updates before these time intervals. But how can we be sure that the public visibility status or the friendship status cannot be forged? We have proved that these statuses can indeed only occur by the standard protocols. These properties (taking 500 LOC), complement our proved confidentiality by a form of accountability: they show that certain statuses can only be forged by identity theft.

6   Related Work

Proof assistants are today's choice for *precise* and *holistic* formal verification of hardware and software systems. Already legendary verification works are the AMD microprocessor floating-point operations [24], the CompCert C compiler [21] and the seL4 operating system kernel [19]. More recent developments include a range of microprocessors [16], Java and ML compilers [20,22], and a model checker [11].

Major "holistic" verification case studies in the area of information flow security are rather scarce, perhaps due to the more complex nature of the involved properties compared to traditional safety and liveness [23]. They include a hardware architecture with information-flow primitives [10] and a separation kernel [9], and noninterference for seL4 [25]. A substantial contribution to web client security is the Quark verified browser [17]. We hope that our line of work, putting CoCon and CoSMed in the spotlight but tuning a general verification framework backstage, will contribute a firm methodology for the holistic verification of server-side confidentiality.

Policy languages for social media platforms have been proposed in the context of Relationship-based Access Control [12], or using epistemic logic [28]. These approaches focus on specifying policies for granting or denying access to data based on the social graph, e.g. friendship relations. While our system implementation does make use of access control, our guarantees go beyond access control to information flow control. A formal connection between these policy languages and BD security would be interesting future work.

Finally, there are quite a few programming languages and tools aimed at supporting information-flow secure programming [2,3,7,30], as well as information-flow tracking tools for the client side of web applications [6,8,14]. We foresee a future where such tools will cooperate with proof assistants to offer light-weight guarantees for free and stronger guarantees (like the ones we proved in this paper) on a need basis.

Conclusion. CoSMed is the first social media platform with verified confidentiality guarantees. Its verification is based on BD security, a framework for information-flow security formalized in Isabelle. CoSMed's specific confidentiality needs require a dynamic topology of declassification bounds and triggers.

Acknowledgements. We are indebted to the reviewers for useful comments and suggestions. We gratefully acknowledge support:
− from Innovate UK through the Knowledge Transfer Partnership 010041 between Caritas Anchor House and Middlesex University: "The Global Noticeboard (GNB): a verified social media platform with a charitable, humanitarian purpose"
− from EPSRC through grant EP/N019547/1, Verification of Web-based Systems (VOWS)
− from DFG through grants Hu 737/5-2, MORES − Modelling and Refinement of Security Requirements on Data and Processes and Ni 491/13-3, Security Type Systems and Deduction.

References

1. OWASP top ten project. www.owasp.org/index.php/Top10#OWASP_Top_10_for_2013
2. Jif: Java + information flow (2014). http://www.cs.cornell.edu/jif
3. SPARK (2014). http://www.spark-2014.org
4. Caritas Anchor House (2016). http://caritasanchorhouse.org.uk/
5. Bauereiß, T., Gritti, A.P., Popescu, A., Raimondi, F.: The CoSMed website (2016). https://cosmed.globalnoticeboard.com
6. Bichhawat, A., Rajani, V., Garg, D., Hammer, C.: Information flow control in WebKit's javascript bytecode. In: Abadi, M., Kremer, S. (eds.) POST 2014 (ETAPS 2014). LNCS, vol. 8414, pp. 159–178. Springer, Heidelberg (2014)
7. Chlipala, A.: Ur/Web: a simple model for programming the web. In: POPL, pp. 153–165 (2015)
8. Chugh, R., Meister, J.A., Jhala, R., Lerner, S.: Staged information flow for javascript. In: PLDI, pp. 50–62 (2009)
9. Dam, M., Guanciale, R., Khakpour, N., Nemati, H., Schwarz, O.: Formal verification of information flow security for a simple ARM-based separation kernel. In: CCS, pp. 223–234 (2013)
10. de Amorim, A.A., Collins, N., DeHon, A., Demange, D., Hritcu, C., Pichardie, D., Pierce, B.C., Pollack, R., Tolmach, A.: A verified information-flow architecture. In: POPL, pp. 165–178 (2014)
11. Esparza, J., Lammich, P., Neumann, R., Nipkow, T., Schimpf, A., Smaus, J.-G.: A fully verified executable LTL model checker. In: Sharygina, N., Veith, H. (eds.) CAV 2013. LNCS, vol. 8044, pp. 463–478. Springer, Heidelberg (2013)
12. Fong, P.W.L., Anwar, M., Zhao, Z.: A privacy preservation model for facebook-style social network systems. In: Backes, M., Ning, P. (eds.) ESORICS 2009. LNCS, vol. 5789, pp. 303–320. Springer, Heidelberg (2009)
13. Goguen, J.A., Meseguer, J.: Unwinding and inference control. In: IEEE Symposium on Security and Privacy, pp. 75–87 (1984)
14. Groef, W.D., Devriese, D., Nikiforakis, N., Piessens, F.: FlowFox: a web browser with flexible and precise information flow control. In: CCS, pp. 748–759 (2012)
15. Haftmann, F., Nipkow, T.: Code generation via higher-order rewrite systems. In: Blume, M., Kobayashi, N., Vidal, G. (eds.) FLOPS 2010. LNCS, vol. 6009, pp. 103–117. Springer, Heidelberg (2010)
16. Hardin, D.S., Smith, E.W., Young, W.D.: A robust machine code proof framework for highly secure applications. In: Manolios, P., Wilding, M. (eds.) ACL2, pp. 11–20 (2006)
17. Jang, D., Tatlock, Z., Lerner, S.: Establishing browser security guarantees through formal shim verification. In: USENIX Security, pp. 113–128 (2012)
18. Kanav, S., Lammich, P., Popescu, A.: A conference management system with verified document confidentiality. In: Biere, A., Bloem, R. (eds.) CAV 2014. LNCS, vol. 8559, pp. 167–183. Springer, Heidelberg (2014)
19. Klein, G., Andronick, J., Elphinstone, K., Heiser, G., Cock, D., Derrin, P., Elkaduwe, D., Engelhardt, K., Kolanski, R., Norrish, M., Sewell, T., Tuch, H., Winwood, S.: seL4: formal verification of an operating-system kernel. Commun. ACM **53**(6), 107–115 (2010)
20. Kumar, R., Myreen, M.O., Norrish, M., Owens, S.: CakeML: a verified implementation of ML. In: POPL, pp. 179–192 (2014)

21. Leroy, X.: Formal verification of a realistic compiler. Commun. ACM **52**(7), 107–115 (2009)
22. Lochbihler, A.: Java and the java memory model — a unified, machine-checked formalisation. In: Seidl, H. (ed.) Programming Languages and Systems. LNCS, vol. 7211, pp. 497–517. Springer, Heidelberg (2012)
23. Mantel, H.: Information flow and noninterference. In: van Tilborg, H.C.A., Jajodia, S. (eds.) Encyclopedia of Cryptography and Security, 2nd edn, pp. 605–607. Springer, Heidelberg (2011)
24. Moore, J.S., Lynch, T.W., Kaufmann, M.: A mechanically checked proof of the $\text{amd5}_k\text{86}^{\text{tm}}$ floating point division program. IEEE Trans. Comput. **47**(9), 913–926 (1998)
25. Murray, T.C., Matichuk, D., Brassil, M., Gammie, P., Bourke, T., Seefried, S., Lewis, C., Gao, X., Klein, G.: seL4: from general purpose to a proof of information flow enforcement. In: Security and Privacy, pp. 415–429 (2013)
26. Nipkow, T., Klein, G.: Concrete Semantics: With Isabelle/HOL. Springer, Heidelberg (2014)
27. Nipkow, T., Paulson, L.C., Wenzel, M.: Isabelle/HOL: A Proof Assistant for Higher-Order Logic. LNCS, vol. 2283. Springer, Heidelberg (2002)
28. Pardo, R., Schneider, G.: A formal privacy policy framework for social networks. In: Giannakopoulou, D., Salaün, G. (eds.) SEFM 2014. LNCS, vol. 8702, pp. 378–392. Springer, Heidelberg (2014)
29. Sutherland, D.: A model of information. In: 9th National Security Conference, pp. 175–183 (1986)
30. Yang, J., Yessenov, K., Solar-Lezama, A.: A language for automatically enforcing privacy policies. In: POPL, pp. 85–96 (2012)

Mechanical Verification of a Constructive Proof for FLP

Benjamin Bisping, Paul-David Brodmann, Tim Jungnickel,
Christina Rickmann, Henning Seidler, Anke Stüber, Arno Wilhelm-Weidner,
Kirstin Peters, and Uwe Nestmann[✉]

Technische Universität Berlin, Berlin, Germany
uwe.nestmann@tu-berlin.de

Abstract. The impossibility of distributed consensus with one faulty
process is a result with important consequences for real world distributed systems e.g., commits in replicated databases. Since proofs are not
immune to faults and even plausible proofs with a profound formalism
can conclude wrong results, we validate the fundamental result named
FLP after Fischer, Lynch and Paterson by using the interactive theorem prover Isabelle/HOL. We present a formalization of distributed
systems and the aforementioned consensus problem. Our proof is based
on Hagen Völzer's paper *A constructive proof for FLP*. In addition to
the enhanced confidence in the validity of Völzer's proof, we contribute
the missing gaps to show the correctness in Isabelle/HOL. We clarify
the proof details and even prove fairness of the infinite execution that
contradicts consensus. Our Isabelle formalization may serve as a starting
point for similar proofs of properties of distributed systems.

Keywords: Formalization · Isabelle/HOL · Verification · FLP ·
Consensus · Distributed systems

1 Introduction

Many informal proofs have been found to be incorrect and even plausible ones
come to invalid results. One example is the "Effective Implementation for the
Generalized Input-Output Construct of CSP" [3] where a later attempt of a
proof produced a counterexample [9]. Hence, increasing the confidence in the
correctness of fundamental results is a good idea in general.

The impossibility of fault-tolerant distributed consensus in asynchronous systems is a fundamental result in computer science. Originally, it has been established and proved by Fischer, Lynch and Paterson (FLP) [6]. Since the result
implies major consequences in distributed computing, it is worth verifying its
correctness to achieve the highest level of certainty. To our knowledge, it has not
yet been verified mechanically. Due to its relevance we checked the claim in the
interactive theorem prover Isabelle/HOL.

We base our formalization on Völzer's paper "A Constructive Proof for
FLP" [13]. Compared to the original proof in [6], Völzer is more precise and more

© Springer International Publishing Switzerland 2016
J.C. Blanchette and S. Merz (Eds.): ITP 2016, LNCS 9807, pp. 107–122, 2016.
DOI: 10.1007/978-3-319-43144-4_7

extensive in defining the model and in the subproofs leading to the crucial contradiction. Völzer argues in [13], it is "not only showing that a non-terminating execution does exist but also how it can be constructed". Due to its higher degree of detail and constructiveness, we chose Völzer's proof as foundation for proving FLP. Accordingly, all design decisions are based on the motivation to stay as close as possible to Völzer. We extend the proof by providing a higher level of precision and clarify the proof details. Moreover, we formalize a notion of fairness and prove that the constructed execution is fair, which Völzer states without proof or proof sketch. Ultimately, we show that FLP's result is correct, up to the correctness of Isabelle/HOL. Our complete formalization is available in the Archive of Formal Proofs [2].

Overview. We start in Sect. 2 with a brief introduction into distributed systems, consensus and Isabelle/HOL. In Sect. 3 we sketch our Isabelle formalization of the model and the proven properties. Section 4 gives an overview of our formalization of the FLP proof and we discuss our observations. We conclude with a brief summary and an outlook in Sect. 5.

Related Work. In "Nonblocking Consensus Procedures Can Execute Forever" [4], Constable shows another way of proving the FLP result. He uses the idea of nonblocking consensus protocols and states that similar work has been formally verified using Nuprl [5] up to a certain point. In contrast, we fully formalized it using Isabelle/HOL [11].

More recently, there is a project[1] formalizing the original FLP paper in the theorem prover Coq [1], which has not yet been finished.

Contributions. Our main contributions are (1) the formalization of a generic model for distributed systems, (2) the constructive proof in the Isabelle/HOL theorem prover, (3) an additional thorough proof of fairness of the infinite execution and (4) the discussion of the differences between the proofs. We obtain an Isabelle/HOL model for distributed consensus that can be reused for future proofs. The theorem prover forced us to be very precise, e.g., when lifting definitions from finite to infinite objects. In the proofs, this precision allowed us to find a tiny error in the paper where the same symbol was used for similar but evidently different configurations. Völzer states without proof that the execution constructed in the paper is fair. We formally proved this claim. Additionally, we obtained a precise list of all preconditions and the dependencies of the previous lemmata for the individual proofs we formalized. We now have the certainty that the proof is correct up to the correctness of the Isabelle/HOL theorem prover and the correctness of our formalization of the model.

2 Preliminaries

We define the (binary) consensus problem as presented by Fischer, Lynch and Paterson in [6]. Then we present a short introduction to the interactive theorem prover Isabelle/HOL [11].

---

[1] https://github.com/ConsensusResearch/flp.

2.1   Distributed Systems and Consensus

Following [6], a *distributed system* is a collection of finitely many nodes, called processes, that communicate asynchronously via messages. In general a message is a pair (p, m), where p is the name of the destination process and m is a so-called message value. Processes communicate asynchronously by sending and receiving messages over reliable channels, i.e., no message is lost but it may be delayed for an arbitrary (finite) amount of time. We assume that this kind of message transfer is the only way for processes to exchange information. Processes are modeled as automata by states and a deterministic transition function that maps a state and a message to its successor state and outgoing messages. The global state of a distributed system is the collection of the local states of its processes together with the collection of messages in transit. The system evolves non-deterministically by choosing one of the messages in transit to be received by one of the processes and changes according to the local transition of this single process. A more detailed description of distributed systems is provided by their formalization described in Sect. 3. Similar to [13], a process is *correct* if it eventually consumes every message directed to it and *faulty* otherwise.

Intuitively, consensus is the problem of whether a distributed system is able to reach a global state such that all processes agree on some condition. More formally, a *consensus algorithm* is a distributed system such that all processes are initialized with a boolean value and, after a number of transition steps, a process may *decide* on a boolean value as output. In [6] initialization of the processes is modeled by a one-bit input register for each process containing its initial boolean value in the initial state and a process decides on a boolean value by writing it into its one-bit output register. Alternatively processes can be enhanced with variables to carry the input and output values, or this information may be carried by input and output messages (as described in Sect. 3).

A *correct consensus algorithm* satisfies the following three conditions:

Agreement: No two processes decide differently.
Validity: The output value of each process is the input value of some process.
Termination: Each correct process eventually decides.

Fischer et al. [6] show that in the presence of a faulty process there are no correct consensus algorithms, i.e., no correct consensus algorithm can tolerate a faulty process. More precisely, they prove that in this case the assumption of agreement and validity implies the existence of an infinite execution violating the termination property.

2.2   Isabelle/HOL

The interactive theorem prover Isabelle/HOL, initially created by Tobias Nipkow and Lawrence Paulson [11], supports machine-checked proofs and thus provides additional confidence in the correctness of proofs. It allows for the reuse of already defined structures and verified lemmata, which are organized in *theories*. Theories are the basic mechanism to organize and encapsulate formalization.

Established theories can be imported into new theories in order to build upon the described models and to reuse facts proven within these theories. *Locales* [7] provide an additional way to structure formalizations. They introduce local contexts, which can fix a number of local assumptions. Within these scopes definitions are based on local assumptions and proofs are carried out depending on them. Locales may also contain local constants and provide syntax rewriting mechanism. They can be organized hierarchically such that the extending structures inherit the assumptions and contents of their sublocales and add content or impose further assumptions.

3 Model

Our model for asynchronous distributed systems is based on the model described by Völzer [13]. In order to compare the formalization with the paper proof in [13], we tried to stay as close as possible to the formalization of Völzer.

The definitions are organized into two locales `AsynchronousSystem` and `Execution`. The first contains a basic branching-time model of concurrent processes communicating by messages. The second adds finite and infinite executions and a concept of fairness.

The formalization of the model is decoupled from later proofs concerning distributed consensus. So these theories can be reused for other proofs about distributed systems. There are however some design decisions—due to our attempt to stay close to [13]—that can be considered inefficient for some purposes.

3.1 Asynchronous System

The theory `AsynchronousSystem` models processes as named automata communicating asynchronously by messages addressing processes by names. It is parametrized over types for process identifiers `'p`, process state space `'s` and communication message values `'v`.

The model makes no assumption on the number of processes although the later proof of the FLP-theorem will only be conducted for finitely many processes.

The communication between processes takes place via messages. Their content is either a value of a (possibly infinite) type `'v` or a Boolean. We explicitly distinguish messages carrying a Boolean to ensure that the messages used to initialize the system are of this type.

```
datatype 'v messageValue =
  Bool bool
| Value 'v
```

According to [13], a message consists of a message value (of type `'v` or `bool`) and a receiver of the message identified by an element of `'p`. We distinguish three types of messages: input messages, output messages and regular inter-process messages. Input and output messages are special messages used to differentiate the input and output of the distributed algorithm. Input messages contain

Boolean start values for the system, i.e., we require that initially there is exactly one input message for each process. The output messages do not have a receiver as they can be seen as the Boolean return value of the system. In the consensus setting every correct process (Termination) sends its decision in exactly one output message and the values of all these messages have to be the same (Agreement). The inner communication of the system is realized with regular messages, whose type is entirely up to the user. Therefore, the model can easily be adapted to various settings.

```
datatype ('p, 'v) message =
  InMsg 'p bool   ("<_, inM _>")
| OutMsg bool     ("<⊥, outM _>")
| Msg 'p 'v        ("<_, _>")
```

A system configuration describes the global state of the system at one instant. It consists of the current states of all processes (a mapping from the set of processes to the set of states) and a multiset of all messages in transit, i.e., messages that have been sent but not yet received; we call these messages **enabled**. The predicate **enabled** checks that a message is part of the messages in transit of a given configuration. We added a small theory to implement multisets.

```
record ('p, 'v, 's) configuration =
  states :: "'p ⇒ 's"
  msgs :: "(('p, 'v) message) multiset"
```

The locale **asynchronousSystem** consists of three parameters, a transition function **trans** for each process, a message function **sends** for each process and an initial state **start** for each process. The transition function describes the local state changes of a process. The message function determines what messages are sent in reaction to a received message.

```
locale asynchronousSystem =
fixes
  trans :: "'p ⇒ 's ⇒ 'v messageValue ⇒ 's" and
  sends :: "'p ⇒ 's ⇒ 'v messageValue ⇒ ('p, 'v) message multiset" and
  start :: "'p ⇒ 's"
```

We assume input enabled processes, i.e., every process can accept any message in transit (directed to the process) at any time. A step in the system consumes exactly one message, changes the state of the receiving process and adds the new messages sent by the process to the multisets of messages in transit. Since we describe an asynchronous distributed system, the executing process does not wait for the reception of its sent messages and the states of all other processes remain unchanged.

```
primrec steps ::
  "('p, 'v, 's) configuration
    ⇒ ('p, 'v) message
    ⇒ ('p, 'v, 's) configuration
    ⇒ bool"
    ("_ ⊢ _ ↦ _" [70,70,70])
where
  StepInMsg: "cfg1 ⊢ <p, inM v> ↦ cfg2 = (
    (∀ s. ((s = p) ⟶ states cfg2 p = trans p (states cfg1 p) (Bool v))
      ∧ ((s ≠ p) ⟶ states cfg2 s = states cfg1 s))
```

```
∧ enabled cfg1 <p, inM v>
∧ msgs cfg2 = (sends p (states cfg1 p) (Bool v)
              ∪# (msgs cfg1 -# <p, inM v>)))"
| StepMsg: "cfg1 ⊢ <p, v> ↦ cfg2 = (
  (∀ s. ((s = p) ⟶ states cfg2 p = trans p (states cfg1 p) (Value v))
     ∧ ((s ≠ p) ⟶ states cfg2 s = states cfg1 s))
  ∧ enabled cfg1 <p, v>
  ∧ msgs cfg2 = (sends p (states cfg1 p) (Value v)
                ∪# (msgs cfg1 -# <p, v>)))"
| StepOutMsg: "cfg1 ⊢ <⊥,outM v> ↦ cfg2 =
  False"
```

The lemma NoReceivingNoChange proves that the states of the other processes do not change during a step and lemma OtherMessagesOnlyGrowing checks that no messages, besides the consumed one, disappear. Lemma ExistsMsg shows that only existing messages can be consumed in a step. The set of outgoing messages can only grow (lemma OutOnlyGrowing). Messages are enabled persistently, i.e., they remain enabled as long as they are not consumed in a step (lemma OnlyOccurenceDisables).

The transitive closure of the step relation defines the reachability of states in the system.

```
inductive reachable ::
  " ('p, 'v, 's) configuration
  ⇒ ('p, 'v, 's) configuration
  ⇒ bool"
where
  init: "reachable cfg1 cfg1"
| step: "⟦ reachable cfg1 cfg2; (cfg2 ⊢ msg ↦ cfg3) ⟧
          ⟹ reachable cfg1 cfg3"
```

Transitivity of the predicate reachable is proved in lemma ReachableTrans.

A configuration cfg3 is qReachable from the configuration cfg1 with the set of processes Q if only processes from the set Q take steps to reach the configuration.

```
inductive qReachable ::
  "('p,'v,'s) configuration
  ⇒ 'p set
  ⇒ ('p,'v,'s) configuration
  ⇒ bool"
where
  initQ: "qReachable cfg1 Q cfg1"
| stepQ: "⟦ qReachable cfg1 Q cfg2; (cfg2 ⊢ msg ↦ cfg3) ;
           ∃ p ∈ Q . isReceiverOf p msg ⟧
           ⟹ qReachable cfg1 Q cfg3"
```

qReachable is transitive (lemma QReachableTrans) and the states of processes outside of Q do not change (lemma NotInQFrozenQReachability). Additionally no messages addressed to processes not in Q are lost (lemma NoActivityNo-MessageLoss).

Dual to qReachable, a configuration is withoutQReachable if only processes not from the set Q take steps. Dual versions of the above lemmata were shown.

With qReachable and withoutQReachable we formalize the confluence property Diamond of our model. As visualized in Fig. 1, if configuration cfg1 is qReachable and another configuration cfg2 is withoutQReachable for the same Q, there is a configuration cfg' which is withoutQReachable from cfg1 and qReachable from cfg2.

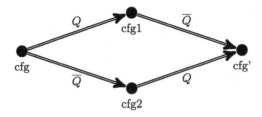

Fig. 1. Diamond property

```
lemma Diamond:
fixes
   cfg cfg1 cfg2 :: "('p,'v,'s) configuration" and
   Q :: "'p set"
assumes
   QReach: "qReachable cfg Q cfg1" and
   WithoutQReach: "withoutQReachable cfg Q cfg2"
shows
   "∃ cfg'. withoutQReachable cfg1 Q cfg'
        ∧ qReachable cfg2 Q cfg'"
```

The proof of lemma `Diamond` is constructed via two auxiliary lemmata. `Diamond-One` shows the confluence property for two single steps, `DiamondTwo` for a number of steps on the one side and a single step on the other side.

3.2 Executions

The Locale Execution: To model an execution of the system, we extend the locale `asynchronousSystem`. Völzer defines an execution as an alternating sequence of configurations and messages. We split such executions into a list `exec` of configurations and a list `trace` of messages.

```
locale execution =
   asynchronousSystem trans sends start
for
   trans :: "'p ⇒ 's ⇒ 'v messageValue ⇒ 's" and
   sends :: "'p ⇒ 's ⇒ 'v messageValue ⇒ ('p, 'v) message multiset" and
   start :: "'p ⇒ 's"
+
fixes
   exec :: "('p, 'v, 's ) configuration list" and
   trace :: "('p, 'v) message list"
assumes
   notEmpty: "length exec ≥ 1" and
   length: "length exec - 1 = length trace" and
   base: "initial (hd exec)" and
   step: "⟦ i < length exec - 1 ; cfg1 = exec ! i ; cfg2 = exec ! (i + 1) ⟧
        ⟹ ((cfg1 ⊢ trace ! i ↦ cfg2)) "
```

For every execution $exec = c_0 c_1 c_2 \ldots$ and $trace = m_0 m_1 m_2 \ldots$, there is a step between c_i and c_{i+1} that consumes m_i. For instance, m_0 is the message that is consumed in the step from c_0 to c_1. Additionally, we require that in c_0 all processes are in their initial state.

For every execution, we introduce the property `minimalEnabled`, which is true for some message `msg` if it has been enabled for the longest time without being consumed. It is defined as a predicate, and not as a function, since multiple messages might be `minimalEnabled`. We provide a proof that if at least one message is enabled, then such a minimal enabled message exists.

We define the `firstOccurrence` of a message, telling that `msg` has been enabled, but not consumed, for n steps. As before, we provide a proof that for any enabled message such a number n exists.

These concepts are used in the construction of the infinite execution to ensure its fairness.

Extensions of Executions: To model infinite executions, we construct an increasing chain of executions. Therefore, we show some lemmata concerning the extension of an execution. In `expandExecutionStep` we show that if we can perform some step from the last configuration of our execution, then we can expand the execution by that step to gain a new execution. We extend this statement to all configurations that are reachable from the last configuration of the considered execution. Finally we show that, if we can reach a configuration `cfg` by consuming the message `msg`, then we can extend the execution such that `msg` is in the extension of the trace.

While the executions discussed so far are only finite lists, the concepts of *fairness* and *correctness* require infinite executions. Our model defines infinite executions to be pairs of:

– a map from $\mathbb{N}$ to `configuration list`
– a map from $\mathbb{N}$ to `message list`

both forming strictly increasing infinite chains with respect to the (strict) prefix relation on lists `prefixList`.

```
definition (in asynchronousSystem) infiniteExecution ::
  "(nat ⇒ (('p, 'v, 's) configuration list))
  ⇒ (nat ⇒ (('p, 'v) message list)) ⇒ bool"
where
  "infiniteExecution fe ft ≡
    ∀ n . execution trans sends start (fe n) (ft n) ∧
      prefixList (fe n) (fe (n+1)) ∧
      prefixList (ft n) (ft (n+1))"
```

Another common concept for representing possibly infinite executions are lazy lists. They are defined as the codatatype corresponding to lists and can be both finite and infinite. We, however, decided to follow Völzer's design. He focuses on finite executions, creating a series of expansions. Keeping finite and infinite definitions separate also helps us in the cases of definitions that require finite executions. Especially for the construction in the main proof, we use finite executions only.

Within a concrete infinite execution, we consider a process as *crashed* if there is a point in time such that there are messages addressed to the process, but the process does not consume any further messages. A process is *correct* in a given execution if it does not crash. Every correct process that still has messages addressed to it has to eventually take a step.

```
definition (in asynchronousSystem) correctInfinite ::
  "(nat ⇒ (('p, 'v, 's) configuration list))
  ⇒ (nat ⇒ (('p, 'v) message list)) ⇒ 'p ⇒ bool"
where
  "correctInfinite fe ft p ≡
   infiniteExecution fe ft
   ∧ (∀ n . ∀ n0 < length (fe n). ∀ msg .
     (enabled ((fe n) ! n0) msg)
      ∧ isReceiverOf p msg
       ⟶ (∃ msg' . ∃ n' ≥ n . ∃ n0' ≥ n0 .isReceiverOf p msg'
         ∧ n0' < length (fe n') ∧ (msg' = ((ft n') ! n0'))))"
```

Finally we define the concept of fairness. An infinite execution is fair, if for each enabled message having a receiver, there is some later point at which this message, or a copy of it, is consumed. We do not differentiate between multiple instances of the same message.

```
definition (in asynchronousSystem) fairInfiniteExecution ::
  "(nat ⇒ (('p, 'v, 's) configuration list))
  ⇒ (nat ⇒ (('p, 'v) message list)) ⇒ bool"
where
  "fairInfiniteExecution fe ft ≡
   infiniteExecution fe ft
   ∧ (∀ n . ∀ n0 < length (fe n). ∀ p . ∀ msg .
     ((enabled ((fe n) ! n0) msg)
      ∧ isReceiverOf p msg ∧ correctInfinite fe ft p )
       ⟶ (∃ n' ≥ n . ∃ n0' ≥ n0 . n0' < length (ft n')
         ∧ (msg = ((ft n') ! n0'))))"
```

4   FLP Formalization

Our formalization of the FLP result is structured in two locales flpSystem and flpPseudoConsensus. The first extends the locale asynchronousSystem by the notion of consensus. The second combines the results of the theory FLPSystem with the concept of fair infinite executions and culminates in showing the impossibility of a consensus algorithm in the proposed setting.

4.1   FLPSystem

Völzer defines consensus in terms of the classical notions of agreement, validity and termination. The proof relies on a weaker version called pseudo consensus. The proof also applies a weakened notion of termination, which we refer to as "pseudo termination". The theory FLPSystem contains our formalization of the consensus properties, i.e., agreement, validity and pseudo termination. Similar to Völzer we define a concept of non-uniformity regarding pending decision possibilities, where non-uniform configurations can always reach other non-uniform ones. It contains all the lemmata and propositions presented by Völzer except for the final construction of the infinite execution, which is contained in FLPTheorem together with the fairness proof and the final contradiction.

The locale flpSystem extends the locale asynchronousSystem with restrictions on the set of processes and the message function. We assume that there are at least two processes in the system, as one process could always decide on its own. Furthermore, we assume that there are only finitely many processes in the system and that every process sends only finitely many messages in each step. To be as general as possible, we make no assumptions on the number of messages a process sends to another one or other restrictions on the communication between the processes. The last assumption states that processes cannot send input messages. Since input messages are special system messages to describe the start values of the system, this does not restrict the inter-processes communication. Although not intentionally created for it, several parts of flpSystem can be reused to solve similar problems in distributed systems.

```
locale flpSystem =
  asynchronousSystem trans sends start
    for trans :: "'p ⇒ 's ⇒ 'v messageValue ⇒'s"
    and sends :: "'p ⇒ 's ⇒ 'v messageValue ⇒ ('p, 'v) message multiset"
    and start :: "'p ⇒ 's" +
assumes finiteProcs: "finite Proc"
    and minimalProcs: "card Proc ≥ 2"
    and finiteSends: "finite {v. v ∈# (sends p s m)}"
    and noInSends: "sends p s m <p2, inM v> = 0"
begin
```

A configuration is **vDecided** with value **v** if it is reachable from an arbitrary initial configuration and there is an output message with value **v** in transit. We call **v** the decision value.

```
abbreviation vDecided ::
  "bool ⇒ ('p, 'v, 's) configuration ⇒ bool"
where
  "vDecided v cfg ≡ initReachable cfg ∧ (<⊥, outM v> ∈# msgs cfg)"
```

The *pSilentDecisionValues* of a process **p** and a configuration **c** are the decision values that are possible if the process **p** no longer executes steps.

```
definition pSilDecVal ::
  "bool ⇒ 'p ⇒ ('p, 'v, 's) configuration ⇒ bool"
where
  "pSilDecVal v p c ≡ initReachable c ∧
    (∃ c'::('p, 'v, 's) configuration . (withoutQReachable c {p} c')
    ∧ vDecided v c')"
```

```
abbreviation pSilentDecisionValues ::
  "'p ⇒ ('p, 'v, 's) configuration ⇒ bool set" ("val[_,_]")
where
  "val[p, c] ≡ {v. pSilDecVal v p c}"
```

A configuration that is reachable from an arbitrary initial configuration is called **vUniform** if, regardless which process stops, **v** is the only decision value possible.

```
definition vUniform ::
  "bool ⇒ ('p, 'v, 's) configuration ⇒ bool"
where
  "vUniform v c ≡ initReachable c ∧ (∀p. val[p,c] = {v})"
```

A configuration is **nonUniform** if it is neither **vUniform** for **True** nor **vUniform** for **False**.

```
abbreviation nonUniform ::
  "('p, 'v, 's) configuration ⇒ bool"
where
  "nonUniform c ≡ initReachable c ∧
    ¬(vUniform False c) ∧
    ¬(vUniform True c)"
```

The three main properties of the consensus problem are agreement, validity and (pseudo) termination.

Agreement states that no two processes decide differently. In our formalization a configuration satisfies this property if all system output messages in transit carry the same value.

```
definition agreement ::
  "('p, 'v, 's) configuration ⇒ bool"
where
  "agreement c ≡
    (∀ v1. (<⊥, outM v1> ∈# msgs c)
      ⟶ (∀ v2. (<⊥, outM v2> ∈# msgs c)
        ⟷ v2 = v1))"
```

Validity holds if the output value is the input value of some process. Therefore, the validity of a configuration depends on the initial configuration. A configuration c that is reachable from an initial configuration i satisfies validity if every value of an output message in c already is the value of an input message in the initial configuration i.

```
definition validity ::
  "('p, 'v, 's) configuration ⇒ ('p, 'v, 's) configuration ⇒ bool"
where
  "validity i c ≡
    initial i ∧ reachable i c ⟶
      (∀ v. (<⊥, outM v> ∈# msgs c)
        ⟶ (∃ p. (<p, inM v> ∈# msgs i)))"
```

A configuration c that is reachable from an initial configuration satisfies pseudo termination if a decided configuration can be reached without the participation of at most t processes and by steps of processes from Q.

```
definition terminationPseudo ::
  "nat ⇒ ('p, 'v, 's) configuration ⇒ 'p set ⇒ bool"
where
  "terminationPseudo t c Q ≡ ((initReachable c ∧ card Q + t ≥ card Proc)
    ⟶ (∃ c'. qReachable c Q c' ∧ decided c'))"
```

4.2 FLPTheorem

The locale flpPseudoConsensus contains the construction of the infinite nondeciding execution, the respective fairness proof and the final contradiction of the consensus assumptions. Parts of the first and the latter two are not contained in the proof of Völzer. The assumptions for the locale are agreement and termination. Agreement is required in the form of **agreementInit** that requires agreement for all configurations c that are **reachable** form an initial configuration.

```
locale flpPseudoConsensus =
  flpSystem trans sends start
for
  trans :: "'p ⇒ 's ⇒ 'v messageValue ⇒'s" and
  sends :: "'p ⇒ 's ⇒ 'v messageValue ⇒ ('p, 'v) message multiset" and
  start :: "'p ⇒ 's" +
assumes
  Agreement: "⋀ i c . agreementInit i c" and
  PseudoTermination: "⋀cc Q . terminationPseudo 1 cc Q"
```

The final contradiction is shown in `ConsensusFails`. It uses all previous lemmata, directly or indirectly. An infinite execution is said to be a `terminationFLP`-execution if at some point it contains a decision message or none of the processes consumes any further messages.

```
theorem ConsensusFails:
assumes
  Termination:
    "⋀ fe ft . (fairInfiniteExecution fe ft ⟹ terminationFLP fe ft)" and
  Validity: "∀ i c . validity i c" and
  Agreement: "∀ i c . agreementInit i c"
shows
  "False"
```

4.3 Proof Structure

Our proof follows the general structure of Völzer's proof. In this section we focus on the ideas and differences to Völzer's proof instead of concentrating on details.

The main argument of the proof is the construction of a *non-uniform infinite configuration*. Non-uniformity implies that no decision has yet been made or is predetermined and thus violates termination.

The proof uses non-uniformity as an invariant. The first part is the lemma `InitialNonUniformCfg` (*Lemma 1* in [13]) which shows the existence of a non-uniform initial configuration. This lemma is proved by constructing a series of initial configurations such that two consecutive configurations differ only in the input of one process. Consequently, there are two configurations such that one is 0-uniform and the other is not 0-uniform. The proof then shows that both can decide on 0 which means the latter is also not 1-uniform and therefore a non-uniform initial configuration.

```
lemma InitialNonUniformCfg:
assumes
  Termination: "⋀cc Q . terminationPseudo 1 cc Q" and
  Validity: "∀ i c . validity i c" and
  Agreement: "∀ i c . agreementInit i c"
shows
  "∃ cfg . initial cfg ∧ nonUniform cfg"
```

Then we have `NonUniformCanReachSilentBivalence`, which corresponds to the *Lemma 2* in [13]. Given some non-uniform configuration and any process p, we can reach a configuration c' such that if p stops both decisions are possible. This lemma is proved by a distinction of cases. Either, if p fails, both values are possible and there is nothing more to show or, if the failure of p would lead to a decided configuration for 0 or 1, we construct an extension where the failure of **p** leads to a decided configuration for the respective other value. In lemma `SilentDecisionValueNotInverting` we have shown that in such a case we can not reach the latter configuration from the former in one step. There must be an intermediate configuration where if **p** fails, both values are possible. So this—non-uniform—configuration must exist.

```
lemma NonUniformCanReachSilentBivalence:
fixes
  p:: 'p and
  c:: "('p, 'v, 's) configuration"
assumes
  NonUni: "nonUniform c" and
  PseudoTermination: "⋀cc Q . terminationPseudo 1 cc Q" and
  Agree: "⋀ cfg . reachable c cfg ⟶ agreement cfg"
shows
  "∃ c' . reachable c c' ∧ val[p,c'] = {True, False}"
```

In `NonUniformExecutionsConstructable` we use these lemmata to construct the extended execution according to the invariant. We assume an execution, that ends in some non-uniform configuration, and chose an arbitrary enabled message **msg**. Then we construct an extension of this execution that consumes **msg** and again ends in some non-uniform configuration.

```
lemma NonUniformExecutionsConstructable:
fixes
  exec :: "('p, 'v, 's ) configuration list " and
  trace :: "('p, 'v) message list" and
  msg :: "('p, 'v) message" and
  p :: 'p
assumes
  MsgEnabled: "enabled (last exec) msg" and
  PisReceiverOf: "isReceiverOf p msg" and
  ExecIsExecution: "execution trans sends start exec trace" and
  NonUniformLexec: "nonUniform (last exec)" and
  Agree: "⋀ cfg . reachable (last exec) cfg ⟶ agreement cfg"
shows
  "∃ exec' trace' . (execution trans sends start exec' trace')
    ∧ nonUniform (last exec')
    ∧ prefixList exec exec' ∧ prefixList trace trace'
    ∧ (∀ cfg . reachable (last exec') cfg ⟶ agreement cfg)
    ∧ stepReachable (last exec) msg (last exec')
    ∧ (msg ∈ set (drop (length trace) trace')))"
```

Figure 2 visualizes the effect of these lemmata. A configuration placed on the 0- or 1-line would mean having decided on the respective value. The infinite execution starts with a non-uniform configuration which we know to exist from *Lemma 1*. With the next step we reach a configuration that moves closer to a decision for one value. From *Lemma 2* we know that from such a configuration we can reach another one which is also non-uniform, i.e., a configuration that is a again not directed towards either of the decision values. By applying this construction repeatedly we never decide. This infinite execution violates the consensus property *termination* and shows that this setting of distributed consensus possibly never decides.

Our construction allows us, whenever we are in a non-uniform configuration, to pic a specific message that is in transit. We choose one that satisfies the predicate `minimalEnabled`, while performing these steps. Hence, to prove the fairness of the constructed execution, we expand the previous lemma by requiring

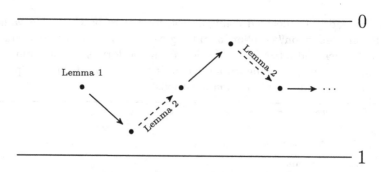

Fig. 2. Concept of the infinite execution

that the extension consumes one of the messages that have been enabled for the longest time.

This construction produces a non-terminating execution, *Theorem 1* in [13]. The resulting infinite execution consists only of non-decided configurations and therefore violates the termination property.

4.4   Discussion

In the following we compare our proof and the one of Völzer [13].

(1) Formalizing the proof in a theorem prover requires a high level of detail. From our formalization we therefore gained precise assumptions for all lemmata. One example for this is *Proposition 2* in [13]. It states for a reachable configuration that each process can decide (a) and that all processes in a decided configuration agree on one value (b). The proof in [13] solely states that "the claims follow directly from the definitions". Our proof in Isabelle/HOL, however, required in part (a) the definition of a decided process and termination and in part (b) it was also necessary to have a decided reachable configuration and agreement for all reachable configurations in addition to the definitions.

(2) This need for precision allowed us to find a small error in the paper where the same name was used for similar but evidently different configurations. This error is located in Völzer's proof of *Lemma 1*, where he uses one name for two configurations that necessarily differ in the input of the same process. He states this himself in the proof, giving the wrong impression these similar configurations are actually the same configuration.

(3) Völzer's construction of the final execution lacks any proof of fairness. It is simply states that "[we] obtain a fair execution where all processes are correct [...]". Although the construction is in fact fair, this lack of proof for fairness is typical of many papers. In contrast, we proved the fairness of our construction. This turned out to be very time-consuming. We had to consider the aspect of fairness throughout the whole proof, which changed several already proven lemmata in retrospect. These changes and the actual fairness proof extended our formalization by about a quarter.

(4) Völzer represents messages in transit as a multiset but the messages sent by a process as a set. We do not make this restriction, i.e. we allow processes to send the same message multiple times to the same process in one step.

(5) In contrast to the four pages that Völzer needs in [13] to describe his model and proof, our formalization of his proof has about 4000 lines of code. Of these, about 1600 are part of defining the model in `AsynchronousSystem` and `Execution`. The remaining 2400 lines consist of the formalization of the FLP result in `flpSystem` and `flpPseudoConsensus`. The proofs considering the so-called *diamond property*, for which [13] simply states that it is easy to verify, took us 300 lines of code. *Lemma 1* took us about 400 lines (`InitialNonUniformCfg`). The proof of the lemma `FairNonUniformExecution` that a fair non-uniform execution exists which is, as mentioned above, stated without proof in [13] took us about 800 lines of code. Of course, since minimality was not our goal, there may be ways to reduce the number of lines considerably, e.g. by using variables to store input and output values instead of the messages used in [13]. Please note, however, the striking discrepancy in the amount of work related to cover fairness: approximately half of the lines of the proof in the formalization instead of a short sentence in the paper.

(6) Finally let us discuss some points one might want to change in order to reuse the formalization. The initial choice, to stay as close to Völzer as possible, resulted in some design decisions which lead to unnecessary long and complicated proofs. Neglecting the design of Völzer, we recommend to model the output of the system as a variable for each process instead of messages. This helps to identify decided processes, leading e.g. to a simpler definition of `terminationFLP` and a less complicated proof of the final contradiction. Alternatively, if you want to keep output messages, it helps to add the sender to each message to simplify some of the proofs. Note that, if you model input values as variables, it is necessary to revise the definition of processes or to add special trigger messages, because processes only perform steps when receiving messages. In either way, we recommend a more differentiated typing of messages. Such a typing could be used e.g. to prevent the sending of input messages or the consumption of output messages. This would also have the advantage that the message value type of interprocess messages does not have to contain Booleans and would be therefore more generic. Both cases avoid the need to consider output messages for the message consumed in steps. Furthermore, assumptions to exclude output messages become unnecessary for several lemmata, leading to clearer proofs.

5 Conclusion

The impossibility of distributed consensus with one faulty process leads to essential restrictions for distributed systems. In [6] Fischer et al. are arguing that "solutions" for real-world consensus problems need more refined models that better reflect realistic assumptions for the system components that are used. With our Isabelle/HOL proof, we increase the confidence that such real-world distributed systems in fact need to align their assumptions and guarantees to the implied restrictions of FLP's result. We see the invention of consensus algorithms like Paxos [10] or RAFT [12] as an evidence of this claim. A framework to formally analyze such algorithms can be found e.g. in [8]. By providing a precise list of assumptions for Völzer's proof based on a mechanical verification, we enable the development of new approaches to solve the consensus problem in distributed systems.

With our Isabelle/HOL formalization, we completed Völzer's proof in [13] by adding the missing proof details and providing a precise list of the necessary assumptions. Moreover we extended Völzer's proof with a formalization of fairness and showed that even fair executions cannot ensure consensus.

The lessons learned in Sect. 4.4 should encourage more people to question and verify more fundamental results with the help of a theorem prover.

References

1. Bertot, Y., Castéran, P.: Interactive Theorem Proving and Program Development: Coq'Art: The Calculus of Inductive Constructions. Texts in Theoretical Computer Science An EATCS Series. Springer, Heidelberg (2013)
2. Bisping, B., Brodmann, P.D., Jungnickel, T., Rickmann, C., Seidler, H., Stüber, A., Wilhelm-Weidner, A., Peters, K., Nestmann, U.: A Constructive Proof for FLP. Archive of Formal Proofs (2016). http://isa-afp.org/entries/FLP.shtml. Formal proof development
3. Buckley, G.N., Silberschatz, A.: An effective implementation for the generalized input-output construct of CSP. ACM Trans. Program. Lang. Syst. (TOPLAS) 5(2), 223–235 (1983)
4. Constable, R.L.: Effectively Nonblocking Consensus Procedures can Execute Forever - a Constructive Version of FLP. Tech. Rep. 11513, Cornell University (2011)
5. Constable, R.L., Allen, S.F., Bromley, H.M., Cleaveland, W.R., Cremer, J.F., Harper, R.W., Howe, D.J., Knoblock, T.B., Mendler, N.P., Panangaden, P., Sasaki, J.T., Smith, S.F.: Implementig Mathematics with the Nuprl Proof Development System. Prentice-Hall, Upper Saddle River (1986)
6. Fischer, M.J., Lynch, N.A., Paterson, M.S.: Impossibility of distributed consensus with one faulty process. J. ACM (JACM) 32(2), 374–382 (1985)
7. Kammüller, F., Wenzel, M., Paulson, L.C.: Locales - a sectioning concept for Isabelle. In: Bertot, Y., Dowek, G., Hirschowitz, A., Paulin, C., Théry, L. (eds.) TPHOLs 1999. LNCS, vol. 1690, pp. 149–165. Springer, Heidelberg (1999)
8. Küfner, P., Nestmann, U., Rickmann, C.: Formal verification of distributed algorithms. In: Baeten, J.C.M., Ball, T., de Boer, F.S. (eds.) TCS 2012. LNCS, vol. 7604, pp. 209–224. Springer, Heidelberg (2012)
9. Kumar, D., Silberschatz, A.: A counter-example to an algorithm for the generalized input-output construct of CSP. Inform. Proc. Lett. 61(6), 287 (1997)
10. Lamport, L.: The part-time parliament. ACM Trans. Comput. Syst. (TOCS) 16(2), 133–169 (1998)
11. Nipkow, T., Paulson, L.C., Wenzel, M. (eds.): Isabelle/HOL: A Proof Assistant for Higher-Order Logic. LNCS, vol. 2283. Springer, Heidelberg (2002)
12. Ongaro, D., Ousterhout, J.: In Search of an Understandable Consensus Algorithm. In: Proceedings of USENIX, pp. 305–320 (2014)
13. Völzer, H.: A constructive proof for FLP. Inform. Proc. Lett. 92(2), 83–87 (2004)

Visual Theorem Proving with the Incredible Proof Machine

Joachim Breitner[(✉)]

Karlsruhe Institute of Technology, Karlsruhe, Germany
breitner@kit.edu

Abstract. The Incredible Proof Machine is an easy and fun to use program to conduct formal proofs. It employs a novel, intuitive proof representation based on port graphs, which is akin to, but even more natural than, natural deduction. In particular, we describe a way to determine the scope of local assumptions and variables implicitly. Our practical classroom experience backs these claims.

1 Introduction

How can we introduce high-school students to the wonderful world of formal logic and theorem proving?

Manual proofs on paper are tedious and not very rewarding: The students have to learn the syntax first, and whatever they produced, they would have to show it to a teacher or tutor before they knew if it was right or wrong.

Interactive theorem provers can amend some of these problems: These computer programs give immediate feedback about whether a proof is faulty or correct, allow free exploration and can be somewhat addictive – they have been called "the world's geekiest computer game" for a reason. Nevertheless, the students still have to learn the syntax first, and beginners without any background in either logic or programming, initially face a motivationally barren phase.

Therefore we built an interactive theorem prover that allows the students to start conducting proofs immediately and without learning syntax first. With *The Incredible Proof Machine* (http://incredible.pm/) the student just drags blocks – which represent assumptions, proof rules and conclusions – onto a canvas and wires them up, using only the mouse or a touch interface. A unification-based algorithm infers the propositions to label the connections with. Once everything is connected properly, such a graph constitutes a rigorous, formal proof.

If one thinks of assumptions as sources of propositions, conclusions as consumers of propositions, and proof rules as little machines that transform propositions to other propositions, then the connections become conveyor belts that transport truth. This not only justifies the name of the software, but is – in our opinion – a very natural representation of how the human mind approaches proving.

Another way of thinking about the Incredible Proof Machine is that it is the result of taking a graphical programming language (e.g. LabView's G [8]) and mangling it through the Curry–Howard correspondence.

© Springer International Publishing Switzerland 2016
J.C. Blanchette and S. Merz (Eds.): ITP 2016, LNCS 9807, pp. 123–139, 2016.
DOI: 10.1007/978-3-319-43144-4_8

The contributions of this paper are:

- We introduce a visual and natural representation of proofs as graph, which is generic in the set of proof rules. In contrast to previous approaches, it supports locally scoped variables and hence predicate logics.
- We infer the scope of local assumptions and variables implicitly from the graph structure, using post-dominators, instead of expecting an explicit declaration. This is a novel way to implement the usual freshness side-conditions.
- We give a formal description of such graphs, define when such a graph constitutes a valid formal proof, and sketch its relation to conventional natural deduction.
- The Incredible Proof Machine provides an intuitive and beginner-friendly way to learn about logic and theorem proving. We describe its interface design and its implementation.
- We report on our practical experience with the tool, including the results of a standard usability questionnaire.

2   Proof Graphs

We begin with a user-level introduction to graphical proofs, as they are used in the Incredible Proof Machine. We put the focus on motivating and explaining the elements of such a proof and giving an intuition about them and defer a rigorous treatment to the subsequent section.

2.1   Conclusion and Assumption

What is the intuitive essence of a proof? We assume certain propositions to be true. From these assumption, we conclude that further propositions are true, using the rules of the logic at hand. Eventually we construct a proposition that matches what we want to prove, i.e. the conclusion. In the simplest case, the conclusion is among the assumptions, and the proof is trivial.

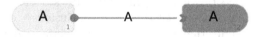

Fig. 1. A very trivial proof (Color figure online)

If we depict such a proof, the picture in Fig. 1 might come up: A *block* representing the assumption, a second block representing the conclusion, and a line between them to draw the *connection*. Both blocks are labelled with the proposition they provide resp. expect, namely A, and the line is also labelled with the proposition. This is a valid proof, and the conclusion turns green.

It is worth pointing out that in these proof graphs, the train of thought runs from left to right. Hence, assumptions have *ports* (the little grey circles where connections can be attached to) on their right, and conclusions on their left.

Fig. 2. A very wrong proof (Color figure online)

Fig. 3. A very incomplete proof (Color figure online)

Such outgoing and incoming ports also have different shapes. The system does not allow connections between two outgoing or two incoming ports.

A wrong proof is shown in Fig. 2, where the proposition of the assumption (B) differs from the proposition of the conclusion (A). Thus, these blocks cannot legally be connected, the false connection is red, and a scary symbol explains the problem. Needless to say, the conclusion is not green.

Similarly, the conclusion in Fig. 3 is not green, as the proof is incomplete. This is indicated by a red port. In general, anything red indicates that the proof is either incomplete or wrong.

2.2 Rule Blocks

To conduct more than just trivial proofs, we need more blocks. These correspond to the inference rule of the underlying logic. Figure 4 shows some typical proof blocks and the corresponding natural deduction rule(s) in conventional inference rule format (antecedents above, consequent below the line).

$$\frac{X \quad Y}{X \wedge Y} \qquad \frac{X \wedge Y}{X} \quad \frac{X \wedge Y}{Y} \qquad \frac{X \to Y \quad X}{Y}$$

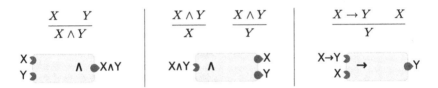

Fig. 4. Some natural deduction rules and their proof block counterparts

Again, incoming ports (on the left) indicate prerequisites of a rule, while outgoing ports (on the right) correspond to the conclusions of a rule. In contrast to usual inference rules, rule blocks can have multiple conclusions, so both conjunction projection rules are represented by just one block. If only one of the

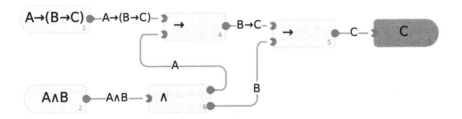

Fig. 5. A more complex proof

conclusions is needed, the other outgoing ports would simply be left unconnected. Unlike unconnected prerequisites, this does in no way invalidate a proof.

The graph in Fig. 5 shows a proof that from $A \wedge B$ and $A \to B \to C$ we can conclude C. Note the connection labels, which indicate the proposition that is "transported" by a connection.

2.3   Local Hypotheses

Figure 4 shows both the introduction and elimination rule for conjunction, as well as the elimination form for implication (*modus ponens*) – clearly we are missing a block that allows us to introduce the implication. Such a block would produce output labelled $A \to B$ if given an input labelled B, where the proof for this B may make use of A. But that

Fig. 6. Implication introduction

local hypothesis A must not be used elsewhere! This restriction is hinted at by the shape of the implication introduction block in Fig. 6, where the dent in the top edge of the block suggests that this block will encompass a subproof. To support this, the block – colloquially called a "sliding calliper" – can be horizontally expanded as needed.

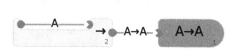

Fig. 7. Implication done right

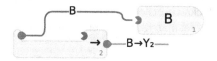

Fig. 8. Implication done wrong (Color figure online)

The graph in Fig. 7 shows the simplest proof using the calliper: By connecting the port of the local hypothesis A with the assumption A, we obtain a valid proof of $A \to A$.

The graph next to it (Fig. 8) shows an invalid use of the implication introduction block: The hypothesis is not used locally to prove the assumption of the block, but is instead connected directly to the conclusion of the proof. The Incredible Proof Machine allows the user to make that connection, but complains about it by colouring it in red.

In this picture you can see that despite the proof being in an invalid state, the system determined that the implication produced by this block would have B as the assumption, and a not yet determined proposition Y_2 as the conclusion. The ability to work with partial and even wrong proofs is an important ingredient to a non-frustrating user experience.

A block can have more than one local hypothesis, with different scoping rules. An example for that is the elimination block for disjunction, shown in Fig. 9. In this case, the conclusion of the block (P) is the same as the local goal on each side of the block. This seems to be a bit redundant, but is necessary to delimit

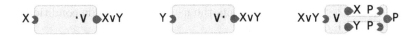

Fig. 9. Disjunction introduction and elimination rules

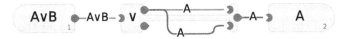

Fig. 10. A local hypotheses of the disjunction block used wrongly.

the scope of the two local hypotheses, respectively, and to keep the two apart – after all, using the local hypothesis from one side in the proof of the other side leads to unsoundness (Fig. 10).

2.4 Predicate Logic

So far, the user can only conduct proofs in propositional logic, which is a good start for beginners, but gets dull eventually. Therefore the Incredible Proof Machine also supports predicate logic. This opens a whole new can of worms, as the system has to keep track of the scope of local, fixed variables.

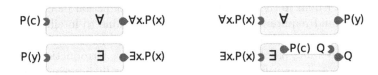

Fig. 11. Blocks for quantifiers

  The additional rules are shown in Fig. 11. The introduction rule for the existential quantifier (bottom left) and the elimination rule for the universal quantifier (top right) are straight forward: If one can prove $P(y)$ for some term y, then $\exists x.P(x)$ holds, and conversely if one has $\forall x.P(x)$, then $P(y)$ holds for some term y.

  At the first glance, it seems strange that the introduction rule for the universal quantifier (top left) has the same shape as the one for the existential quantifier. But there is a small difference, visible only from the naming convention: To obtain $\forall x.P(x)$ the user has to prove $P(c)$ for an (arbitrary but fixed) *constant c*.

  Furthermore, and not visible from the shape of the block, is that this constant c is available only locally, in the proof of $P(c)$. To enforce this, the Incredible Proof Machine identifies those proof blocks from where all paths pass through the universal quantifier introduction block on their way to a conclusion, and only the free variables of these blocks are allowed to be instantiated by a term that mentions c. This restriction implements the usual freshness side condition in an inference rule with explicit contexts:

$$\frac{\Gamma \vdash P(c) \qquad c \text{ does not occur in } \Gamma}{\Gamma \vdash \forall x.P(x)}$$

Such a local constant is also used in the elimination rule for the existential quantifier (bottom right), where in order to prove a proposition Q, we may use that $P(c)$ holds for some constant c, but this constant may only occur in this part of the proof, and moreover the proposition $P(c)$ is a local hypothesis (Sect. 2.3) and may not escape this scope.

In this formulation of predicate logic, the universe is unspecified, but not empty. In particular, it is valid to derive $\exists x.P(x)$ from $\forall x.P(x)$ (Fig. 12).

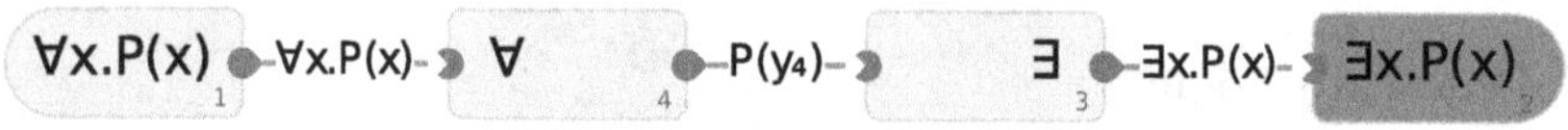

Fig. 12. A proof that $\forall x.P(x)$ entails $\exists x.P(x)$.

The asymmetry in Fig. 11 is striking, and the question arises why the elimination block for the existential quantifier would not just produce $P(c)$ as its output, forming a proper dual to the universal quantifier introduction block. This could work, but it would require the Incredible Proof Machine to intelligently determine a scope for c; in particular it had to ensure that scopes nest properly. With some scopes extending backwards (universal quantifier introduction) and some forwards (existential quantifier elimination), automatically inferring sensible and predictable scoping becomes tricky, so we chose to use a block shape that makes the scope explicit. More on scopes in Sect. 3.2.

2.5   Helper Block

With full-scale theorem provers such as Isabelle or Coq it is quite helpful to break down a proof into smaller steps and explicitly state intermediate results. The same holds for the Incredible Proof Machine, and is made possible using the so-called helper block, shown in Fig. 13. Once placed in the proof area, the user can click on it and enter a proposition, which is then both assumed and produced by this block. Logically, this corresponds to a use of the cut rule.

Fig. 13. The helper block

The block is also useful if the desired proposition is not inferred, which can be the case with partial proofs, especially if quantifiers are involved.

2.6   Custom Blocks

After performing a few proofs with the Incredible Proof Machine, the user soon notices that some patterns appear repeatedly. One such pattern would be a proof by contradiction, which consists of the three blocks highlighted in Fig. 14: Tertium non datur, disjunction elimination and ex falso quodlibet. (Note that the negation of X is expressed as $X \to \bot$.)

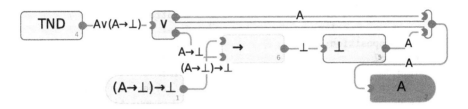

Fig. 14. A primitive proof of double negation elimination

When the user has selected a part of the proof this way (by shift-clicking), he can create a custom block that represents the selected proof fragment. In this case, the custom block would look as in Fig. 15, and with that block, which now directly represents a proof by contradiction, the whole proof is greatly simplified (Fig. 16). This mechanism corresponds to the lemma command in, say, Isabelle.

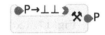

Fig. 15. A custom block

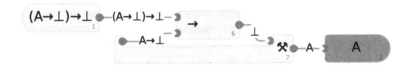

Fig. 16. A shorter proof of double negation elimination

2.7   Custom Logics

The rule blocks, and hence the underlying logic, are not baked into the Incredible Proof Machine, but read from a simple text file. Figure 17 shows the declaration of the first disjunction introduction block, the implication introduction block and the universal quantifier introduction block.

Each rule needs to have an identifier (`id`), but may specify a more readable description (`desc`), which includes a hint towards what side of the block the description should be aligned. The next two fields specify which variables in the

```
rules:                              proposition: X
- id: disjI1                        in:
  desc:                               type: assumption
    intro: ·∨                         proposition: "Y"
  free: ["X","Y"]                   out:
  ports:                              type: conclusion
    in:                               proposition: X→Y
      type: assumption          - id: allI
      proposition: X              desc:
    out:                            intro: ∀
      type: conclusion            free: ["P"]
      proposition: X∨Y            local: ["c"]
- id: impI                         ports:
  desc:                             in1:
    intro: →                          type: assumption
  free: ["X","Y"]                     proposition: P(c)
  ports:                              scoped: ["c"]
    hyp:                            out2:
      type: local hypothesis          type: conclusion
      consumedBy: in                  proposition: ∀x.P(x)
```

Fig. 17. Extract of `predicate.yaml`, where the rule blocks are defined

following propositions are `free`, i.e. may be instantiated upon connecting the blocks, and which are `local`, i.e. different for each instance of the block.

Then the list of `ports` is indexed by an arbitrary identifier (`in`, `out`, ...). A port has a `type`, which is `assumption`, `conclusion` or `local hypothesis`. In the latter case, this port may only be used towards proving the port specified in the `consumedBy` field. Every port specifies the `proposition` that is produced resp. expected by this block. A local constant (such as the `c` in rule `allI`) is usually `scoped` by a port of type `assumption` (see Sects. 2.4 and 3.2).

It is simple to experiment with completely different logics, without changing the code. For example, we have implemented a Hilbert-style system for propositional logic (one rule block for modus ponens and three rules blocks for the axioms) and the typing derivations of the simply typed lambda calculus.

A similar file specifies the pre-defined tasks, which can be grouped into sessions, and each session can use a different logic, or – for an educational progression between the sessions – can show only a subset of a logic's rules.

Naturally, none of these files are user-visible. They would, however, provide the mechanism by which an educator who wants to use the Incredible Proof Machine in his course, simply by editing the rules and tasks therein as desired.

3 Theory

The previous section has (intentionally) only scratched the surface of the Incredible Proof Machine, and avoided some more fundamental questions such as: What precisely makes up a proof graph (and what is just cosmetic frill)? When is it valid? And what does it actually prove?

These questions are answered in this section with some level of formalism. In particularly, we define when the shape of a proof graph is valid (e.g. no cycles, local hypotheses wired up correctly) and when scoped variables are used correctly. While we use the language of *port graphs* as introduced in [2], the notion of a well-shaped graph is a new contribution.

3.1 Port Graphs

In contrast to those in [2], our port graphs are *directed*.

Definition 1 (Port Graph Signature). *A (directed) port graph signature ∇ over a set $\mathcal{N}$ of node names and a set $\mathcal{P}$ of port names consists of the two functions $\_\nabla : \mathcal{N} \to 2^{\mathcal{P}}$ and $\nabla\!\_ : \mathcal{N} \to 2^{\mathcal{P}}$, which associate to a node name the names of its incoming resp. outgoing ports.*

Definition 2 (Port Graph). *A (directed) port graph G over a signature ∇ is a tuple (V, n, E) consisting of*

- *a set V of vertices,*
- *a function $n : V \to \mathcal{N}$ to associate a node to each vertex, and*
- *a multiset $E \subseteq (V \times \mathcal{P}) \times (V \times \mathcal{P})$ of edges such that for every edge $(v_1, p_1)\!-\!(v_2, p_2) \in E$ we have $p_1 \in \nabla\!\_(n(v_1))$ and $p_2 \in \_\nabla(n(v_2))$.*

The notation $s\!-\!t$ used for an edge is just syntax for the tuple (s, t).

We need a number of graph-theoretic definitions.

Definition 3 (Path). *A path in G is a sequence of edges $(v_1, p_1')\!-\!(v_2, p_2)$, $(v_2, p_2')\!-\!(v_3, p_3)$, $\ldots$, $(v_{n-1}, p_{n-1}')\!-\!(v_n, p_n) \in E$. The path begins in (v_1, p_1') (or just v_1) and ends in (v_n, p_n) (or just v_n).*

Definition 4 (Terminal Node, Pruned Graph). *A node $n \in \mathcal{N}$ is called a terminal node, if $\nabla\!\_(n) = \{\}$, and a vertex $v \in V$ is called a terminal vertex if $n(v)$ is a terminal node. A graph is called* pruned *if every vertex v is either a terminal vertex, or there is a path from v to a terminal vertex.*

3.2 Scopes

The key idea to support both local assumptions and scoped variables is that the scope of a local proof can be implicitly inferred from the shape of the graph.

It is desirable to have as large as possible scopes, so that as many proof graphs as possible are well-scoped. On the other hand, the scopes must be small enough to still be valid. This motivates

Definition 5 (Scope). *In a port graph $G = (V, n, E)$, the* scope *of an incoming port (v, p) with $p \in \_\nabla(n(v))$ is the set $S(v, p) \subseteq V$ of vertices post-dominated by the port. More precisely: $v' \in S(v, p)$ iff v' is not a terminal vertex and every path that begins in v' and ends in a terminal vertex passes through (v, p).*

As an indication that this is a sensible definition, we show that the scopes nest the way one would expect them to. We restrict this to pruned graphs; pruning a graph removes only unused and hence irrelevant parts of the proof.

Lemma 1 (Scopes Nest). *Let $G = (V, n, E)$ be a pruned graph. For any two (v_1, p_1) and (v_2, p_2) with $v_i \in V$ and $p_i \in \_\!\nabla(n(v_i))$ $(i = 1, 2)$, we have $S(v_1, p_1) \subseteq S(v_2, p_2)$ or $S(v_2, p_2) \subseteq S(v_1, p_1)$ or $S(v_1, p_1) \cap S(v_2, p_2) = \{\}$.*

Proof. We show that $S(v_1, p_1) \cap S(v_2, p_2) \neq \{\}$ implies $S(v_1, p_1) \subseteq S(v_2, p_2)$ or $S(v_2, p_2) \subseteq S(v_1, p_1)$.

Let $v \in S(v_1, p_1) \cap S(v_2, p_2)$. The vertex v is not terminal, so there is a path from v to a terminal node, and it necessarily passes through (v_1, p_1) and (v_2, p_2). W.l.o.g. assume (v_1, p_1) occurs before (v_2, p_2) on that path. Then all paths from (v_1, p_1) to a terminal node go through (v_2, p_2), as otherwise we could construct a path from v to a terminal node that does not go through (v_2, p_2).

Now consider a $v' \in S(v_1, p_1)$. All paths to a terminal node go through (v_1, p_1), and hence also through (v_2, p_2), and we obtain $S(v_1, p_1) \subseteq S(v_2, p_2)$.

3.3 Graph Shapes

The above definition of scopes allows us to say how a local hypothesis needs to be wired up. We also need a relaxed definition of acyclicity.

Definition 6 (Local Hypothesis). *A local hypothesis specification for a graph signature ∇ is a partial function $h_n : \mathcal{P} \rightharpoonup \mathcal{P}$ for every $n \in \mathcal{N}$ such that $h_n(p) = p'$ implies $p \in \_\!\nabla(n)$ and $p' \in \_\!\nabla(n)$. In that case, p is a local hypothesis of n and p' defines its scope.*

Definition 7 (Well-Scoped Graph). *A port graph $G = (V, n, E)$ with a local hypothesis specification h is well-scoped if for every edge (v_1, p_1)—(v_2, p_2) where $h_{n(v_1)}(p_1) = p'$ we have $(v_2, p_2) = (v_1, p')$ or $v_2 \in S(v_1, p')$.*

Definition 8 (Acyclic Graph). *A port graph $G = (V, n, E)$ with a local hypothesis specification h is acyclic if there is no path connecting a node to itself, disregarding paths that pass by some local hypothesis, i.e. where there is a (v, p) on the path with $p \in \mathrm{dom}\, h_{n(v)}$.*

Definition 9 (Saturated Graph). *A port graph $G = (V, n, E)$ is saturated if every (v, p) with $p \in \_\!\nabla(n(v))$ is incident to an edge.*

To summarise when a graph is in a good shape to form a proof, we give

Definition 10 (Well-Shaped Graph). *A port graph G is well-shaped if it is well-scoped, acyclic and saturated.*

3.4 Propositions

So far we have described the shape of the graphs; it is about time to give them meaning in terms of logical formulas. We start with propositional logic (no binders and no scoped variables) first.

Definition 11 (Formulas). *Let $\mathcal{X}$ be a set of variables, and $\mathcal{F}_\mathcal{X}$ a set of formulas with variables in X.*

Definition 12 (Labelled Signature). *A port graph signature ∇ is labelled by formulas $l : \mathcal{N} \times \mathcal{P} \rightarrow \mathcal{F}_\mathcal{X}$.*

For two vertices v_1, v_2 with the same node name, the free variables of the formulas need to be distinct. So in the context of a specific graph, we annotate the variables with the vertex they originate from:

$$l' : V \times \mathcal{P} \rightarrow \mathcal{F}_{\mathcal{X} \times V}$$
$$l'(v, p) = l(n(v), p)[x_v/x \mid x \in \mathcal{X}]$$

We use the subscript syntax x_v to denote the tuple (x, v).

The Incredible Proof Machine employs a unification algorithm to make sure the formulas expected on either side of an edge match, if possible. Here, we abstract over this and simply require a unifying substitution, which we model as a function.

Definition 13 (Instantiation). *An* instantiation *for a port graph G with a labelled signature is, for every vertex $v \in N$, a function $\theta_v : \mathcal{F}_{\mathcal{X} \times N} \rightarrow \mathcal{F}_{\mathcal{X} \times N}$.*

Definition 14 (Solution). *An instantiation θ for a port graph G with a labelled signature is a solution if for every edge $(v_1, p_1)-(v_2, p_2) \in E$ we have $\theta_{v_1}(l'(v_1, p_1)) = \theta_{v_2}(l'(v_2, p_2))$.*

Definition 15 (Proof Graph). *A proof graph is a well-shaped port graph with a solution.*

3.5 Scoped Variables

To support binders and scoped variables, we need to define which variables are scoped (a property of the signature), and then ensure that the scopes are adhered to (a property of the graph). For the latter we need – a bit vaguely, to stay abstract in the concrete structure of terms – the notion of the range of an instantiation, ran $\theta_i \subseteq \mathcal{X} \times N$, which is the set of free variables of the formulas that the substitution substitutes for.

Definition 16 (Scoped Variables). *A port graph signature ∇ can be annotated with* variable scopes *by a partial function $s_n : \mathcal{X} \rightharpoonup \mathcal{P}$, for every $n \in \mathcal{N}$, where $s_n(x) = p$ implies $p \in \_\nabla(n)$.*

Definition 17 (Well-Scoped Instantiation). *An instantiation θ for a port graph G with a labelled signature and scoped variables is* well-scoped *if for every scoped variable x, i.e. $s_{n(v)}(x) = p$ for some vertex $v \in V$, $x \in \operatorname{ran} \theta_{v'}$ implies that $v' \in S(v, p)$.*

In the presence of scoped variables, we extend Definition 15 to

Definition 18 (Proof Graph). *A* proof graph *is a well-shaped port graph with a well-scoped solution.*

3.6   Example

After this flood of definitions, let us give a complete and comprehensive example.

To prove that $\exists x.(\mathbf{P}(x) \wedge \mathbf{Q}(x))$ entails $\exists x.\mathbf{P}(x)$ (where the bold $\mathbf{P}$ indicates a constant, not a variable in the sense of $\mathcal{X}$), we would use node names $\mathcal{N} = \{\texttt{a}, \texttt{c}, \texttt{exE}, \texttt{conjE}, \texttt{exI}\}$ and port names $\mathcal{P} = \{\texttt{in}, \texttt{out}, \texttt{in2}, \texttt{out2}\}$. The (labelled and scoped) signature is given by

$$\nabla(\texttt{a}) = \{\} \qquad\qquad \sqsubseteq(\texttt{a}) = \{\texttt{out}\}$$
$$\nabla(\texttt{c}) = \{\texttt{in}\} \qquad\qquad \sqsubseteq(\texttt{a}) = \{\}$$
$$\nabla(\texttt{exE}) = \{\texttt{in}, \texttt{in2}\} \qquad\qquad \sqsubseteq(\texttt{exE}) = \{\texttt{out}, \texttt{out2}\}$$
$$\nabla(\texttt{conJ}) = \{\texttt{in}\} \qquad\qquad \sqsubseteq(\texttt{conjE}) = \{\texttt{out}, \texttt{out2}\}$$
$$\nabla(\texttt{exI}) = \{\texttt{in}\} \qquad\qquad \sqsubseteq(\texttt{exI}) = \{\texttt{out}\}$$

$$l(\texttt{a}, \texttt{out}) = \exists x.(\mathbf{P}(x) \wedge \mathbf{Q}(x)) \qquad\qquad l(\texttt{c}, \texttt{in}) = \exists x.\mathbf{P}(x)$$
$$l(\texttt{exE}, \texttt{in}) = \exists x.\mathbf{P}(x) \qquad\qquad l(\texttt{exE}, \texttt{out}) = P(c)$$
$$l(\texttt{exE}, \texttt{in2}) = Q \qquad\qquad l(\texttt{exE}, \texttt{out2}) = Q$$
$$l(\texttt{conjE}, \texttt{in}) = X \wedge Y \qquad\qquad l(\texttt{conjE}, \texttt{out}) = X$$
$$l(\texttt{conjE}, \texttt{out2}) = Y$$
$$l(\texttt{exI}, \texttt{in}) = P(y) \qquad\qquad l(\texttt{exI}, \texttt{out}) = \exists x.\mathbf{P}(x)$$

$$h_{\texttt{exE}}(\texttt{out}) = \texttt{in2} \qquad\qquad s_{\texttt{exE}}(c) = \texttt{in2}.$$

A well-shaped proof graph for this signature – also shown in Fig. 18 – is given by $N = \{1..5\}$, $n(1) = \texttt{a}$, $n(2) = \texttt{c}$, $n(3) = \texttt{exE}$, $n(4) = \texttt{conjE}$, $n(5) = \texttt{exI}$ and

$$E = \{(1, \texttt{out})\text{—}(3, \texttt{in}), (3, \texttt{out})\text{—}(4, \texttt{in}),$$
$$(4, \texttt{out})\text{—}(5, \texttt{in}), (5, \texttt{out})\text{—}(3, \texttt{in2}), (3, \texttt{out2})\text{—}(2, \texttt{in})\}.$$

A solution for this graph is given by these higher-order substitutions:

$$\theta_3 = [(\lambda x.\mathbf{P}(x) \wedge \mathbf{Q}(x))/P_3, \; \exists x.\mathbf{P}(x)/Q_3]$$
$$\theta_4 = [\mathbf{P}(c_3)/X_4, \; \mathbf{Q}(c_3)/Y_4]$$
$$\theta_5 = [(\lambda x.\mathbf{P}(x))/P_5, \; c_3/y_5]$$

Note that this is well-scoped: We have $c_3 \in \operatorname{ran} \theta_i$ only for $i \in S(3, \texttt{in2}) = \{4, 5\}$.

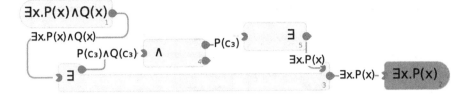

Fig. 18. A comprehensive example

3.7 Proof Conclusions

In order to relate a proof graph with a proof in a given logic, we assume a partition of the nodes $\mathcal{N}$ into assumptions $\mathcal{N}_A$, conclusions $\mathcal{N}_C$ and rules $\mathcal{N}_R$, where $\mathcal{N}_A$ contains only assumptions (no input and precisely one output) and $\mathcal{N}_C$ only conclusions (no output and precisely one input). For a vertex $v \in V$ with $n(v) \in \mathcal{N}_A \cup \mathcal{N}_C$ let $l'(v) = l'(v, p)$ where p is the single outgoing resp. incoming port of $n(v)$.

If we assume that the nodes in $\mathcal{N}_R$ faithfully implement the inference rules of a natural deduction-style implementation of a given logic, we can state

Theorem 1 (Soundness and Completeness). *The existence of a proof graph, with a vertex for every conclusion ($\mathcal{N}_C \subseteq n(V)$), implies that from the set of formulas $\{l'(v) \mid n(v) \in \mathcal{N}_A\}$, all formulas in $\{l'(n) \mid n(v) \in \mathcal{N}_C\}$ are derivable by natural deduction, and vice versa.*

A mechanized proof of this theorem, built using the interactive theorem prover Isabelle, can be found in the Archive of Formal Proofs [6].

4 Implementation

The Incredible Proof Machine is based on web technologies (HTML, JavaScript, SVG) and runs completely in the web browser. Once it is loaded, no further internet connection is required – this was useful when the workshop WiFi turned out to be unreliable. This also simplifies hosting customised versions. It adjusts to the browser's configured language, currently supporting English and German.

The logical core is implemented in Haskell, which we compile to JavaScript using GHCJS. It uses the unbound library [16] to handle local names, and a translation of Nipkow's higher-order pattern unification algorithm [13]. There is little in the way of a LCF-style trusted core, and the system can easily be tricked from the browser's JavaScript console.

The Incredible Proof Machine greets its users with a list of tasks to prove (Fig. 19, left). Attempted tasks are highlighted yellowishly; completed tasks in green. The main working view (Fig. 19, right) consist of a left pane, listing the current task and the various blocks which can be dragged onto the main pane. The interface supports undo/redo, zooming and can save the proof as an SVG graphic.

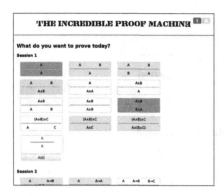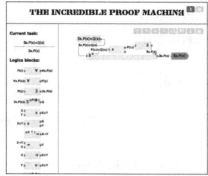

Fig. 19. The Incredible Proof Machine, task selection and proving (Color figure online)

The system continuously checks and annotates the proof, even in the presence of errors, supporting an incremental work flow. This currently happens synchronously and it gets a little sluggish with larger proofs.

Custom blocks are also listed in the left pane, where they can be created and deleted. The overview page allows users to quickly define new tasks.

Custom blocks, new tasks and the state of all proofs are stored in the browser (by way of Web Storage), so a returning user can continue where he left.

An educator would customise the Incredible Proof Machine by adjusting the files that contain the logic definition (Sect. 2.7) and the tasks.

All code is liberally licensed Free Software, and contributions at http://github.com/nomeata/incredible are welcome.

5   Evaluation

The Incredible Proof Machine has been used in practice, and our experience shows that it does indeed achieve the desired goal of providing an entertaining low barrier entry to logic and formal proofs.

5.1   Classroom Experience

Development of the Incredible Proof Machine was initiated when the author was given the possibility to hold a four day workshop with high school students to a topic of his choosing. The audience consisted of 13 students ages 13 to 20, all receiving a scholarship by the START-Stiftung for motivated students with migration background. Prior knowledge in logic, proofs or programming was not expected, and in most cases, not present.

Within the 14 h spent exclusively working with the Incredible Proof Machine, the class covered the propositional content; two very apt students worked quicker and managed most of the predicate proofs as well.

After a very quick introduction to the user interface, the procedure was to let the students explore the next session, which was unlocked using a password we gave them to keep everyone on the same page, on their own. They should experiment and come up with their own mental picture of the next logical connective, before we eventually discussed it together, explained the new content, and handed out a short text for later reference.

We evaluated the user experience using a standardised usability questionnaire (UEQ, [9]). The students answered 26 multiple-choice questions, which are integrated into a score in six categories. The evaluation tool compares the score against those of 163 product evaluations. It confirms an overall positive user experience, with the Incredible Proof Machine placed in the top quartile in the categories Attractiveness, Dependability and Novelty and beating the averages in Perspicuity, Efficiency and Novelty. Additional free-form feedback also confirmed that most students enjoyed the course – some greatly – and that some found the difficulty level rather high.

We have since used the Incredible Proof Machine two further times as a 90 min taster course addressing high-school students interested in the mathematics and computer science courses at our university.

5.2 Online Reception

The Incredible Proof Machine is free to use online, and "random people from the internet" have played with it. While we cannot provide representative data on how it was perceived, roughly two hundreds posts on online fora (Twitter, Reddit, Hacker News)[1] are an encouraging indication. Users proudly posted screenshots of their solutions, called the Incredible Proof Machine "addictive" and one even reported that his 11-year old daughter wants to play with it again.

A German science podcast ran a 3-h feature presenting the Incredible Proof Machine [5].

6   Future Directions

We plan to continue developing the Incredible Proof Machine into a versatile and accessible tool to teach formal logic and rigorous proving. For that we want to make it easier to use, more educational and more powerful.

To improve usability, we envision a better visualisation of the inferred scopes, to ease proofs in predicate logic.

To make it more educational, we plan to add an interactive tutorial mode targeting self-learners. A simultaneous translation of the proof graph in to a (maybe bumpy) natural language proof, with a way to explore how the respective components correspond, will also greatly improve the learning experience.

---

[1] https://twitter.com/nomeata/status/647056837062324224, https://reddit.com/mbt k2, https://reddit.com/3m7li1, https://news.ycombinator.com/item?id=10276160, https://twitter.com/d_christiansen/status/647117704764256260, https://twitter. com/mjdominus/status/675673521255788544, https://twitter.com/IlanGodik/sta tus/716258636566290432.

A powerful, yet missing, feature is the ability to abstract not only over proofs (lemmas), but also over terms (definitions). Inspired by ML's sealing of abstract types [12], we'd make, for a given proof, a given definition (such as $\neg A :=$ $A \to \bot$) either transparent or abstract. This would encourage and teach a more disciplined approach to abstraction than if a definition could be unfolded locally whenever convenient.

Proof graphs might be worthwhile to use also in full interactive theorem provers: Consider a local **proof** with intermediate results in a typical Isar proof. It would be quite natural to free the user from having to place them into a linear order, to give names and to refer to these names when he could just draw lines! The statefulness of Isabelle code (e.g. attribute changes, simplifier setup, etc.) pose some interesting challenges in implementing this idea.

7   Related Work

Given how intuitive it appears to us to write proofs as graphs, we were surprised to find little prior work on that. Closest to our approach is [1], which identified (undirected) port graphs as defined in [2] as the right language to formulate these ideas in, and covers intuitionistic propositional logic. We develop their approach further by deducing scopes from the graph structure and by supporting predicate logic as well, and we believe that *directed* port graphs, as used in this work, are more suitable to represent proofs.

The inner workings of the Incredible Proof Machine, as well as our implementation of predicate logic, were obviously influenced by Isabelle's [14].

We looked into existing graphical and/or educational approaches to formal logic. We particularly like *Domino on Acid* [4], which represents proofs as domino pieces. It provides a very game-like experience, but is limited to propositional proofs with $\to$ and $\bot$ only. The graphical interactive tools *Polymorphic Blocks* [10] and *Clickable Proofs* [15] support all the usual propositional connectives, but none of these, though, support predicate logic.

There is a greater variety in tools that allow mouse-based editing of more textual proof representations. Examples are *Logitext* [17], which sports a slick interface and provides a high assurance due to the Coq [7] back end, and *KeY* [3], a practical system for program verification. *Easyprove* [11] sticks out as it allows the user to click their way to proper, though clumsy, English proofs. With all these tools the user usually loses a part of his proof when he needs to change a step done earlier, while the Incredible Proof Machine allows him to edit anything at any time, and broken or partial proof fragments can stay around.

8   Conclusions

We lowered the entry barrier to formal logic and theorem proving by offering an intuitive graphical interface to conduct proofs. We have used our program in practice and found that this approach works: Young students with no prior knowledge can work with the tool, and actually enjoy the puzzle-like experience.

We therefore conclude that the non-linear, graphical proof representation, as presented in this work, has advantages over more conventional text-based approach in learning logic.

Acknowledgements. We thank Denis Lohner, Richard Molitor, Martin Mohr and Nicole Rauch for their contributions to the Incredible Proof Machine, and Andreas Lochbihler and Sebastian Ritterbusch for helpful comments on a draft of this paper. Furthermore, I thank the anonymous referees for the encouraging review and the list of feature requests.

References

1. Alves, S., Fernández, M., Mackie, I.: A new graphical calculus of proofs. In: TER-MGRAPH. EPTCS, vol. 48 (2011)
2. Andrei, O., Kirchner, H.: A rewriting calculus for multigraphs with ports. ENTCS **219**, 67–82 (2008)
3. Beckert, B., Hähnle, R., Schmitt, P.H. (eds.): Verification of Object-Oriented Software. The KeY Approach. LNCS (LNAI), vol. 4334. Springer, Heidelberg (2007)
4. Benkmann, M.: Visualization of natural deduction as a game of dominoes. http://www.winterdrache.de/freeware/domino/data/article.html
5. Breitner, J.: Incredible proof machine. Conversation with Sebastian Ritterbusch, Modellansatz Podcast, Episode 78, Karlsruhe Institute of Technology (2016). http://modellansatz.de/incredible-proof-machine
6. Breitner, J., Lohner, D.: The meta theory of the incredible proof machine. Arch. Form. Proofs (2016). Formal proof development. http://isa-afp.org/entries/Incredible_Proof_Machine.shtml
7. Coq Development Team. The Coq proof assistant reference manual. LogiCal Project (2004). version 8.0. http://coq.inria.fr
8. Johnson, G.W.: LabVIEW Graphical Programming. McGraw-Hill, New York (1997)
9. Laugwitz, B., Held, T., Schrepp, M.: Construction and evaluation of a User Experience Questionnaire. In: Holzinger, A. (ed.) USAB 2008. LNCS, vol. 5298, pp. 63–76. Springer, Heidelberg (2008)
10. Lerner, S., Foster, S.R., Griswold, W.G.: Polymorphic blocks: formalism-inspired UI for structured connectors. In: CHI. ACM (2015)
11. Materzok, M.: Easyprove: a tool for teaching precise reasoning. In: TTL. Université de Rennes 1 (2015)
12. Mitchell, J.C., Plotkin, G.D.: Abstract types have existential type. TOPLAS **10**(3), 470–502 (1988)
13. Nipkow, T.: Functional unification of higher-order patterns. In: LICS (1993)
14. Nipkow, T., Paulson, L.C., Wenzel, M.: Isabelle/HOL. LNCS, vol. 2283. Springer, Heidelberg (2002)
15. Selier, T.: A Propositionlogic-, naturaldeduction-proof app(lication). Bachelor's thesis, Utrecht University (2013)
16. Weirich, S., Yorgey, B.A., Sheard, T.: Binders unbound. In: ICFP. ACM (2011)
17. Yang, E.Z.: Logitext. http://logitext.mit.edu/

Proof Pearl: Bounding Least Common Multiples with Triangles

Hing-Lun Chan[1]([✉]) and Michael Norrish[2]

[1] Australian National University, Canberra, Australia
joseph.chan@anu.edu.au
[2] Canberra Research Laboratory, NICTA / Data61, Australian National University,
Canberra, Australia
Michael.Norrish@data61.csiro.au

Abstract. We present a proof of the fact that $2^n \leq \text{lcm}\{1, 2, 3, \ldots, (n+1)\}$. This result has a standard proof *via* an integral, but our proof is purely number theoretic, requiring little more than list inductions. The proof is based on manipulations of a variant of Leibniz's Harmonic Triangle, itself a relative of Pascal's better-known Triangle.

1 Introduction

The least common multiple of the consecutive natural numbers has a lower bound[1]:

$$2^n \leq \text{lcm}\{1, 2, 3, \ldots, (n+1)\}$$

This result is a minor (though important) part of the proof of the complexity of the "PRIMES is in P" AKS algorithm (see below for more motivational detail). A short proof is given by Nair [10], based on a sum expressed as an integral. That paper ends with these words:

> It also seems worthwhile to point out that there are different ways to prove the identity implied [...], for example, [...] by using the difference operator.

Nair's remark indicates the possibility of an elementary proof of the above number-theoretic result. Nair's integral turns out to be an expression of the beta-function, and there is a little-known relationship between the beta-function and Leibniz's harmonic triangle [2]. The harmonic triangle can be described as the difference table of the harmonic sequence: $1, \frac{1}{2}, \frac{1}{3}, \frac{1}{4}, \frac{1}{5}, \ldots$ (*e.g.*, as presented in [3]).

Exploring this connection, we work out an interesting proof of this result that is both clear and elegant. Although the idea has been sketched in various sources (*e.g.*, [9]), we put the necessary pieces together in a coherent argument, and prove it formally in HOL4.

NICTA is funded by the Australian Government through the Department of Communications and the Australian Research Council through the ICT Centre of Excellence Program.

[1] We use $(n + 1)$ here since we allow $n = 0$.

J.C. Blanchette and S. Merz (Eds.): ITP 2016, LNCS 9807, pp. 140–150, 2016.
DOI: 10.1007/978-3-319-43144-4_9

Overview. We find that the rows of denominators in Leibniz's harmonic triangle provide a trick to enable an estimation of the lower bound of least common multiple (LCM) of consecutive numbers. The route from this row property to the LCM bound is subtle: we exploit an LCM property of triplets of neighboring elements in the denominator triangle. We shall show how this property gives a wonderful proof of the LCM bound for consecutive numbers in HOL4:

Theorem 1. *Lower bound for LCM of consecutive numbers.*

$\vdash 2^n \leq$ list_lcm $[1 \mathrel{..} n + 1]$

where list_lcm *is the obvious extension of the binary* lcm *operator to a list of numeric arguments. This satisfies, for example, the following properties:*

$\vdash$ list_lcm $(h::t) =$ lcm h (list_lcm t)
$\vdash$ list_lcm $(l_1 \frown l_2) =$ lcm (list_lcm l_1) (list_lcm l_2)
$\vdash$ list_lcm (REVERSE ℓ) $=$ list_lcm ℓ

Motivation. This work was initiated as part of our mechanization of the AKS algorithm [1], the first unconditionally deterministic polynomial-time algorithm for primality testing. As part of its initial action, the algorithm searches for a parameter k satisfying a condition dependent on the input number. The major part of the AKS algorithm then involves a for-loop whose count depends on the size of k.

In our first paper on the correctness (but not complexity) of the AKS algorithm [4], we proved the existence of such a parameter k on general grounds, but did not give a bound. Now wanting to also show the complexity result for the AKS algorithm, we must provide a tight bound on k. As indicated in the AKS paper [1, Lemma 3.1], the necessary bound can be derived from a lower bound on the LCM of consecutive numbers.

Historical Notes. Pascal's arithmetic triangle ($c1654$) is well-known, but Leibniz's harmonic triangle (1672) has been comparatively neglected. As reported by Massa Esteve and Delshams [5], Pietro Mengoli investigated certain sums of special form in 1659, using a combinatorial triangle identical to the harmonic triangle. Those same sums are the basis of Euler's beta-function (1730) defined by an integral.

In another vein, Hardy and Wright's *Theory of Numbers* [7] related the LCM bound of consecutive numbers to the Prime Number Theorem, which work was followed up by Nair [10], giving the bound in Theorem 1 through application of the beta-function.

Our approach to prove Theorem 1 is inspired by Farhi [6], in which a binomial coefficient identity, equivalent to our Theorem 6, was established using Kummer's theorem. A direct computation to relate both results of Nair and Farhi was given by Hong [8].

Paper Structure. The rest of this paper is devoted to explaining the mechanised proof of this result. We give some background to Pascal's and Leibniz's triangles in Sect. 2. Section 3 discusses two forms of the Leibniz's triangle: the harmonic form and the denominator form, and proves the important LCM property for our Leibniz triplets. Section 4 shows how paths in the denominator triangle can make use of an LCM exchange property, eventually proving that both the consecutive numbers and a row of the denominator triangle share the same LCM. In Sect. 5, we apply this LCM relationship to give a proof of Theorem 1, and conclude in Sect. 6.

HOL4 Notation. All statements starting with a turnstile ($\vdash$) are HOL4 theorems, automatically pretty-printed to LaTeX from the relevant theory in the HOL4 development. Generally, our notation allows an appealing combination of quantifiers ($\forall$, $\exists$), logical connectives ($\land$ for "and", $\Rightarrow$ for "implies", and $\iff$ for "if and only if"). Lists are enclosed in square-brackets [], with members separated by semicolon (;), using infix operators :: for "cons", $\frown$ for append, and . . for inclusive range. Common list operators are: LENGTH, SUM, REVERSE, MEM for list member, and others to be introduced as required. Given a binary relation $\mathcal{R}$, its reflexive and transitive closure is denoted by $\mathcal{R}^*$.

HOL4 Sources. Our proof scripts, one for the Binomial Theory and one for the Triangle Theory, can be found at https://bitbucket.org/jhlchan/hol/src/, in the sub-folder `algebra/lib`.

2   Background

2.1   LCM Lower Bound for a List

The following observation is simple:

Theorem 2. *The least common multiple of a list of positive numbers equals at least its average.*

$$\vdash (\forall x.\ \mathsf{MEM}\ x\ \ell \Rightarrow 0 < x) \Rightarrow \mathsf{SUM}\ \ell \leq \mathsf{LENGTH}\ \ell \times \mathsf{list\_lcm}\ \ell$$

Proof. For a list ℓ, since every element is nonzero, list_lcm ℓ is also nonzero. There are LENGTH ℓ elements, and each element $x \leq$ list_lcm ℓ. Therefore adding together LENGTH ℓ copies of list_lcm ℓ cannot be smaller than their sum, which is SUM ℓ. $\square$

A naïve application of this theorem to the list of consecutive numbers gives a trivial and disappointing LCM lower bound. For an ingenious application of the theorem to obtain the better LCM lower bound in Theorem 1, we turn to Leibniz's Triangles, close relatives of Pascal's Triangle.

2.2 Pascal's Triangle

Pascal's well-known triangle (first in Fig. 1) can be constructed as follows:

- Each boundary entry: always 1.
- Each inside entry: sum of two immediate parents.

The entries of Pascal's triangle (the k-th element on n-th row) are binomial coefficients $\binom{n}{k}$, with the n-th row sum: $\sum_{k=0}^{n} \binom{n}{k} = 2^n$.

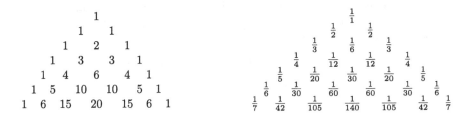

Fig. 1. Pascal's and Leibniz's Triangles

Since Leibniz's triangle (see Sect. 2.3 below) will be defined using Pascal's triangle, we include the binomials as a foundation in our HOL4 implementation, proving the above result:

Theorem 3. *Sum of a row in Pascal's Triangle.*

$\vdash$ SUM $(\mathcal{P}_{\text{row}}\ n) = 2^n$

We use $(\mathcal{P}_{\text{row}}\ n)$ to represent the n-th row of the Pascal's triangle, counting n from 0.

2.3 Leibniz's Harmonic Triangle

Leibniz's harmonic triangle (second in Fig. 1) can be similarly constructed:

(a) Each boundary entry: $\dfrac{1}{(n+1)}$ for the n-th row, with n starting from 0.

(b) Each entry (inside or not): sum of two immediate children.

With the boundary entries forming the harmonic sequence, this Leibniz's triangle is closely related to Pascal's triangle. Denoting the harmonic triangle entries (also the k-th element on n-th row) by $\begin{bmatrix} n \\ k \end{bmatrix}$, then it is not hard to show (e.g., [2]) from the construction rules that:

(a) $\begin{bmatrix} n \\ k \end{bmatrix} = \dfrac{1}{(n+1)\binom{n}{k}}$

(b) $\displaystyle\sum_{k=0}^{n} \binom{n}{k} \begin{bmatrix} n \\ k \end{bmatrix} = 1$

Therefore all entries of the harmonic triangle are unit fractions. So, we choose to work with Leibniz's "Denominator Triangle", by picking only the denominators of the entries. This allows us to deal with whole numbers rather than rational numbers in HOL4.

3   Leibniz's Denominator Triangle and Its Triplets

Taking the denominators of each entry of Leibniz's Harmonic Triangle to form Leibniz's Denominator Triangle, denoted by $\mathcal{L}$, we define its entries in HOL4 *via* the binomial coefficients:

Definition 1. *Denominator form of Leibniz's triangle: k-th entry at n-th row.*

$\vdash \mathcal{L}\ n\ k = (n + 1) \times \binom{n}{k}$

Table 1. Leibniz's denominator triangle. A typical triplet is marked.

row n \ column k	$k=0,$	$k=1,$	$k=2,$	$k=3,$	$k=4,$	$k=5,$	$k=6, \cdots$
$n=0$	1						
$n=1$	2	2					
$n=2$	3	6	3				
$n=3$	4	12	12	4			
$n=4$	5	20	30	20	5		
$n=5$	6	30	60	60	30	6	
$n=6$	7	42	105	140	105	42	7

The first few rows of the denominator triangle are shown (Table 1) in a vertical-horizontal format. Evidently from Definition 1, the n-th horizontal row is just a multiple of the n-th row in Pascal's triangle by a factor $(n+1)$, and the left vertical boundary consists of consecutive numbers:

$\vdash \mathcal{L}\ n\ 0 = n + 1$

Within this vertical-horizontal format, we identify L-shaped "Leibniz triplets" rooted at row n and column k, involving three entries:

– the top of the triplet being α_{nk}, and
– its two child entries as β_{nk} and γ_{nk} on the next row.

In other words, we can define the constituents of a typical Leibniz triplet as:

Table 2. The Leibniz triplet

	$\cdots$			$\cdots$	
row $\cdots$	$\cdots\;\cdots$			$\cdots$	

Denominator Triangle   Harmonic Triangle

$$\vdash \alpha_{nk} \;=\; \mathcal{L}\; n\; k$$

$$\vdash \beta_{nk} \;=\; \mathcal{L}\; (n\,+\,1)\; k \qquad \vdash \gamma_{nk} \;=\; \mathcal{L}\; (n\,+\,1)\; (k\,+\,1)$$

Note that the values α_{nk}, β_{nk} and γ_{nk} occur as denominators in Leibniz's original harmonic triangle, corresponding to the situation that the entry $\dfrac{1}{\alpha_{nk}}$ has immediate children $\dfrac{1}{\beta_{nk}}$ and $\dfrac{1}{\gamma_{nk}}$ (refer to Table 2). By the construction rule of harmonic triangle, we should have:

$$\frac{1}{\alpha_{nk}} = \frac{1}{\beta_{nk}} + \frac{1}{\gamma_{nk}}, \quad \text{or} \quad \frac{1}{\gamma_{nk}} = \frac{1}{\alpha_{nk}} - \frac{1}{\beta_{nk}}$$

which, upon clearing fractions, becomes:

$$\alpha_{nk} \times \beta_{nk} = \gamma_{nk} \times (\beta_{nk} - \alpha_{nk})$$

Indeed, it is straightforward to show that our definition of $(\mathcal{L}\; n\; k)$ satisfies this property:

Theorem 4. *Property of a Leibniz triple in Denominator Triangle.*

$$\vdash \alpha_{nk} \times \beta_{nk} = \gamma_{nk} \times (\beta_{nk} - \alpha_{nk})$$

This identity for a Leibniz triplet is useful for computing the entry γ_{nk} from previously calculated entries α_{nk} and β_{nk}. Indeed, the entire Denominator Triangle can be constructed directly out of such overlapping triplets:

– Each left boundary entry: $(n + 1)$ for the n-th row, with n starting from 0.
– Each Leibniz triplet: $\gamma_{nk} = \dfrac{\alpha_{nk} \times \beta_{nk}}{\beta_{nk} - \alpha_{nk}}$.

This is also the key for the next important property of the triplet.

3.1 LCM Exchange

A Leibniz triplet has an important property related to least common multiple:

Theorem 5. *In a Leibniz triplet, the vertical pair $[\beta_{nk};\ \alpha_{nk}]$ and the horizontal pair $[\beta_{nk};\ \gamma_{nk}]$ both share the same least common multiple.*

$\vdash\ \text{lcm}\ \beta_{nk}\ \alpha_{nk}\ =\ \text{lcm}\ \beta_{nk}\ \gamma_{nk}$

Proof. Let $a = \alpha_{nk}$, $b = \beta_{nk}$, $c = \gamma_{nk}$. Recall from Theorem 4 that: $ab = c(b - a)$.

$$
\begin{aligned}
&\text{lcm}\ b\ c\\
&= bc \div \gcd(b, c) && \text{by definition}\\
&= abc \div (a \times \gcd(b, c)) && \text{introduce factor } a \text{ above and below division}\\
&= bac \div \gcd(ab, ca) && \text{by common factor } a, \text{commutativity}\\
&= bac \div \gcd(c(b - a), ca) && \text{by Leibniz triplet property, Theorem 4}\\
&= bac \div (c \times \gcd(b-a, a)) && \text{extract common factor } c\\
&= ba \div \gcd(b, a) && \text{apply GCD subtraction and cancel factor } c\\
&= \text{lcm}\ b\ a && \text{by definition.}
\end{aligned}
$$
□

Table 3. A column and a row intersecting at a left boundary entry of denominator triangle

row n \ column k	$k = 0,$	$k = 1,$	$k = 2,$	$k = 3,$	$k = 4,$	$k = 5,$	$k = 6, \cdots$
$n = 0$	1						
$n = 1$	2	2					
$n = 2$	3	6	3				
$n = 3$	4	12	12	4			
$n = 4$	5	20	30	20	5		
$n = 5$	6	30	60	60	30	6	
$n = 6$	7	42	105	140	105	42	7

We shall make good use of this LCM invariance through swapping vertical and horizontal pairs in Leibniz triplets to establish an "enlarged" L-shaped LCM invariance involving columns and rows, as shown in Table 3. Theorem 1 will be deduced from this extended LCM invariance.

4   Paths Through Triangles

Our theorem requires us to capture the notion of the least common multiple of a list of elements (a path within the Denominator Triangle). We formalize paths as lists of numbers, without requiring the path to be connected. However, the paths we work with will be connected and include (refer to Table 3):

– $(\mathcal{L}_{\text{down}}\ n)$: the list $[1\ ..\ n + 1]$, which happens to be the first $n+1$ elements of the leftmost column of the Denominator Triangle, reading down;

– ($\mathcal{L}_{up}$ n): the reverse of $\mathcal{L}_{down}$ n, or the leftmost column of the triangle reading up; and
– ($\mathcal{L}_{row}$ n): the n-th row of the Denominator Triangle, reading from the left.

Then, due to the possibility of LCM exchange within a Leibniz triplet (Theorem 5), we can prove the following:

Theorem 6. *In the Denominator Triangle, consider the first element (at left boundary) of the n-th row. Then the least common multiple of the column of elements above it is equal to the least common multiple of elements in its row.*

$$\vdash \; \mathsf{list\_lcm} \; (\; \mathcal{L}_{down} \; n) \; = \; \mathsf{list\_lcm} \; (\; \mathcal{L}_{row} \; n)$$

The proof is done *via* a kind of zig-zag transformation, see Fig. 2. In the Denominator Triangle, we represent the entries for LCM consideration as a path of black discs, and indicate the Leibniz triplets by discs marked with small gray dots. Recall that, by Theorem 5, the vertical pair of a Leibniz triplet can be swapped with its horizontal pair without affecting the least common multiple.

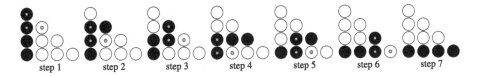

Fig. 2. Transformation of a path from vertical to horizontal in the Denominator Triangle, stepping from left to right. The path is indicated by entries with black discs. The 3 gray-dotted discs in L-shape indicate the Leibniz triplet, which allows LCM exchange. Each step preserves the overall LCM of the path.

It takes a little effort to formalize such a transformation. We use the following approach in HOL4.

4.1 Zig-Zag Paths

If a path happens to have a vertical pair, we can match the vertical pair with a Leibniz triplet and swap with its horizontal pair to form another path, its zig-zag equivalent, which keeps the list LCM of the path.

Definition 2. *Zig-zag paths are those transformable by a Leibniz triplet.*

$$\vdash \; p_1 \rightsquigarrow p_2 \; \Longleftrightarrow \;$$
$$\exists n \; k \; x \; y. \; p_1 = x \; \frown \; [\beta_{nk}; \; \alpha_{nk}] \; \frown \; y \land p_2 = x \; \frown \; [\beta_{nk}; \; \gamma_{nk}] \; \frown \; y$$

Basic properties of zig-zag paths are:

Theorem 7. *Zig-zag path properties.*

$$\vdash \; p_1 \rightsquigarrow p_2 \Rightarrow \forall x. \; [x] \; \frown \; p_1 \rightsquigarrow [x] \; \frown \; p_2 \qquad \text{zig-zag a congruence wrt (::)}$$
$$\vdash \; p_1 \rightsquigarrow p_2 \Rightarrow \mathsf{list\_lcm} \; p_1 = \mathsf{list\_lcm} \; p_2 \qquad \text{preserving LCM by exchange } via \text{ triplet}$$

4.2   Wriggle Paths

A path can *wriggle* to another path if there are zig-zag paths in between to facilitate the transformation. Thus, wriggling is the reflexive and transitive closure of zig-zagging, giving the following:

Theorem 8. *Wriggle path properties.*

$\vdash p_1 \rightsquigarrow^* p_2 \Rightarrow \forall x.\ [x] \frown p_1 \rightsquigarrow^* [x] \frown p_2$ wriggle a congruence wrt (::)

$\vdash p_1 \rightsquigarrow^* p_2 \Rightarrow \mathsf{list\_lcm}\ p_1 = \mathsf{list\_lcm}\ p_2$ preserves LCM by zig-zags

4.3   Wriggling Inductions

We use wriggle paths to establish a key step[2]:

Theorem 9. *In the Denominator Triangle, a left boundary entry with the entire row above it can wriggle to its own row.*

$\vdash [\mathcal{L}\ (n + 1)\ 0] \frown \mathcal{L}_{\mathrm{row}}\ n \rightsquigarrow^* \mathcal{L}_{\mathrm{row}}\ (n + 1)$

Proof. We prove a more general result by induction, with the step case given by the following lemma:

$\vdash k \le n \Rightarrow$
$\qquad \mathsf{TAKE}\ (k + 1)\ (\mathcal{L}_{\mathrm{row}}\ (n + 1)) \frown \mathsf{DROP}\ k\ (\mathcal{L}_{\mathrm{row}}\ n) \rightsquigarrow$
$\qquad \mathsf{TAKE}\ (k + 2)\ (\mathcal{L}_{\mathrm{row}}\ (n + 1)) \frown \mathsf{DROP}\ (k + 1)\ (\mathcal{L}_{\mathrm{row}}\ n)$

where the list operators TAKE and DROP extract, respectively, prefixes and suffixes of our paths.

In other words: in the Denominator Triangle, the two partial rows TAKE $(k + 1)$ $(\mathcal{L}_{\mathrm{row}}\ (n + 1))$ and DROP k $(\mathcal{L}_{\mathrm{row}}\ n)$ can zig-zag to a longer prefix of the lower row, with the upper row becoming one entry shorter. This is because there is a Leibniz triplet at the zig-zag point (see, for example, Step 5 of Fig. 2), making the zig-zag condition possible. The subsequent induction is on the length of the upper partial row.   □

With this key step, we can prove the whole transformation illustrated in Fig. 2.

Theorem 10. *In the Denominator Triangle, for any left boundary entry: its upward vertical path wriggles to its horizontal path.*

$\vdash \mathcal{L}_{\mathrm{up}}\ n \rightsquigarrow^* \mathcal{L}_{\mathrm{row}}\ n$

Proof. By induction on the path length n.
For the basis $n = 0$, both $(\mathcal{L}_{\mathrm{up}}\ 0)$ and $(\mathcal{L}_{\mathrm{row}}\ 0)$ are [1], hence they wriggle trivially.

---

[2] This is illustrated in Fig. 2 from the middle (step 4) to the last (step 7).

For the induction step, note that the head of $(\mathcal{L}_{up}\ (n+1))$ is $(\mathcal{L}\ (n+1)\ 0)$. Then,

$$
\begin{aligned}
&\mathcal{L}_{up}\ (n+1) \\
=\ &[\mathcal{L}\ (n+1)\ 0] \frown \mathcal{L}_{up}\ n &&\text{by taking apart head and tail} \\
\rightsquigarrow^*\ &[\mathcal{L}\ (n+1)\ 0] \frown \mathcal{L}_{row}\ n &&\text{by induction hypothesis and tail wriggle (Theorem 8)} \\
\rightsquigarrow^*\ &\mathcal{L}_{row}\ (n+1) &&\text{by key step of wriggling (Theorem 9).} \qquad\square
\end{aligned}
$$

Now we can formally prove the LCM transform of Theorem 6.

$\vdash$ list_lcm $(\mathcal{L}_{down}\ n) =$ list_lcm $(\mathcal{L}_{row}\ n)$

Proof. Applying path wriggling of Theorem 10 in the last step,

$$
\begin{aligned}
&\text{list\_lcm}\ (\mathcal{L}_{down}\ n) \\
=\ &\text{list\_lcm}\ (\mathcal{L}_{up}\ n) &&\text{by reverse paths keeping LCM} \\
=\ &\text{list\_lcm}\ (\mathcal{L}_{row}\ n) &&\text{by wriggle paths keeping LCM (Theorem 8).} \quad\square
\end{aligned}
$$

5   LCM Lower Bound

Using the equality of least common multiples just proved for Theorem 6, here is the proof of Theorem 1:

$\vdash 2^n \leq$ list_lcm $[1\ ..\ n+1]$

Proof. Recall from Sect. 3 that the left vertical boundary of Leibniz's Denominator Triangle consists of consecutive numbers, thus $(\mathcal{L}_{down}\ n) = [1\ ..\ n+1]$. Also, the horizontal $(\mathcal{L}_{row}\ n)$ is just a multiple of $(\mathcal{P}_{row}\ n)$ by a factor $(n+1)$. Therefore,

$$
\begin{aligned}
&\text{list\_lcm}\ [1\ ..\ n+1] \\
=\ &\text{list\_lcm}\ (\mathcal{L}_{down}\ n) &&\text{as asserted} \\
=\ &\text{list\_lcm}\ (\mathcal{L}_{row}\ n) &&\text{by LCM transform (Theorem 6)} \\
=\ &(n+1) \times \text{list\_lcm}\ (\mathcal{P}_{row}\ n) &&\text{by LCM common factor} \\
=\ &\text{LENGTH}\ (\mathcal{P}_{row}\ n) \times \text{list\_lcm}\ (\mathcal{P}_{row}\ n) &&\text{by length of horizontal row} \\
\geq\ &\text{SUM}\ (\mathcal{P}_{row}\ n) &&\text{by Theorem 2} \\
=\ &2^n &&\text{by binomial sum (Theorem 3).} \quad\square
\end{aligned}
$$

6   Conclusion

We have proved a lower bound for the least common multiple of consecutive numbers, using an interesting application of Leibniz's Triangle in denominator form. By elementary reasoning over natural numbers and lists, we have not just mechanized what we believe to be a cute proof, but now have a result that will be useful in our ongoing work on the mechanization of the AKS algorithm.

References

1. Agrawal, M., Kayal, N., Saxena, N.: PRIMES is in P. Ann. Math. **160**(2), 781–793 (2004)
2. Ayoub, A.B.: The harmonic triangle and the beta function. Math. Mag. **60**(4), 223–225 (1987)
3. Bicknell-Johnson, M.: Diagonal sums in the harmonic triangle. Fibonacci Q. **19**(3), 196–199 (1981)
4. Chan, H.-L., Norrish, M.: Mechanisation of AKS algorithm: Part 1–the main theorem. In: Urban, C., Zhang, X. (eds.) ITP 2015. LNCS, vol. 9236, pp. 117–136. Springer, Heidelberg (2015)
5. Esteve, M.R.M., Delshams, A.: Euler's beta function in Pietro Mengoli's works. Arch. Hist. Exact Sci. **63**(3), 325–356 (2009)
6. Farhi, B.: An identity involving the least common multiple of binomial coefficients and its application. Am. Math. Mon. **116**(9), 836–839 (2009)
7. Hardy, G.H., Wright, E.M.: An Introduction to the Theory of Numbers, 6th edn. Oxford University Press, USA (2008). ISBN: 9780199219865
8. Hong, S.: Nair's and Farhi's identities involving the least common multiple of binomial coefficients are equivalent, July 2009. http://arxiv.org/pdf/0907.3401
9. Grigory, M.: Answer to: is there a direct proof of this LCM identity? Question 1442 on Math Stack Exchange, August 2010. http://math.stackexchange.com/questions/1442/
10. Nair, M.: On Chebyshev-type inequalities for primes. Am. Math. Mon. **89**(2), 126–129 (1982)

Two-Way Automata in Coq

Christian Doczkal[(⊠)] and Gert Smolka[(⊠)]

Saarland University, Saarbrücken, Germany
{doczkal,smolka}@ps.uni-saarland.de

Abstract. We formally verify translations from two-way automata to one-way automata based on results from the literature. Following Vardi, we obtain a simple reduction from nondeterministic two-way automata to one-way automata that leads to a doubly-exponential increase in the number of states. By adapting the work of Shepherdson and Vardi, we obtain a singly-exponential translation from nondeterministic two-way automata to DFAs. The translation employs a constructive variant of the Myhill-Nerode theorem. Shepherdson's original bound for the translation from deterministic two-way automata to DFAs is obtained as a corollary. The development is formalized in Coq/Ssreflect without axioms and makes extensive use of countable and finite types.

1 Introduction

Two-way finite automata are a representation for regular languages introduced by Rabin and Scott [15]. Unlike one-way automata, two-way automata may move back and forth on the input word and may be seen as read-only Turing machines without memory.

Both deterministic two-way automata (2DFAs) and nondeterministic two-way automata (2NFAs) exactly accept regular languages [15,17,20]. However, some languages have 2DFAs that are exponentially smaller than the minimal DFA; for instance the languages $I_n := (a + b)^* a (a + b)^n$ from [14]. It is known that the cost (in terms of the number of states) of simulating both 2DFAs and 2NFAs with DFAs is exponential [17,20]. Whether the cost of simulating NFAs and 2NFAs using 2DFAs is also exponential is still an open problem [14,16].

As is frequently the case with language-theoretic results, the proofs in the literature are described in a fairly informal manner. When carried out in detail, the constructions are delicate and call for formalization. We are the first to provide constructive and machine-checked proofs of the following results:

1. For every n-state 2NFA M there exists an NFA with at most 2^{2n} states accepting the complement of the language of M.
2. For every n-state 2DFA there is an equivalent DFA with at most $(n+1)^{(n+1)}$ states.
3. For every n-state 2NFA there is an equivalent DFA with at most 2^{n^2+n} states.

Our proofs mostly refine the proofs given by Shepherdson [17] and Vardi [20]. Result (1) is easiest to show. It establishes that the languages accepted by 2NFAs

© Springer International Publishing Switzerland 2016
J.C. Blanchette and S. Merz (Eds.): ITP 2016, LNCS 9807, pp. 151–166, 2016.
DOI: 10.1007/978-3-319-43144-4_10

(and therefore also 2DFAs) are regular. Our proof is based on a construction in [20]. If one wants to obtain an automaton for the original language, using (1) leads to a doubly exponential increase in the number of states. A singly-exponential bound can be obtained using a construction from Shepherdson [17] originally used to establish (2). Building on ideas from [20], we adapt Shepherdson's construction to 2NFAs. That this is possible appears to be known [14], but to the best of our knowledge the construction for 2NFAs has never been published. Once we have established (3), we obtain (2) by showing that if the input automaton is deterministic, the constructed DFA has at most $(n+1)^{(n+1)}$ states. This allows us to get both results with a single construction.

The reduction to DFAs makes use of the Myhill-Nerode theorem. We employ a constructive variant where Myhill-Nerode relations are represented as functions we call classifiers that are supplemented with decidability assumptions to provide for a constructive proof. When constructing DFAs from 2NFAs, the decidability requirements are easily satisfied. The application of the constructive Myhill-Nerode theorem to the reduction from 2NFAs to DFAs demonstrates that the construction is useful.

We formalize our results in Coq [18] using the Ssreflect [9] extension. The formalization accompanying this paper[1] extends and revises previous work [6] and contains a number of additional results. The development makes extensive use of finite and countable types [7,8] as provided by Ssreflect. In particular, we use finite types to represent states for finite automata.

Various aspects of the theory of regular languages have been formalized in different proof assistants. In addition to executable certified decision methods [2,3,5,12,19] based on automata or regular expressions, there are a number of purely mathematical developments. Constable et al. [4] formalize automata theory in Nuprl, including the Myhill-Nerode theorem. Wu et al. [22] give a proof of the Myhill-Nerode theorem based on regular expressions. Recently, Paulson [13] has formalized the Myhill-Nerode theorem and Brzozowski's minimization algorithm in Isabelle.

The paper is organized as follows. Sections 2 and 3 recall some type theoretic constructions underlying our proofs and describe how the usual language theoretic notions are represented in type theory. In Sect. 4 we define one-way automata. In Sect. 5 we prove the constructive variant of the Myhill-Nerode theorem. Section 6 defines two-way automata. Section 7 presents the reduction from 2NFAs to NFAs (for the complement) and Sect. 8 the reductions from 2NFAs and 2DFAs to DFAs.

2   Type Theory Preliminaries

We formalize our results in the constructive type theory of the proof assistant Coq [18]. In this setting, decidability properties are of great importance. We call a proposition *decidable*, if it is equivalent to a boolean expression. Similarly,

---

[1] www.ps.uni-saarland.de/extras/itp16-2FA.

we call a predicate decidable, if it is equivalent to a boolean predicate. In the mathematical presentation, we will not distinguish between decidable propositions and the associated boolean expressions.

In type theory, operations such as boolean equality tests and choice operators are not available for all types. Nevertheless, there are certain classes of types for which these operations are definable. For our purposes, three classes of types are of particular importance. These are discrete types, countable types, and finite types [8].

We call a type X *discrete* if equality on (elements of) X is decidable. The type of booleans $\mathbb{B}$ and the type of natural numbers $\mathbb{N}$ are both discrete.

We call a type X *countable* if there are functions $f : X \to \mathbb{N}$ and $g : \mathbb{N} \to X_\perp$ such that $g(f\,x) = \mathsf{Some}\,x$ for all $x : X$, where $X_\perp$ is the option type over X. All countable types are also discrete. We will make use of the fact that surjective functions from countable types to discrete types have right inverses.

Lemma 1. *Let X be countable, Y be discrete, and $f : X \to Y$ be surjective. Then there exists a function $f^{-1} : Y \to X$ such that $f(f^{-1}y) = y$ for all y.*

Proof. The countable type X is equipped with a choice operator

$$\mathsf{xchoose}_X : \forall p : X \to \mathbb{B}.\, (\exists x : X.\, p\,x) \to X$$

satisfying $p(\mathsf{xchoose}_X\, p\, E)$ for all $E : (\exists x : X.\, p\,x)$. Given some $y : Y$, we construct $f^{-1}y$ using the choice operator with $p := \lambda x : X.\, f\,x = y$. □

A *finite type* is a type X together with a list enumerating all elements of X. If X is finite, we write $|X|$ for the number of elements of X. For our purposes, the most important property of finite types is that quantification over finite types preserves decidability.

Discrete, countable, and finite types are closed under forming product types $X \times Y$, sum types $X + Y$, and option types $X_\perp$. Moreover, all three classes of types are closed under building subtypes with respect to decidable predicates. Let $p : X \to \mathbb{B}$. The Σ-type $\{x : X \mid p\,x\}$, whose elements are dependent pairs of elements $x : X$ and proofs of $p\,x = \mathsf{true}$, can be treated as a subtype of X. In particular, the first projection yields an injection from $\{x : X \mid p\,x\}$ to X since $p\,x = \mathsf{true}$ is proof irrelevant [10].

Finite types also come with a power operator. That is, if X and Y are finite types then there is a finite type Y^X whose $|Y|^{|X|}$ elements represent the functions from X to Y up to extensionality. We write 2^X for the finite type of (extensional) finite sets with decidable membership represented as $\mathbb{B}^X$. If a finite type X appears as a set, it is to be read as the full set over X.

3   Languages in Type Theory

For us, an *alphabet* is a finite type. For simplicity, we fix some alphabet Σ throughout the paper and refer to its elements as *symbols*. The type of lists over Σ, written Σ^*, is a countable type. We refer to terms of this type as *words*.

The letters a, b always denote symbols. The letters x, y, and z always denote words and ε denotes the empty word. We write $|x|$ to denote the *length* of the word x and xy or $x \cdot y$ (if this increases readability) for the concatenation of x and y. We also write $x[n, m]$ for the subword from position n (inclusive) to m (exclusive), e.g., $x = x[0, j] \cdot x[j, |x|]$.

A *language* is a predicate on words, i.e., a function of type $\Sigma^* \to \mathsf{Prop}$ (or $\Sigma^* \to \mathbb{B}$ for decidable languages). This yields an intensional representation. We write $L_1 \equiv L_2$ to denote that L_1 and L_2 are equivalent (i.e., extensionally equal). The absence of extensionality causes no difficulties since all our constructions respect language equivalence. To increase readability, we employ the usual set-theoretic notations for languages. In particular, we write $\overline{L}$ for the complement of the language L.

4   One-Way Automata

Deterministic one-way automata (DFAs) can be seen as the most basic operational characterization of regular languages. In addition to DFAs, we also define nondeterministic finite automata (NFAs) since both will serve as targets for our translations from two-way automata to one-way automata.

Definition 2. *A* deterministic finite automaton (DFA) *is a structure* (Q, s, F, δ) *where*

- Q *is a finite type of* states.
- $s : Q$ *is the* starting state.
- $F : Q \to \mathbb{B}$ *determines the* final states.
- $\delta : Q \to \Sigma \to Q$ *is the* transition function.

In Coq, we represent DFAs using dependent records:

$$\mathsf{dfa} := \{\ \mathsf{state} : \mathsf{finType}$$
$$\mathsf{start} : \mathsf{state}$$
$$\mathsf{final}\ : \mathsf{state} \to \mathbb{B}$$
$$\mathsf{trans} : \mathsf{state} \to \Sigma \to \mathsf{state}\}$$

Here, $\mathsf{state} : \mathsf{finType}$ restricts the type states to be a finite type. Finite types provide for a formalization of finite automata that is very convenient to work with. In particular, finite types have all the closure properties required for the usual constructions on finite automata [6].

Let $A = (Q, s, F, \delta)$ be a DFA. We extend δ to a function $\hat{\delta} : Q \to \Sigma^* \to Q$ by recursion on words:

$$\hat{\delta}\, q\, \varepsilon := q$$
$$\hat{\delta}\, q(a :: x) := \hat{\delta}\, (\delta\, q\, a)\, x$$

We say that a state q of A *accepts* a word x if $\hat{\delta}\,q\,x \in F$. The *language of A*, written $\mathcal{L}(A)$, is then defined as the collection of words accepted by the starting state:

$$\mathcal{L}(A) := \{\, x \in \Sigma^* \mid \hat{\delta}\,s\,x \in F \,\}$$

Note that is a decidable language.

Definition 3. *We say that a DFA A accepts the language L if $L \equiv \mathcal{L}(A)$. We call L* regular *if it is accepted by some DFA.*

Nondeterministic finite automata differ from DFAs in that the transition function is replaced with a relation. Moreover, we allow multiple stating states.

Definition 4. *A* nondeterministic finite automation (NFA) *is a structure (Q, S, F, δ) where:*

- Q *is a finite type of states.*
- $S : 2^Q$ *is the set of* starting *states.*
- $F : 2^Q$ *is the set of* final *states.*
- $\delta : Q \to \Sigma \to Q \to \mathbb{B}$ *is the* transition relation.

Let $A = (Q, S, F, \delta)$ be an NFA. Similar to DFAs, we define acceptance for every state of an NFA by structural recursion on the input word.

$$\mathsf{accept}\,p\,\varepsilon := p \in F$$
$$\mathsf{accept}\,p\,(a :: x) := \exists q \in Q.\, \delta\,p\,a\,q \wedge \mathsf{accept}\,q\,x$$

The *language* of an NFA is then the union of the languages accepted by its starting states.

$$\mathcal{L}(A) := \{\, x \in \Sigma^* \mid \exists s \in S.\, \mathsf{accept}\,s\,x \,\}$$

Note that since S is finite, this is also a decidable language. As with DFAs, acceptance of languages is defined up to language equivalence.

NFAs can be converted to DFAs using the well-known powerset construction.

Fact 5. *For every n-state NFA A, there exists a DFA with at most 2^n states accepting $\mathcal{L}(A)$.*

5 Classifiers and Myhill-Nerode

We now introduce classifiers as an abstract characterization of DFAs. For us, classifiers play the role of Myhill-Nerode relations (cf. [11]). Classifiers differ from Myhill-Nerode relations mainly in that they include decidability assumptions required for constructive proofs. Classifiers have a cut-off property which yields a number of useful decidability properties. Further, classifiers provide a sufficient criterion for the existence of DFAs that is useful for the translation from two-way automata to one-way automata.

Definition 6. *Let Q be a type and let $f : \Sigma^* \to Q$. Then f is called* right congruent *if $fx = fy$ implies $f(xa) = f(ya)$ for all $x, y : \Sigma^*$ and all $a : \Sigma$.*

Definition 7. *A function $f : \Sigma^* \to Q$ is called a* classifier *if it is right congruent and Q is a finite type. If L is a decidable language, a* classifier for L *is a classifier that refines L, i.e., that satisfies $\forall x\, y.\, f\,x = f\,y \to (x \in L \leftrightarrow y \in L)$.*

Fact 8. *If $A = (Q, s, F, \delta)$ is a DFA, then $\hat{\delta}s$ is a classifier for $\mathcal{L}(A)$.*

If $f : \Sigma^* \to Q$ is a classifier, the congruence property of f allows us to decide whether a certain element of Q is in the image of f.

Theorem 9 (Cut-Off). *Let $f : \Sigma^* \to Q$ be a classifier and let $P : Q \to \mathsf{Prop}$. Then*

$$\exists x.\, P(f\,x) \iff \exists x.\, |x| \le |Q| \wedge P(f\,x)$$

Proof. The direction from right to left is trivial. For the other direction let x such that $P(f\,x)$. We proceed by induction on $|x|$. If $|x| \le |Q|$ the claim is trivial. Otherwise, there exist $i < j < |x|$ such that $f(x[0, i]) = f(x[0, j])$. Since f is right congruent, we have $f\,x = f\,(x[0, i] \cdot x[j, |x|])$ and the claim follows by induction hypothesis. □

Corollary 10. *Let $f : \Sigma^* \to Q$ be a classifier. Then $\exists x.\, p(f\,x)$ and $\forall x.\, p(f\,x)$ are decidable for all decidable predicates $p : Q \to \mathbb{B}$.*

Proof. Decidability of $\exists x.\, p(f\,x)$ follows with Theorem 9, since there are only finitely many words of length at most $|Q|$. Decidability of $\forall x.\, p(f\,x)$ then follows from decidability of $\exists x.\, \neg p(f\,x)$. □

Corollary 11. *Language emptiness for DFAs is decidable.*

Proof. Let $A = (Q, s, F, \delta)$ be a DFA. Then $\mathcal{L}(A)$ is empty iff $\hat{\delta}\,s\,x \notin F$ for all $x : \Sigma^*$. Since $\hat{\delta}s$ is a classifier, this is a decidable property (Corollary 10). □

Remark 12. The proof of Corollary 11 is essentially the proof of decidability of emptiness given by Rabin and Scott [15].

As mentioned above, every DFA yields a classifier for its language. We now show that a classifier for a given decidable language L contains all the information required to construct a DFA accepting L.

Lemma 13. *Let $f : \Sigma^* \to Q$ be a classifier. Then the image of f can be constructed as a subtype of Q.*

Proof. By Corollary 10, we have that $\exists x \in \Sigma^*.\, f\,x = q$ is decidable for all q. Hence, we can construct the subtype $\{\, q : Q \mid \exists x.\, f\,x = q \,\}$. □

If $f : \Sigma^* \to Q$ is a classifier, we write $f(\Sigma^*)$ for the subtype of Q corresponding to the image of f.

Theorem 14 (Myhill-Nerode). *Let L be decidable and let $f : \Sigma^* \to Q$ be a classifier for L. Then one can construct a DFA accepting L that has at most $|Q|$ states.*

Proof. By casting the results of f from Q to $f(\Sigma^*)$, we obtain a surjective classifier $g : \Sigma^* \to f(\Sigma^*)$ for L (Lemma 13). Since g is surjective, it has a right inverse g^{-1} (Lemma 1). It is straightforward to verify that the DFA $(f(\Sigma^*), s, F, \delta)$ where

$$
\begin{aligned}
s &:= g\,\varepsilon \\
F &:= \{\, q \mid g^{-1}q \in L \,\} \\
\delta\,q\,a &:= g((g^{-1}q) \cdot a)
\end{aligned}
$$

$\square$

accepts the language L.

We remark that in order to use Theorem 14 for showing that a language is regular, one first has to show that the language is decidable. It turns out that this restriction is unavoidable in a constructive setting. Let P be some independent proposition. Then $P \vee \neg P$ is not provable. Now consider the language $L := \{\, w \in \Sigma^* \mid P \,\}$. Save for the decidability requirement on L, the constant function from Σ^* into the unit type is a regular classifier for L. If Theorem 14 were to apply, the resulting DFA would allow us to decide $\varepsilon \in L$ and consequently obtain a proof of $P \vee \neg P$.

For the translation from two-way automata to one-way automata, the restriction to decidable languages poses no problem since the language of a two-way automaton is easily shown to be decidable.

6 Two-Way Finite Automata

A two-way finite automaton (2FA) is essentially a read-only Turing machine, i.e., a machine with a finite state control and a read head that may move back and forth on the input word. One of the fundamental results about 2FAs is that the ability to move back and forth does not increase expressiveness [15]. That is, two-way automata are yet another representation of the class of regular languages. As for one-way automata, we consider both the deterministic and the nondeterministic variant.

In the literature, two-way automata appear in a number of variations. Modern accounts of two-way automata [14] usually consider automata with *end-markers*. That is, on input x the automaton is run on the string $\triangleright x \triangleleft$, where $\triangleright$ and $\triangleleft$ are marker symbols that do not occur in Σ and allow the automaton to detect the word boundaries. These marker symbols are not present in early work on two-way automata [15,17,20]. Marker symbols allow the detection of the word boundaries and allow for the construction of more compact automata for some languages. In fact, the emptiness problem for nondeterministic two-way automata with only one endmarker over a singleton alphabet is polynomial while the corresponding problem for two-way automata with two endmarkers is NP-complete [21].

Definition 15. *A* nondeterministic two-way automaton (2NFA) *is a structure* $M = (Q, s, F, \delta, \delta_\triangleright, \delta_\triangleleft)$ *where*

– Q *is a finite type of* states
– $s : Q$ *is the* starting state
– $F : 2^Q$ *is the set of* final states
– $\delta : Q \to \Sigma \to 2^{Q \times \{L,R\}}$ *is the* transition function *for symbols*
– $\delta_{\triangleright} : Q \to 2^{Q \times \{L,R\}}$ *is the* transition function *for the left marker*
– $\delta_{\triangleleft} : Q \to 2^{Q \times \{L,R\}}$ *is the* transition function *for the right marker*

Let $M = (Q, s, F, \delta, \delta_{\triangleright}, \delta_{\triangleleft})$ be a 2NFA. On an input word $x : \Sigma^*$ the *configurations of M on x*, written C_x, are pairs $(p, i) \in Q \times \{0, \ldots, |x| + 1\}$ where i is the position of the read head. We take $i = 0$ to mean that the head is on the left marker and $i = |x| + 1$ to mean that the head is on the right marker. Otherwise, the head is on the i-th symbol of x (counting from 1). In particular, we do not allow the head to move beyond the end-markers. In following, we write $x[i]$ for the i-th symbol of x. The *step relation* $\to_x : C_x \to C_x \to \mathbb{B}$ updates state and head position according to the transition function for the current head position:

$$
\dot{\delta}\, p\, i := \begin{cases} \delta_{\triangleright}\, p & i = 0 \\ \delta\, p\, (x[i]) & 0 < i \leq |x| \\ \delta_{\triangleleft}\, p & i = |x| + 1 \end{cases}
$$

$$
(p, i) \xrightarrow[x]{} (q, j) := (q, \mathsf{L}) \in \dot{\delta}\, p\, i \wedge i = j + 1 \;\vee\; (q, \mathsf{R}) \in \dot{\delta}\, p\, i \wedge i + 1 = j
$$

We write $\to_x^*$ for the reflexive transitive closure of $\to_x$. The *language* of M is then defined as follows:

$$
\mathcal{L}(M) := \{\, x \mid \exists q \in F.\, (s, 1) \xrightarrow[x]{}^* (q, |x| + 1)\, \}
$$

That is, M accepts the word x if it can reach the right end-marker while being in a final state.

In Coq, we represent C_x as the finite type $Q \times \mathrm{ord}(|x| + 2)$, where $\mathrm{ord}\, n := \{\, m : \mathbb{N} \mid m < n\, \}$. This allows us to represent $\to_x$ as well as $\to_x^*$ as decidable relations on C_x.[2] Hence, $\mathcal{L}(M)$ is a decidable language. In the mathematical presentation, we treat $\mathrm{ord}\, n$ like $\mathbb{N}$ and handle the bound implicitly. In Coq, we use a conversion function inord : $\forall n.\, \mathbb{N} \to \mathrm{ord}(n + 1)$ which behaves like the 'identity' on numbers in the correct range and otherwise returns 0. This allows us to sidestep most of the issues arising from the dependency of the type of configurations on the input word.

Definition 16. *A deterministic two-way automaton (2DFA) is a 2NFA $(Q, s, F, \delta, \delta_{\triangleright}, \delta_{\triangleleft})$ where $|\delta_{\triangleleft}\, q| \leq 1$, $|\delta_{\triangleright}\, q| \leq 1$, and $|\delta\, q\, a| \leq 1$ for all $q : Q$ and $a : \Sigma$.*

Fact 17. *For every n-state DFA there is an n-state 2DFA that accepts the same language and only moves its head to the right.*

---

[2] That the transitive closure of a decidable relation is decidable is established in the Ssreflect libraries using depth-first search.

Remark 18. While Fact 17 is obvious from the mathematical point of view, the formal proof is somewhat cumbersome due to the mismatch between the acceptance condition for DFAs, which is defined by recursion on the input word, and the acceptance condition for 2FAs, where the word remains constant throughout the computation.

The rest of the paper is devoted to the translation of two-way automata to one-way automata. There are several such translations in the literature. Vardi [20] gives a simple construction that takes as input some 2NFA M and yields an NFA accepting $\overline{\mathcal{L}(M)}$. This establishes that deterministic and nondeterministic two-way automata accept exactly the regular languages. The size of the constructed NFA is exponential in the size of M. Consequently, if one wants to obtain an automaton for the input language, rather than its complement, the construction incurs a doubly exponential blowup in the number of states. Shepherdson [17] gives a translation from 2DFAs to DFAs that incurs only an exponential blowup. Building on ideas from [20], we adapt the construction to 2NFAs.

We first present the translation to NFAs since it is conceptually simpler. We then give a direct translation from 2NFAs to DFAs. We also show that when applied to 2DFAs, the latter construction yields the bounds on the size of the constructed DFA established in [17].

7   Vardi Construction

Let $M = (Q, s, F, \delta, \delta_{\triangleright}, \delta_{\triangleleft})$ be a 2NFA. We construct an NFA accepting $\overline{\mathcal{L}(M)}$. Vardi [20] formulates the proof for 2NFAs without markers. We adapt the proof to 2NFAs with markers. The main idea is to define certificates for the non-acceptance of a string x by M. The proof then consists of two parts:

1. proving that these negative certificates are sound and complete
2. constructing an NFA whose accepting runs correspond to negative certificates

Definition 19. *A negative certificate for a word x is a set $\mathcal{C} \subseteq C_x$ satisfying:*

N1. $(s, 1) \in \mathcal{C}$
N2. If $(p, i) \in \mathcal{C}$ and $(p, i) \to_x (q, j)$, then $(q, j) \in \mathcal{C}$.
N3. If $q \in F$ then $(q, |x| + 1) \notin \mathcal{C}$.

The first two conditions ensure that the negative certificates for x overapproximate the configurations M can reach on input x. The third condition ensures that no accepting configuration is reachable.

Lemma 20. *Let $x \in \Sigma^*$. There exists a negative certificate for x iff $x \notin \mathcal{L}(M)$.*

Proof. Let $R := \{\, (q, j) \mid (s, 1) \to_x^* (q, j) \,\}$. If there exists a negative certificate $\mathcal{C}$ for x, then $R \subseteq \mathcal{C}$ and, therefore, $x \notin \mathcal{L}(M)$. Conversely, if $x \notin \mathcal{L}(M)$, then R is a negative certificate for x.   □

Let x be a word and let $\mathcal{C}$ be a negative certificate for x. The certificate $\mathcal{C}$ can be viewed as $|x|+2$-tuple over 2^Q where the i-th component, written $\mathcal{C}_i$, is the set $\{ q \mid (q, i) \in \mathcal{C} \}$.

We define an NFA whose accepting runs correspond to this tuple view of negative certificates. For this, condition (N2) needs to be rephrased into a collection of local conditions, i.e., conditions that no longer mention the head position.

Definition 21. *Let $U, V, W : 2^Q$ and $a : \Sigma$. We say that*

- (U, V) *is $\triangleright$-closed if $q \in V$ whenever $p \in U$ and $(q, \mathsf{R}) \in \delta_\triangleright\, p$.*
- (U, V) *is $\triangleleft$-closed if $q \in U$ whenever $p \in V$ and $(q, \mathsf{L}) \in \delta_\triangleleft\, p$.*
- (U, V, W) *is a-closed if for all $p \in V$ we have*
  1. $q \in U$ *whenever* $(q, \mathsf{L}) \in \delta\, p\, a$
  2. $q \in W$ *whenever* $(q, \mathsf{R}) \in \delta\, p\, a$

We define an NFA $N = (Q', S', F', \delta')$ that incrementally checks the closure conditions defined above:

$$Q' := 2^Q \times 2^Q$$
$$S' := \{(U, V) \mid s \in V \text{ and } (U, V) \text{ is } \triangleright\text{-closed}\}$$
$$F' := \{(U, V) \mid F \cap V = \emptyset \text{ and } (U, V) \text{ is } \triangleleft\text{-closed}\}$$
$$\delta'\, (U, V)\, a\, (V', W) := (V = V' \wedge (U, V, W) \text{ is } a\text{-closed})$$

Note that transition relation requires the two states to overlap. Hence, the runs of N on some word x, which consist of $|x|$ transitions, define $|x|+2$-tuples. We will show that the accepting runs of N correspond exactly to negative certificates.

For this we need to make the notion of accepting runs (of N) explicit. For many results on NFAs this is not necessary since the recursive definition of acceptance allows for proofs by induction on the input word. However, for two-way automata, the word remains static throughout the computation. Having a matching declarative acceptance criterion for NFAs makes it easier to relate the two automata models.

We define an inductive relation run $: \Sigma^* \to Q \to Q^* \to \mathsf{Prop}$ relating words and non-empty sequences of states:

$$\frac{q \in F'}{\mathsf{run}\, \varepsilon\, q\, []} \qquad\qquad \frac{\delta'\, p\, a\, q \qquad \mathsf{run}\, x\, q\, l}{\mathsf{run}\, (ax)\, p\, (q :: l)}$$

An *accepting run for x* is a sequence of states $(s :: l)$ such that $s \in S'$ and $\mathsf{run}\, x\, s\, l$. Note that accepting runs for x must have length $|x|+1$. In the following we write $(r_i)_{i \leq |x|}$ for runs of length $|x| + 1$ and r_i for the i-th element (counting from 0).

Lemma 22. $x \in \mathcal{L}(N)$ *iff there exists an accepting run for x.*

Lemma 23. $x \in \mathcal{L}(N)$ *iff there exists a negative certificate for x.*

Proof. By Lemma 22, it suffices to show that there exists an accepting run iff there exists a negative certificate.

"⇒" Let $(r_i)_{i \leq |x|}$ be an accepting run of N on x. We define a negative certificate $\mathcal{C}$ for x where $\mathcal{C}_0 := (r_0).1$ and $\mathcal{C}_{i+1} := (r_i).2$.

"⇐" If $\mathcal{C}$ is a negative certificate for x we can define a run $(r_i)_{i \leq |x|}$ for x on M where $r_0 := (\mathcal{C}_0, \mathcal{C}_1)$ and $r_{i+1} := (\mathcal{C}_i, \mathcal{C}_{i+1})$. □

Remark 24. The formalization of the lemma above is a straightforward but tedious proof of about 60 lines.

Lemma 25. $\mathcal{L}(N) = \overline{\mathcal{L}(M)}$.

Proof. Follows immediately with Lemmas 20 and 23. □

Theorem 26. *For every n-state 2NFA M there exists an NFA accepting $\overline{\mathcal{L}(M)}$ and having at most 2^{2n} states.*

If one wants to obtain a DFA for $\mathcal{L}(M)$ using this construction, one needs to determinize N before complementing it. Since N is already exponentially larger than M, the resulting DFA then has a size that is doubly exponential in $|Q|$.

8 Shepherdson Construction

We now give a second proof that the language accepted by a 2NFA is regular. The proof follows the original proof of Shepherdson [17]. In [17], the proof is given for 2DFAs without end-markers. Building on ideas form Vardi [20], we adapt the proof to 2NFAs with end-markers.

We fix some 2NFA $M = (Q, s, F, \delta, \delta_\triangleright, \delta_\triangleleft)$ for the rest of this section. In order to construct a DFA for $\mathcal{L}(M)$, it suffices to construct a classifier for $\mathcal{L}(M)$ (Theorem 14). For this, we need to come up with a finite type X and a function $T : \Sigma^* \to X$ which is right congruent and refines $\mathcal{L}(M)$.

The construction exploits that the input is read-only. Therefore, M can only save a finite amount of information in its finite state control. Consider the situation where M is running on a composite word xz. In order to accept xz, M must move its head all the way to the right. In particular, it must move the read-head beyond the end of x and there is a finite set of states M can possibly be in when this happens for the first time. Once the read head is to the right of x, M may move its head back onto x. However, the only additional information that can be gathered about x is set of states M may be in when returning to z. Since the possible states upon return may depend on the state M is in when entering x form the right, this defines a relation on $Q \times Q$. This is all the information required about x to determine whether $xz \in \mathcal{L}(M)$. This information can be recorded in a finite table. We will define a function

$$T : \Sigma^* \to 2^Q \times 2^{Q \times Q}$$

returning the table for a given word. Note that $2^Q \times 2^{Q \times Q}$ is a finite type. To show that $\mathcal{L}(M)$ is regular, it suffices to show that T is right-congruent and refines $\mathcal{L}(M)$.

To formally define T, we need to be able to stop M once its head reaches a specified position. We define the k-stop relation on x:

$$(p,i) \xrightarrow[x]{k} (q,i) := (p,i) \xrightarrow[x]{} (q,j) \wedge i \neq k$$

Note that for $k \geq |x| + 2$ the stop relation coincides with the step relation. The function T is then defined as follows:

$$T x := (\{\ q\ \mid (s,1) \xrightarrow[x]{|x|+1}{}^* (q, |x| + 1)\ \},$$
$$\{(p,q) \mid (p, |x|) \xrightarrow[x]{|x|+1}{}^* (q, |x| + 1)\ \})$$

Note that T returns a pair of a set and a relation. We write $(T x).1$ for the first component of $T x$ and $(T x).2$ for the second component.

Before we can show that T is a classifier for $\mathcal{L}(M)$, we need a number of properties of the stop relation. The first lemma captures the intuition, that for composite words xz, all the information M can gather about x is given by $T x$.

Lemma 27. Let $p,q : Q$ and let $x, z : \Sigma^*$. Then

1. $q \in (T x).1 \iff (s,1) \xrightarrow[xz]{|x|+1}{}^* (q, |x| + 1)$

2. $(p,q) \in (T x).2 \iff (p, |x|) \xrightarrow[xz]{|x|+1}{}^* (q, |x| + 1)$

Since for composite words xz everything that can be gathered about x is provided by $T x$, M behaves the same on xz and yz whenever $T x = T y$. To show this, we need to exploit that M moves its head only one step at a time. This is captured by the lemma below.

Lemma 28. Let $i \leq k \leq j$ and let l be a $\xrightarrow[x]{k'}$-path from (p,i) to (q,j). Then there exists some state p' such that l can be split into a $\xrightarrow[x]{k}$-path from (p,i) to (p',k) and a $\xrightarrow[x]{k'}$-path from (p',k) to (q,j).

Proof. By induction on the length of the $\xrightarrow[x]{k'}$-path from (p,i) to (q,j).

Lemma 28 can be turned into an equivalence if $k' \geq k$. We state this equivalence in terms in terms of transitive closure since for most parts of the development the concrete path is irrelevant.

Lemma 29. Let $i \leq k \leq j$ and let $k' \geq k$. Then $(p,i) \xrightarrow[x]{k'}{}^* (q,j)$ iff there exists some p' such that $(p,i) \xrightarrow[x]{k}{}^* (p',k) \xrightarrow[x]{k'}{}^* (q,j)$.

We now show that for runs of M that start and end on the right part of a composite word xz, x can be replaced with y whenever $T\,x = T\,y$.

Lemma 30. *Let $p, q : Q$ and let $x, y, z : \Sigma^*$ such that $T\,x = T\,y$. Then for all $k > 1$, $i \leq |z| + 1$, and $1 \leq j \leq |z| + 1$, we have*

$$(p, |x| + i) \xrightarrow[xz]{|x|+k}\!\!^* (q, |x| + j) \iff (p, |y| + i) \xrightarrow[yz]{|y|+k}\!\!^* (q, |y| + j)$$

Proof. By symmetry, it suffices to show the direction from left to right. We proceed by induction on the length of the path from $(p, |x| + i)$ to $(q, |x| + j)$. There are two cases to consider:

$i = 0$. According to Lemma 28 the path can be split such that:

$$(p, |x|) \xrightarrow[xz]{|x|+1}\!\!^* (p', |x| + 1) \xrightarrow[xz]{|x|+k}\!\!^* (q, |x| + j)$$

Thus, $(p, p') \in (T\,x).2$ by Lemma 27. Applying Lemma 27 again, we obtain $(p, |y|) \xrightarrow[yz]{|y|+1}\!\!^* (p', |y| + 1)$. The claim then follows by induction hypothesis since the path from $(p, |x|)$ to $(p', |x| + 1)$ must make at least one step.

$i > 0$. The path from $(p, |x|+i)$ to $(q, |x|+j)$ is either trivial and the claim follows immediately or there exist p' and i' such that $(p, |x| + i) \xrightarrow[xz]{|x|+k} (p', |x| + i')$. But then $(p, |y| + i) \xrightarrow[yz]{|y|+k} (p', |y| + i')$ and the claim follows by induction hypothesis. $\qquad\square$

Now we have everything we need to show that T is a classifier for $\mathcal{L}(M)$.

Lemma 31. *T refines $\mathcal{L}(M)$.*

Proof. Fix $x, y : \Sigma^*$ and assume $T\,x = T\,y$. By symmetry, it suffices to show $y \in \mathcal{L}(M)$ whenever $x \in \mathcal{L}(M)$. If $x \in \mathcal{L}(M)$, then $(s, 1) \xrightarrow[x]{|x|+2}\!\!^* (p, |x| + 1)$ for some $p \in F$. We show $y \in \mathcal{L}(M)$ by showing $(s, 1) \xrightarrow[y]{|y|+2}\!\!^* (p, |y| + 1)$. By Lemma 29, there exists a state q such that:

$$(s, 1) \xrightarrow[x]{|x|+1}\!\!^* (q, |x| + 1) \xrightarrow[x]{|x|+2}\!\!^* (p, |x| + 1)$$

We can simulate the first part on y using Lemma 27 and the second part using Lemma 30. $\qquad\square$

Lemma 32. *T is right congruent*

Proof. Fix words $x, y : \Sigma^*$ and some symbol $a : \Sigma$ and assume $T\,x = T\,y$. We need to show $T\,xa = T\,ya$. We first show $(T\,xa).2 = (T\,ya).2$. Let $(p, q) \in Q \times Q$. We have to show

$$(p, |xa|) \xrightarrow[xa]{|xa|+1}\!\!^* (q, |xa| + 1) \implies (p, |ya|) \xrightarrow[ya]{|ya|+1}\!\!^* (q, |ya| + 1)$$

Since $|xa| + 1 = |x| + 2$ this follows immediately with Lemma 30. It remains to show $(T\,xa).1 = (T\,ya).1$. By symmetry, it suffices to show:

$$(s,1) \xrightarrow[xa]{|xa|+1}{}^* (q, |xa| + 1) \Longrightarrow (s,1) \xrightarrow[ya]{|ya|+1}{}^* (q, |ya| + 1)$$

By Lemma 29, there exists a state p such that:

$$(s,1) \xrightarrow[xa]{|x|+1}{}^* (p, |x| + 1) \xrightarrow[xa]{|xa|+1}{}^* (q, |xa| + 1)$$

Thus, we have $p \in (T\,x).1$ (and therefore also $p \in (T\,y).1$) and $(p,q) \in (T\,xa).2$. Since we have shown above that $(T\,xa).2 = (T\,ya).2$, the claim follows with Lemma 27.   □

Note that Lemma 30 is used very differently in the proofs of Lemmas 31 and 32. In the first case we are interested in acceptance and set k to $|x| + 2$ so we never actually stop. In the second case we set k to $|xa| + 1$ to stop on the right marker.

Using Theorem 14 and the two lemmas above, we obtain:

Theorem 33. *Let M be a 2NFA with n states. Then there exists a DFA with at most $2^{n^2 + n}$ states accepting $\mathcal{L}(M)$.*

We now show that for deterministic two-way automata, the bound on the size of the constructed DFA can be improved from $2^{n^2 + n}$ to $(n + 1)^{(n+1)}$.

Fact 34. *Let $M = (Q, s, F, \delta, \delta_\triangleright, \delta_\triangleleft)$ be a 2DFA. Then $\xrightarrow[x]{k}$ is functional for all $k : \mathbb{N}$ and $x : \Sigma^*$.*

Corollary 35. *Let M be a 2DFA with n states. Then there exists a DFA with at most $(n + 1)^{(n+1)}$ states accepting $\mathcal{L}(M)$.*

Proof. Let $M = (Q, s, F, \delta, \delta_\triangleright, \delta_\triangleleft)$ be deterministic and let $T : \Sigma^* \to 2^Q \times 2^{Q \times Q}$ be defined as above. Since T is right-congruent (Lemma 32) we can construct the type $T(\Sigma^*)$ (Lemma 13). By Theorem 14, it suffices to show that $T(\Sigma^*)$ has at most $(|Q| + 1)^{(|Q|+1)}$ elements.

Let $(A, R) : T(\Sigma^*)$. Then $T\,x = (A, R)$ for some $x : \Sigma^*$. We show that A has at most one element. Assume $p, q \in A$. By the definition of T, we have

$$(s,1) \xrightarrow[x]{|x|+1}{}^* (p, |x| + 1) \qquad \text{and} \qquad (s,1) \xrightarrow[x]{|x|+1}{}^* (q, |x| + 1)$$

Since $\xrightarrow[x]{|x|+1}$ is functional and both $(p, |x| + 1)$ and $(p, |x| + 1)$ are terminal, we have $p = q$. A similar argument yields that R is a functional relation. Consequently, we can construct an injection

$$i : T(\Sigma^*) \to Q_\perp \times (Q_\perp)^Q$$

Given some $(A, R) : T(\Sigma^*)$, $(i(A, R)).1$ is defined to be the unique element of A or $\perp$ if $A = \emptyset$. The definition of $(i(A, R)).2$ is analogous. The claim then follows since $Q_\perp \times (Q_\perp)^Q$ has exactly $(|Q| + 1)^{(|Q|+1)}$ elements.   □

9   Conclusion

We have shown how results about two-way automata can be formalized in Coq with reasonable effort. The translation from 2NFAs to DFAs makes use of a constructive variant of the Myhill-Nerode theorem that is interesting in its own right. When spelled out in detail, the constructions involved become fairly delicate. The formalization accompanying this paper matches the paper proofs fairly closely and provides additional detail.

Even though both Shepherdson [17] and Vardi [20] consider two-way automata without end-markers, the changes required to handle two-way automata with end-markers are minimal. Perhaps surprisingly, the translation to NFAs for the complement becomes simpler and more 'symmetric' when end-markers are added. The original construction [20] uses $2^Q + 2^Q \times 2^Q$ as the type of states while the construction in Sect. 7 gets along with the type $2^Q \times 2^Q$. States from the type 2^Q are required to check beginning and end of a negative certificate in the absence of end-markers.

Automata are a typical example of a dependently typed mathematical structure. Our representation of finite automata relies on dependent record types and on finite types being first-class objects. Paulson [13] formalizes finite automata in Isabelle/HOL using heriditarily finite (HF) sets to represent states. Like finite types, HF sets have all the closure properties required for the usual constructions on finite automata. Due to the absence of dependent types, the definition of DFAs in [13] is split into a type that overapproximates DFAs and a predicate that checks well-formedness conditions (e.g., that the starting state is a state of the automaton). Thus, the natural typing of DFAs is lost.

We also use dependent types in the representation of two-way automata. The possible configurations of a two-way automaton are represented as a word-indexed collection of finite types. The truncation of natural numbers to bounded natural numbers mentioned in Sect. 6 allows us to recover the separation between stating lemmas (e.g. Lemma 30) and establishing that all indices stay within the correct bounds, thus avoiding many of the problems commonly associated with using dependent types.

Acknowledgments. We thank Jan-Oliver Kaiser, who was involved in our previous work on one-way automata and also in some of the early experiments with two-way automata. We also thank one of the anonymous referees for his helpful comments.

References

1. Berghofer, S., Nipkow, T., Urban, C., Wenzel, M. (eds.): TPHOLs 2009. LNCS, vol. 5674. Springer, Heidelberg (2009)
2. Berghofer, S., Reiter, M.: Formalizing the logic-automaton connection. In: Berghofer et al. [1], pp. 147–163
3. Braibant, T., Pous, D.: Deciding Kleene algebras in Coq. Log. Meth. Comp. Sci. **8**(1:16), 1–42 (2012)

4. Constable, R.L., Jackson, P.B., Naumov, P., Uribe, J.C.: Constructively formalizing automata theory. In: Plotkin, G.D., Stirling, C., Tofte, M. (eds.) Proof, Language, and Interaction, pp. 213–238. The MIT Press, Cambridge (2000)
5. Coquand, T., Siles, V.: A decision procedure for regular expression equivalence in type theory. In: Jouannaud, J.-P., Shao, Z. (eds.) CPP 2011. LNCS, vol. 7086, pp. 119–134. Springer, Heidelberg (2011)
6. Doczkal, C., Kaiser, J.-O., Smolka, G.: A constructive theory of regular languages in Coq. In: Gonthier, G., Norrish, M. (eds.) CPP 2013. LNCS, vol. 8307, pp. 82–97. Springer, Heidelberg (2013)
7. Garillot, F., Gonthier, G., Mahboubi, A., Rideau, L.: Packaging mathematical structures. In: Berghofer et al. [1], pp. 327–342
8. Gonthier, G., Mahboubi, A., Rideau, L., Tassi, E., Théry, L.: A modular formalisation of finite group theory. In: Schneider, K., Brandt, J. (eds.) TPHOLs 2007. LNCS, vol. 4732, pp. 86–101. Springer, Heidelberg (2007)
9. Gonthier, G., Mahboubi, A., Tassi, E.: A Small Scale Reflection Extension for the Coq system. Rapport de recherche RR-6455, INRIA (2008)
10. Hedberg, M.: A coherence theorem for Martin-Löf's type theory. J. Funct. Program. 8(4), 413–436 (1998)
11. Kozen, D.: Automata and Computability. Undergraduate Texts in Computer Science. Springer, New York (1997)
12. Nipkow, T., Traytel, D.: Unified decision procedures for regular expression equivalence. In: Klein, G., Gamboa, R. (eds.) ITP 2014. LNCS, vol. 8558, pp. 450–466. Springer, Heidelberg (2014)
13. Paulson, L.C.: A formalisation of finite automata using hereditarily finite sets. In: Felty, A.P., Middeldorp, A. (eds.) CADE-25. LNCS(LNAI), vol. 9195, pp. 231–245. Springer, Heidelberg (2015)
14. Pighizzini, G.: Two-way finite automata: old and recent results. Fundam. Inform. 126(2–3), 225–246 (2013)
15. Rabin, M.O., Scott, D.: Finite automata and their decision problems. IBM J. Res. Dev. 3(2), 114–125 (1959)
16. Sakoda, W.J., Sipser, M.: Nondeterminism and the size of two way finite automata. In: Lipton, R.J., Burkhard, W.A., Savitch, W.J., Friedman, E.P., Aho, A.V. (eds.) Proceedings of the 10th Annual ACM Symposium on Theory of Computing, pp. 275–286. ACM (1978)
17. Shepherdson, J.: The reduction of two-way automata to one-way automata. IBM J. Res. Develp. 3, 198–200 (1959)
18. The Coq Development Team. http://coq.inria.fr
19. Traytel, D., Nipkow, T.: Verified decision procedures for MSO on words based on derivatives of regular expressions. J. Funct. Program. 25, e18 (30 pages) (2015)
20. Vardi, M.Y.: A note on the reduction of two-way automata to one-way automata. Inf. Process. Lett. 30(5), 261–264 (1989)
21. Vardi, M.Y.: Endmarkers can make a difference. Inf. Process. Lett. 35(3), 145–148 (1990)
22. Wu, C., Zhang, X., Urban, C.: A formalisation of the Myhill-Nerode theorem based on regular expressions. J. Autom. Reasoning 52(4), 451–480 (2014)

Mostly Automated Formal Verification of Loop Dependencies with Applications to Distributed Stencil Algorithms

Thomas Grégoire[1] and Adam Chlipala[2]([⊠])

[1] ÉNS Lyon, Lyon, France
thomas.gregoire@ens-lyon.fr
[2] MIT CSAIL, Cambridge, MA, USA
adamc@csail.mit.edu

Abstract. The class of *stencil* programs involves repeatedly updating elements of arrays according to fixed patterns, referred to as stencils. Stencil problems are ubiquitous in scientific computing and are used as an ingredient to solve more involved problems. Their high regularity allows massive parallelization. Two important challenges in designing such algorithms are cache efficiency and minimizing the number of communication steps between nodes. In this paper, we introduce a mathematical framework for a crucial aspect of formal verification of both sequential and distributed stencil algorithms, and we describe its Coq implementation. We present a domain-specific embedded programming language with support for automating the most tedious steps of proofs that nested loops respect dependencies, applicable to sequential and distributed examples. Finally, we evaluate the robustness of our library by proving the dependency-correctness of some real-world stencil algorithms, including a state-of-the-art cache-oblivious sequential algorithm, as well as two optimized distributed kernels.

1 Introduction

Broadly speaking, in this paper we are interested in verifying, within a proof assistant, the correctness of a class of algorithms in which some matrices are computed, with some matrix cells depending on others. The aim is to check that these quantities are computed *in the right order*. This archetypical style of calculation arises in such situations as computing solutions of partial differential equations using finite-difference methods. The algorithms used in this setting are usually referred to as *stencil codes*, and they are the focus of the framework that we present in this paper.

A *stencil* defines a value for each element of a d-dimensional spatial grid at time t as a function of neighboring elements at times $t-1, t-2, \ldots, t-k$, for some fixed $k, d \in \mathbb{N}^+$. Figure 1 defines a stencil and gives its graphical representation.

Stencil problems naturally occur in scientific-computing and engineering applications. For example, consider the two-dimensional heat equation:

$$\frac{\partial u}{\partial t} - \alpha \left(\frac{\partial^2 u}{\partial x^2} + \frac{\partial^2 u}{\partial y^2} \right) = 0$$

© Springer International Publishing Switzerland 2016
J.C. Blanchette and S. Merz (Eds.): ITP 2016, LNCS 9807, pp. 167–183, 2016.
DOI: 10.1007/978-3-319-43144-4_11

$$\begin{cases} \text{Space} = \{0, 1, \ldots, N\}^2, \\ u_0[x, y] = \alpha_{x,y}, \\ u_{t+1}[x, y] = F(u_t[x, y], u_t[x + 1, y], u_t[x - 1, y], \\ \qquad\qquad\qquad u_t[x, y + 1], u_t[x, y - 1]), \\ N \in \mathbb{N}, \quad \alpha_{x,y} \in \mathbb{R}, \quad F : \mathbb{R}^4 \to \mathbb{R} \end{cases}$$

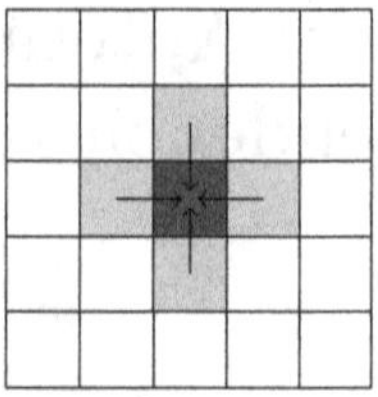

Fig. 1. A two-dimensional Jacobi stencil

We might try to solve it by discretizing both space and time using a finite-difference approximation scheme as follows:

$$\frac{\partial u}{\partial t} \approx \frac{u(t + \Delta t, x, y) - u(t, x, y)}{\Delta t}$$

$$\frac{\partial^2 u}{\partial x^2} \approx \frac{u(t, x + \Delta x, y) - 2u(t, x, y) + u(t, x - \Delta x, y)}{\Delta x^2}$$

Proceeding similarly for $\frac{\partial^2 u}{\partial y^2}$, we obtain the two-dimensional stencil depicted by Fig. 1. More generally, stencil computations are used when solving partial differential equations using finite-difference methods on regular grids [4,11], in iterative methods solving linear systems [11], but also in simulations of cellular automata, such as Conway's game of life [8].

In this paper, we will focus on stencil codes where all the values of the grid are required at the end of the computation–a common situation when partial differential equations are used to simulate the behavior of a real-world system, and the user is interested not only in the final result but also in the dynamics of the underlying process.

Although writing stencil code might seem very simple at first glance–a program with nested loops that respects the dependencies of the problem is enough–there are in fact many different optimizations that have a huge impact on performance, especially when the grid size grows. For stencil code running on a single processor or core, changing the order in which computations are performed can significantly increase cache efficiency, hence dramatically lowering computation times. In the case of multicore implementations, reducing the number of synchronizations between the cores is important since communication is usually the main bottleneck. All these optimizations *reorder* computations. It is therefore crucial to check that these reorderings *do not break the dependencies implied by the underlying stencil.* This is the overall goal of this paper.

We describe a formal framework to define stencils, encode their sequential and distributed implementations, and prove their correctness. We show how using a domain-specific language adapted to stencil code allows a very effective form of *symbolic execution* of programs, supporting verification without annotations like loop invariants that are common in traditional approaches. As for distributed algorithms, we show how the *synchronousness* of stencil kernels impacts verification. In particular, we show how verification of distributed stencil code boils down to verification of *sequential* algorithms. Finally, we showcase the robustness of our theory by applying it to some real-world examples, including a state-of-the-art sequential cache-oblivious algorithm, as well as a communication-efficient distributed kernel.

We have implemented our framework and example verifications within the Coq proof assistant[1]. Because everything is formalized in Coq, we will, in this paper, tend toward relatively informal explanations, to help develop the reader's intuition.

So, to summarize, our contributions are *the first mechanized proof of soundness of a dependency-verification framework for loopy programs over multidimensional arrays*, in addition to *a set of Coq tactics that support use of the framework with reasonable effort* plus a set of case-study verifications showing the framework in action for both sequential and distributed programs.

1.1   Related Work

To the authors' knowledge, this is the first attempt at designing a verification framework for dependencies in stencil code.

Stencils have drawn some attention in the formal-proof community, since finite-difference schemes are among the simplest (yet most powerful) methods available to solve differential equations in low dimension. Therefore, recent work has been more focused on proving stability and convergence of a given discretization scheme [2,3,9], rather than investigating different ways to solve a given stencil. In a different direction, we also mention Ypnos [16], a domain-specific language that enforces indexing safety guarantees in stencil code through type checking, therefore eliminating the need for run-time bounds checking.

As mentioned earlier in this introduction, stencil code can suffer from poor cache performance. This has led to intensive research on cache-oblivious stencil algorithms [6,7]. Writing such optimized stencil code can be very tedious and error-prone, and code might have to be rewritten from scratch when switching architecture. Recently, different techniques have been proposed to automate difficult parts of stencil implementation, including sketching [18], program synthesis [21], or compiling a domain-specific stencil language [19]. In our new work, we show how to verify *a posteriori* that such generated code respects dependencies, for infinite input domains.

One of the most natural approaches to generate efficient parallel stencil code is to use the *polyhedral model* (see [5,10]), to represent and manipulate loop nests. The goal of our framework is different from that of a vectorizing compiler, since we are not trying to generate code, but rather check that it does not violate any dependencies. Nevertheless, our approach is reminiscent of the line of work on the polyhedral model and parallelizing compilers (see, *e.g.*, [14,15,17]), in the sense that we produce an algebraic representation of the current state of the program and use it to check that dependencies are satisfied. Notice that, since we are working within a proof assistant, we can represent arbitrary mathematical

---

[1] https://github.com/mit-plv/stencils.

sets, which implies that we are not limited to linear integer programs; and indeed our evaluation includes codes that employ nonlinear arithmetic.

Closely related to this paper, there has been recent work on the formal verification of GPU kernels. In this direction, we refer the reader to PUG [12], GKLEE [13], and GPUverify [1]. Our approach is somewhat orthogonal to this line of work: we study a smaller class of programs, which is also big enough to have many applications, supporting first-principles proofs at low human cost.

2   Verifying Stencil Code

2.1   Stencils and Their Representation

A *stencil* is defined on a set $\mathcal{G}$, which represents a spatial *grid*. Its elements are called *cells*, and most stencils of interest are based on $\mathcal{G} = \mathbb{Z}^d$ for some small integer d.

We assign a numerical value $u_t[c]$ to every cell $c \in \mathcal{G}$ at every time $t \in \mathbb{N}$ (or any time $0 \leq t \leq t_{\max}$, for some $t_{\max} \in \mathbb{N}$). A stencil is then defined by some initial conditions $u_0[c] = \alpha_c$, with $c \in \mathcal{G}$, $\alpha_c \in \mathbb{R}$, and a pattern

$$u_t[c] = F(u_{t-1}[d_1(c)], \ldots, u_{t-1}[d_k(c)]),$$

for some function F. The cells $d_1(c), \ldots, d_k(c)$ are called c's *neighbors*. See Fig. 1 for an example of such a formal stencil definition.

Let us turn to the representation of stencils within the proof assistant.

Definition 1. *A stencil is defined by:*

- *A type* cell *representing grid elements;*
- *A term* space : set cell, *indicating which finite subset of the grid we will be computing on;*
- *A term* target : set cell, *indicating which grid elements have to be computed by the end of the execution of the algorithm;*
- *A term* dep : cell $\rightarrow$ list cell *representing the dependencies of each cell, that is, its neighbors.*

Figure 2 shows how we formalize the Jacobi 2D stencil we introduced in Fig. 1. We will conclude this section with a few comments on this definition:

Remark 1. Here, set A denotes "mathematical" sets of elements of type A, implemented in Coq as $A \rightarrow$ Prop.

Remark 2. Notice that this is an *abstract* notion of stencil. In particular, we do not specify the function applied at each step–it is seen as a black box–nor do we give the initial conditions. Moreover, the formulation is much more general than the schematic equation given above and encompasses *e.g.* Gauss-Seidel iterations or box-blur filtering.

```
Parameters T I J : Z.

Module Jacobi2D <: (PROBLEM Z3).
  Local Open Scope aexpr.

  Definition space := [[0, T]]×[[0, I]]×[[0, J]].
  Definition target := [[0, T]]×[[0, I]]×[[0, J]].
  Definition dep c :=
    match c with
    | (t,i,j) ⇒ [(t−1,i,j); (t−1,i−1,j); (t−1,i+1,j); (t−1,i,j−1); (t−1,i,j+1)]
    end.
End Jacobi2D.
```

Fig. 2. Coq representation of the two-dimensional Jacobi stencil

Remark 3. The `space` parameter is used to encode *boundary conditions*. For example, in the case of the Jacobi 2D stencil, we might want to ensure that $u_t[(i,j)] = 0$ as soon as $i < 0$, $i > I$, $j < 0$, or $j > J$, for some parameters I and J. In this case, we would pick[2] `space` $= [[0,I]] \times [[0,J]]$.

2.2   Programs: Syntax, Semantics, and Correctness

Now that we have a way to describe stencils, we turn to the representation and correctness of programs. Let us consider the following trivial program solving the Jacobi 2D stencil:

for t=0 **to** T **do**
  for i=0 **to** I **do**
    for j=0 **to** J **do**
      Compute $u_t[i,j]$

Verifying the correctness of this program amounts to proving that

1. It does not violate any dependency. That is, $u_t[i,j]$ is never computed before $u_{t-1}[i,j]$, $u_{t-1}[i+1,j]$, $u_{t-1}[i-1,j]$, $u_{t-1}[i,j+1]$, or $u_{t-1}[i,j-1]$;
2. It is complete, in the sense that it computes all the values of cells in `target`.

Therefore, we see programs as *agents*, having some *knowledge*. The state of a program is a *set of cells*, those with values known by the program.

For now, we will only discuss requirement 1, and we will see how to prove requirement 2 in the next section. Our starting point is a basic imperative language, which we extend with a **flag** c command, which adds cell c to the current state, as well as **assert** c, which checks that c belongs to the current state. If not, the program halts abnormally.

To verify requirement 1 for the trivial program above, we would therefore have to prove the normal termination of the following program:

---

[2] In this paper, for all $a, b \in \mathbb{Z}$, we write $[[a, b]]$ for $\{n \in \mathbb{Z} : a \leq n \leq b\}$.

$$p ::= \mathbf{nop} \mid p;p \mid \mathbf{flag}\ c \mid \mathbf{assert}\ c \mid \mathbf{if}\ b\ \mathbf{then}\ p\ \mathbf{else}\ p \mid \mathbf{for}\ v = e\ \mathbf{to}\ e\ \mathbf{do}\ p$$
$$e ::= k \mid v \mid e + e \mid e - e \mid e \times e \mid e/e \mid e\ \mathrm{mod}\ e$$
$$b ::= \epsilon \mid \mathbf{not}\ b \mid b\ \mathbf{or}\ b \mid b\ \mathbf{and}\ b \mid e = e \mid e \neq e \mid e \leq e \mid e \geq e \mid e < e \mid e > e$$
$$k \in \mathbb{Z},\ \epsilon \in \{\top, \bot\}$$

Fig. 3. Syntax of arithmetic and Boolean expressions and programs

```
for  t=0 to T do
   for  i=0 to I do
      for  j=0 to J do
         assert (t − 1, i, j);   assert (t − 1, i + 1, j);
         assert (t − 1, i − 1, j);   assert (t − 1, i, j + 1);
         assert (t − 1, i, j − 1);   flag (t, i, j)
```

The precise syntax of expressions and programs is given in Fig. 3. Notice that there is no assignment command $x := e$, but that a statement like $x := e; p$ can be simulated by the program **for** $x = e$ **to** e **do** p. Our programs are *effect-free*, in the sense that variables are bound to values functionally within loops. Moreover, our framework is parameterized by a type and evaluation function for cell expressions.

The operational semantics of our programming language is given by a judgment $\rho \vdash (C_1, p) \Downarrow C_2$, where ρ assigns an integer value to every variable, C_1 and C_2 are sets of cells, and p is a program. The intended meaning is that, in a state where the program knows the values of the cells in C_1 (and these values only), and where variables are set according to ρ, the execution of p terminates (without any assertion failing) and the final knowledge of the program is described by C_2.

The semantics is given in Fig. 4. There, $[\![e]\!]_\rho$ denotes the evaluation of expression e in environment ρ, when e is an arithmetic, Boolean, or cell expression. $\rho[x \leftarrow i]$ denotes the environment obtained by setting x to $i \in \mathbb{Z}$ in ρ.

$$\text{Nop:} \frac{}{\rho \vdash (C, \mathbf{nop}) \Downarrow C}$$

$$\text{Seq:} \frac{\rho \vdash (C_1, p_1) \Downarrow C_2 \quad \rho \vdash (C_2, p_2) \Downarrow C_3}{\rho \vdash (C_1, p_1; p_2) \Downarrow C_3}$$

$$\text{If (1):} \frac{[\![b]\!]_\rho = \top \quad \rho \vdash (C_1, p_1) \Downarrow C_2}{\rho \vdash (C_1, \mathbf{if}\ b\ \mathbf{then}\ p_1\ \mathbf{else}\ p_2) \Downarrow C_2}$$

$$\text{If (2):} \frac{[\![b]\!]_\rho = \bot \quad \rho \vdash (C_1, p_2) \Downarrow C_2}{\rho \vdash (C_1, \mathbf{if}\ b\ \mathbf{then}\ p_1\ \mathbf{else}\ p_2) \Downarrow C_2}$$

$$\text{Assert:} \frac{[\![c]\!]_\rho \in C \vee [\![c]\!]_\rho \notin \mathbf{space}}{\rho \vdash (C, \mathbf{assert}\ c) \Downarrow C}$$

$$\text{Flag:} \frac{}{\rho \vdash (C, \mathbf{flag}\ c) \Downarrow C \cup \{[\![c]\!]_\rho\}}$$

$$\text{For:} \frac{\forall(S_k).\forall(U_k). \quad A = [\![a]\!]_\rho \quad B = [\![b]\!]_\rho \quad \forall i.\ U_i = C \cup \bigcup_{k \in [\![A,i]\!]} S_k \quad \forall A \leq i \leq B.\ \rho[x \leftarrow i] \vdash (U_{i-1}, p) \Downarrow U_i}{\rho \vdash (C, \mathbf{for}\ v = a\ \mathbf{to}\ b\ \mathbf{do}\ p) \Downarrow U_B}$$

Fig. 4. Operational semantics of programs

Remark 4. As mentioned earlier, assertions check that a cell's value is known *only if this cell belongs to* **space**. Therefore, in the Jacobi 2D example above, **assert** $(−1,0,0)$ would not fail.

Remark 5. In the loop rule, S_i represents the set of cells computed by the program at iteration i, while U_i represents the knowledge of the program since its execution started, until the beginning of iteration i.

2.3 Verification Conditions and Symbolic Execution

Proving the correctness of a program (at least for requirement 1) amounts to proving a statement of the form $\rho \vdash (C_1, p) \Downarrow C_2$. Can we automate this process?

The key insight here is that our domain-specific language is very simple. It is not Turing-complete. We take advantage of this fact to simplify the verification scheme.

We can in fact perform some *symbolic execution* of programs. For an environment ρ and a program p, we define a set $\mathrm{Shape}_\rho(p)$ that intuitively corresponds to the knowledge that the program will acquire after its execution *if it does not fail*. The rules defining $\mathrm{Shape}_\rho(p)$ are purely *syntactic* and given in Fig. 5.

$$\mathrm{Shape}_\rho(\mathbf{nop}) := \emptyset, \quad \mathrm{Shape}_\rho(\mathbf{flag}\ c) := \{[\![c]\!]_\rho\}$$

$$\mathrm{Shape}_\rho(\mathbf{if}\ b\ \mathbf{then}\ p_1\ \mathbf{else}\ p_2) := \begin{cases} \mathrm{Shape}_\rho(p_1)\ \text{if}\ [\![b]\!]_\rho = \top \\ \mathrm{Shape}_\rho(p_2)\ \text{otherwise} \end{cases}$$

$$\mathrm{Shape}_\rho(p_1; p_2) := \mathrm{Shape}_\rho(p_1) \cup \mathrm{Shape}_\rho(p_2)$$

$$\mathrm{Shape}_\rho(\mathbf{assert}\ c) := \emptyset$$

$$\mathrm{Shape}_\rho(\mathbf{for}\ x = a\ \mathbf{to}\ b\ \mathbf{do}\ p) := \bigcup_{k \in [\![A,B]\!]} \mathrm{Shape}_{\rho[x \leftarrow k]}(p), \quad A = [\![a]\!]_\rho, B = [\![b]\!]_\rho$$

Fig. 5. Symbolic execution of programs

Note that $\mathrm{Shape}_\rho(p)$ will also allow us to prove requirement 2 from the correctness statement: the latter can be reformulated as $\mathtt{target} \subseteq \mathrm{Shape}_\rho(p)$. To be more precise, let us now formalize the notion of correctness for sequential stencil algorithms.

Definition 2. *Consider a stencil* ($\mathtt{space}, \mathtt{target}, \mathtt{dep}$). *For every cell* c, *let* **fire** $c \equiv$ **assert** d_1; ...; **assert** d_n; **flag** c, *where* $\mathbf{dep}(c) = \{d_1, \ldots, d_n\}$. *Here and after, it is assumed that the user uses exclusively the* **fire** c *command, and not* **flag** c.

Let ρ_0 *denote an empty environment. A program* p *is correct with respect to the aforementioned stencil if* $\rho_0 \vdash (\emptyset, p) \Downarrow \mathrm{Shape}_{\rho_0}(p)$ *and* $\mathrm{Shape}_{\rho_0}(p) \subseteq \mathtt{target}$.

Now that we have a means to symbolically evaluate programs, we can write our verification-condition generator. The symbolic-execution step is very important, since it allows us to *synthesize loop invariants*, without the need for any human intervention. The verification-condition generator is defined in Fig. 6.

Theorem 1 proves the correctness of the verification-condition generator.

Theorem 1. *Let* p *be a program,* ρ *an environment, and* C *a set of cells. If* $VC_{\rho,C}(p)$ *holds, then* $\rho \vdash (C, p) \Downarrow (C \cup \mathrm{Shape}_\rho(p))$.

$$\mathrm{VC}_{\rho,C}(\mathbf{nop}) := \top, \quad \mathrm{VC}_{\rho,C}(\mathbf{flag}\ c) := \top$$

$$\mathrm{VC}_{\rho,C}(\mathbf{if}\ b\ \mathbf{then}\ p_1\ \mathbf{else}\ p_2) := \begin{cases} \mathrm{VC}_{\rho,C}(p_1)\ \text{if}\ [\![b]\!]_\rho = \top \\ \mathrm{VC}_{\rho,C}(p_2)\ \text{otherwise} \end{cases}$$

$$\mathrm{VC}_{\rho,C}(p_1;p_2) := \mathrm{VC}_{\rho,C}(p_1) \wedge \mathrm{VC}_{\rho,C\cup\mathrm{Shape}_\rho(p_1)}(p_2)$$

$$\mathrm{VC}_{\rho,C}(\mathbf{assert}\ c) := [\![c]\!]_\rho \in C \vee [\![c]\!]_\rho \nsubseteq \mathbf{space}$$

$$\mathrm{VC}_{\rho,C}(\mathbf{for}\ x = a\ \mathbf{to}\ b\ \mathbf{do}\ p) := \forall A \leq i \leq B.\ \mathrm{VC}_{\rho[x \leftarrow i],D}(p)$$

$$A = [\![a]\!]_\rho,\ B = [\![b]\!]_\rho,\ D = C \cup \mathrm{Shape}_\rho(\mathbf{for}\ x = a\ \mathbf{to}\ i - 1\ \mathbf{do}\ p)$$

Fig. 6. Verification-condition generator

3   Verifying Distributed Stencil Algorithms

We now turn to the problem of verifying *distributed* stencil algorithms. We will start with an informal description of our programming model.

3.1   Reduction to the Sequential Case

Stencil problems are inherently regular. Therefore, they are susceptible to substantial parallelization. More importantly, distributed stencil code is in general *synchronous*. The program alternates between *computation steps*, where each thread computes some cell values, and *communication steps*, during which threads send some of these values to other threads. Figure 7 gives a graphical representation of such an algorithm's execution, with three threads. Each of the three threads is assigned a strip in the plane, depicted by the dark dashed line. There is:

– A computation step, where each thread computes a "triangle";
– A communication step, during which each thread sends the edges of its "triangle" to its left and right neighbors;
– Another computation step, where each thread computes two "trapezoids."

We will use the phrase *time step* to refer to the combination of one computation step and one communication step. Our model, inspired by that of Xu et. al [21], also shares similarities with the Bulk-Synchronous Parallel (BSP) model [20].

Let us write a pseudo-code implementation of this simple algorithm, to give a flavor of what our formalization will eventually look like.

Computation step	Communication step

`if T=0 then`
`  for t=0 to 3 do`
`    for ` $i = t$ ` to ` $7 - t$ ` do`
`      fire ` $(8 \times \mathrm{id} + i, t)$
`else  (* T=1 *)`
`  for t=1 to 3 do`
`    for ` $i = -t$ ` to ` $t - 1$ ` do`
`      fire ` $(8 \times \mathrm{id} + i, t)$
`    for ` $i = -t$ ` to ` $t - 1$ ` do`
`      fire ` $(8 \times \mathrm{id} + 8 + i, t)$
	`if T=0 then`
	`  if ` $\mathrm{to} = \mathrm{id} - 1$ ` then`
	`    for t=0 to 3 do`
	`      fire ` $(8 \times \mathrm{id} + t, t)$
	`  else if ` $\mathrm{to} = \mathrm{id} + 1$ ` then`
	`    for t=0 to 3 do`
	`      fire ` $(8 \times \mathrm{id} + 4 + t, 3 - t)$

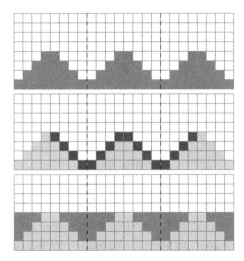

Fig. 7. Example of two computation steps (green) and one communication step (red) (Color figure online)

On the left is the "computation kernel." Notice that every thread can access its unique identifier through the variable "id" and the current time step through variable "T". We also implicitly assume that "fire" commands involving a cell that is out of the rectangle depicted in Fig. 7 have no effect. This is why we included **space** in our definition of stencils. The communication step is given on the right. This time, besides the variables "id" and "T", every thread has access to the variable "to", which contains the unique identifier of the thread it is currently sending data to. Now "fire" corresponds to sending a cell's local value to a neighbor thread.

For the reader familiar with verification of more complicated distributed code, we would like to emphasize that there is *no data race* here: threads have their own separate memories, and communication between threads only happens via message passing followed by barriers waiting for all threads to receive all messages sent to them. As a result, it does not make sense to talk about different threads racing on reads or writes to grid cells.

3.2 Distributed Kernels: Syntax, Semantics, and Correctness

The syntax of distributed code is given in Fig. 8. We use the *same* syntax for both computation steps (in which case **fire** c means "compute the value of cell c") and for communication steps (in which case **fire** c means "send the value of cell c to the thread we are currently communicating with"). More formally, we give two different translations to the programs defined in Fig. 8. The first one, Comp.denote, compiles **fire** c into the "fire" command of Definition 2, which *checks that all dependencies of c are satisfied*. The second semantics, Comm.denote, simply compiles it into a **flag** c command, *without checking any* *dependencies.* This trick allows us to factor computation and communication steps into the same framework: a distributed kernel is a pair of sequential codes, one for each step. The translation is formally given in Fig. 9.

$$p ::= \mathbf{nop} \mid p; p \mid \mathbf{fire}\ c \mid \mathbf{if}\ b\ \mathbf{then}\ p\ \mathbf{else}\ p \mid \mathbf{for}\ v = e\ \mathbf{to}\ e\ \mathbf{do}\ p$$

Fig. 8. Syntax of distributed code

$$F \in \{\mathbf{fire}, \mathbf{flag}\}$$

$$\text{denote}_F(\mathbf{nop}) := \mathbf{nop}, \text{denote}_F(\mathbf{fire}\ c) := F(c)$$

$$\text{denote}_F(\mathbf{if}\ b\ \mathbf{then}\ p_1\ \mathbf{else}\ p_2) := \mathbf{if}\ b\ \mathbf{then}\ \text{denote}_F(p_1)\ \mathbf{else}\ \text{denote}_F(p_2)$$

$$\text{denote}_F(p_1; p_2) := \text{denote}_F(p_1); \text{denote}_F(p_2)$$

$$\text{denote}_F(\mathbf{for}\ x = a\ \mathbf{to}\ b\ \mathbf{do}\ p) := \mathbf{for}\ x = a\ \mathbf{to}\ b\ \mathbf{do}\quad \text{denote}_F(p)$$

$$\text{Comp.denote}(p) := \text{denote}_{\mathbf{fire}}(p), \quad \text{Comm.denote}(p) := \text{denote}_{\mathbf{flag}}(p)$$

Fig. 9. Translation of distributed code. The arguments are distributed programs, but the values returned by denote are sequential programs.

Now that we are equipped with a language describing distributed stencil kernels and the associated semantics, we can formalize the correctness of such algorithms. A few intuitive comments are given right below the definitions, and the reader might find them useful in order to interpret our formalization.

Definition 3. *Suppose we fix a number* $\text{id}_{\max}$, *such that threads are indexed over* $[\![0, \text{id}_{\max}]\!]$, *and let us fix a maximum execution time* $T_{\max}$.

A trace is a triple (beforeComp, afterComp, sends), *where* beforeComp *and* afterComp *have type* time × thread → set cell *and* sends *has type* time × thread × thread → set cell.

A distributed kernel is a pair (comp, comm) *of distributed codes. It is* correct *if there exists a trace satisfying the following properties:*

– *Initially, nothing is known:*

$$\forall 0 \le i \le \text{id}_{\max}.\ \text{beforeComp}(0, i) = \emptyset;$$

– *We go from* beforeComp(T, i) *to* afterComp(T, i) *through a computation step:* $\forall 0 \le i \le \text{id}_{\max}.\ \forall 0 \le T \le T_{\max}.$

$$\rho_0[\text{``id''} \leftarrow i, \text{``T''} \leftarrow T] \vdash (\text{beforeComp}(T, i), Comp.denote(\text{comp})) \Downarrow \text{afterComp}(T, i);$$

– sends(T, i, j) *represents what is sent by thread* i *to thread* j *at step* T:

$$\forall 0 \le i, j \le \text{id}_{\max}.\ \forall 0 \le T \le T_{\max}.$$

$$\rho_0[\text{``id''} \leftarrow i, \text{``to''} \leftarrow j, \text{``T''} \leftarrow T] \vdash (\emptyset, Comm.denote(\text{comm})) \Downarrow \text{sends}(T, i, j);$$

– *A thread cannot send a value it does not know:*

$$\forall 0 \leq i, j \leq \mathrm{id_{max}}. \ \forall 0 \leq T \leq T_{\max}. \ \mathtt{sends}(T, i, j) \subseteq \mathtt{afterComp}(T, i);$$

– *"Conservation of knowledge": what a thread knows at time $T + 1$ comes from what it knew at time T and what other threads sent to it:*

$$\forall 0 \leq i \leq \mathrm{id_{max}}. \ \forall 0 \leq T \leq T_{\max}.$$

$$\mathtt{beforeComp}(T + 1, i) \subseteq \mathtt{afterComp}(T, i) \cup \bigcup_{j \in [\![0, \mathrm{id_{max}}]\!]} \mathtt{sends}(T, j, i);$$

– *Completeness: when we reach time step $T_{\max} + 1$, all the required values have been computed:*

$$\mathtt{target} \subseteq \bigcup_{i \in [\![0, \mathrm{id_{max}}]\!]} \mathtt{beforeComp}(T_{\max} + 1, i).$$

Remark 6. $\mathtt{beforeComp}(T, i)$ represents the set of cells whose values are known by thread i at the beginning of the Tth time step. Similarly, $\mathtt{afterComp}(T, i)$ represents the knowledge of thread i right after the computation step of time step T. Finally, $\mathtt{sends}(T, i, j)$ represents the set of cells whose values are sent from thread i to thread j at the end of time step T.

3.3 Trace Generation

The definition of correctness involves proving a lot of different properties. Nevertheless, we will now show how the tools developed in the previous section, the symbolic-execution engine and the verification-condition generator, can be used to support mostly automated verification. In particular, we will *synthesize the trace that the program would follow if it does not fail*. The trace generator is given in Fig. 10.

$$[\mathtt{sends}](k, \mathrm{id_{max}}, T, i, j) := \mathrm{Shape}_{\rho_0[\text{"id"} \leftarrow i, \text{"to"} \leftarrow j, \text{"T"} \leftarrow T]}(\mathrm{Comm.denote}(k))$$

$$[\mathtt{computes}](k, \mathrm{id_{max}}, T, i) := \mathrm{Shape}_{\rho_0[\text{"id"} \leftarrow i, \text{"T"} \leftarrow T]}(\mathrm{Comp.denote}(k))$$

$$[\mathtt{beforeComp}](k, \mathrm{id_{max}}, T, i) := \bigcup_{t \in [\![0, T-1]\!]} [\mathtt{computes}](k, \mathrm{id_{max}}, t, i) \cup \bigcup_{\substack{t \in [\![0, T-1]\!] \\ j \in [\![0, \mathrm{id_{max}}]\!]}} [\mathtt{sends}](k, \mathrm{id_{max}}, t, j, i)$$

$$[\mathtt{afterComp}](k, \mathrm{id_{max}}, T, i) := \bigcup_{t \in [\![0, T]\!]} [\mathtt{computes}](k, \mathrm{id_{max}}, t, i) \cup \bigcup_{\substack{t \in [\![0, T-1]\!] \\ j \in [\![0, \mathrm{id_{max}}]\!]}} [\mathtt{sends}](k, \mathrm{id_{max}}, t, j, i)$$

Fig. 10. Trace generator

Of course, the trace generator would be useless without a proof of its correctness. Theorem 2 is the key result of this paper: it shows that, thanks to the trace generator, verification of distributed kernels *boils down to verifying two sequential programs*, proving a law of "conservation of knowledge" and a set inclusion that encodes completeness.

Theorem 2. *Let $k = (\mathtt{comp}, \mathtt{comm})$ be a kernel. If the following conditions hold:*

- $\forall 0 \leq i \leq \mathrm{id}_{\max}.\ \forall 0 \leq T \leq T_{\max}.\ VC_{\rho_{i,T},D}(Comp.denote(\mathtt{comp}))$, *where* $\rho_{i,T} = \rho_0[\text{``id''} \leftarrow i, \text{``T''} \leftarrow T]$ *and* $D = [\mathtt{beforeComp}](k, \mathrm{id}_{\max}, T, i)$;
- $\forall 0 \leq i,j \leq \mathrm{id}_{\max}.\ \forall 0 \leq T \leq T_{\max}.\ VC_{\rho_{i,j,T},\emptyset}(Comm.denote(\mathtt{comm}))$, *where* $\rho_{i,j,T} = \rho_0[\text{``id''} \leftarrow i, \text{``to''} \leftarrow j, \text{``T''} \leftarrow T]$;
- $\forall 0 \leq id,to \leq \mathrm{id}_{\max}.\ \forall 0 \leq T \leq T_{\max}.\ [\mathtt{sends}](k, \mathrm{id}_{\max}, T, id, to) \subseteq [\mathtt{afterComp}](k, \mathrm{id}_{\max}, T, id)$;
- $\mathtt{target} \subseteq \bigcup_{i \in [0,\mathrm{id}_{\max}]} [\mathtt{beforeComp}](k, \mathrm{id}_{\max}, T_{\max} + 1, i)$.

Then, k is correct, with trace

$$([\mathtt{beforeComp}](k, \mathrm{id}_{\max}), [\mathtt{afterComp}](k, \mathrm{id}_{\max}), [\mathtt{sends}](k, \mathrm{id}_{\max})).$$

Notice that we get very close here to the way a human being would write the proof of correctness: we have to prove that the dependencies are satisfied at any point, and the current state of the program is synthesized for us. Then, we need to prove that the final set of values contains all those that had to be computed.

4   Implementation and Experimental Results

The framework described in this paper has been implemented in Coq. In this section, we show how our library can be used to prove a very simple stencil algorithm for the two-dimensional Jacobi stencil introduced in Fig. 1, and whose Coq definition is given in Fig. 2.

4.1   A Simple Example

Let us come back to our straightforward sequential program:

```
Definition naive_st :=
  (For"t" From 0 To T Do
    For "i" From 0 To I Do
      For "j" From 0 To J Do
        Fire ("t":aexpr, "i":aexpr, "j":aexpr))%prog.
```

Let us start by stating the correctness of this algorithm:

```
Fact naive_st_correct : correct naive_st.
Proof.
  split.
```

We obtain two subgoals. The first corresponds to the verification conditions and can be simplified by using the symbolic-execution engine. We then clean up the goal.

```
* decide i=0; [bruteforce | bruteforce' [i-1; i0; i1]].
```

The case $i = 0$ is special and deserves special treatment. Both cases are handled by our automation. The bruteforce tactic discharges the first subgoal, while its sister, bruteforce', which takes as argument a list of candidates for existential-variable instantiation, discharges the second one.

The four other subgoals are handled similarly.

```
* decide i=0; [bruteforce | ].
  decide i0=0; [bruteforce | bruteforce' [i-1; i0-1; i1]].
* decide i=0; [bruteforce | ].
  decide i0=I; [bruteforce | bruteforce' [i-1; i0+1; i1]].
* decide i=0; [bruteforce | ].
  decide i1=0; [bruteforce | bruteforce' [i-1; i0; i1-1]].
* decide i=0; [bruteforce | ].
  decide i1=J; [bruteforce | bruteforce' [i-1; i0; i1+1]].
```

Although syntactically different, the computer-checked proof is very similar to the one a human being would write: case analysis to tackle boundaries, and the remaining proofs are "easy."

The second part of the proof of correctness is completeness:

$$\text{Shape}_{\rho_0}(\texttt{naive\_st}) \subseteq \texttt{target},$$

which is easily discharged by our automation, this time with the forward tactic. Contrary to bruteforce, the latter tries to "make progress" on the goal, without failing if it cannot discharge it completely.

```
- unfold target; simpl; simplify sets with ceval.
  forward. subst; forward. simpl; forward.
Qed.
```

4.2   Automation: Sets and Nonlinear Arithmetic

In this section, we describe the different tactics that we designed to reduce the cost of verifying stencil code.

Sets are represented as predicates: given a universe $\mathcal{U}$, a set has type $\mathcal{U} \longrightarrow$ Prop. The empty set is $\emptyset : u \mapsto \bot$, while for example the union of two sets A and B is defined as $A \cup B : u \mapsto A(u) \vee B(u)$. These definitions are here to give an experience to the user as close as possible to a handwritten proof, and they are automatically unfolded and simplified though first-order reasoning when using the tactic library. This process is implemented by two tactics, simplify_hyps and simplify_goal, which respectively clean up the current context and goal.

Programs are Coq terms. Therefore, the symbolic execution and trace synthesis are purely *syntactic*. We provide a tactic symbolic execution that unfolds all the required definitions.

This choice brings one inconvenience: their output has to be cleaned up. Some variables may be inferred automatically thanks to symbolic execution, which leads to expressions of the form **if** $0=1$ **then** A **else** B, where A and B are sets. Or, some of the verification conditions may look like $c \in \emptyset \cup \emptyset \cup \emptyset \cup \{c\}$, where c is a given cell. While the first one is pretty harmless, the second one can reduce the efficiency of automation: to prove that an element belongs to the union of two sets, we have to try and prove that it belongs to the first one, and if that fails, that it belongs to the second one. Therefore, we have implemented a *rewriting system* that tries to simplify the goal heuristically. For example, we use the following simplification rules:

$$A \cup \emptyset = \emptyset \cup A = A \qquad\qquad \bigcup_{c \in [\![a,b]\!]} A \times \{c\} = A \times [\![a,b]\!]$$

Moreover, we proved that set-theoretic operations are "morphisms," which in Coq's jargon means that we can apply the simplification rules to *subterms*. The rewriting system is implemented as a tactic called `simplify sets`.

The goals we obtain are set-theoretic: we usually have to prove that an element belongs to a set (*e.g.*, to show that a cell's value has already been computed) or that a set is a subset of another one (*e.g.*, what we need at this step was already known from the previous step). Most of this is first-order reasoning and is dealt with by the `forward` tactic, which repeatedly applies `simplify_goal`, `simplify_hyps`, and some first-order reasoning to the goal and context until no progress is made.

The next obstacles are goals of the following forms: $x \in A \cup B$ and $x \in \bigcup_{t \in [\![a,b]\!]} A_t$. We have already mentioned how the first one could be handled, contingent on the number of unions being "not too large." The second one can be tackled by taking as input a list of candidate variables, which we use to instantiate unknown parameters like t in the above expression. This more aggressive automation is implemented as the `bruteforce'` tactic, which takes a list of variables as input. `bruteforce` is a shortcut for `bruteforce' ∅`.

The challenge for full automation in stencil-code verification is that set-theoretic reasoning has to be followed up by a final arithmetic step. Indeed, a goal like $t \in [\![a, b]\!]$ is equivalent to $a \leq t \leq b$. But most of the time, stencil code acts on *blocks*, or subregions within grids, which are parameterized by some integers. For example, a typical goal might be $t \in [\![N \cdot a, (N + 1) \cdot a - 1]\!]$. This leads to very nonlinear systems of inequalities. In our experience, most of these goals can be discharged using Coq's `nia` tactic, an (incomplete) proof procedure for integer nonlinear arithmetic.

Unfortunately, in our experience, `nia` is somewhat slow to fail, when given an unprovable goal. This is the main obstruction to a fully automated framework. We built a tactic to accumulate a list of all variables available in the current context and use it to enumerate expression trees that could be used as instantiation candidates in goals involving parameterized unions. However, in practice for interactive proving, we found it unusable with `bruteforce` due to the combinatorial explosion, combined with `nia`'s slowness to fail. It may still be cost-effective in overnight proof-search runs, for a program that is not likely to need much further debugging. In that case, we achieve full automation for a variety of stencil algorithms.

4.3   More Examples

We have implemented a few stencil algorithms and proved their correctness. They come from different areas, including simulation of a differential equation, computational finance, and computational biology. Table 1 shows the number of lines of code needed to prove their correctness. The framework scales well and allows to prove optimized and optimal algorithms of various kinds.

Table 1. Stencils implemented using our framework

	Type	Lines of proof
Heat Equation, 2D	Naive	30
American Put Stock Options	Naive	25
American Put Stock Options	Optimized	25
Distributed American Put Stock Options	Naive	65
Distributed American Put Stock Options	Optimized	150
Pairwise Sequence Alignment	Dynamic programming	20
Distributed Three-Point Stencil	Naive	60
Distributed Three-Point Stencil	Optimized	160
Universal Three-Point Stencil Algorithm	Optimal	300

Examples come in four different groups.

- The *two-dimensional Jacobi kernel* was introduced at the beginning of this paper (see Fig. 1). We verified a naive sequential algorithm that is often used as a textbook example for finite-difference methods, applied to the Heat Equation.
- We verified a cache-oblivious sequential algorithm as well as a communication-efficient distributed kernel for *three-point stencils*.
- We also verified a cache-oblivious sequential algorithm and a communication-efficient distributed kernel for *American put stock options pricing*. This example is interesting since dependencies go backward in time: the price of an option depends on the price of the underlying asset *in the future*.
- The *Pairwise Sequence Alignment* problem is different from the other examples. It shows that our framework can be used to prove the correctness of algorithms based on *dynamic programming*.

5   Conclusions and Future Work

In this paper, we have shown how dependencies for both sequential and distributed stencil algorithms could be formally verified, and how to design automation to drastically reduce the cost of proving the correctness of such programs.

By focusing on a restricted class of problems and working with a domain-specific language adapted to this class, we were able to *symbolically execute* algorithms, which allowed us to *synthesize* program states, therefore avoiding the need to manually write any kind of loop invariants. A natural and interesting extension of this work could be to add symbolic tracking of cache-relevant behavior.

We also showed how to verify *distributed* stencil algorithms. Here, the key result is that verifying *synchronous* algorithms, when their program states can be *synthesized*, actually boils down to the verification of *several sequential programs*. An interesting extension to this work would be to design an extraction mechanism able to translate our DSL into MPI code or conversely, to get a program in our DSL from MPI code.

Acknowledgments. This work was supported in part by the U.S. Department of Energy, Office of Science, Advanced Scientific Computing Research Program, under Award Number DE-SC0008923; and by National Science Foundation grant CCF-1253229. We also thank Shoaib Kamil for feedback on this paper.

References

1. Betts, A., Chong, N., Donaldson, A., Qadeer, S., Thomson, P.: GPUVerify: a verifier for GPU kernels. In: Proceedings of the OOPSLA, pp. 113–132. ACM (2012)
2. Boldo, S., Clément, F., Filliâtre, J.-C., Mayero, M., Melquiond, G., Weis, P.: Formal proof of a wave equation resolution scheme: the method error. In: Kaufmann, M., Paulson, L.C. (eds.) ITP 2010. LNCS, vol. 6172, pp. 147–162. Springer, Heidelberg (2010)
3. Boldo, S., Clément, F., Filliâtre, J.C., Mayero, M., Melquiond, G., Weis, P.: Wave equation numerical resolution: a comprehensive mechanized proof of a C program. J. Autom. Reasoning **50**(4), 423–456 (2013)
4. Epperson, J.F.: An Introduction to Numerical Methods and Analysis. Wiley, New York (2014)
5. Feautrier, P.: Automatic parallelization in the polytope model. In: Perrin, G.-R., Darte, A. (eds.) The Data Parallel Programming Model. LNCS, vol. 1132, pp. 79–103. Springer, Heidelberg (1996)
6. Frigo, M., Strumpen, V.: Cache oblivious stencil computations. In: Proceedings of the Supercomputing, pp. 361–366. ACM (2005)
7. Frigo, M., Strumpen, V.: The cache complexity of multithreaded cache oblivious algorithms. Theory Comput. Syst. **45**(2), 203–233 (2009)
8. Gardner, M.: Mathematical games - The fantastic combinations of John Conway's new solitaire game "Life". Sci. Am. **223**(4), 120–123 (1970)
9. Immler, F., Hölzl, J.: Numerical analysis of ordinary differential equations in Isabelle/HOL. In: Beringer, L., Felty, A. (eds.) ITP 2012. LNCS, vol. 7406, pp. 377–392. Springer, Heidelberg (2012)
10. Kelly, W., Pugh, W.: A unifying framework for iteration reordering transformations. In: Proceedings of the ICAPP, vol. 1, pp. 153–162. IEEE (1995)
11. LeVeque, R.J.: Finite Difference Methods for Ordinary and Partial Differential Equations: Steady-State and Time-Dependent Problems, vol. 98. SIAM, Philadelphia (2007)

12. Li, G., Gopalakrishnan, G.: Scalable SMT-based verification of GPU kernel functions. In: Proceedings of the FSE, pp. 187–196. ACM (2010)
13. Li, G., Li, P., Sawaya, G., Gopalakrishnan, G., Ghosh, I., Rajan, S.P.: GKLEE: concolic verification and test generation for GPUs. In: PPOPP, pp. 215–224 (2012)
14. Loechner, V.: PolyLib: a library for manipulating parameterized polyhedra(1999). http://camlunity.ru/swap/Library/Conflux/Techniques%20-%20Code%20Analy sis%20and%20Transformations%20(Polyhedral)/Free%20Libraries/polylib.ps
15. Maydan, D.E., Hennessy, J.L., Lam, M.S.: Efficient and exact data dependence analysis. In: Proceedings of the PLDI, PLDI 1991, pp. 1–14. ACM (1991)
16. Orchard, D., Mycroft, A.: Efficient and correct stencil computation via pattern matching and static typing. arXiv preprint arXiv:1109.0777 (2011)
17. Pugh, W.: The Omega test: a fast and practical integer programming algorithm for dependence analysis. In: Proceedings of the Supercomputing, pp. 4–13. ACM (1991)
18. Solar-Lezama, A., Arnold, G., Tancau, L., Bodik, R., Saraswat, V., Seshia, S.: Sketching stencils. In: Proceedings of the PLDI, pp. 167–178. ACM (2007)
19. Tang, Y., Chowdhury, R.A., Kuszmaul, B.C., Luk, C.K., Leiserson, C.E.: The Pochoir stencil compiler. In: Proceedings of the SPAA, pp. 117–128. ACM (2011)
20. Valiant, L.G.: A bridging model for parallel computation. Commun. ACM **33**(8), 103–111 (1990)
21. Xu, Z., Kamil, S., Solar-Lezama, A.: MSL: a synthesis enabled language for distributed implementations. In: Proceedings of the International Conference for High Performance Computing, Networking, Storage and Analysis, pp. 311–322. IEEE Press (2014)

The Flow of ODEs

Fabian Immler[✉] and Christoph Traut

Institut für Informatik, Technische Universität München, Munich, Germany
immler@in.tum.de

Abstract. Formal analysis of ordinary differential equations (ODEs) and dynamical systems requires a solid formalization of the underlying theory. The formalization needs to be at the correct level of abstraction, in order to avoid drowning in tedious reasoning about technical details. The *flow* of an ODE, i.e., the solution depending on initial conditions, and a dedicated type of bounded linear functions yield suitable abstractions. The dedicated type integrates well with the type-class based analysis in Isabelle and we prove advanced properties of the flow, most notably, differentiable dependence on initial conditions via the variational equation and a rigorous numerical algorithm to solve it.

1 Introduction

Ordinary differential equations (ODEs) are ubiquitous for modeling continuous problems in e.g., physics, biology, or economics. A formalization of the theory of ODEs allows us to verify algorithms for the analysis of such systems. A popular example, where a verified algorithm is highly relevant, is Tucker's proof on the topic of a strange attractor for the Lorenz equations [9]. This proof relies on the output of a computer program, that computes bounds for analytical properties of the so-called *flow* of an ODE.

The flow is the solution as a function depending on an initial condition. We formalize the flow and prove conditions for analytical properties like continuity of differentiability (the derivative is of particular importance in Tucker's proof). Most of these properties seem very "natural", as Hirsch, Smale and Devaney call them in their textbook [2]. However, despite being "natural" properties and fairly standard results, they are delicate to prove: In the textbook, the authors present these properties rather early, but

> "postpone all of the technicalities [...], primarily because understanding this material demands a firm and extensive background in the principles of real analysis."

In this paper, we show that it is feasible to cope with these technicalities in a formal setting and confirm that Isabelle/HOL supplies a sufficient background of real analysis.

F. Immler—Supported by the DFG RTG 1480 (PUMA).

J.C. Blanchette and S. Merz (Eds.): ITP 2016, LNCS 9807, pp. 184–199, 2016.
DOI: 10.1007/978-3-319-43144-4_12

We present our Isabelle/HOL library for reasoning about the flow of ODEs. The main results are formalizations of continuous and differentiable dependence on initial conditions. The differentiable dependence is characterized by a particular ODE, the variational equation, and we show how to use existing rigorous numerical algorithms to solve it (Sect. 4). The variational equation is posed on the space of linear functions. We introduce a separate type for this space in order to profit from the type class based formalization of mathematics in Isabelle/HOL.

We are not aware of any other formalization that covers this foundational part of the theory of ODEs in similar detail.

2 Overview

We will first (in Sect. 3) present the "interface" to our theory, i.e., the definitions and assumptions that are needed for formalizing our main results. Any potential user of the library needs in principle only know about these concepts. Because the general topic is very theoretical and foundational work, we present a practical application right afterwards in Sect. 4.

Only then, we go into the details of the techniques that we used to make this formalization possible. Mathematics and analysis is formalized in Isabelle mostly based on type classes and filters, as has been presented earlier in earlier work [3]. We follow this path to formalize the foundations of our work:

Several proofs needed the notion of a uniform limit. We cast this notion into the "Isabelle/HOL approach to limits": we define it using a filter. This gives a versatile formalization and one can profit from the existing infrastructure for filters in limits. This will be presented in Sect. 5.

The derivative of the flow is a linear function. The space of linear functions forms a Banach space. In order to profit from the structure and properties that hold in a Banach space (which is a type class in Isabelle/HOL), we needed to introduce a type of bounded linear functions. We will present this type and further applications of its formalization in Sect. 6.

In Sect. 7, we present the technical lemmas that are needed to prove continuity and differentiability of the Flow in order to give an impression of the kind of reasoning that is required.

All of the theorems we present here and in the following are formalized in Isabelle/HOL [8], the source code can be found in the development version of the Archive of Formal Proof[1].

3 The Flow of a Differential Equation

In this section, we introduce the concept of *flow* and *existence interval* (which guarantees that the flow is well-defined) and present our main results (without proofs at first, we will present some of the lemmas leadings to the proofs in Sect. 7).

---

[1] http://www.isa-afp.org/devel-entries/Ordinary_Differential_Equations.shtml.

The claim we want to make in this section is the flow as definition is a suitable abstraction for initial value problems. But beware: do not get deceived by simplicity of statements: as already mentioned in the introduction, these are all "natural" properties, but the proofs (also in the textbook) require many technical lemmas.

First of all, let us introduce the concepts we are interested in. We consider open sets T, X and an autonomous[2] ODE with right hand side f

$$\dot{x}(t) = f(x(t)), \text{ where } f : \mathbb{R}^n \to \mathbb{R}^n \text{ is a function from } X \text{ to } X \qquad (1)$$

Under mild assumptions (which we will make more precise later in Definition 27), there exists a solution $\varphi(t)$, which is unique for an initial condition $x(t_0) = x_0$. To emphasize the dependence on the initial condition, we write $\varphi(x_0, t)$ for the solution of Eq. (1). This solution depending on initial conditions is called the *flow* of the differential equation:

Definition 1 (Flow). *The flow $\varphi(x_0, t)$ is the (unique) solution of the ODE (1) with initial condition $x\,0 = x_0$*

The solution does not necessarily exist for every $t \in T$. For example, solutions can *explode* in finite time s: if $\lim_{t \to s} \varphi(x_3, t) = \infty$, then the flow is only defined for $t < s$ as is illustrated in Fig. 1 for $\varphi(x_3, \_)$. We therefore need to define a notion of (maximal) existence interval.

Definition 2 (Maximal Existence Interval). *The maximal existence interval of the ODE (1) is the open interval*

$$\text{ex-ivl}\,(x_0) := \,]\alpha; \beta[$$

for $\alpha, \beta \in \mathbb{R} \cup \{\infty, -\infty\}$, such that $\varphi(x_0, t)$ is a solution for $t \in$ ex-ivl. Moreover for every other interval I and every solution $\psi(x_0, t)$ for $t \in I$, one has $I \subseteq J$ and $\forall t \in I$. $\psi(x_0, t) = \varphi(x_0, t)$.

We claim that the flow φ (together with ex-ivl, which guarantees the flow to be well-defined) is a very nice way to talk about solutions, because after guaranteeing that they are well-defined, these constants have many nice properties, which can be stated without further assumptions.

3.1   Composition of Solutions

A first nice property is the abstract property of the generic notion of flow. This notion makes it possible to easily state composition of solutions and to algebraically reason about them. As illustrated in Fig. 1, flowing from x_1 for time $s + t$ is equivalent to first flowing for time s, and from there flowing for time t.

This only works if the flow is defined also for the intermediate times (the theorem can not be true for $\varphi(x_0, t + (-t))$ if $t \notin$ ex-ivl).

Theorem 3 (Flow property)

$$\{s, t, s + t\} \subseteq \text{ex-ivl}\,(x_0) \implies \varphi(x_0, s + t) = \varphi(\varphi(x_0, s), t)$$

---

[2] This means that f does not depend on t. Many of our results are also formalized for non-autonomous ODEs, but the presentation is clearer, and reduction is possible.

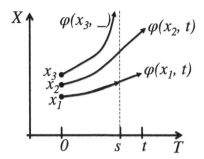

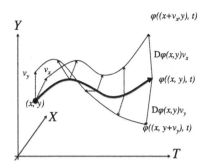

Fig. 1. The flow for different initial values

Fig. 2. Illustration of the derivative of the flow

3.2 Continuity of the Flow

In the previous lemma, the assumption that the flow is defined (i.e., that the time is contained in the existence interval) was important. Let us now study the domain $\Omega = \{(x,t) \mid t \in \text{ex-ivl}(x)\} \subseteq X \times T$ of the flow in more detail. Ω is called the *state space*.

For the first "natural" property, we consider an element in the state space. $(t,x) \in \Omega$ means that we can follow a solution starting at x for time t. It is "natural" to expect that solutions starting close to x can be followed for times that are close to t. In topological parlance, the state space is open.

Theorem 4 (Open State Space). *open* Ω

In the previous theorem, the state space allows us to reason about the fact that solutions are *defined* for close times and initial values. For quantifying how deviations in the initial values are propagated by the flow, Grönwall's lemma is an important tool that is used in several proofs. Because of its importance in the theory of dynamical systems, we list it here as well, despite it being a rather technical result. Starting from an *implicit* inequality $g\,t \le C + K \cdot \int_0^t g(s)\,\mathrm{d}s$ involving a continuous, nonnegative function $g : \mathbb{R} \to \mathbb{R}$, it allows one to deduce an *explicit* bound for g:

Lemma 5 (Grönwall)

$$0 < C \implies 0 < K \implies \textit{continuous-on } [0;a]\ g \implies$$

$$\forall t.\ 0 \le g(t) \le C + K \cdot \int_0^t g(s)\,\mathrm{d}s \implies$$

$$\forall t \in [0;a].\ g(t) \le C \cdot e^{K \cdot t}$$

Grönwall's lemma can be used to show that solutions deviate *at most* exponentially fast: $\exists K.\ |\varphi(x,t) - \varphi(y,t)| < |x-y| \cdot e^{K \cdot |t|}$ (see also Lemma 30). Therefore, by choosing x and y close enough, one can make the distance of the solutions arbitrarily small. In other words, the flow is a continuous function on the state space:

Theorem 6 (Continuity of Flow). *continuous-on* Ω φ

3.3 Differentiability of the Flow

Continuity just states that small deviation in the initial values result in small deviations of the flow. But one can be more precise on how initial deviations propagate. Consider Fig. 2, which depicts a solution starting at (x, y) and its evolution up to time t, as well as two other solutions evolving from initial values that have been perturbed via vectors v_x and v_y, respectively. A nice property of the flow is that it is differentiable: the way initial deviations propagate can be approximated by a linear function. So instead of solving the ODE for perturbed initial values, one can approximate the resulting perturbation with the linear function: $D\varphi \cdot v \approx \varphi((x, y), t) - \varphi((x, y) + v, t)$. More formally, our main result is the formalization of the fact that the derivative of the flow exists and is continuous.

Theorem 7 (Differentiability of the Flow). *For every* $(x, t) \in \Omega$ *There exists a linear function* $W(x, t)$, *which is the derivative of the flow at* (x, t):

$$\exists W. \ D\varphi|_{(x,t)} = W(x, t) \wedge \text{continuous-on} \ \Omega \ W$$

4   Rigorous Numerics for the Derivative of the Flow

In this section, we show that the formalization is not something abstract and detached, but something that can actually be computed with: The derivative W of the flow can be characterized as the solution of a linear, matrix-valued ODE, a byproduct of the (constructive) proof of differentiability in Lemma 36: The derivative with respect to x, written W_x, is the solution to the following ODE[3]

$$\dot{W}(t) = Df|_{\varphi(x_0, t)} \cdot W(t)$$

with initial condition $W(0)$ is the identity matrix.

We encode this matrix valued variational equation into a vector valued one, use an existing rigorous numerical algorithm for solving ODEs in Isabelle [5] to compute bounds on the solutions. Re-interpreting the result as bounds on matrices, we obtain bounds on the solution of the variational equation. As a concrete example, we use the van der Pol system: $\dot{x} = y$; $\dot{y} = (1 - x^2)y - x$ for initial condition $(x_0, y_0) = (1.25, 2.27)$.

The overall setup for the computation is as follows: We have an executable specification of the Euler method (which is formally verified to produce rigorous enclosures for the solution of an ODE) and use Isabelle's code generator [1] to generate SML code from this specification. We chose to compute the evolution

---

[3] Here, $\cdot$ stands for matrix multiplication.

until time $t = 2$ with a discrete grid of 500 time steps. The computation takes about 3 min on an average laptop computer. As a result, we get the following inclusion for the variational equation:

Theorem 8

$$W(2) \in \begin{pmatrix} [0.18; 0.23] & [0.41; 0.414] \\ [-0.048; -0.041] & [0.26; 0.27] \end{pmatrix}$$

The left column of the matrix shows the propagation of a deviation in the x direction: a $(1, 0)$ deviation is propagated to a $([0.18; 0.23], [-0.048; -0.041])$ deviation: it gets smaller but remains mostly in the x direction. For the right column, a deviation in the y direction $(0, 1)$ is propagated to a $([0.41; 0.414]; [0.26; 0.27])$ deviation: it contracts as well, but it gets rotated towards the x direction.

5 Uniform Limit as Filter

Filters have proved to be useful to describe all kinds of limits and convergence [3]. We use filters to define uniform convergence. For details about filters, please consider the source code and the paper [3]. In the formalization, the uniform limit *uniform-limit* X f l F is parameterized by a filter F, here we just present the explicit formulations for the *sequentially* and *at* filters.

A sequence of functions $f_n : \alpha \to \beta$ for $n \in \mathbb{N}$ is said to converge uniformly on $X : \mathcal{P}(\alpha)$ against the *uniform limit* $l : \alpha \to \beta$, if

Definition 9

> *uniform-limit* X f l *sequentially* $:=$
> $\forall \varepsilon > 0.\ \exists N.\ \forall x \in X.\ \forall n \geq N.\ |f_n\ x - l\ x| < \varepsilon$

Note the difference to pointwise convergence, where one would exchange the order of the quantifiers $\exists N$ and $\forall x \in X$.

With the (*at* z) filter, we can also handle uniform convergence of a family of functions $f_y : \alpha \to \beta$ as y approaches z:

Definition 10

> *uniform-limit* X f l (*at* z) $:=$
> $\forall \varepsilon > 0.\ \exists \delta > 0.\quad \forall y.\ |y - z| < \delta \implies (\forall x \in X.\ \text{dist}\ (f_y\ x)\ (l\ x) < \varepsilon)$

The advantage of the filter approach is that many important lemmas can be expressed for arbitrary filters, for example the uniform limit theorem, which states that the uniform limit of a (via filter F generalized) sequence f_n of continuous functions is continuous.

Theorem 11 (Uniform Limit Theorem)

> $(\forall n \in F.\ \textit{continuous-on}\ X\ f_n) \implies \textit{uniform-limit}\ X\ f\ l\ F \implies$
> *continuous-on* X f_n

A frequently used criterion to show that an infinite series of functions converges uniformly is the Weierstrass M-test. Assuming majorants M_n for the functions f_n and assuming that the series of majorants converges, it allows one to deduce uniform convergence of the partial sums towards the series.

Lemma 12 (Weierstrass M-Test)

$$\forall n.\ \forall x \in X.\ |f_n\ x| \leq M_n \implies \sum_{n \in \mathbb{N}} M_n < \infty \implies$$

$$\textit{uniform-limit}\ X\ (n \mapsto x \mapsto \sum_{i \leq n} f_i\ x)\ (x \mapsto \sum_{i \in \mathbb{N}} f_i\ x)\ \textit{sequentially}$$

6   Bounded Linear Functions

We introduce a type of bounded linear functions (or equivalently *continuous* linear functions) in order to be able to profit from the hierarchy of mathematical type classes in Isabelle/HOL.

6.1   Type Classes for Mathematics in Isabelle/HOL

In Isabelle/HOL, many of the mathematical concepts (in particular spaces with a certain structure) are formalized using type classes.

The advantage of type class based reasoning is that most of the reasoning is generic: formalizations are carried out in the context of type classes and can then be used for all types inhabiting that type class. For generic formalizations, we use Greek letters α, β, γ and name their type class constraints in prose (i.e., if we write that we "consider a topological space" α, then this result is formalized generically for every type α that fulfills the properties of a topological space).

The spaces we consider are topological spaces with **open** sets, (real) vector spaces with addition $+ : \alpha \to \alpha \to \alpha$ and scalar multiplication $\cdot : \mathbb{R} \to \alpha \to \alpha$. Normed vector spaces come with a norm $|(\_)| : \alpha \to \mathbb{R}$. A vector space with multiplication $* : \alpha \to \alpha \to \alpha$ that is compatible with addition $(a + b) * c = a * c + b * c$ is an algebra and can also be endowed with a norm. Complete normed vector spaces are called Banach spaces.

6.2   A Type of Bounded Linear Functions

An important concept is that of a linear function. For vector spaces α and β, a linear function is a function $f : \alpha \to \beta$ that is compatible with addition and scalar multiplication.

Definition 13

$$\textit{linear}\ f := \forall x\,y\,c.\ f(c \cdot x + y) = c \cdot f(x) + f(y)$$

We need topological properties of linear functions, we therefore now assume normed vector spaces α and β. One usually wants linear functions to be continuous, and if α and β are vector spaces of finite dimension, any linear function $\alpha \to \beta$ is continuous. In general, this is not the case, and one usually assumes *bounded* linear functions. The norm of the result of a bounded linear function is linearly bounded by the norm of the argument:

Definition 14

$$\textit{bounded-linear } f := \textit{linear } f \wedge \exists K. \ \forall x. \ |f(x)| \leq K * |x|$$

We now cast bounded linear functions $\alpha \to \beta$ as a type $\alpha \to_{bl} \beta$ in order to make it an instance of Banach space.

Definition 15

$$\textbf{typedef } \alpha \to_{bl} \beta := \{f : \alpha \to \beta \mid \textit{bounded-linear } f\}$$

6.3 Instantiations

For defining operations on type $\alpha \to_{bl} \beta$, the Lifting and Transfer package [4] is an essential tool: operations on the plain function type $\alpha \to \beta$ are automatically lifted to definitions on the type $\alpha \to_{bl} \beta$ when supplied with a proof that functions in the result are *bounded-linear* under the assumption that argument functions are *bounded-linear*. We write application of a bounded linear function $f : \alpha \to_{bl} \beta$ with an element $x : \alpha$ as follows.

Definition 16 (application of bounded linear functions)

$$(f \cdot x) : \beta$$

We present the definitions of operations involving the type $\alpha \to_{bl} \beta$ by presenting them in an extensional form using $\cdot$. Bounded linear functions with pointwise addition and pointwise scalar multiplication form a vector space.

Definition 17 (Vector Space Operations). *For $f, g : \alpha \to_{bl} \beta$ and $c : \mathbb{R}$,*

$$(f + g) \cdot x := f \cdot x + g \cdot x$$

$$(c \cdot f) \cdot x := c \cdot (f \cdot x)$$

The usual choice of a norm for bounded linear functions is the operator norm: the maximum of the image of the bounded linear function on the unit ball. With this norm, $\alpha \to_{bl} \beta$ forms a normed vector space and we prove that it is Banach if α and β are Banach.

Definition 18 (Norm in Banach Space). *For $f : \alpha \to_{bl} \beta$,*

$$|f| := \max \{|f \cdot y| \mid |y| \leq 1\}$$

One can also compose bounded linear functions according to $(f \circ g) \cdot x = f \cdot (g \cdot x)$. Bounded linear operators—that is bounded linear functions $\alpha \to_{bl} \alpha$ from one type α into itself—form a Banach algebra with composition as multiplication and the identity function as neutral element:

Definition 19 (Banach Algebra of Bounded Linear Operators)
For $f, g : \alpha \to_{bl} \alpha$,

$$(f * g) \cdot x := (f \circ g) \cdot x$$

$$1 \cdot x := x$$

6.4   Applications

Now we can profit from many of the developments that are available for Banach spaces or algebras. Here we present some useful applications: The exponential function is defined generically for *banach-algebra* and can therefore be used for bounded linear functions as well. Furthermore, the type of bounded linear functions can be used to describe derivatives in arbitrary vector spaces and therefore allows one to naturally express (and conveniently prove) basic results from analysis: the Leibniz rule for differentiation under the integral sign and conditions for (total) differentiability of multidimensional functions. Note that not everything in this section is directly necessary for the formalizations of our main results, it is rather intended to show the versatile use of a separate type for bounded linear functions in Isabelle/HOL.

Exponential of Operators. The exponential function for bounded linear functions is a useful concept and important for the analysis linear ODEs. Here we present that the solution of linear autonomous homogeneous differential equations can be expressed using the exponential function. For a Banach algebra α, the exponential function is defined using the usual power series definition (B^k is a k fold multiplication $B * \cdots * B$):

Definition 20 (Exponential Function). *For a Banach algebra α and $B : \alpha$,*

$$e^B := \sum_{k=0}^{\infty} \frac{1}{k!} \cdot B^k$$

We prove the following rule for the derivative of the exponential function

Lemma 21 (Derivative of Exponential). $\frac{\mathrm{d}\, e^{x \cdot A}}{\mathrm{d}x} = e^{x \cdot A} \cdot A$

Proof. After unfolding the definition of derivative $\frac{\mathrm{d}\, e^{x \cdot A}}{\mathrm{d}x} = \lim_{h \to 0} \frac{e^{(x+h) \cdot A} - e^{x \cdot A}}{h}$, the crucial step in the proof is to exchange the two limits (one is explicit in $\lim_{h \to 0}$, and the other one is hidden as the limit of the series Definition 20 of the exponential). Exchange of limits can be done similar to Theorem 11, while uniform convergence is guaranteed according to the Weierstrass M-Test from Lemma 12. □

With this rule for the derivative and an obvious calculation for the initial value, one can show the following

Lemma 22. (Solution of linear initial value problem)
$\varphi_{x_0,t_0}(t) := \left(e^{(t-t_0) \cdot A}\right)(x_0)$ *is the unique solution to the ODE* $\dot{\varphi}\, t = A\, (\varphi\, t)$ *with initial condition* $\varphi(t_0) = x_0$.

Total Derivatives. The total derivative (or Fréchet derivative) is a generalization of the ordinary derivative (of functions $\mathbb{R} \to \mathbb{R}$) for arbitrary normed vector spaces. To illustrate this generalization, recall that the ordinary derivative yields the slope of the function: if $f'(x) = m$, then

$$\lim_{h \to 0} \frac{f(x+h) - f(x)}{h} = m \tag{2}$$

Moving the m under the limit, one sees that the (linear) function $h \mapsto h \cdot m$ is a good approximation for the difference of the function value at nearby points x and $x + h$:

$$\lim_{h \to 0} \frac{f(x+h) - f(x) - h \cdot m}{h} = 0$$

This concept can be generalized by replacing $h \mapsto h \cdot m$ with an arbitrary (bounded) linear function A. In the following equation, A is a good linear approximation.

$$\lim_{h \to 0} \frac{f(x+h) - f(x) - A \cdot h}{|h|} = 0 \tag{3}$$

Note that in the previous equation, we can (just formally) drop many of the restrictions on the type of f. We started with $f : \mathbb{R} \to \mathbb{R}$ in Eq. 2, but the last equation still makes sense for $f : \alpha \to \beta$ for normed vector spaces α, β. We call $A : \alpha \to_{bl} \beta$ the total derivative Df of f at a point x:

Definition 23 (Total Derivative). *For* $A : \alpha \to_{bl} \beta$ *in Eq. 3, we write*

$$Df|_x = A$$

The total derivative is important for our developments as it is for example the derivative W of the flow in Theorem 7. It is only due to the fact that the resulting type $\alpha \to_{bl} \alpha$ is a normed vector space, that makes it possible to express continuity of the derivative or to express higher derivatives.

Another example, where interpreting the derivative as bounded linear function $\alpha \to_{bl} \beta$ is helpful, is when deducing the total derivative of a function f by looking at its partial derivatives f_1 and f_2 (that is, the derivatives w.r.t. one variable while fixing the other). One needs the assumption that the partial derivatives are continuous.

Lemma 24 (Total Derivative via Continuous Partial Derivatives)
For $f : \alpha \to \beta \to \gamma$, $f_1 : \alpha \to \beta \to (\alpha \to_{bl} \gamma)$, $f_2 : \alpha \to \beta \to (\beta \to_{bl} \gamma)$

$$\forall x. \ \forall y. \ \mathsf{D}(x \mapsto f \ x \ y)|_x = f_1 \ x \ y \implies$$
$$\forall x. \ \forall y. \ \mathsf{D}(y \mapsto f \ x \ y)|_y = f_2 \ x \ y \implies$$
$$continuous\,((x, y) \mapsto f_1 \ x \ y) \implies$$
$$continuous\,((x, y) \mapsto f_2 \ x \ y) \implies$$
$$\mathsf{D}((x, y) \mapsto f \ x \ y)|_{(x,y)} \cdot (t_1, t_2) = (f_1 \ x \ y) \cdot t_1 + (f_2 \ x \ y) \cdot t_2$$

Leibniz Rule. Another example is a general formulation of the Leibniz rule. The following rule is a generalization of e.g., the rule formalized by Lelay and Melquiond [7] to general vector spaces. Here $[[a; b]]$ is a hyperrectangle in Euclidean space $\mathbb{R}^n$. The rule allows one to differentiate under the integral sign: the derivative of the parameterized integral $\int_a^b f \ x \ t \, dt$ with respect to x can be expressed as the integral of the derivative of f. Note that the integral on the right is in the Banach space of bounded linear functions.

Lemma 25 (Leibniz rule). *For Banach spaces* α, β *and* $f : \alpha \to \mathbb{R}^n \to \beta$, $f_1 : \alpha \to \mathbb{R}^n \to (\alpha \to_{bl} \beta)$,

$$\forall t. \ \mathsf{D}(x \mapsto f \ x \ t)|_x = f_1 \ x \ t \implies$$
$$\forall x. \ (f \ x) \ integrable\text{-}on \ [[a; b]]$$
$$\forall xt. \ t \in [[a; b]] \implies continuous\,((x, t) \mapsto f \ x \ t)$$

$$\mathsf{D}\left(x \mapsto \int_a^b f \ x \ t \, dt\right)|_x = \int_a^b f_1 \ x \ t \, dt$$

7 Proofs About the Flow

We will now go into the technical details of the proofs leading towards continuity and differentiability of the flow (Theorems 6 and 7). We still do not present the proofs: their structure is very similar to the textbook [2] proofs. Nevertheless, we want to present the detailed statements of the propositions, as they give a good impression on the kind of reasoning that was required.

7.1 Criteria for Unique Solution

First of all, we specify the common assumptions to guarantee existence of a unique solution for an initial value problem and therefore a condition for the flow in Definition 1 to be well-defined.

We assume that f is locally Lipschitz continuous in its second argument: for every $(t, x) \in T \times X$ there exist ε-neighborhoods $U_\varepsilon(t)$ and $U_\varepsilon(t)$ around t and x, in which f is Lipschitz continuous w.r.t. the second argument (uniformly w.r.t. the first): the distance of function values is bounded by a constant times the distance of argument values:

Definition 26

local-lipschitz T X $f :=$
$\forall t \in T.\ \forall x \in X.$
$\quad \exists \varepsilon > 0.\ \exists L.$
$\qquad \forall t' \in U_\varepsilon(t).\ \forall x_1, x_2 \in U_\varepsilon(x).\ |f\ t'\ x_1 - f\ t'\ x_2| \le L \cdot |x_1 - x_2|$

Now the only assumptions that we need to prove continuity of the flow are open sets for time and phase space and a locally Lipschitz continuous right-hand side f that is continuous in t:

Definition 27 (Conditions for unique solution)

1. T *is an open set*
2. X *is an open set*
3. *f is locally Lipschitz continuous on X:* local-lipschitz T X f
4. *for every $x \in X$, $t \mapsto f(t, x)$ is continuous on T.*

These assumptions (the detailed proofs that these assumptions guarantee the existence of a unique solution for initial value problems has been presented in Theorem 3 of earlier work [6]).

7.2   The Frontier of the State Space

It is important to study the behavior of the flow at the frontier of the state space (e.g., as time or the solution tend to infinity). From this behavior, one can deduce conditions under which solutions can be continued. This yields techniques to gain more precise information on the existence interval ex-ivl.

If the solution only exists for finite time, it has to explode (i.e., leave every compact set):

Lemma 28 (Explosion for Finite Existence Interval)

$$\text{ex-ivl}\,(x_0) = \,]\alpha, \beta[\implies \beta < \infty \implies \text{compact } K \implies$$
$$\exists t \ge 0.\ t \in \text{ex-ivl}\,(x_0) \wedge \varphi(x_0, t) \notin K$$

This lemma can be used to prove a condition on the right-hand side f of the ODE, to certify that the solution exists for the whole time. Here the assumption guarantees that the solution stays in a compact set.

Lemma 29 (Global Existence of Solution)

$$(\forall s \in T.\ \forall u \in T.\ \exists L.\ \exists M.\ \forall t \in [s; u].\ \forall x \in X.\ |f\ t\ x| \le M + L \cdot |x|)$$
$$\implies \text{ex-ivl}\,(x_0) = T$$

7.3   Continuity of the Flow

The following lemmas are all related to continuity of the flow. With the help of Grönwall's Lemma 5, one can show that when two solutions (starting from different initial conditions x_0 and y_0) both exist for a time t and are restricted to some set Y on which the right-hand side f satisfies a (global) Lipschitz condition K, then the distance between the solutions grows at most exponentially with increasing time:

Lemma 30 (Exponential Initial Condition for Two Solutions)

$$t \in \text{ex-ivl}(x_0) \implies t \in \text{ex-ivl}(y_0) \implies$$
$$x_0 \in Y \implies y_0 \in Y \implies Y \subseteq X \implies$$
$$\forall s \in [0; t].\ \varphi(x_0, s) \in Y \implies$$
$$\forall s \in [0; t].\ \varphi(y_0, s) \in Y \implies$$
$$\forall s \in [0; t].\ \text{lipschitz } Y\ (f\ s)\ K \implies$$
$$|\varphi(x_0, t) - \varphi(y_0, t)| \leq |x_0 - y_0| \cdot e^{K \cdot t}$$

Note that it can be hard to establish the assumptions of this lemma, in particular the assumption that both solutions from x_0 and y_0 exist for the same time t. Consider Fig. 1: not all solutions (e.g., from x_3) do necessarily exist for the same time s. One can choose, however, a neighborhood of x_1, such that all solutions starting from within this neighborhood exist for at least the same time, and with the help of the previous lemma, one can show that the distance of these solutions increases at most exponentially:

Lemma 31 (Exponential Initial Condition of Close Solutions)

$$a \in \text{ex-ivl}(x_0) \implies b \in \text{ex-ivl}(x_0) \implies a \leq b$$
$$\exists \delta > 0.\ \exists K > 0.\ U_\delta(x_0) \subseteq X \ \wedge$$
$$(\forall y \in U_\delta(x_0).\ \forall t \in [a; b].$$
$$t \in \text{ex-ivl}(y) \wedge |\varphi(x_0, t) - \varphi(y, t)| \leq |x_0 - y| \cdot e^{K \cdot |t|})$$

Using this lemma is the key to showing continuity of the flow (Theorem 6).

A different kind of continuity is not with respect to the initial condition, but with respect to the right-hand side of the ODE.

Lemma 32 (Continuity with respect to ODE). *Assume two right-hand sides f, g defined on X and uniformly close $|f\ x - g\ x| < \varepsilon$. Furthermore, assume a global Lipschitz constant K for f on X. Then the deviation of the flows φ_f and φ_g can be bounded:*

$$|\varphi_f(x_0, t) - \varphi_g(x_0, t)| \leq \frac{\varepsilon}{K} \cdot e^{K \cdot t}$$

7.4   Differentiability of the Flow

The proof for the differentiability of the flow incorparates many of the tools that we have presented up to now, we will therefore go a bit more into the details of this proof.

Assumptions. The assumptions in Definition 27 are not strong enough to prove differentiability of the flow. However, a continuously differentiable right-hand side $f : \mathbb{R}^n \to \mathbb{R}^n$ suffices. To be more precise:

Definition 33 (Criterion for Continuous Differentiability of the Flow)

$$\exists f' : \mathbb{R}^n \to (\mathbb{R}^n \to_{bl} \mathbb{R}^n). \ (\forall x \in X. \ \mathsf{D}f|_x = f' \ x) \wedge \textit{continuous-on } X \ f'$$

From now on, we denote the derivative along the flow from x_0 with $A_{x_0} : \mathbb{R} \to \mathbb{R}^n$:

Definition 34 (Derivative along the Flow). $A_{x_0}(t) := \mathsf{D}f|_{\varphi(x_0,t)}$

The derivative of the flow is the solution to the so-called variational equation, a non-autonomous linear ODE. The initial condition ξ is supposed to be a perturbation of the initial value (like v_x and v_y in Fig. 2) and in what follows we will prove that the solution to this ODE is a good (linear) approximation of the propagation of this perturbation.

$$\begin{cases} \dot{u}(t) = A_{x_0}(t) \cdot u(t) \\ u(0) = \xi \end{cases} \quad , \tag{4}$$

We will write $u_{x_0}(\xi, t)$ for the flow of this ODE and omit the parameter x_0 and/or the initial value ξ if they can be inferred from the context.

As a prerequisite for the next proof, we begin by proving that $u_{x_0}(\xi, t)$ is linear in ξ, a property that holds because u is the solution of a linear ODE (this is often also called the "superposition principle").

Lemma 35 (Linearity of $u_{x_0}(\xi, t)$ in ξ)

$$\alpha \cdot u_{x_0,a}(t) + \beta \cdot u_{x_0,b}(t) = u_{x_0,\alpha \cdot a + \beta \cdot b}(t).$$

Because $\xi \mapsto u_{x_0}(\xi, t) : \mathbb{R}^n \to \mathbb{R}^n$ is linear on Euclidean space, it is also bounded linear, so we will identify this function with the corresponding element of type $\mathbb{R}^n \to_{bl} \mathbb{R}^n$. The main efforts go into proving the following lemma, showing that the aforementioned function is the derivative of the flow $\varphi(x_0, t)$ in x_0.

Lemma 36 (Space Derivative of the Flow). *For $t \in$ ex-ivl(x_0),*

$$(\mathsf{D}(x \to \varphi(x,t))|_{x_0}) \cdot \xi = u_{x_0}(\xi, t)$$

Proof. The proof starts out with the integral identities of the flow, the perturbed flow, and the linearized propagation of the perturbation:

$$\varphi(x_0, t) = x_0 + \int_0^t f(\varphi(x_0, s)) \, ds$$

$$\varphi(x_0 + \xi, t) = x_0 + \xi + \int_0^t f(\varphi(x_0 + \xi, s)) \, ds$$

$$u_{x_0}(\xi, t) = \xi + \int_0^t A_{x_0}(s) \cdot u_{x_0}(\xi, s) \, ds$$

$$= \xi + \int_0^t f'(\varphi(x_0, s)) \cdot u_{x_0}(\xi, s) \, ds$$

Then, for any fixed ε, after a sequence of estimations (3 pages in the textbook proof) involving e.g., uniform convergence (Sect. 5) of the first-order remainder term of the Taylor expansion of f, continuity of the flow (Theorem 6), and linearity of u (Lemma 35) one can prove the following inequality.

$$\frac{\|\varphi(x_0 + \xi, t) - \varphi(x_0, t) - u_{x_0}(\xi, t)\|}{\|\xi\|} \leq \varepsilon$$

This shows that $u_{x_0}(\xi, t)$ is indeed a good approximation for the propagation of the initial perturbation ξ and exactly the definition for the space derivative of the flow. □

Note that $u_{x_0}(\xi, t)$ yields the space derivative in direction of the vector ξ. The total space derivative of the flow is then the linear function $\xi \mapsto u_{x_0,\xi}(t)$. But this derivative can also be described as the solution of the following "matrix-valued" variational equation:

$$\begin{cases} \dot{W}_{x_0}(t) = A_{x_0}(t) \circ W_{x_0}(t) \\ W_{x_0}(0) = \mathsf{Id} \end{cases} \tag{5}$$

This initial value problem is defined for linear operators of type $\mathbb{R}^n \to_{bl} \mathbb{R}^n$. Thanks to Lemma 29, one can show that it is defined on the same existence interval as the flow φ. The solution W_{x_0} is related to solutions of the variational IVP as follows:

$$u_{x_0}(\xi, t) = W_{x_0}(t) \cdot \xi$$

The derivative of the flow φ at (x_0, t) with respect to t is given directly by the ODE, namely $f(\varphi(x_0, t))$. Therefore and according to Lemma 24 the total derivative of the flow is characterized as follows:

Theorem 37. (Derivative of the Flow)

$$D\varphi|_{(x_0,t)} \cdot (\xi, \tau) = W_{x_0}(t) \cdot \xi + \tau \cdot f(\varphi(x_0, t))$$

7.5 Continuity of Derivative

Regarding the continuity of the derivative $D\varphi|_{(x_0,t)} \cdot (\xi, \tau)$ with respect to (x_0, t): $\tau \cdot f\left(\varphi(x_0, t)\right)$ is continuous because of Definition 27 and Theorem 6.

$W_{x_0}(t)$ is continuous with respect to t, so what remains to be shown is continuity of the space derivative regarding x_0. The proof of this statement relies on Theorem 32, because for different values of x_0, W_{x_0} is the solution to ODEs with slightly different right-hand sides. A technical difficulty here is to establish the assumption of *global* Lipschitz continuity for Theorem 32.

8 Conclusion

To conclude, our formalization contains essentially all lemmas and proofs of at least 22 pages (Chap. 17) of the textbook by Hirsch *et al.* [2] and additionally required some more general-purpose background to be formalized, in particular uniform limits and the Banach space of (bounded) linear functions. The separate type for bounded linear functions was a minor complication that was necessary because of the type class based library for analysis in Isabelle/HOL. We showed the concrete usability of our results by verifying the connection of the abstract formalization with a concrete rigorous numerical algorithm.

References

1. Haftmann, F.: Code generation from specifications in higher-order logic. Dissertation, Technische Universität München, München (2009)
2. Hirsch, M.W., Smale, S., Devaney, R.L.: Differential Equations, Dynamical Systems, and an Introduction to Chaos. Elsevier Academic Print, Waltham (2013)
3. Hölzl, J., Immler, F., Huffman, B.: Type classes and filters for mathematical analysis in Isabelle/HOL. In: Blazy, S., Paulin-Mohring, C., Pichardie, D. (eds.) ITP 2013. LNCS, vol. 7998, pp. 279–294. Springer, Heidelberg (2013). doi:10.1007/978-3-642-39634-2_21
4. Huffman, B., Kunčar, O.: Lifting and transfer: a modular design for quotients in Isabelle/HOL. In: Gonthier, G., Norrish, M. (eds.) CPP 2013. LNCS, vol. 8307, pp. 131–146. Springer, Heidelberg (2013). doi:10.1007/978-3-319-03545-1_9
5. Immler, F.: Formally verified computation of enclosures of solutions of ordinary differential equations. In: Badger, J.M., Rozier, K.Y. (eds.) NFM 2014. LNCS, vol. 8430, pp. 113–127. Springer, Heidelberg (2014)
6. Immler, F., Hölzl, J.: Numerical analysis of ordinary differential equations in Isabelle/HOL. In: Beringer, L., Felty, A. (eds.) ITP 2012. LNCS, vol. 7406, pp. 377–392. Springer, Heidelberg (2012)
7. Lelay, C., Melquiond, G.: Différentiabilité et intégrabilité en Coq. application à la formule de d'Alembert. In: JFLA - Journées Francophone des Langages Applicatifs, Carnac, France (2012). https://hal.inria.fr/hal-00642206
8. Nipkow, T., Paulson, L.C., Wenzel, M.: Isabelle/HOL: A Proof Assistant for Higher-order Logic. LNCS. Springer, Berlin (2002)
9. Tucker, W.: A rigorous ODE solver and Smale's 14th problem. Found. Comput. Math. **2**(1), 53–117 (2002)

From Types to Sets by Local Type Definitions in Higher-Order Logic

Ondřej Kunčar[1(✉)] and Andrei Popescu[2,3]

[1] Fakultät für Informatik, Technische Universität München, Munich, Germany
kuncar@in.tum.de
[2] Department of Computer Science, School of Science and Technology,
Middlesex University, London, UK
[3] Institute of Mathematics Simion Stoilow of the Romanian Academy,
Bucharest, Romania

Abstract. Types in Higher-Order Logic (HOL) are naturally interpreted as nonempty sets—this intuition is reflected in the type definition rule for the HOL-based systems (including Isabelle/HOL), where a new type can be defined whenever a nonempty set is exhibited. However, in HOL this definition mechanism cannot be applied *inside proof contexts*. We propose a more expressive type definition rule that addresses the limitation and we prove its soundness. This higher expressive power opens the opportunity for a HOL tool that relativizes type-based statements to more flexible set-based variants in a principled way. We also address particularities of Isabelle/HOL and show how to perform the relativization in the presence of type classes.

1 Motivation

The proof assistant community is mainly divided in two successful camps. One camp, represented by provers such as Agda [7], Coq [6], Matita [5] and Nuprl [10], uses expressive type theories as a foundation. The other camp, represented by the HOL family of provers (including HOL4 [2], HOL Light [14], HOL Zero [3] and Isabelle/HOL [26]), mostly sticks to a form of classic set theory typed using simple types with rank-1 polymorphism. (Other successful provers, such as ACL2 [19] and Mizar [12], could be seen as being closer to the HOL camp, although technically they are not based on HOL.)

According to the HOL school of thought, a main goal is to acquire a sweet spot: keep the logic as simple as possible while obtaining *sufficient expressiveness*. The notion of sufficient expressiveness is of course debatable, and has been debated. For example, PVS [29] includes dependent types (but excludes polymorphism), HOL-Omega [16] adds first-class type constructors to HOL, and Isabelle/HOL adds ad hoc overloading of polymorphic constants. In this paper, we want to propose a gentler extension of HOL. We do not want to promote new "first-class citizens," but merely to give better credit to an old and venerable HOL citizen: the notion of types emerging from sets.

© Springer International Publishing Switzerland 2016
J.C. Blanchette and S. Merz (Eds.): ITP 2016, LNCS 9807, pp. 200–218, 2016.
DOI: 10.1007/978-3-319-43144-4_13

The problem that we address in this paper is best illustrated by an example. Let lists : α set $\rightarrow$ α list set be the constant that takes a set A and returns the set of lists whose elements are in A, and P : α list $\rightarrow$ bool be another constant (whose definition is not important here). Consider the following statements, where we extend the usual HOL syntax by explicitly quantifying over types at the outermost level:

$$\forall \alpha. \, \exists xs_{\alpha \text{ list}}. \, \mathsf{P} \, xs \tag{1}$$

$$\forall \alpha. \, \forall A_{\alpha \text{ set}}. \, A \neq \emptyset \longrightarrow (\exists xs \in \mathsf{lists} \, A. \, \mathsf{P} \, xs) \tag{2}$$

The formula (2) is a relativized form of (1), quantifying not only over all types α, but also over all their nonempty subsets A, and correspondingly relativizing the quantification over all lists to quantification over the lists built from elements of A. We call theorems such as (1) *type based* and theorems such as (2) *set based*.

Type-based theorems have obvious advantages compared to the set-based ones. First, they are more concise. Moreover, automatic proof procedures work better for them, thanks to the fact that they encode properties more rigidly and more implicitly, namely, in the HOL types (such as membership to α list) and not via formulas (such as membership to the set lists A). On the downside, type-based theorems are less flexible, and therefore unsuitable for some developments. Indeed, when working with mathematical structures, it is often the case that they have the desired property only on a proper subset of the whole type. For example, a function f from τ to σ may be injective or continuous only on a subset of τ. When wishing to apply type-based theorems from the library to deal with such situations, users are forced to produce ad hoc workarounds for relativizing them from types to sets. In the most striking cases, the relativization is created manually. For example, in Isabelle/HOL there exists the constant inj-on $A \, f = (\forall x \, y \in A. \, f \, x = f \, y \longrightarrow x = y)$ together with a small library about functions being injective only on a subset of a type. In summary, while it is easier to reason about type-based statements such as (1), the set-based statements such as (2) are more general and easier to apply.

An additional nuance to this situation is specific to Isabelle/HOL, which allows users to annotate types with Haskell-like type-class constraints. This provides a further level of implicit reasoning. For example, instead of explicitly quantifying a statement over an associative operation $*$ on a type σ, one marks σ as having class semigroup (which carries implicitly the assumptions). This would also need to be reversed when relativizing from types to sets. If (1) made the assumption that α is a semigroup, as in $\forall(\alpha_{\text{semigroup}}). \, \exists xs_{\alpha \text{ listt}}. \, \mathsf{P} \, xs$, then the statement (2) would need to quantify universally not only over A, but also over a binary operation on A, and explicitly assume it to be associative.

The aforementioned problem, of the mismatch between type-based theorems from libraries and set-based versions needed by users, shows up regularly in requests posted on the Isabelle community mailing lists. Here is a concrete example [33]: *Various lemmas* [from the theory Finite_Set] *require me to show that f* [commutes with ∘] *for all x and y. This is a too strong requirement for me. I can show that it holds for all x and y in A, but not for all x and y in general.*

Often, users feel the need to convert entire libraries from type-based theorems to set-based ones. For example, our colleague Fabian Immler writes about his large formalization experience [18, Sect. 5.7]: *The main reason why we had to introduce this new type* [of finite maps] *is that almost all topological properties are formalized in terms of type classes, i.e., all assumptions have to hold on the whole type universe. It feels like a cleaner approach* [would be] *to relax all necessary topological definitions and results from types to sets because other applications might profit from that, too.*

A prophylactic alternative is of course to develop the libraries in a set-based fashion from the beginning, agreeing to pay the price in terms of verbosity and lack of automation. And numerous developments in different HOL-based provers do just that [4, 8, 9, 15, 23].

In this paper, we propose an alternative that gets the best of both worlds: *prove easily and still be flexible.* More precisely, develop the libraries type based, but export the results set based. We start from the observation that, from a set-theoretic semantics standpoint, the theorems (1) and (2) are equivalent: they both state that, for every nonempty collection of elements, there exists a list of elements from that collection for which P holds. Unfortunately, the HOL logic in its current form is blind to one direction of this equivalence: assuming that (1) is a theorem, one cannot prove (2). Indeed, in a proof attempt of (2), one would fix a nonempty set A and, to invoke (1), one would need to define a new type corresponding to A—an action not currently allowed inside a HOL proof context. In this paper, we propose a gentle eye surgery to HOL (and to Isabelle/HOL) to enable proving such equivalences, and show how this can be used to leverage user experience as outlined above.

The paper is organized as follows. In Sect. 2, we recall the logics of HOL and Isabelle/HOL. In Sect. 3, we describe the envisioned extension of HOL: adding a new rule for simulating type definitions in proof contexts. In Sect. 4, we demonstrate how the new rule allows us to relativize type-based theorems to set-based ones in HOL. Due to the presence of type classes, we need to extend Isabelle/HOL's logic further to achieve the relativization—this is the topic of Sect. 5. Finally, in Sect. 6 we outline the process of performing the relativization in a principled and automated way.

We created a website [1] associated to the paper where we published the Isabelle implementation of the proposed logical extensions and the Isabelle proof scripts showing examples of applying the new rules to relativize from types to sets (including this paper's introductory example).

2   HOL and Isabelle/HOL Recalled

In this section, we briefly recall the logics of HOL and Isabelle/HOL mostly for the purpose of introducing some notation. For more details, we refer the reader to standard textbooks [11, 25]. We distinguish between the *core logic* and the *definitional mechanisms*.

2.1   Core Logic

The core logic is common to HOL and Isabelle/HOL: it is classical Higher-Order Logic with rank-1 polymorphism, Hilbert choice and the Infinity axioms. A HOL signature consists of a collection of type constructor symbols $k \in K$, which include the binary function type constructor $\rightarrow$ and the nullary bool and ind (for representing the booleans and an infinite type, respectively). The types σ, τ are built from type variables α and type constructors. The signature also contains a collection of constants $c \in C$ together with an indication of their types, $c : \tau$. Among these, we have equality, $= : \alpha \rightarrow \alpha \rightarrow$ bool, and implication, $\longrightarrow : $ bool $\rightarrow$ bool $\rightarrow$ bool. The terms t, s are built using typed (term) variables x_σ, constant instances c_σ, application and λ-abstraction. When writing concrete terms, types of variables and constants will be omitted when they can be inferred. HOL typing assigns types to terms, $t : \sigma$, in a standard way. The notation $\sigma \leq \tau$ means that σ is an instance of τ, e.g., bool list is an instance of α list, which itself is an instance of α. A formula is a term of type bool. The formula connectives and quantifiers are defined in a standard way starting from equality and implication.

In HOL, types represent "rigid" collections of elements. More flexible collections can be obtained using sets. Essentially, a set on a type σ, also called a subset of σ, is given by a predicate $S : \sigma \rightarrow$ bool. Then membership of an element a to S is given by $S\ a$ being true. HOL systems differ in the details of representing sets: some consider sets as syntactic sugar for predicates, others use a specialized type constructor for wrapping predicates, yet others consider the "type of subsets of a type" unary type constructor as a primitive. All these approaches yield essentially the same notion.

HOL deduction is parameterized by an underlying theory D. It is a system for inferring formulas starting from the formulas in D and HOL axioms (containing axioms for equality, infinity, choice, and excluded middle) and applying deduction rules (introduction and elimination of $\longrightarrow$, term and type instantiation and extensionality).

2.2   Definitional Mechanisms of HOL

Most of the systems implementing HOL follow the tradition to discourage their users from using arbitrary underlying theories D and to promote merely *definitional ones*, containing definitions of constants and types.

A *HOL constant definition* is a formula $c_\sigma = t$, where:

- c is a fresh constant of type σ
- t is a term that is closed (i.e., has no free term variables) and whose type variables are included in those of σ

HOL type definitions are more complex entities. They are based on the notion of a newly defined type β being embedded in an existing type α, i.e., being isomorphic to a given nonempty subset S of α via mappings *Abs* and *Rep*. Let $_\alpha(\beta \approx S)^{Abs}_{Rep}$ denote the formula expressing this:

$$(\forall x_\beta.\ Rep\ x \in S) \wedge (\forall x_\beta.\ Abs\ (Rep\ x) = x) \wedge (\forall y_\alpha.\ y \in S \longrightarrow Rep\ (Abs\ y) = y)$$

When the user issues a command `typedef` $\tau = S_{\sigma\ \text{set}}$, they are required to discharge the goal $S \neq \emptyset$, after which the system introduces a new type τ and two constants $\text{Abs}^\tau : \sigma \to \tau$ and $\text{Rep}^\tau : \tau \to \sigma$ and adds the axiom $_\sigma(\tau \approx S)^{\text{Abs}^\tau}_{\text{Rep}^\tau}$ to the theory.

2.3 Definitional Mechanisms of Isabelle/HOL

While a member of the HOL family, Isabelle/HOL is special w.r.t. constant definitions. Namely, a constant is allowed to be declared with a given type σ and then "overloaded" on various types τ less general than σ and mutually orthogonal. For example, we can have d declared to have type α, and then d_{bool} defined to be True and $\text{d}_{\alpha\ \text{list}}$ defined to be $[\text{d}_\alpha]$. We shall write Δ_c for the collection of all types where c has been overloaded. In the above example, $\Delta_{\text{d}} = \{\text{bool}, \alpha\ \text{list}\}$.

The mechanism of overloaded definitions offers broad expressive power. But with power also comes responsibility. The system has to make sure that the defining equations cannot form a cycle. To guarantee that, a binary constant/type dependency relation $\rightsquigarrow$ on types and constants is maintained, where $u \rightsquigarrow v$ holds true iff one of the following holds:

1. u is a constant c that was declared with type σ and v is a type in σ
2. u is a constant c defined as $c = t$ and v is a type or constant in t
3. u is a type σ defined as $\sigma = A$ and v is a type or constant in A

We write $\rightsquigarrow^\downarrow$ for (type-)substitutive closure of the constant/type dependency relation, i.e., if $p \rightsquigarrow q$, the type instances of p and q are in $\rightsquigarrow^\downarrow$. The system accepts only overloaded definitions for which $\rightsquigarrow^\downarrow$ does not contain an infinite chain.

In addition, Isabelle supports user-defined *axiomatic type classes*, which are essentially predicates on types. They effectively improve the type system with the ability to carry implicit assumptions. For example, we can define the type class $\text{finite}(\alpha)$ expressing that α has a finite number of inhabitants. Then, we are allowed to annotate type variables by such predicates, e.g., α_{finite} or $\alpha_{\text{semigroup}}$ from Sect. 1. Finally, we can substitute a type τ for α_{finite} only if τ has been previously proved to fulfill $\text{finite}(\tau)$.

The axiomatic type classes become truly useful when we use overloaded constants for their definitions. This combination allows the use of Haskell-style type classes. E.g., we can reason about arbitrary semigroups by declaring a global constant $* : \alpha \to \alpha \to \alpha$ and defining the HOL predicate $\text{semigroup}(\alpha)$ stating that $*$ is associative on α.

In this paper, we are largely concerned with results relevant for the entire HOL family of provers, but also take special care with the Isabelle/HOL maverick. Namely, we show that our local typedef proposal can be adapted to cope with Isabelle/HOL's type classes.

3   Proposal of a Logic Extension: Local Typedef

To address the limitation described in Sect. 1, we propose extending the HOL logic with a new rule for type definition with the following properties:

- It enables type definitions to be emulated inside proofs while avoiding the introduction of dependent types by a simple syntactic check.[1]
- It is natural and sound w.r.t. the standard HOL semantics à la Pitts [27] as well as with the logic of Isabelle/HOL.

To motivate the formulation of the new rule and to understand the intuition behind it, we will first look deeper into the idea behind type definitions in HOL. Let us take a purely semantic perspective and ignore the rank-1 polymorphism for a minute. Then the principle behind type definitions simply states that for all types α and nonempty subsets A of them, there exists a type β isomorphic to A:

$$\forall \alpha.\ \forall A_{\alpha\ \mathsf{set}}.\ A \neq \emptyset \longrightarrow \exists \beta.\ \exists Abs_{\alpha \to \beta}\ Rep_{\beta \to \alpha}.\ \alpha(\beta \approx A)^{Abs}_{Rep} \qquad (\star)$$

The typedef mechanism can be regarded as the result of applying a sequence of standard rules for connectives and quantifiers to $(\star)$ in a more expressive logic (notationally, we use Gentzen's sequent calculus):

1. Left $\forall$ rule of α and A with given type σ and term $S_{\sigma\ \mathsf{set}}$ (both provided by the user), and left implication rule:

$$\frac{\Gamma \vdash S \neq \emptyset \qquad \dfrac{\Gamma, \exists \beta\ Abs\ Rep.\ _\sigma(\beta \approx S)^{Abs}_{Rep} \vdash \varphi}{\Gamma, (\star) \vdash \varphi}\ \forall_L,\ \forall_L,\ \longrightarrow_L}{\Gamma \vdash \varphi}\ \text{Cut of } (\star)$$

2. Left $\exists$ rule for β, Abs and Rep, introducing some new/fresh type τ, and functions Abs^τ and Rep^τ:

$$\frac{\Gamma \vdash S \neq \emptyset \qquad \dfrac{\dfrac{\Gamma, _\sigma(\tau \approx S)^{\mathsf{Abs}^\tau}_{\mathsf{Rep}^\tau} \vdash \varphi}{\Gamma, \exists \beta\ Abs\ Rep.\ _\sigma(\beta \approx S)^{Abs}_{Rep} \vdash \varphi}\ \exists_L,\ \exists_L,\ \exists_L}{\Gamma, (\star) \vdash \varphi}\ \forall_L,\ \forall_L,\ \longrightarrow_L}{\Gamma \vdash \varphi}\ \text{Cut of } (\star)$$

The user further discharges $\Gamma \vdash S \neq \emptyset$, and therefore the overall effect of this chain is the sound addition of $_\sigma(\tau \approx S)^{\mathsf{Abs}^\tau}_{\mathsf{Rep}^\tau}$ as an extra assumption when trying to prove an arbitrary fact φ.

What we propose is to use a variant of the above (with fewer instantiations) as an actual rule:

---

[1] Dependent type theory has its own pluses and minuses. Moreover, even if we came to the conclusion that the pluses prevail, we do not know how to combine dependent types with higher-order logic and the tools built around it. Hence the avoidance of the dependent types.

- In step 1. we do not ask the user to provide concrete σ and $S_{\sigma\,\text{set}}$, but work with a type σ and a term $A_{\sigma\,\text{set}}$ that can contain type and term *variables*.
- In step 2., we only apply the left $\exists$ rule to the type β and introduce a fresh type *variable* β.

We obtain:

$$\cfrac{\Gamma \vdash A \neq \emptyset \qquad \cfrac{\cfrac{\Gamma, \exists \text{Abs Rep.}\,_\sigma (\beta \approx A)^{Abs}_{Rep} \vdash \varphi}{\Gamma, \exists \beta\ \text{Abs Rep.}\,_\sigma (\beta \approx A)^{Abs}_{Rep} \vdash \varphi}\ [\beta \text{ fresh}]\ \exists_L}{\Gamma, (\star) \vdash \varphi}\ \forall_L, \forall_L, \longrightarrow_L}{\Gamma \vdash \varphi}\ \text{Cut of } (\star)$$

To conclude, the overall rule, written (LT) as in "Local Typedef", looks as follows:

$$\cfrac{\Gamma \vdash A \neq \emptyset \qquad \Gamma \vdash (\exists \text{Abs Rep.}\,_\sigma (\beta \approx A)^{Abs}_{Rep}) \longrightarrow \varphi}{\Gamma \vdash \varphi}\ [\beta \notin A, \varphi, \Gamma]\ \text{(LT)}$$

This rule allows us to locally assume that there is a type β isomorphic to an arbitrary nonempty set A. The syntactic check $\beta \notin A, \varphi, \Gamma$ prevents an introduction of a dependent type (since A can contain term variables in general).

The above discussion merely shows that (LT) is morally correct and more importantly *natural* in the sense that it is an instance of a more general principle, namely the rule $(\star)$.

As for any extension of a logic, we have to make sure that the extension is correct.

Proposition 1. *HOL extended by the (LT) rule is consistent.*

This means that using rules of the HOL deduction system together with the (LT) rule cannot produce a proof of False. The same property holds for Isabelle/HOL.

Proposition 2. *Isabelle/HOL extended by the (LT) rule is consistent.*

The justification of both Propositions can be found in the extended version of this paper [1]. The soundness argument of the (LT) rule in HOL uses the standard HOL semantics à la Pitts [27] and the soundness of the rule in the context of Isabelle/HOL's overloading is based on our new work on proving Isabelle/HOL's consistency [21].

In the next section we will look at how the (LT) rule helps us to achieve the transformation from types to sets in HOL.

4   From Types to Sets in HOL

Let us look again at the motivating example from Sect. 1 and see how the rule (LT) allows us to achieve the relativization from a type-based theorem to a set-based theorem in HOL or Isabelle/HOL without type classes. We assume (1) is a theorem, and wish to prove (2). We fix α and $A_{\alpha\,\text{set}}$ and assume $A \neq \emptyset$. Applying (LT), we obtain a type β (represented by a fresh type variable) such that $\exists Abs\ Rep.\ {}_\alpha(\beta \approx A)^{Abs}_{Rep}$, from which we obtain Abs and Rep such that ${}_\alpha(\beta \approx A)^{Abs}_{Rep}$. From this, (1) with α instantiated to β, and the definition of lists, we obtain

$$\exists xs_{\beta\,\text{list}} \in \text{lists}\,(\text{UNIV}_{\beta\,\text{set}}).\ \mathsf{P}_{\beta\,\text{list}\rightarrow\text{bool}}\ xs.$$

Furthermore, using that Abs and Rep are isomorphisms between $A_{\alpha\,\text{set}}$ and $\text{UNIV}_{\beta\,\text{set}}$, we obtain

$$\exists xs_{\alpha\,\text{list}} \in \text{lists}\,A_{\alpha\,\text{set}}.\ \mathsf{P}_{\alpha\,\text{list}\rightarrow\text{bool}}\ xs,$$

as desired.[2]

We will consider a general case now. Let us start with a type-based theorem

$$\forall \alpha.\ \varphi[\alpha], \tag{3}$$

where $\varphi[\alpha]$ is a formula containing α. We fix α and $A_{\alpha\,\text{set}}$, assume $A \neq \emptyset$ and "define" a new type β isomorphic to A. Technically, we fix a fresh type variable β and assume

$$\exists Abs\ Rep.\ {}_\alpha(\beta \approx A)^{Abs}_{Rep}. \tag{4}$$

From the last formula, we can obtain the isomorphism Abs and Rep between β and A. Having the isomorphisms, we can carry out the relativization along them and prove

$$\varphi[\beta] \longleftrightarrow \varphi^{\text{on}}[\alpha, A_{\alpha\,\text{set}}], \tag{5}$$

where $\varphi^{\text{on}}[\alpha, A_{\alpha\,\text{set}}]$ is the relativization of $\varphi[\beta]$. In the motivational example:

$$\varphi[\beta] = \exists xs_{\beta\,\text{list}}.\ \mathsf{P}\ xs$$
$$\varphi^{\text{on}}[\alpha, A_{\alpha\,\text{set}}] = \exists xs_{\alpha\,\text{list}} \in \text{lists}\,A.\ \mathsf{P}\ xs$$

We postpone the discussion how to derive φ^{on} from φ in a principled way and how to automatically prove the equivalence between them until Sect. 6. We only appeal to the intuition here: for example, if φ contains the universal quantification $\forall x_\beta$, we replace it by the related bounded quantification $\forall x_\alpha \in A$ in φ^{on}. Or if φ contains the predicate inj $f_{\beta\rightarrow\gamma}$, we replace it by the related notion of $\text{inj}^{\text{on}}\,A_{\alpha\,\text{set}}\,f_{\alpha\rightarrow\gamma}$ in φ^{on}.

---

[2] We silently assume parametricity of the quantifier $\exists$ and P.

Since the left-hand side of the equivalence (5) is an instance of (3), we discharge the left-hand side and obtain $\varphi^{on}[\alpha, A_{\alpha\,set}]$, which does not contain the locally "defined" type β anymore. Thus we can discard β. Technically, we use the (LT) rule and remove the assumption (4). Thus we obtain the final result:

$$\forall\alpha. \; \forall A_{\alpha\,set}. \; A \neq \emptyset \longrightarrow \varphi^{on}[\alpha, A]$$

This theorem is the set-based version of $\forall\alpha. \; \varphi[\alpha]$.

We will move to Isabelle/HOL in the next section and explore how the isomorphic journey between types and sets proceeds in the environment where we are allowed to restrict type variables by type-class annotations.

5   From Types to Sets in Isabelle/HOL

Isabelle/HOL goes beyond traditional HOL and extends it by axiomatic type classes and overloading. We will explain in this section how these two features are in conflict with the algorithm described in Sect. 4 and how to circumvent these complications.

5.1   Local Axiomatic Type Classes

The first complication is the implicit assumptions on types given by the axiomatic type classes. Let us recall that α_{finite} means that α can be instantiated only with a type that we proved to fulfill the conditions of the type class finite, namely that the type must contain finitely many elements.

To explain the complication on an example, let us modify (3) to speak about types of class finite:

$$\forall\alpha_{finite}. \; \varphi[\alpha_{finite}] \tag{6}$$

Clearly, the set that is isomorphic to α_{finite} must be some nonempty set A that is *finite*. Thus as a modification of the algorithm from Sect. 4, we fix a set A and assume that it is nonempty *and* finite. As previously, we locally define a new type β isomorphic to A. Although β fulfills the condition of the type class finite, we cannot add the type into the type class since this action is allowed only at the global theory level in Isabelle and not locally in a proof context.

On the other hand, without adding β into finite we cannot continue since we need to instantiate β for α_{finite} to prove the analog of the equivalence (5). Our solution is to internalize the type-class assumption in (6) and obtain

$$\forall\alpha. \; finite(\alpha) \longrightarrow \varphi[\alpha], \tag{7}$$

where $finite(\alpha)$ is a term of type bool, which is true if and only if α is a finite type.[3] Now we can instantiate α by β and get $finite(\beta) \longrightarrow \varphi[\beta]$. Using the fact

---

[3] This is Wenzel's approach [32] to represent axiomatic type classes by internalizing them as predicates on types, i.e., constants of type $\forall\alpha.$ bool. As this particular type is not allowed in Isabelle, Wenzel uses instead α itself $\rightarrow$ bool, where α itself is a singleton type.

that the relativization of finite(β) is finite A, we apply the isomorphic translation between β and A and obtain

$$\text{finite } A \longrightarrow \varphi^{\text{on}}[\alpha, A].$$

Quantifying over the fixed variables and adding the assumptions yields the final result, the set-based version of (6):

$$\forall \alpha.\ \forall A_{\alpha\ \text{set}}.\ A \neq \emptyset \longrightarrow \text{finite } A \longrightarrow \varphi^{\text{on}}[\alpha, A]$$

The internalization of type classes (inferring (7) from (6)) is already supported by the kernel of Isabelle—thus no further work is required from us. The rule for internalization of type classes is a result of the work by Haftmann and Wenzel [13,32].

5.2 Local Overloading

In the previous section we addressed implicit assumptions on types given by axiomatic type classes and showed how to reduce the relativization of such types to the original translation algorithm by internalizing the type classes as predicates on types. As we explained in Sect. 2.3, the mechanism of Haskell-like type classes in Isabelle is more general than the notion of axiomatic type classes since additionally we are allowed to associate operations with every type class. In this respect, the type class finite is somewhat special since there are no operations associated with it.

Therefore we use semigroups as the running example in this section since semigroups require an associated operation—multiplication. A general specification of a semigroup would contain a nonempty set $A_{\alpha\ \text{set}}$, a binary operation $f_{\alpha\to\alpha\to\alpha}$ such that A is closed under f, and a proof of the specific property of semigroups that f is associative on A. We capture the last property by the predicate

$$\text{semigroup}^{\text{on}}_{\text{with}}\ A\ f = (\forall x\ y\ z \in A.\ f\ (f\ x\ y)\ z = f\ x\ (f\ y\ z)),$$

which we read along the paradigm: *a structure on the set A with operations $f_1, \ldots, f_n$.*

The realization of semigroups by type classes in Isabelle is somewhat more specific. The type σ can belong to the type class semigroup if semigroup(σ) is provable, where

$$\text{semigroup}(\alpha) \text{ iff } \forall x_\alpha\ y_\alpha\ z_\alpha.\ (x * y) * z = x * (y * z). \tag{8}$$

Notice that the associated multiplication operation is represented by the *global* overloaded constant $*_{\alpha\to\alpha\to\alpha}$, which will cause the complication.

Let us relativize $\forall \alpha_{\text{semigroup}}.\ \varphi[\alpha_{\text{semigroup}}]$ now. We fix a nonempty set A, a binary f such that A is closed under f and assume $\text{semigroup}^{\text{on}}_{\text{with}}\ A\ f$. As before, we locally define β to be isomorphic to A and obtain the respective isomorphisms Abs and Rep.

Having defined β, we want to prove that β belongs into semigroup. Using the approach from the previous section, this goal translates into proving semigroup(β), which requires that the overloaded constant $*_{\beta \to \beta \to \beta}$ used in the definition of semigroup (see (8)) must be isomorphic to f on A. In other words, we have to locally define $*_{\beta \to \beta \to \beta}$ to be a projection of f onto β, i.e., $x_\beta * y_\beta$ must equal $\mathsf{Abs}(f\ (\mathsf{Rep}\ x)\ (\mathsf{Rep}\ y))$. Although we can locally "define" a new constant (fix a fresh term variable c and assume $c = t$), we cannot overload the global symbol $*$ locally for β. This is not supported by Isabelle.

We will cope with the complication by compiling out the overloaded constant $*$ from

$$\forall \alpha.\ \mathsf{semigroup}(\alpha) \longrightarrow \varphi[\alpha] \qquad (9)$$

by the dictionary construction as follows: whenever $c = \ldots * \ldots$ (i.e., c was defined in terms of $*$ and thus depends implicitly on the overloaded meaning of $*$), define $c_{\mathsf{with}}\ f = \ldots f \ldots$ and use it instead of c. The parameter f plays a role of the dictionary here: whenever we want to use c_{with}, we have to explicitly specify how to perform multiplication in c_{with} by instantiating f. That is to say, the implicit meaning of $*$ in c was made explicit by f in c_{with}. Using this approach, we obtain:

$$\forall \alpha.\ \forall f_{\alpha \to \alpha \to \alpha}.\ \mathsf{semigroup}_{\mathsf{with}}\ f \longrightarrow \varphi_{\mathsf{with}}[\alpha, f], \qquad (10)$$

where $\mathsf{semigroup}_{\mathsf{with}}\ f_{\alpha \to \alpha \to \alpha} = (\forall x_\alpha\ y_\alpha\ z_\alpha.\ f\ (f\ x\ y)\ z = f\ x\ (f\ y\ z))$ and similarly for φ_{with}. For now, we assume that (10) is a theorem and look at how it helps us to finish the relativization and later we will explain how to derive (10) as a theorem.

Given (10), we will instantiate α with β and obtain

$$\forall f_{\beta \to \beta \to \beta}.\ \mathsf{semigroup}_{\mathsf{with}}\ f \longrightarrow \varphi_{\mathsf{with}}[\beta, f].$$

Recall that the quantification over all functions of type $\beta \to \beta \to \beta$ is isomorphic to the bounded quantification over all functions of type $\alpha \to \alpha \to \alpha$ under which $A_{\alpha\ \mathsf{set}}$ is closed.[4] The difference after compiling out the overloaded constant $*$ is that now we are isomorphically relating two bounded (local) variables from the quantification and not a global constant $*$ to a local variable.

Thus we reduced the relativization once again to the original algorithm and can obtain the set-based version

$$\forall \alpha.\ \forall A_{\alpha\ \mathsf{set}}.\ A \neq \emptyset \longrightarrow$$
$$\forall f_{\alpha \to \alpha \to \alpha}.\ (\forall x_\alpha\ y_\alpha \in A.\ f\ x\ y \in A) \longrightarrow \mathsf{semigroup}^{\mathsf{on}}_{\mathsf{with}}\ A\ f \longrightarrow \varphi^{\mathsf{on}}_{\mathsf{with}}[\alpha, A, f].$$

Let us get back to the dictionary construction. Its detailed description can be found, for example, in the paper by Krauss and Schropp [20]. We will outline the

---

[4] Let us recall that $\forall x.\ P\ x$ is a shorthand for $\mathsf{All}\ (\lambda x.\ P\ x)$ and $\forall x \in A.\ P\ x$ for $\mathsf{Ball}\ A\ (\lambda x.\ P\ x)$, where All and Ball are the HOL combinators for quantification. Thus the statement about isomorphism between the two quantifications means isomorphism between All and $\mathsf{Ball}\ A$.

process only informally here. Our task is to compile out an overloaded constant $*$ from a term s. As a first step, we transform s into $s_{\mathsf{with}}[*/f]$ such that $s = s_{\mathsf{with}}[*/f]$ and such that unfolding the definitions of all constants in s_{with} does not yield $*$ as a subterm. We proceed for every constant c in s as follows: if c has no definition, we do not do anything. If c was defined as $c = t$, we first apply the construction recursively on t and obtain t_{with} such that $t = t_{\mathsf{with}}[*/f]$; thus $c = t_{\mathsf{with}}[*/f]$. Now we define a new constant c_{with} $f = t_{\mathsf{with}}$. As c_{with} $* = c$, we replace c in s by c_{with} $*$. At the end, we obtain $s = s_{\mathsf{with}}[*/f]$ as a theorem. Notice that this procedure produces s_{with} that does not semantically depends on $*$ only if there is no type in s that depends on $*$.

Thus the above-described step applied to (9) produces

$$\forall \alpha.\ \mathsf{semigroup}_{\mathsf{with}}\ *_{\alpha\to\alpha\to\alpha} \longrightarrow \varphi_{\mathsf{with}}[\alpha,\ f_{\alpha\to\alpha\to\alpha}][*_{\alpha\to\alpha\to\alpha}/f_{\alpha\to\alpha\to\alpha}].$$

To finish the dictionary construction, we replace every occurrence of $*_{\alpha\to\alpha\to\alpha}$ by a universally quantified variable $f_{\alpha\to\alpha\to\alpha}$ and obtain (10). This derivation step is not currently allowed in Isabelle. The idea why this is a sound derivation is as follows: since $*_{\alpha\to\alpha\to\alpha}$ is a type-class operation, there exist overloaded definitions only for strict instances of $*$ (such as $*_{\mathsf{nat}\to\mathsf{nat}\to\mathsf{nat}}$) but never for $*_{\alpha\to\alpha\to\alpha}$; thus the meaning of $*_{\alpha\to\alpha\to\alpha}$ remains unrestricted. That is to say, $*_{\alpha\to\alpha\to\alpha}$ permits any interpretation and hence it must behave as a term variable. We will formulate a rule (an extension of Isabelle's logic) that allows us to perform the above-described derivation.

First, let us recall that $\leadsto^\downarrow$ is the substitutive closure of the constant/type dependency relation $\leadsto$ from Sect. 2.3 and Δ_c is the set of all types for which c was overloaded. The notation $\sigma \not\leq S$ means that σ is not an instance of any type in S. We shall write R^+ for the transitive closure of R. Now we can formulate the Unoverloading Rule (UO):

$$\frac{\varphi[c_\sigma/x_\sigma]}{\forall x_\sigma.\ \varphi}\ [\neg(u \leadsto^{\downarrow +} c_\sigma)\ \text{for any type or constant } u \text{ in } \varphi;\ \sigma \not\leq \Delta_c] \quad (\mathrm{UO})$$

This means that we can replace occurrences of the constant c_σ in φ by the universally quantified variable x_σ under the following two side conditions:

1. All types and constant instances in φ do not semantically depend on c_σ through a chain of constant and type definitions. The constraint is fulfilled in the first step of the dictionary construction since for example $\varphi_{\mathsf{with}}[\alpha, *]$ does not contain any hidden $*$s due to the construction of φ_{with}.[5]
2. There is no matching definition for c_σ. In our use case, c_σ is always a type-class operation with its most general type (e.g., $*_{\alpha\to\alpha\to\alpha}$). As already mentioned, we overload a type-class operation only for strictly more specific types (such as $*_{\mathsf{nat}\to\mathsf{nat}\to\mathsf{nat}}$) and never for its most general type and thus the condition $\sigma \not\leq \Delta_c$ must be fulfilled.

---

[5] Unless there is a type depending on $*$.

Proposition 3. Isabelle/HOL extended by the (UO) rule is consistent.[6]

Notice that the (UO) rule suggests that even in presence of *ad hoc* overloading, the polymorphic overloaded constants retain parametricity under some conditions.

In the next section, we will look at a concrete example of relativization of a formula with type classes.

5.3   Example: Relativization of Topological Spaces

We will show an example of relativization of a type-based theorem with type classes in a set-based theorem from the field of topology (addressing Immler's concern discussed in Sect. 1). The type class in question will be a topological space, which has one associated operation open : α set $\rightarrow$ bool, a predicate defining the open subsets of α. We require that the whole space is open, finite intersections of open sets are open, finite or infinite unions of open sets are open and that every two distinct points can be separated by two open sets that contain them. Such a topological space is called a T2 space and therefore we call the respective type class T2-space.

One of the basic properties of T2 spaces is the fact that every compact set is closed:

$$\forall \alpha_{\text{T2-space}}. \ \forall S_{\alpha \text{ set}}. \ \text{compact } S \longrightarrow \text{closed } S \tag{11}$$

A set is compact if every open cover of it has a finite subcover. A set is closed if its complement is open. i.e., closed S = open $(-S)$. Recall that our main motivation is to solve the problem when we have a T2 space on a proper subset of α. Let us show the translation of (11) into a set-based variant, which solves the problem. We will observe what happens to the predicate closed during the translation.

We will first internalize the type class T2-space and then abstract over its operation open via the first step of the dictionary construction. As a result, we obtain

$$\forall \alpha. \ \text{T2-space}_{\text{with}} \text{ open} \longrightarrow \forall S_{\alpha \text{ set}}. \ \text{compact}_{\text{with}} \text{ open } S \longrightarrow \text{closed}_{\text{with}} \text{ open } S,$$

where closed$_{\text{with}}$ $open$ S = $open$ $(-S)$. Let us apply (UO) and generalize over open:

$$\forall \alpha. \ \forall open_{\alpha \text{ set} \rightarrow \text{bool}}. \\ \text{T2-space}_{\text{with}} \ open \longrightarrow \forall S_{\alpha \text{ set}}. \ \text{compact}_{\text{with}} \ open \ S \longrightarrow \text{closed}_{\text{with}} \ open \ S \tag{12}$$

The last formula is a variant of (11) after we internalized the type class T2-space and compiled out its operation. Now we reduced the task to the original algorithm (using Local Typedef) from Sect. 4. As always, we fix a nonempty set

---

[6] Again, the rigorous justification of this result is based on our work on Isabelle/HOL's consistency [21] and can be found in the extended version of this paper [1].

$A_{\alpha\,\text{set}}$, locally define β to be isomorphic to A and transfer the β-instance of (12) onto the $A_{\alpha\,\text{set}}$-level:

$$\forall\alpha.\,\forall A_{\alpha\,\text{set}}.\,A \neq \emptyset \longrightarrow \forall open_{\alpha\,\text{set}\to\text{bool}}.\,\text{T2-space}^{\text{on}}_{\text{with}}\,A\,open \longrightarrow$$
$$\forall S_{\alpha\,\text{set}} \subseteq A.\,\text{compact}^{\text{on}}_{\text{with}}\,A\,open\,S \longrightarrow \text{closed}^{\text{on}}_{\text{with}}\,A\,open\,S$$

This is the set-based variant of the original theorem (11). Let us show what happened to $\text{closed}_{\text{with}}$: its relativization is defined as $\text{closed}^{\text{on}}_{\text{with}}\,A\,open\,S = open\,(-S \cap A)$. Notice that we did not have to restrict $open$ while moving between β and A (since the function does not produce any values of type β), whereas S is restricted since subsets of β correspond to subsets of A.

5.4 General Case

Having seen a concrete example, let us finally aim for the general case. Let us assume that Υ is a type class depending on the overloaded constants $*_1, \ldots, *_n$, written $\overline{*}$. We write $A \downarrow \overline{f}$ to mean that A is closed under operations $f_1, \ldots, f_n$.

The following derivation tree shows how we derive, from the type-based theorem $\vdash \forall\alpha_\Upsilon.\,\varphi[\alpha_\Upsilon]$ (the topmost formula in the tree), its set-based version (the bottommost formula). Explanation of the derivation steps follows after the tree.

$$\cfrac{\cfrac{\cfrac{\cfrac{\cfrac{\cfrac{\cfrac{\cfrac{\vdash \forall\alpha_\Upsilon.\,\varphi[\alpha_\Upsilon]}{\vdash \forall\alpha.\,\Upsilon(\alpha) \longrightarrow \varphi[\alpha]}\,(1)}{\vdash \forall\alpha.\,\Upsilon_{\text{with}}\,\overline{*}[\alpha] \longrightarrow \varphi_{\text{with}}[\alpha,\overline{f}][\overline{*}/\overline{f}]}\,(2)}{\vdash \forall\alpha.\,\forall\overline{f}[\alpha].\,\Upsilon_{\text{with}}\,\overline{f} \longrightarrow \varphi_{\text{with}}[\alpha,\overline{f}]}\,(3)}{A_{\alpha\,\text{set}} \neq \emptyset,\,_\alpha(\beta \approx A)^{Abs}_{Rep} \vdash \forall\alpha.\,\forall\overline{f}[\alpha].\,\Upsilon_{\text{with}}\,\overline{f} \longrightarrow \varphi_{\text{with}}[\alpha,\overline{f}]}\,(4)}{A_{\alpha\,\text{set}} \neq \emptyset,\,_\alpha(\beta \approx A)^{Abs}_{Rep} \vdash \forall\overline{f}[\beta].\,\Upsilon_{\text{with}}\,\overline{f} \longrightarrow \varphi_{\text{with}}[\beta,\overline{f}]}\,(5)}{A_{\alpha\,\text{set}} \neq \emptyset,\,_\alpha(\beta \approx A)^{Abs}_{Rep} \vdash \forall\overline{f}[\alpha].\,A \downarrow \overline{f} \longrightarrow \Upsilon^{\text{on}}_{\text{with}}\,A\,\overline{f} \longrightarrow \varphi^{\text{on}}_{\text{with}}[\alpha,A,\overline{f}]}\,(6)}{A_{\alpha\,\text{set}} \neq \emptyset \vdash \forall\overline{f}[\alpha].\,A \downarrow \overline{f} \longrightarrow \Upsilon^{\text{on}}_{\text{with}}\,A\,\overline{f} \longrightarrow \varphi^{\text{on}}_{\text{with}}[\alpha,A,\overline{f}]}\,(7)}{\vdash \forall\alpha.\,\forall A_{\alpha\,\text{set}}.\,A \neq \emptyset \longrightarrow \forall\overline{f}[\alpha].\,A \downarrow \overline{f} \longrightarrow \Upsilon^{\text{on}}_{\text{with}}\,A\,\overline{f} \longrightarrow \varphi^{\text{on}}_{\text{with}}[\alpha,A,\overline{f}]}\,(8)$$

Derivation steps:

(1) The class internalization from Sect. 5.1.
(2) The first step of the dictionary construction from Sect. 5.2.
(3) The Unoverloading rule (UO) from Sect. 5.2.
(4) We fix fresh α, $A_{\alpha\,\text{set}}$ and assume that A is nonempty. We locally define a new type β to be isomorphic to A; i.e., we fix fresh β, $Abs_{\alpha\to\beta}$ and $Rep_{\beta\to\alpha}$ and assume $_\alpha(\beta \approx A)^{Abs}_{Rep}$.
(5) We instantiate α in the conclusion with β.
(6) Relativization along the isomorphism between β and A—see Sect. 6.
(7) Since Abs and Rep are present only in $_\alpha(\beta \approx A)^{Abs}_{Rep}$, we can existentially quantify over them and replace the hypothesis with $\exists Abs\ Rep.\,_\alpha(\beta \approx A)^{Abs}_{Rep}$, which we discharge by the Local Typedef rule from Sect. 3, as β is not present elsewhere either (the previous step (6) removed all occurrences of β in the conclusion).

(8) We move all hypotheses into the conclusion and quantify over all fixed variables.

As previously discussed, step (2), the dictionary construction, cannot be performed for types depending on overloaded constants unless we want to compile out such types too. In the next section, we will explain the last missing piece: the relativization step (6).

Note that our approach addresses one of the long-standing user complaints: the impossibility to provide two different orders for the same type when using the type class of orders. With our approach, users can still enjoy the advantages of type classes while proving abstract properties about orders, and then only export the final product as a set-based theorem (which quantifies over all possible orders).

6   Transfer: Automated Relativization

In this section, we will describe a procedure that automatically achieves relativization of the type-based theorems. Recall that we are facing the following problem: we have two types β and α such that β is isomorphic to some (nonempty) set $A_{\alpha\,\text{set}}$, a proper subset of α, via two isomorphisms $\mathsf{Abs}_{\alpha\to\beta}$ and $\mathsf{Rep}_{\beta\to\alpha}$. In this setting, given a formula $\varphi[\beta]$, we want to find its isomorphic counterpart $\varphi^{\mathsf{on}}[\alpha, A]$ and prove $\varphi[\beta] \longleftrightarrow \varphi^{\mathsf{on}}[\alpha, A]$. Thanks to the previous work in which the first author of this paper participated [17], we can use Isabelle's Transfer tool, which automatically synthesizes the relativized formula $\varphi^{\mathsf{on}}[\alpha, A]$ and proves the equivalence with the original formula $\varphi[\beta]$.

We will sketch the main principles of the tool on the following example, where the formula (14) is a relativization of the formula (13):

$$\forall f_{\beta\to\gamma}\ xs_{\beta\ \text{list}}\ ys_{\beta\ \text{list}}.\ \mathsf{inj}\ f \longrightarrow (\mathsf{map}\ f\ xs = \mathsf{map}\ f\ ys) \longleftrightarrow (xs = ys) \qquad (13)$$

$$\forall f_{\alpha\to\gamma}.\ \forall xs\ ys \in \mathsf{lists}\ A_{\alpha\ \text{set}}.\ \mathsf{inj}_{\mathsf{on}}\ A\ f \longrightarrow (\mathsf{map}\ f\ xs = \mathsf{map}\ f\ ys) \longleftrightarrow (xs = ys) \qquad (14)$$

First of all, we reformulate the problem a little bit. We will not talk about isomorphisms Abs and Rep but express the isomorphism between A and β by a binary relation $\mathsf{T}_{\alpha\to\beta\to\text{bool}}$ such that $\mathsf{T}\ x\ y = (\mathsf{Rep}\ y = x)$. We call T a transfer relation.

To make transferring work, we require some setup. First of all, we assume that there exists a relator for every nonnullary type constructor in φ. Relators lift relations over type constructors: Related data structures have the same shape, with pointwise-related elements (e.g., the relator list_all2 for lists), and related functions map related input to related output. Concrete definitions follow:

list_all2 : $(\alpha \to \beta \to \text{bool}) \to \alpha\ \text{list} \to \beta\ \text{list} \to \text{bool}$

$(\text{list\_all2}\ R)\ xs\ ys \equiv (\text{length}\ xs = \text{length}\ ys) \wedge (\forall(x, y) \in \text{set}\ (\text{zip}\ xs\ ys).\ R\ x\ y)$

$\Mapsto : (\alpha \to \gamma \to \text{bool}) \to (\beta \to \delta \to \text{bool}) \to (\alpha \to \beta) \to (\gamma \to \delta) \to \text{bool}$

$(R \Mapsto S)\ f\ g \equiv \forall x\ y.\ R\ x\ y \longrightarrow S\ (f\ x)\ (g\ y)$

Moreover, we need a transfer rule for every constant present in φ. The transfer rules express the relationship between constants on β and α. Let us look at some examples:

$$((\mathsf{T} \Mapsto =) \Mapsto =) \ (\mathsf{inj}_{\mathsf{on}} \ A) \ \mathsf{inj} \tag{15}$$

$$((\mathsf{T} \Mapsto =) \Mapsto =) \ (\forall_{-} \in A) \ (\forall) \tag{16}$$

$$((\mathsf{list\_all2} \ \mathsf{T} \Mapsto =) \Mapsto =) \ (\forall_{-} \in \mathsf{lists} \ A) \ (\forall) \tag{17}$$

$$((\mathsf{T} \Mapsto =) \Mapsto \mathsf{list\_all2} \ \mathsf{T} \Mapsto \mathsf{list\_all2} =) \ \mathsf{map} \ \mathsf{map} \tag{18}$$

$$(\mathsf{list\_all2} \ \mathsf{T} \Mapsto \mathsf{list\_all2} \ \mathsf{T} \Mapsto =) \ (=) \ (=) \tag{19}$$

As already mentioned, the universal quantification on β corresponds to a bounded quantification over A on α ($\forall_{-} \in A$). The relation between the two constants is obtained purely syntactically: we start with the type (e.g., $(\beta \to \gamma) \to \mathsf{bool}$ for inj) and replace every type that does not change (γ and bool) by the identity relation $=$, every nonnullary type constructor by its corresponding relator ($\to$ by $\Mapsto$ and list by list_all2) and every type that changes by the corresponding transfer relation (β by T).

To derive the equivalence theorem between (13) and (14), we use the above-stated transfer rules (15)–(19) (they are leaves in the derivation tree) and combine them with the following three rules (for a bound variable, application and lambda abstraction):

$$\frac{R \, x \, y \in \Gamma}{\Gamma \vdash R \, x \, y} \qquad \frac{\Gamma_1 \vdash (R \Mapsto S) \, f \, g \quad \Gamma_2 \vdash R \, x \, y}{\Gamma_1 \cup \Gamma_2 \vdash S \, (f \, x) \, (g \, y)} \qquad \frac{\Gamma, \, R \, x \, y \vdash S \, (f \, x) \, (g \, y)}{\Gamma \vdash (R \Mapsto S) \, (\lambda x. \, f \, x) \, (\lambda y. \, g \, y)}$$

Similarity of the rules to those for typing of the simply typed lambda calculus is not a coincidence. A typing judgment here involves two terms instead of one, and a binary relation takes the place of a type. The environment Γ collects the local assumptions for bound variables. Thus since (13) and (14) are of type bool, the procedure produces (13) = (14) as the corresponding relation for bool is $=$. Having all appropriate transfer rules for all the involved constants (such as (15)–(19)), we can derive the equivalence theorem for any closed lambda term.

Of course, it is impractical to provide transfer rules for every instance of a given constant and for every particular transfer relation (T, in our example). In general, we are solving the transfer problem for some relation $R_{\alpha \to \beta \to \mathsf{bool}}$ such that R is right-total ($\forall y. \, \exists x. \, R \, x \, y$), right-unique ($\forall x \, y \, z. \, R \, x \, y \longrightarrow R \, x \, z \longrightarrow y = z$) and left-unique ($\forall x \, y \, z. \, R \, x \, z \longrightarrow R \, y \, z \longrightarrow x = y$). Notice that our concrete T fulfills all these three conditions. Instead of requiring specific transfer rules (such as (15)–(19)), we automatically derive them from general parametrized transfer rules[7] talking about basic polymorphic constants of HOL.

---

[7] These rules are related to Reynolds's relational parametricity [28] and Wadler's free theorems [31]. The Transfer tool is a working implementation of Mitchell's representation independence [24] and it demonstrates that transferring of properties across related types can be organized and largely automated using relational parametricity.

For example, we obtain (16) and (19) from the following rules:

$$\text{right\_total } R \longrightarrow ((R \mapsto =) \mapsto =) \ (\forall_{-} \in (\text{Domain } R)) \ (\forall)$$
$$\text{left\_unique } R \longrightarrow \text{right\_unique } R \longrightarrow (R \mapsto R \mapsto =) \ (=) \ (=)$$

These rules are part of Isabelle's library. Notice that, in the Transfer tool, we cannot regard type constructors as mere sets of elements, but need to impose an additional structure on them. Indeed, we required a relator structure for the involved type constructors. In addition, for standard type constructors such as list we implicitly used some ad hoc knowledge, e.g., that "lists whose elements are in A" can be expressed by lists A. For space constraints, we cannot describe the structure in detail here. We only note that the Transfer tool generates automatically the structure for every type constructor that is a natural functor (sets, finite sets, all algebraic datatypes and codatatypes) [30]. More can be found in the first author's thesis [22, Sect. 4].

Overall, the tool is able to perform the relativization completely automatically.

7   Conclusion

In this paper, we proposed extending Higher-Order Logic with a Local Typedef (LT) rule. We showed that the rule is not an ad hoc, but a natural addition to HOL in that it incarnates a semantic perspective characteristic to HOL: for every nonempty set A, there must be a type that is isomorphic to A. At the same time, (LT) is careful not to introduce dependent types since it is an open question how to integrate them into HOL.

We demonstrated how the rule allows for more flexibility in the proof development: with (LT) in place, the HOL users can enjoy succinctness and proof automation provided by types during the proof activity, while still having access to the more widely applicable, set-based theorems.

Being natural, semantically well justified and useful, we believe that the Local Typedef rule is a good candidate for HOL citizenship. We have implemented this extension in Isabelle/HOL, but its implementation should be straightforward and noninvasive in any HOL prover. And in a more expressive prover, such as HOL-Omega [16], this rule could simply be added as an axiom in the user space.

In addition, we showed that our method for relativizing theorems is applicable to types restricted by type classes as well, provided we extend the logic by a rule for compiling out overloading constants (UO). With (UO) in place, the Isabelle users can reason abstractly using type classes, while at the same time having access to different instances of the relativized result.

All along according to the motto: *Prove easily and still be flexible.*

Acknowledgements. We are indebted to the reviewers for useful comments and suggestions. We gratefully acknowledge support from DFG through grant Ni 491/13-3 and from EPSRC through grant EP/N019547/1.

References

1. From Types to Sets - Associated Web Page. http://www21.in.tum.de/~kuncar/documents/types-to-sets/
2. The HOL4 Theorem Prover. http://hol.sourceforge.net/
3. Adams, M.: Introducing HOL Zero. In: Fukuda, K., Hoeven, J., Joswig, M., Takayama, N. (eds.) ICMS 2010. LNCS, vol. 6327, pp. 142–143. Springer, Heidelberg (2010)
4. Aransay, J., Ballarin, C., Rubio, J.: A mechanized proof of the basic perturbation lemma. J. Autom. Reason. **40**(4), 271–292 (2008)
5. Asperti, A., Ricciotti, W., Sacerdoti Coen, C., Tassi, E.: The Matita interactive theorem prover. In: Bjorner, N., Sofronie-Stokkermans, V. (eds.) CADE 2011. LNCS, vol. 6803, pp. 64–69. Springer, Heidelberg (2011)
6. Bertot, Y., Castéran, P.: Interactive Theorem Proving and Program Development - Coq'Art: The Calculus of Inductive Constructions. Texts in Theoretical Computer Science. An EATCS Series. Springer, Berlin (2004)
7. Bove, A., Dybjer, P., Norell, U.: A brief overview of Agda – a functional language with dependent types. In: Berghofer, S., Nipkow, T., Urban, C., Wenzel, M. (eds.) TPHOLs 2009. LNCS, vol. 5674, pp. 73–78. Springer, Heidelberg (2009)
8. Chan, H., Norrish, M.: Mechanisation of AKS algorithm: part 1 - the main theorem. In: Urban, C., Zhang, X. (eds.) ITP 2015. LNCS, vol. 9236, pp. 117–136. Springer, New York (2015)
9. Coble, A.R.: Formalized information-theoretic proofs of privacy using the HOL4 theorem-prover. In: Borisov, N., Goldberg, I. (eds.) PETS 2008. LNCS, vol. 5134, pp. 77–98. Springer, Heidelberg (2008)
10. Constable, R.L., Allen, S.F., Bromley, H.M., Cleaveland, W.R., Cremer, J.F., Harper, R.W., Howe, D.J., Knoblock, T.B., Mendler, N.P., Panangaden, P., Sasaki, J.T., Smith, S.F.: Implementing Mathematics with the Nuprl Proof Development System. Prentice-Hall Inc, Upper Saddle River (1986)
11. Gordon, M.J.C., Melham, T.F. (eds.): Introduction to HOL: A Theorem Proving Environment for Higher Order Logic. Cambridge University Press, Cambridge (1993)
12. Grabowski, A., Kornilowicz, A., Naumowicz, A.: Mizar in a nutshell. J. Formalized Reason. **3**(2), 153–245 (2010)
13. Haftmann, F., Wenzel, M.: Constructive type classes in Isabelle. In: Altenkirch, T., McBride, C. (eds.) TYPES 2006. LNCS, vol. 4502, pp. 160–174. Springer, Heidelberg (2007)
14. Harrison, J.: HOL Light: a tutorial introduction. In: Srivas, K., Camilleri, M.A.J. (eds.) FMCAD 1996. LNCS, vol. 1166, pp. 265–269. Springer, Heidelberg (1996)
15. Hölzl, J., Heller, A.: Three chapters of measure theory in Isabelle/HOL. In: van Eekelen, M., Geuvers, H., Schmaltz, J., Wiedijk, F. (eds.) ITP 2011. LNCS, vol. 6898, pp. 135–151. Springer, Heidelberg (2011)
16. Homeier, P.V.: The HOL-Omega logic. In: Berghofer, S., Nipkow, T., Urban, C., Wenzel, M. (eds.) TPHOLs 2009. LNCS, vol. 5674, pp. 244–259. Springer, Heidelberg (2009)
17. Huffman, B., Kunčar, O.: Lifting and Transfer: a modular design for quotients in Isabelle/HOL. In: Gonthier, G., Norrish, M. (eds.) CPP 2013. LNCS, vol. 8307, pp. 131–146. Springer, Heidelberg (2013)
18. Immler, F.: Generic Construction of Probability Spaces for Paths of Stochastic Processes. Master's thesis, Institut für Informatik, Technische Universität München (2012)

19. Kaufmann, M., Manolios, P., Moore, J.S.: Computer-Aided Reasoning: An Approach. Kluwer Academic Publishers, Boston (2000)
20. Krauss, A., Schropp, A.: A mechanized translation from higher-order logic to set theory. In: Kaufmann, M., Paulson, L.C. (eds.) ITP 2010. LNCS, vol. 6172, pp. 323–338. Springer, Heidelberg (2010)
21. Kunčar, O., Popescu, A.: Comprehending Isabelle/HOL's Consistency, Draft. http://andreipopescu.uk/HOLC.html
22. Kunčar, O.: Types, Abstraction and Parametric Polymorphism in Higher-Order Logic. Ph.D. thesis, Fakultät für Informatik, Technische Universität München (2016). http://www21.in.tum.de/~kuncar/documents/kuncar-phdthesis.pdf
23. Maggesi, M.: A formalisation of metric spaces in HOL Light. In: Presented at the workshop formal mathematics for mathematicians, CICM 2015 (2015). http://www.cicm-conference.org/2015/fm4m/FMM_2015_paper_3.pdf
24. Mitchell, J.C.: Representation independence and data abstraction. In: POPL 1986, pp. 263–276. ACM (1986)
25. Nipkow, T., Paulson, L.C., Wenzel, M.: Isabelle/HOL–A Proof Assistant for Higher-Order Logic. LNCS, vol. 2283. Springer, Heidelberg (2002)
26. Nipkow, T., Paulson, L.C., Wenzel, M.: Isabelle/HOL – A Proof Assistant for Higher-Order Logic. Part of the Isabelle 2015 distribution (2015). https://isabelle.in.tum.de/dist/Isabelle2015/doc/tutorial.pdf
27. Pitts, A.: The HOL Logic. In: Gordon and Melham [11], pp. 191–232 (1993)
28. Reynolds, J.C.: Types, Abstraction and Parametric Polymorphism. In: IFIP Congress, pp. 513–523 (1983)
29. Shankar, N., Owre, S., Rushby, J.M.: PVS Tutorial. Computer Science Laboratory, SRI International (1993)
30. Traytel, D., Popescu, A., Blanchette, J.C.: Foundational, compositional (co)datatypes for higher-order logic: category theory applied to theorem proving. In: LICS 2012, pp. 596–605. IEEE (2012)
31. Wadler, P.: Theorems for Free! In: FPCA 1989, pp. 347–359. ACM (1989)
32. Wenzel, M.: Type classes and overloading in higher-order logic. In: Gunter, E.L., Felty, A.P. (eds.) TPHOLs 1997. LNCS, vol. 1275, pp. 307–322. Springer, Heidelberg (1997)
33. Wickerson, J.: Isabelle Users List, February 2013. https://lists.cam.ac.uk/mailman/htdig/cl-isabelle-users/2013-February/msg00222.html

Formalizing the Edmonds-Karp Algorithm

Peter Lammich$^{(\boxtimes)}$ and S. Reza Sefidgar$^{(\boxtimes)}$

Technische Universität München, Munich, Germany
{lammich,sefidgar}@in.tum.de

Abstract. We present a formalization of the Ford-Fulkerson method for computing the maximum flow in a network. Our formal proof closely follows a standard textbook proof, and is accessible even without being an expert in Isabelle/HOL — the interactive theorem prover used for the formalization. We then use stepwise refinement to obtain the Edmonds-Karp algorithm, and formally prove a bound on its complexity. Further refinement yields a verified implementation, whose execution time compares well to an unverified reference implementation in Java.

1 Introduction

Computing the maximum flow of a network is an important problem in graph theory. Many other problems, like maximum-bipartite-matching, edge-disjoint-paths, circulation-demand, as well as various scheduling and resource allocating problems can be reduced to it. The Ford-Fulkerson method [10] describes a class of algorithms to solve the maximum flow problem. An important instance is the Edmonds-Karp algorithm [9], which was one of the first algorithms to solve the maximum flow problem in polynomial time for the general case of networks with real-valued capacities.

In this paper, we present a formal verification of the Edmonds-Karp algorithm and its polynomial complexity bound. The formalization is conducted in the Isabelle/HOL proof assistant [27]. Stepwise refinement techniques [1,2,33] allow us to elegantly structure our verification into an abstract proof of the Ford-Fulkerson method, its instantiation to the Edmonds-Karp algorithm, and finally an efficient implementation. The abstract parts of our verification closely follow the textbook presentation of Cormen et al. [7]. Using the Isar [32] proof language, we were able to produce proofs that are accessible even to non-Isabelle experts.

While there exists another formalization of the Ford-Fulkerson method in Mizar [23][1], we are, to the best of our knowledge, the first that verify a polynomial maximum flow algorithm, prove the polynomial complexity bound, or provide a verified executable implementation. Moreover, this paper is a case study on elegantly formalizing algorithms.

The rest of this paper is structured as follows: In Sect. 2 we give a short informal introduction to the Ford-Fulkerson method. In Sect. 3, we report on our formalization of the abstract method. Section 4 gives a brief overview of the Isabelle

---

[1] Section 8.1 provides a detailed discussion.

© Springer International Publishing Switzerland 2016
J.C. Blanchette and S. Merz (Eds.): ITP 2016, LNCS 9807, pp. 219–234, 2016.
DOI: 10.1007/978-3-319-43144-4_14

Refinement Framework [17,22], which supports stepwise refinement based algorithm development in Isabelle/HOL. In Sect. 5, we report on our instantiation of the Ford-Fulkerson method to the Edmonds-Karp algorithm and the proof of its complexity. Section 6 reports on the further refinement steps required to yield an efficient implementation. Section 7 reports on benchmarking our implementation against a reference implementation of the Edmonds-Karp algorithm from Sedgewick et al. [31]. Finally, Sect. 8 gives a conclusion and discusses related and future work. The source code of our formalization is available at http://www21.in.tum.de/~lammich/edmonds_karp/.

2   The Ford-Fulkerson Method

In this section, we give a short introduction to the Ford-Fulkerson method, closely following the presentation by Cormen et al. [7].

A (flow) network is a directed graph over a finite set of vertices V and edges E, where each edge $(u,v) \in E$ is labeled by a positive real-valued capacity $c(u,v) > 0$. Moreover, there are two distinct vertices $s, t \in V$, which are called *source* and *sink*.

A *flow* f on a network is a labeling of the edges with real values satisfying the following constraints: (1) *Capacity constraint*: the flow on each edge is a non-negative value smaller or equal to the edge's capacity; (2) *Conservation constraint*: For all vertices except s and t, the sum of flows over all incoming edges is equal to the sum of flows over all outgoing edges. The value of a flow f is denoted by $|f|$, and defined to be the sum over the outgoing flows of s minus the sum over the incoming flows of s. Given a network G, the maximum flow problem is to find a flow with a maximum value among all flows of the network.

To simplify reasoning about the maximum flow problem, we assume that our network satisfies some additional constraints: (1) the source only has outgoing edges while the sink only has incoming edges; (2) if the network contains an edge (u,v) then there is no *parallel edge* (v,u) in the reverse direction[2]; and (3) every vertex of the network must be on a path from s to t. Note that any network can be transformed to a network with the aforementioned properties and the same maximum flow [7].

An important result is the relation between flows and cuts in a network. A *cut* is a partitioning of the vertices into two sets, such that one set contains the source and the other set contains the sink. The capacity of a cut is the sum of the capacities of all edges going from the source's side to the sink's side of the cut. It is easy to see that the value of any flow cannot exceed the capacity of any cut, as all flow from the source must ultimately reach the sink, and thus go through the edges of the cut. The Ford-Fulkerson theorem tightens this bound and states that the value of the maximum flow is equal to the capacity of the minimum cut.

The Ford-Fulkerson method is a corollary of this theorem. It is based on a greedy approach: Starting from a zero flow, the value of the flow is iteratively

---

[2] With $u = v$, this also implies that there are no self loops.

increased until a maximum flow is reached. In order to increase the overall flow value, it may be necessary to redirect some flow, i.e. to decrease the flow passed through specific edges. For this purpose the Ford-Fulkerson method defines the residual graph, which has edges in the same and opposite direction as the network edges. Each edge is labeled by the amount of flow that can be effectively passed along this edge, by either increasing or decreasing the flow on a network edge. Formally, the residual graph G_f of a flow f is the graph induced by the edges with positive labels according to the following labeling function c_f:

$$c_f(u, v) = \begin{cases} c(u,v) - f(u,v) & \text{if } (u,v) \in E \\ f(v,u) & \text{if } (v,u) \in E \\ 0 & \text{otherwise} \end{cases}$$

In each iteration, the Ford-Fulkerson method tries to find an *augmenting path*, i.e. a simple path from s to t in the residual graph. It then pushes as much flow as possible along this path to increase the value of the current flow. Formally, for an augmenting path p, one first defines the *residual capacity* c_p as the minimum value over all edges of p:

$$c_f(p) = \min\{c_f(u,v) : (u,v) \text{ is on } p\}$$

An augmenting path then yields a residual flow f_p, which is the flow that can be passed along this path:

$$f_p(u, v) = \begin{cases} c_f(p) & \text{if } (u,v) \text{ is on } p \\ 0 & \text{otherwise} \end{cases}$$

Finally, to actually push the flow induced by an augmenting path, we define the *augment* function $f \uparrow f'$, which augments a flow f in the network by any *augmenting flow* f', i.e. any flow in the residual graph:

$$(f \uparrow f')(u, v) = \begin{cases} f(u,v) + f'(u,v) - f'(v,u) & \text{if } (u,v) \in E \\ 0 & \text{otherwise} \end{cases}$$

Note that, for any edge in the network, the augmenting flow in the same direction is added to the flow, while the augmenting flow in the opposite direction is subtracted. This matches the intuition of passing flow in the indicated direction, by either increasing or decreasing the flow of an edge in the network.

The correctness of the Ford-Fulkerson algorithm follows from the Ford-Fulkerson theorem, which is usually stated as the following three statements being equivalent:

1. f is a maximum flow in a network G.
2. there is no augmenting path in the residual graph G_f.
3. there is a cut C in G such that the capacity of C is equal to the value of f.

The Ford-Fulkerson method does not specify how to find an augmenting path in the residual graph. There are several possible implementations with different execution times. The general method is only guaranteed to terminate for networks with rational capacities, while it may infinitely converge against non-maximal flows in the case of irrational edge capacities [10,34]. When always choosing a *shortest* augmenting path, the number of iterations is bound by $O(VE)$, even for the general case of real-valued capacities. Note that we write V and E instead of $|V|$ and $|E|$ for the number of nodes and edges if the intended meaning is clear from the context. A shortest path can be found by breadth first search (BFS) in time $O(E)$, yielding the Edmonds-Karp algorithm [9] with an overall running time of $O(VE^2)$.

3   Formalizing the Ford-Fulkerson Method

In this section, we provide a brief overview of our formalization of the Ford-Fulkerson method. In order to develop theory in the context of a fixed graph or network, we use Isabelle's concept of *locales* [3], which allows us to define named contexts that fix some parameters and assumptions. For example, the graph theory is developed in the locale `Graph`, which fixes the edge labeling function `c`, and defines the set of edges and nodes based on `c`:

```
locale Graph = fixes c :: edge ⇒ capacity begin
   definition E ≡ {(u, v). c (u, v) ≠ 0}
   definition V ≡ {u. ∃v. (u, v) ∈ E ∨ (v, u) ∈ E}
   [...]
```

Moreover, we define basic concepts like (simple, shortest) paths, and provide lemmas to reason about them.

Networks are based on graphs, and add the source and sink nodes, as well as the network assumptions:

```
locale Network = Graph + fixes s t :: node
   assumes no_incoming_s: ∀u. (u, s) ∉ E
   [...]
```

Most theorems presented in this paper are in the context of the `Network` locale.

3.1   Presentation of Proofs

Informal proofs focus on the relevant thoughts by leaving out technical details and obvious steps. In contrast, a formal proof has to precisely specify each step as the application of some inference rules. Although modern proof assistants provide high-level tactics to summarize some of these steps, formal proofs tend to be significantly more verbose than informal proofs. Moreover, formal proofs are conducted in the tactic language of the proof assistant, which is often some dialect of ML. Thus, many formal proofs are essentially programs that instruct the proof assistant how to conduct the proof. They tend to be inaccessible without a deep knowledge of the used proof assistant, in many cases requiring to replay the proof in the proof assistant in order to understand the idea behind it.

For the Isabelle/HOL proof assistant, the Isar proof language [32] allows to write formal proofs that resemble standard mathematical textbook proofs, and are accessible, to a certain extent, even for those not familiar with Isabelle/HOL. We use Isar to present our proof of the Ford-Fulkerson method such that it resembles the informal proof described by Cormen et al. [7].

As an example, consider the proof that for a flow f and a residual flow f', the augmented flow $f \uparrow f'$ is again a valid flow. In particular, one has to show that the augmented flow satisfies the capacity constraint. Cormen et al. give the following proof, which we display literally here, only replacing the references to "Equation 26.4" by "definition of $\uparrow$":

For the capacity constraint, first observe that if $(u, v) \in E$, then $c_f(v, u) = f(u, v)$. Therefore, we have $f'(v, u) \leq c_f(v, u) = f(u, v)$, and hence

$$
\begin{aligned}
(f \uparrow f')(u, v) &= f(u, v) + f'(u, v) - f'(v, u) && \text{(definition of $\uparrow$)} \\
&\geq f(u, v) + f'(u, v) - f(u, v) && \text{(because $f'(v, u) \leq f(u, v)$)} \\
&= f'(u, v) \\
&\geq 0.
\end{aligned}
$$

In addition,

$$
\begin{aligned}
(f \uparrow f')(u, v) &= f(u, v) + f'(u, v) - f'(v, u) && \text{(definition of $\uparrow$)} \\
&\leq f(u, v) + f'(u, v) && \text{(because flows are nonnegative)} \\
&\leq f(u, v) + c_f(u, v) && \text{(capacity constraint)} \\
&= f(u, v) + c(u, v) - f(u, v) && \text{(definition of c_f)} \\
&= c(u, v).
\end{aligned}
$$

In the following we present the corresponding formal proof in Isar:

```
lemma augment_flow_presv_cap:
   shows 0 ≤ (f↑f')(u,v) ∧ (f↑f')(u,v) ≤ c(u,v)
proof (cases (u,v)∈E; rule conjI)
   assume [simp]: (u,v)∈E
   hence f(u,v) = cf(v,u)
     using no_parallel_edge by (auto simp: residualGraph_def)
   also have cf(v,u) ≥ f'(v,u) using f'.capacity_const by auto
   finally have f'(v,u) ≤ f(u,v) .

   have (f↑f')(u,v) = f(u,v) + f'(u,v) - f'(v,u)
     by (auto simp: augment_def)
   also have ... ≥ f(u,v) + f'(u,v) - f(u,v)
     using ⟨f'(v,u) ≤ f(u,v)⟩ by auto
   also have ... = f'(u,v) by auto
   also have ... ≥ 0 using f'.capacity_const by auto
   finally show (f↑f')(u,v) ≥ 0 .

   have (f↑f')(u,v) = f(u,v) + f'(u,v) - f'(v,u)
     by (auto simp: augment_def)
```

```
also have ... ≤ f(u,v) + f'(u,v) using f'.capacity_const by auto
also have ... ≤ f(u,v) + cf(u,v) using f'.capacity_const by auto
also have ... = f(u,v) + c(u,v) - f(u,v)
  by (auto simp: residualGraph_def)
also have ... = c(u,v) by auto
finally show (f↑f')(u, v) ≤ c(u, v) .
qed (auto simp: augment_def cap_positive)
```

The structure of the Isar proof is exactly the same as that of the textbook proof, except that we had to also consider the case $(u, v) \notin E$, which is not mentioned in the informal proof at all, and easily discharged in our formal proof by the auto-tactic after the **qed**. We also use exactly the same justifications as the original proof, except that we had to use the fact that there are no parallel edges to show $c_f(v, u) = f(u, v)$, which is not mentioned in the original proof.

3.2 Presentation of Algorithms

In textbooks, it is common to present algorithms in pseudocode, which captures the essential ideas, but leaves open implementation details. As a formal equivalent to pseudocode, we use the monadic programming language provided by the Isabelle Refinement Framework [17,22]. For example, we define the Ford-Fulkerson method as follows:

```
definition ford_fulkerson_method ≡ do {
  let f = (λ(u,v). 0);

  (f,brk) ← while (λ(f,brk). ¬brk)
    (λ(f,brk). do {
      p ← selectp p. is_augmenting_path f p;
      case p of
        None ⇒ return (f,True)
      | Some p ⇒ return (augment c f p, False)
    })
    (f,False);
  return f
}
```

The code looks quite similar to pseudocode that one would expect in a textbook, but actually is a rigorous formal specification of the algorithm, using nondeterminism to leave open the implementation details (cf. Sect. 4). Note that we had to use the available combinators of the Isabelle Refinement Framework, which made the code slightly more verbose than we would have liked. We leave it to future work to define a set of combinators and appropriate syntax that allows for more concise presentation of pseudocode.

Finally, using the Ford-Fulkerson theorem and the verification condition generator of the Isabelle Refinement Framework, it is straightforward to prove (partial) correctness of the Ford-Fulkerson method, which is stated in Isabelle/HOL by the following theorem:

```
theorem (in Network) ford_fulkerson_method ≤ (spec f. isMaxFlow f)
```

4   Refinement in Isabelle/HOL

After having stated and proved correct an algorithm on the abstract level, the next step is to provide an (efficient) implementation. In our case, we first specialize the Ford-Fulkerson method to use shortest augmenting paths, then implement the search for shortest augmenting paths by BFS, and finally use efficient data structures to represent the abstract objects modified by the algorithm.

A natural way to achieve this formally is *stepwise refinement* [33], and in particular *refinement calculus* [1,2], which allows us to systematically transform an abstract algorithm into a more concrete one, preserving its correctness.

In Isabelle/HOL, stepwise refinement is supported by the Isabelle Refinement Framework [17,22]. It features a refinement calculus for programs phrased in a nondeterminism monad. The monad's type is a set of possible results plus an additional value that indicates a failure:

```
datatype α nres = res α set | fail
```

The operation `return x` of the monad describes the single result x, and the operation `bind m f` nondeterministically picks a result from m and executes f on it. The bind operation fails iff either $m = $ `fail`, or f may fail for a result in m,

We define the *refinement ordering* on α `nres` by lifting the subset ordering with `fail` being the greatest element. Intuitively, $m \leq m'$ means that m is a refinement of m', i.e. all possible results of m are also possible results of m'. Note that the refinement ordering is a complete lattice, and bind is monotonic. Thus, we can define recursion using a fixed-point construction [16]. Moreover, we can use the standard Isabelle/HOL constructs for if, let and case distinctions, yielding a fully fledged programming language, shallowly embedded into Isabelle/HOL's logic. For simpler usability, we define standard loop constructs (while, foreach), a syntax for postcondition specifications, and use a Haskell-like do-notation:

```
spec P ≡ spec x. P x ≡ res {x. P x}
do {x ← m; f x} ≡ bind m f
do {m; m'} ≡ bind m (λ_. m')
```

Correctness of a program m with precondition P and postcondition Q is expressed as $P \implies m \leq$ `spec r`. Q r (or, eta-contracted, just `spec Q`), which means that, if P holds, m does not fail and all possible results of m satisfy Q. Note that we provide different recursion constructs for partial and total correctness: A nonterminating total correct recursion yields `fail`, which satisfies no specification, even if joined with results from other possible runs. On the other hand, a nonterminating partial correct recursion yields `res {}`, which refines any specification and disappears when joined with other results.

The Isabelle Refinement Framework also supports data refinement. The representation of results can be changed according to a *refinement relation*, which relates concrete with abstract results: Given a relation R, $\Downarrow R$ m is the set of concrete results that are related to an abstract result in m by R. If $m = $ `fail`, then also $\Downarrow R$ $m = $ `fail`.

In a typical program development, one first comes up with an initial version m_0 of the algorithm and its specification P, Q, and shows $P \implies m_0 \leq$ spec Q. Then, one iteratively provides refined versions m_i of the algorithm, proving $m_i \leq \Downarrow R_i\, m_{i-1}$. Using transitivity and composability of data refinement, one gets $P \implies m_i \leq \Downarrow R_i \ldots R_1$ spec Q, showing the correctness of the refined algorithm. If no data refinement is performed, R_i is set to the identity relation, in which case $\Downarrow R_i$ becomes the identity function.

Various tools, including a verification condition generator, assist the user in conducting the refinement proofs by breaking them down to statements that do not contain monad constructs any more. In many cases, these verification conditions reflect the core idea of the proof precisely.

Monotonicity of the standard combinators also allows for modular refinement: Replacing a part of a program by a refined version results in a program that refines the original program. This gives us a natural formal model for statements like "we implement shortest path finding by BFS", or "we use arrays to represent the edge labeling".

5   The Edmonds-Karp Algorithm

Specializing the Ford-Fulkerson method to the Edmonds-Karp algorithm is straightforward, as finding a shortest augmenting path is a refinement of finding any augmenting path.

Considerably more effort is required to show that the resulting algorithm terminates within $O(VE)$ iterations. The idea of the proof is as follows: Edges in the opposite direction to an edge on a shortest path cannot lie on a shortest path itself. On every augmentation, at least one edge of the residual graph that lies on a shortest augmenting path is flipped. Thus, either the length of the shortest path increases, or the number of edges that lie on some shortest path decreases. As the length of a shortest path is at most V, there are no more than $O(VE)$ iterations.

Note that Cormen et al. present the same idea a bit differently: They define an edge of the residual graph being *critical* if it lies on a shortest path such that it will be flipped by augmentation. Then, they establish an upper bound of how often an edge can get critical during the algorithm. Our presentation is more suited for a formal proof, as we can directly construct a measure function from it, i.e. a function from flows to natural numbers, which decreases on every iteration and is bounded by $O(VE)$.

Formalizing the above intuitive argument was more tricky than it seemed on first glance: While it is easy to prove that, in a *fixed graph*, an edge and its opposite cannot both lie on shortest paths, generalizing the argument to a graph transformation which may add multiple flipped edges and removes at least one original edge requires some generalization of the statement. Note that a straightforward induction on the length of the augmenting path or on the number of flipped edges fails, as, after flipping the first edge, the path no longer exists.

Having defined the measure function and shown that it decreases on augmentation, it is straightforward to refine the partial correct while loop to a total correct one. Moreover, to make explicit the bound on the number of loop iterations, we instrument the loop to count its iterations, and assert the upper bound after the loop.

6   Refinement to Executable Code

In the previous section, we have presented our abstract formalization of the Edmonds-Karp algorithm, leaving open how to obtain a shortest augmenting path and how to implement the algorithm. In this section, we outline the further refinement steps that were necessary to obtain an efficient implementation.

6.1   Using Breadth First Search

A standard way to find a shortest path in a graph is breadth first search (BFS). Luckily, we had already formalized a BFS algorithm as an example for the Isabelle Refinement Framework. Unfortunately, this algorithm only computed the minimum distance between two nodes, without returning an actual path. For this project, we extended the formalization accordingly, and added an efficient imperative implementation, using the same stepwise refinement techniques as for the main algorithm. Note that the resulting BFS algorithm is independent, and can be reused for finding shortest paths in other applications.

Implementing shortest path finding by BFS in Edmonds-Karp algorithm yields a specification that algorithmically describes all major operations, but still leaves open the data structures used for implementation.

6.2   Manipulating Residual Graphs Directly

Next, we observe that the algorithm is phrased in terms of a flow, which is updated until it is maximal. In each iteration, the augmenting path is searched on the residual graph induced by the current flow. Obviously, computing the complete residual graph in each iteration is a bad idea. One solution to this problem is to compute the edges of the residual graph on the fly from the network and the current flow. Although this solution seems to be common, it has the disadvantage that for each edge of the residual graph, two (or even three) edges of the network and the flow have to be accessed. As edges of the residual graph are accessed in the inner loop, during the BFS, these operations are time critical.

After our profiling indicated a hot spot on accessing the capacity matrices of the network and the flow, we switched to an algorithm that operates on a representation of the residual graph directly. This resulted in a speed-up of roughly a factor of two. As the residual graph uniquely determines the flow (and vice versa), it is straightforward to phrase the operations directly on the residual graph. Performing a data refinement of the flow wrt. the refinement relation $\{(c_f, f) \mid f$ is a flow$\}$ then yields the desired algorithm.

6.3   Implementing Augmentation

In our abstract formalization, which matches the presentation in Sect. 2, we have formulated augmentation by first defining the residual capacity c_p of the augmenting path. Using c_p, we have defined the residual flow f_p, which was finally added to the current flow. In the refinement to operate on residual graphs, we have refined this to augment the residual graph. For the implementation, we compute the residual capacity in a first iteration over the augmenting path, and modify the residual graph in a second iteration. Proving this implementation correct is straightforward by induction on the augmenting path.

6.4   Computing Successors

In order to find an augmenting path, the BFS algorithm has to compute the successors of a node in the residual graph. Although this can be implemented on the edge labeling function by iterating over all nodes, this implementation tends to be inefficient for sparse graphs, where we would have to iterate over many possible successor nodes just to find that there is no edge.

A common optimization is to pre-compute an *adjacency map* from nodes to adjacent nodes in the network. As an edge in the residual graph is either in the same or opposite direction of a network edge, it is enough to iterate over the adjacent nodes in the network, and check whether they are actual successors in the residual graph. It is straightforward to show that this implementation actually returns the successor nodes in the residual graph.

6.5   Using Efficient Data Structures

In a final step, we have to choose efficient data structures for the algorithm.

We implement capacities as (arbitrary precision) integer numbers[3]. Note that an implementation as fixed precision numbers would also be possible, but requires additional checks on the network to ensure that no overflows can happen.

We implement nodes as natural numbers less than an upper bound N, and residual graphs are implemented by their capacity matrices, which, in turn, are realized as arrays of size $N \times N$ with row-major indexing, such that the successors of a node are close together in memory. The adjacency map of the network is implemented as an array of lists of nodes. An augmenting path is represented by a list of edges, i. e. a list of pairs of nodes.

The input network of the algorithm is represented as a function from network edges to capacities, which is tabulated into an array to obtain the initial residual graph. This gives us some flexibility in using the algorithm, as any capacity matrix representation can be converted into a function easily, without losing efficiency for read-access. Similarly, our implementation expects an adjacency

---

[3] Up to this point, the formalization models capacities as *linearly ordered integral domains*, which subsume reals, rationals, and integers. Thus, we could chose any executable number representation here.

map as additional parameter, which is then tabulated into an array. This is convenient in our context, where a preprocessing step computes the adjacency map anyway.

The output flow of the algorithm is represented as the residual graph. The user can decide how to compute the maximum flow from it. For example, in order to compute the maximum flow value, only the outgoing edges of the source node have to be computed, which is typically less expensive than computing the complete flow matrix. The correctness theorem of the algorithm abstractly states how to obtain the maximum flow from the output.

Note that there is still some optimization potential left in the choice of our data structures: For example, the BFS algorithm computes a predecessor map P. It then iterates over P to extract the shortest path as a list of edges. A subsequent iteration over this list computes the residual capacity, and a final iteration performs the augmentation. This calls for a deforestation optimization to get rid of the intermediate list, and iterate only two times over the predecessor map directly. Fortunately, iteration over the shortest path seems not to significantly contribute to the runtime of our implementation, such that we leave this optimization for future work.

Note that we performed the last refinement step using our Sepref tool [19, 20], which provides tool support for refinement from the purely functional programs of the Isabelle Refinement Framework into imperative programs expressed in Imperative/HOL [5]. The formalization of this refinement step consists of setting up the mappings between the abstract and concrete data structures, and then using Sepref to synthesize the Imperative/HOL programs and their refinement proofs. Finally, Imperative/HOL comes with a setup for the Isabelle code generator [14,15] to generate imperative programs in OCaml, SML, Scala, and Haskell.

6.6   Network Checker

Additionally, we implemented an algorithm that takes as input a list of edges, a source node, and a target node. It converts these to a capacity matrix and an adjacency map, and checks whether the resulting graph satisfies our network assumptions. We proved that this algorithm returns the correct capacity matrix and adjacency map iff the input describes a valid network, and returns a failure value otherwise.

Combining the implementation of the Edmonds-Karp algorithm with the network checker yields our final implementation, for which we can export code, and have proved the following theorem:

```
theorem
  fixes el defines c ≡ ln_α el
  shows <emp> edmonds_karp el s t <λ
      None  ⇒ ↑(¬ln_invar el ∨ ¬Network c s t)
    | Some (N, cf) ⇒
      ↑(ln_invar el ∧ Network c s t ∧ Graph.V c ⊆ {0..<N})
    * (∃ₐf. is_rflow c N f cf * ↑(Network.isMaxFlow c s t f))>ₜ
```

Note that this theorem is stated as a Hoare triple, using separation logic [21,30] assertions. There are no preconditions on the input. If the algorithm returns None, then the edge list was malformed or described a graph that does not satisfy the network assumptions. Here, `ln_invar` describes well-formed edge lists, i.e. edge lists that have no duplicate edges and only edges with positive capacity, and `ln_α` describes the mapping from (well-formed) edge lists to capacity matrices (note that we set `c ≡ ln_α el`). If the algorithm returns some number N and residual graph `cf`, then the input was a well-formed edge list that describes a valid network with at most N nodes. Moreover, the returned residual graph describes a flow f in the network, which is maximal. As the case distinction is exhaustive, this theorem states the correctness of the algorithm. Note that Isabelle/HOL does not have a notion of execution, thus total correctness of the generated code cannot be expressed. However, the program is phrased in a heap-exception monad, thus introducing some (coarse grained) notion of computation. On this level, termination can be ensured, and, indeed, the above theorem implies that all the recursions stated by recursion combinators in the monad must terminate. However, it does not guarantee that we have not injected spurious code equations like $f\ x = f\ x$, which is provable by reflexivity, but causes the generated program to diverge.

7   Benchmarking

We have compared the running time of our algorithm in SML against an unverified reference implementation in Java, taken from Sedgewick and Wayne's book on algorithms [31]. We have used MLton 20100608 [26] and OpenJDK Java 1.7.0_95, running on a standard laptop machine with a 2.8 GHz i7 quadcore processor and 16 GiB of RAM.

We have done the comparison on randomly generated sparse and dense networks, the sparse networks having a *density* (= $\frac{E}{V(V-1)}$) of 0.02, and the dense networks having a density of 0.25. Note that the maximum density for networks that satisfy our assumptions is 0.5, as we allow no parallel edges. For sparse networks, we varied the number of nodes between 1000 and 5500, for dense networks between 1000 and 1450. The results are shown in Fig. 1, in a double-logarithmic scale.

We observe that, for sparse graphs, the Java implementation is roughly faster by a factor of 1.6, while for dense graphs, our implementation is faster by a factor of 1.2. Note that the Java implementation operates on flows, while our implementation operates on residual graphs (cf. Sect. 6.2). Moreover, the Java implementation does not store the augmenting path in an intermediate list, but uses the predecessor map computed by the BFS directly (cf. Sect. 6.5). Finally note that a carefully optimized C++ implementation of the algorithm is only slightly faster than the Java implementation for sparse graphs, but roughly one order of magnitude faster for dense graphs. We leave it to future work to investigate this issue, and conclude that we were able to produce a reasonably fast verified implementation.

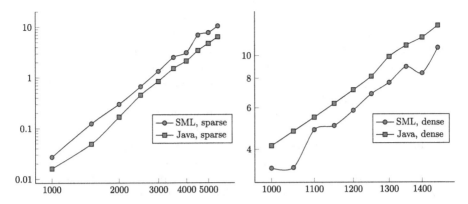

Fig. 1. Benchmark of different implementations. The x-axis shows the number of nodes, the y axis the execution time in seconds.

8   Conclusion

We have presented a verification of the Edmonds-Karp algorithm, using a stepwise refinement approach. Starting with a proof of the Ford-Fulkerson theorem, we have verified the generic Ford-Fulkerson method, specialized it to the Edmonds-Karp algorithm, and proved the upper bound $O(VE)$ for the number of outer loop iterations. We then conducted several refinement steps to derive an efficiently executable implementation of the algorithm, including a verified breadth first search algorithm to obtain shortest augmenting paths. Finally, we added a verified algorithm to check whether the input is a valid network, and generated executable code in SML. The runtime of our verified implementation compares well to that of an unverified reference implementation in Java.

Our formalization has combined several techniques to achieve an elegant and accessible formalization: Using the Isar proof language [32], we were able to provide a completely rigorous but still accessible proof of the Ford-Fulkerson theorem. The Isabelle Refinement Framework [17,22] and the Sepref tool [19,20] allowed us to present the Ford-Fulkerson method on a level of abstraction that closely resembles pseudocode presentations found in textbooks, and then formally link this presentation to an efficient implementation. Moreover, modularity of refinement allowed us to develop the breadth first search algorithm independently, and later link it to the main algorithm. The BFS algorithm can be reused as building block for other algorithms. The data structures are re-usable, too: although we had to implement the array representation of (capacity) matrices for this project, it will be added to the growing library of verified imperative data structures supported by the Sepref tool, such that it can be re-used for future formalizations.

During this project, we have learned some lessons on verified algorithm development:

- It is important to keep the levels of abstraction strictly separated. For example, when implementing the capacity function with arrays, one needs to show that it is only applied to valid nodes. However, proving that, e.g., augmenting paths only contain valid nodes is hard at this low level. Instead, one can protect the application of the capacity function by an assertion — already on a high abstraction level where it can be easily discharged. On refinement, this assertion is passed down, and ultimately available for the implementation. Optimally, one wraps the function together with an assertion of its precondition into a new constant, which is then refined independently.
- Profiling has helped a lot in identifying candidates for optimization. For example, based on profiling data, we decided to delay a possible deforestation optimization on augmenting paths, and to first refine the algorithm to operate on residual graphs directly.
- "Efficiency bugs" are as easy to introduce as for unverified software. For example, out of convenience, we implemented the successor list computation by *filter*. Profiling then indicated a hot-spot on this function. As the order of successors does not matter, we invested a bit more work to make the computation tail recursive and gained a significant speed-up. Moreover, we realized only lately that we had accidentally implemented and verified matrices with column major ordering, which have a poor cache locality for our algorithm. Changing the order resulted in another significant speed-up.

We conclude with some statistics: The formalization consists of roughly 8000 lines of proof text, where the graph theory up to the Ford-Fulkerson algorithm requires 3000 lines. The abstract Edmonds-Karp algorithm and its complexity analysis contribute 800 lines, and its implementation (including BFS) another 1700 lines. The remaining lines are contributed by the network checker and some auxiliary theories. The development of the theories required roughly 3 man month, a significant amount of this time going into a first, purely functional version of the implementation, which was later dropped in favor of the faster imperative version.

8.1   Related Work

We are only aware of one other formalization of the Ford-Fulkerson method conducted in Mizar [25] by Lee. Unfortunately, there seems to be no publication on this formalization except [23], which provides a Mizar proof script without any additional comments except that it "defines and proves correctness of Ford/Fulkerson's Maximum Network-Flow algorithm at the level of graph manipulations". Moreover, in Lee et al. [24], which is about graph representation in Mizar, the formalization is shortly mentioned, and it is clarified that it does not provide any implementation or data structure formalization. As far as we understood the Mizar proof script, it formalizes an algorithm roughly equivalent to our abstract version of the Ford-Fulkerson method. Termination is only proved for integer valued capacities.

Apart from our own work [18,28], there are several other verifications of graph algorithms and their implementations, using different techniques and proof assistants. Noschinski [29] verifies a checker for (non-)planarity certificates using a bottom-up approach. Starting at a C implementation, the AutoCorres tool [12, 13] generates a monadic representation of the program in Isabelle. Further abstractions are applied to hide low-level details like pointer manipulations and fixed size integers. Finally, a verification condition generator is used to prove the abstracted program correct. Note that their approach takes the opposite direction than ours: While they start at a concrete version of the algorithm and use abstraction steps to eliminate implementation details, we start at an abstract version, and use concretization steps to introduce implementation details.

Charguéraud [6] also uses a bottom-up approach to verify imperative programs written in a subset of OCaml, amongst them a version of Dijkstra's algorithm: A verification condition generator generates a *characteristic formula*, which reflects the semantics of the program in the logic of the Coq proof assistant [4].

8.2   Future Work

Future work includes the optimization of our implementation, and the formalization of more advanced maximum flow algorithms, like Dinic's algorithm [8] or push-relabel algorithms [11]. We expect both formalizing the abstract theory and developing efficient implementations to be challenging but realistic tasks.

References

1. Back, R.-J.: On the correctness of refinement steps in program development. Ph.D. thesis, Department of Computer Science, University of Helsinki (1978)
2. Back, R.-J., von Wright, J.: Refinement Calculus - A Systematic Introduction. Springer, New York (1998)
3. Ballarin, C.: Interpretation of locales in Isabelle: theories and proof contexts. In: Borwein, J.M., Farmer, W.M. (eds.) MKM 2006. LNCS (LNAI), vol. 4108, pp. 31–43. Springer, Heidelberg (2006)
4. Bertot, Y., Castran, P., Proving, I.T., Development, P.: Coq'Art The Calculus of Inductive Constructions, 1st edn. Springer (2010)
5. Bulwahn, L., Krauss, A., Haftmann, F., Erkök, L., Matthews, J.: Imperative functional programmingwith Isabelle/HOL. In: Mohamed, O.A., Muñoz, C., Tahar, S. (eds.) TPHOLs 2008. LNCS, vol. 5170, pp. 134–149. Springer, Heidelberg (2008)
6. Charguéraud, A.: Characteristic formulae for the verification of imperative programs. In: ICFP, pp. 418–430. ACM (2011)
7. Cormen, T.H., Leiserson, C.E., Rivest, R.L., Stein, C.: Introduction to Algorithms, 3rd edn. The MIT Press (2009)
8. Dinitz, Y.: Dinitz' algorithm: the original version and Even's version. In: Goldreich, O., Rosenberg, A.L., Selman, A.L. (eds.) Theoretical Computer Science. LNCS, vol. 3895, pp. 218–240. Springer, Heidelberg (2006)
9. Edmonds, J., Karp, R.M.: Theoretical improvements in algorithmic efficiency for network flow problems. J. ACM **19**(2), 248–264 (1972)
10. Ford, L.R., Fulkerson, D.R.: Maximal flow through a network. Can. J. Math. **8**(3), 399–404 (1956)
11. Goldberg, A.V., Tarjan, R.E.: A new approach to the maximum-flow problem. J. ACM **35**(4), 921–940 (1988)
12. Greenaway, D.: Automated proof-producing abstraction of C code. Ph.D. thesis, CSE, UNSW, Sydney, Australia (2015)

13. Greenaway, D., Andronick, J., Klein, G.: Bridging the gap: automatic verified abstraction of C. In: Beringer, L., Felty, A. (eds.) ITP 2012. LNCS, vol. 7406, pp. 99–115. Springer, Heidelberg (2012)

14. Haftmann, F.: Code generation from specifications in higher order logic. Ph.D. thesis, Technische Universität München (2009)

15. Haftmann, F., Nipkow, T.: Code generation via higher-order rewrite systems. In: Blume, M., Kobayashi, N., Vidal, G. (eds.) FLOPS 2010. LNCS, vol. 6009, pp. 103–117. Springer, Heidelberg (2010)

16. Krauss, A.: Recursive definitions of monadic functions. In: Proceedings of PAR, vol. 43, pp. 1–13 (2010)

17. Lammich, P.: Refinement for monadic programs. In: Archive of Formal Proofs, Formal proof development (2012). http://afp.sf.net/entries/Refine_Monadic.shtml

18. Lammich, P.: Verified efficient implementation of Gabow's strongly connected component algorithm. In: Klein, G., Gamboa, R. (eds.) ITP 2014. LNCS, vol. 8558, pp. 325–340. Springer, Heidelberg (2014)

19. Lammich, P.: Refinement to Imperative/HOL. In: Urban, C., Zhang, X. (eds.) ITP 2015. LNCS, vol. 9236, pp. 253–269. Springer, Heidelberg (2015)

20. Lammich, P.: Refinement based verification of imperative data structures. In: CPP, pp. 27–36. ACM (2016)

21. Lammich, P., Meis, R.: A separation logic framework for Imperative HOL. Archive of Formal Proofs, Formal proof development, Nov. 2012. http://afp.sf.net/entries/Separation_Logic_Imperative_HOL.shtml

22. Lammich, P., Tuerk, T.: Applying data refinement for monadic programs to Hopcroft's algorithm. In: Beringer, L., Felty, A. (eds.) ITP 2012. LNCS, vol. 7406, pp. 166–182. Springer, Heidelberg (2012)

23. Lee, G.: Correctnesss of Ford-Fulkersons maximum flow algorithm. Formalized Math. 13(2), 305–314 (2005)

24. Lee, G., Rudnicki, P.: Alternative aggregates in MIZAR. In: Kauers, M., Kerber, M., Miner, R., Windsteiger, W. (eds.) MKM/CALCULEMUS 2007. LNCS (LNAI), vol. 4573, pp. 327–341. Springer, Heidelberg (2007)

25. Matuszewski, R., Rudnicki, P.: Mizar: the first 30 years. Mechanized Math. Appl. 4(1), 3–24 (2005)

26. MLton Standard ML compiler. http://mlton.org/

27. Nipkow, T., Paulson, L.C., Wenzel, M. (eds.): Isabelle/HOL. LNCS, vol. 2283. Springer, Heidelberg (2002)

28. Nordhoff, B., Lammich, P.: Formalization of Dijkstra's algorithm. Archive of Formal Proofs, Formal proof development, Jan. 2012. http://afp.sf.net/entries/Dijkstra_Shortest_Path.shtml

29. Noschinski, L.: Formalizing graph theory and planarity certificates. Ph.D. thesis, Fakultät für Informatik, Technische Universität München, November 2015

30. Reynolds, J.C.: Separation logic: A logic for shared mutable data structures. In: Proceedings of Logic in Computer Science (LICS), pp. 55–74. IEEE (2002)

31. Sedgewick, R., Wayne, K.: Algorithms, 4th edn. Addison-Wesley (2011)

32. Wenzel, M.: Isar - A generic interpretative approach to readable formal proof documents. In: Bertot, Y., Dowek, G., Hirschowitz, A., Paulin, C., Théry, L. (eds.) TPHOLs 1999. LNCS, vol. 1690, pp. 167–184. Springer, Heidelberg (1999)

33. Wirth, N.: Program development by stepwise refinement. Commun. ACM 14(4), 221–227 (1971)

34. Zwick, U.: The smallest networks on which the Ford-Fulkerson maximum flow procedure may fail to terminate. Theor. Comput. Sci. 148(1), 165–170 (1995)

A Formal Proof of Cauchy's Residue Theorem

Wenda Li[(⊠)] and Lawrence C. Paulson

Computer Laboratory, University of Cambridge, Cambridge, UK
{wl302,lp15}@cam.ac.uk

Abstract. We present a formalization of Cauchy's residue theorem and two of its corollaries: the argument principle and Rouché's theorem. These results have applications to verify algorithms in computer algebra and demonstrate Isabelle/HOL's complex analysis library.

1 Introduction

Cauchy's residue theorem — along with its immediate consequences, the argument principle and Rouché's theorem — are important results for reasoning about isolated singularities and zeros of holomorphic functions in complex analysis. They are described in almost every textbook in complex analysis [3,15,16].

Our main motivation of this formalization is to certify the standard quantifier elimination procedure for real arithmetic: *cylindrical algebraic decomposition* [4]. Rouché's theorem can be used to verify a key step of this procedure: Collins' projection operation [8]. Moreover, Cauchy's residue theorem can be used to evaluate improper integrals like

$$\int_{-\infty}^{\infty} \frac{e^{itz}}{z^2 + 1} dz = \pi e^{-|t|}$$

Our main contribution[1] is two-fold:

- Our machine-assisted formalization of Cauchy's residue theorem and two of its corollaries is new, as far as we know.
- This paper also illustrates the second author's achievement of porting major analytic results, such as Cauchy's integral theorem and Cauchy's integral formula, from HOL Light [12].

The paper begins with some background on complex analysis (Sect. 2), followed by a proof of the residue theorem, then the argument principle and Rouché's theorem (3–5). Then there is a brief discussion of related work (Sect. 6) followed by conclusions (Sect. 7).

2 Background

We briefly introduce some basic complex analysis from Isabelle/HOL's Multivariate Analysis library. Most of the material in this section was first formalized in HOL Light by John Harrison [12] and later ported to Isabelle.

---

[1] Source is available from https://bitbucket.org/liwenda1990/src_itp_2016/src.

© Springer International Publishing Switzerland 2016
J.C. Blanchette and S. Merz (Eds.): ITP 2016, LNCS 9807, pp. 235–251, 2016.
DOI: 10.1007/978-3-319-43144-4_15

2.1   Contour Integrals

Given a path γ, a map from the real interval $[0, 1]$ to $\mathbb{C}$, the contour integral of a complex-valued function f on γ can be defined as

$$\oint_\gamma f = \int_0^1 f(\gamma(t))\gamma'(t)dt.$$

Because integrals do not always exist, this notion is formalised as a relation:

definition `has_contour_integral ::`
    `"(complex ⇒ complex) ⇒ complex ⇒ (real ⇒ complex) ⇒ bool"`
        `(infixr "has'_contour'_integral" 50)`
  where `"(f has_contour_integral i) g ≡`
        `((λx. f(g x) * vector_derivative g (at x within {0..1}))`
        `has_integral i) {0..1}"`

We can introduce an operator for the integral to use in situations when we know that the integral exists. This is analogous to the treatment of ordinary integrals, derivatives, etc., in HOL Light [12] as well as Isabelle/HOL.

2.2   Valid Path

In order to guarantee the existence of the contour integral, we need to place some restrictions on paths. A *valid path* is a piecewise continuously differentiable function on $[0..1]$. In plain English, the function must have a derivative on all but finitely many points, and this derivative must also be continuous.

definition `piecewise_C1_differentiable_on`
        `:: "(real ⇒ 'a :: real_normed_vector) ⇒real set ⇒ bool"`
        `(infixr "piecewise'_C1'_differentiable'_on" 50)`
  where `"f piecewise_C1_differentiable_on i ≡`
        `continuous_on i f ∧`
        `(∃s. finite s ∧ (f C1_differentiable_on (i - s)))"`

definition `valid_path :: "(real ⇒ 'a :: real_normed_vector) ⇒ bool"`
  where `"valid_path f ≡ f piecewise_C1_differentiable_on {0..1::real}"`

2.3   Winding Number

The winding number of the path γ at the point z is defined (following textbook definitions) as

$$n(\gamma, z) = \frac{1}{2\pi i} \oint_\gamma \frac{dw}{w - z}$$

A lemma to illustrate this definition is as follows:

lemma `winding_number_valid_path:`
  fixes `γ::"real ⇒ complex" and z::complex`
  assumes `"valid_path γ" and "z ∉ path_image γ"`
  shows `"winding_number γ z`
        `= 1/(2*pi*i) * contour_integral γ (λw. 1/(w - z))"`

2.4   Holomorphic Functions and Cauchy's Integral Theorem

A function is *holomorphic* if it is complex differentiable in a neighborhood of every point in its domain. The Isabelle/HOL version follows that of HOL Light:

definition holomorphic_on :: (**infixl** "(holomorphic'_on)" 50)   **where**
  "f holomorphic_on s ≡ ∀x∈s. f complex_differentiable (at x within s)"

As a starting point to reason about holomorphic functions, it is fortunate that John Harrison has made the effort to prove Cauchy's integral theorem in a rather general form:

theorem Cauchy_theorem_global:
  fixes s::"complex set" **and** f::"complex ⇒ complex"
    and γ::"real ⇒ complex"
  assumes "open s" **and** "f holomorphic_on s"
    and "valid_path γ" **and** "pathfinish γ = pathstart γ"
    and "path_image γ ⊆ s"
    and "⋀w. w ∉ s ⟹ winding_number γ w = 0"
  shows "(f has_contour_integral 0) γ"

Note, a more common statement of Cauchy's integral theorem requires the open set s to be simply connected (connected and without holes). Here, the simple connectedness is encoded by a homologous assumption

$$"⋀w. w ∉ s ⟹ winding\_number γ w = 0"$$

The reason behind this homologous assumption is that a non-simply-connected set s should contain a cycle γ and a point a within one of its holes, such that winding_number γ a is non-zero. Statements of such homologous version of Cauchy's integral theorem can be found in standard texts [1,15].

2.5   Remarks on the Porting Efforts

We have been translating the HOL Light proofs manually in order to make them more general and more legible. In the HOL Light library, all theorems are proved for $\mathbb{R}^n$, where n is a positive integer encoded as a type [14]. The type of complex numbers is identified with $\mathbb{R}^2$, and sometimes the type of real numbers must be coded as $\mathbb{R}^1$. Even worse, the ordered pair (x,y) must be coded, using complicated translations, as $\mathbb{R}^{m+n}$. We are able to eliminate virtually all mention of $\mathbb{R}^n$ in favour of more abstract notions such as topological or metric spaces. Moreover, our library consists of legible structured proofs, where the formal development is evident from the proof script alone.

3   Cauchy's Residue Theorem

As a result of Cauchy's integral theorem, if f is a holomorphic function on a simply connected open set s which contains a closed path γ, then

$$\oint_\gamma f(w) = 0$$

However, if the set s does have a hole, then Cauchy's integral theorem will not apply. For example, consider $f(w) = \frac{1}{w}$ so that f has a pole at $w = 0$, and γ is the circle path $\gamma(t) = e^{2\pi it}$:

$$\oint_{\gamma} \frac{dw}{w} = \int_0^1 \frac{1}{e^{2\pi it}} \left(\frac{d}{dt} e^{2\pi it} \right) dt = \int_0^1 2\pi i \, dt = 2\pi i \neq 0$$

Cauchy's residue theorem applies when a function is holomorphic on an open set except for a finite number of points (i.e. isolated singularities):

lemma `Residue_theorem:`
   fixes `s pts::"complex set"` and `f::"complex ⇒ complex"`
      and `γ::"real ⇒ complex"`
   assumes `"open s"` and `"connected s"` and `"finite pts"` and
         `"f holomorphic_on s - pts"` and
         `"valid_path γ"` and
         `"pathfinish γ = pathstart γ"` and
         `"path_image γ ⊆ s - pts"` and
         `"∀z. (z ∉ s) ⟶ winding_number γ z = 0"`
   shows `"contour_integral γ f`
            `= 2 * pi * i *(∑p∈pts. winding_number γ p * residue f p)"`

where `residue f p` denotes the residue of `f` at `p`, which we will describe in details in the next subsection.

Note, definitions and lemmas described from this section onwards are our original proofs (i.e. not ported from HOL Light) except where clearly noted.

3.1   Residue

A complex function f is defined to have an *isolated singularity* at point z, if f is holomorphic on an open disc centered at z but not at z.

We now define `residue f z` to be the path integral of f (divided by a constant $2\pi i$) along a small circle path around z:

definition `residue::"(complex ⇒ complex) ⇒ complex ⇒ complex"` **where**
   `"residue f z = (SOME int. ∃e>0. ∀ε>0. ε<e`
      `⟶ (f has_contour_integral 2 * pi * i * int) (circlepath z ε))"`

To actually utilize our definition, we need not only to show the existence of such integral but also its invariance when the radius of the circle path becomes sufficiently small.

lemma `base_residue:`
   fixes `s::"complex set"` and `f::"complex ⇒ complex"`
      and `e::real` and `z::complex`
   assumes `"open s"` and `"z ∈ s"` and `"e > 0"`
      and `"f holomorphic_on (s - {z})"` and `"cball z e ⊆ s"`
   shows `"(f has_contour_integral 2 *pi *i *residue f z) (circlepath z e)"`

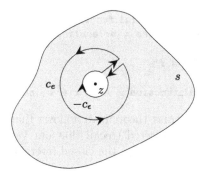

Fig. 1. Circlepath c_e and c_ϵ around an isolated singularity z

Here `cball` denotes the familiar concept of a closed ball:

definition `cball :: "'a::metric_space ⇒ real ⇒ 'a set"`
  where `"cball x e = {y. dist x y ≤ e}"`

Proof. Given two small circle path c_ϵ and c_e around z with radius ϵ and e respectively, we want to show that

$$\oint_{c_\epsilon} f = \oint_{c_e} f$$

Let γ is a line path from the end of c_e to the start of $-c_\epsilon$. As illustrated in Fig. 1, consider the path

$$\Gamma = c_e + \gamma + (-c_\epsilon) + (-\gamma)$$

where $+$ is path concatenation, and $-c_\epsilon$ and $-\gamma$ are reverse paths of c_ϵ and γ respectively. As Γ is a valid closed path and f is holomorphic on the interior of Γ, we have

$$\oint_\Gamma f = \oint_{c_e} f + \oint_\gamma f + \left(-\oint_{c_\epsilon} f\right) + \left(-\oint_\gamma f\right) = \oint_{c_e} f - \oint_{c_\epsilon} f = 0$$

hence

$$\oint_{c_\epsilon} f = \oint_{c_e} f$$

and the proof is completed.   □

3.2   Generalization to a Finite Number of Singularities

The lemma `base_residue` can be viewed as a special case of the lemma `Residue_theorem` where there is only one singularity point and γ is a circle path. In this section, we will describe our proofs of generalizing the lemma `base_residue` to a plane with finite number of singularities.

First, we need the Stone-Weierstrass theorem, which approximates continuous functions on a compact set using polynomial functions.[2]

---

[2] Our formalization is based on a proof by Brosowski and Deutsch [7].

lemma *Stone_Weierstrass_polynomial_function:*
  fixes *f* :: *"'a::euclidean_space ⇒ $\dot{b}$::euclidean_space"*
  assumes *"compact s"*
    and *"continuous_on s f"*
    and *"0 < e"*
  shows *"∃g. polynomial_function g ∧ (∀x ∈ s. norm(f x - g x) < e)"*

From the Stone-Weierstrass theorem, it follows that each open connected set is actually valid path connected (recall that our valid paths are piecewise continuous differentiable functions on the closed interval $[0, 1]$):

lemma *connected_open_polynomial_connected:*
  fixes *s::"'a::euclidean_space set"* **and** *x y::'a*
  assumes *"open s"* **and** *"connected s"* **and** *"x ∈ s"* **and** *"y ∈ s"*
  shows *"∃g. polynomial_function g ∧ path_image g ⊆ s ∧*
               pathstart g = x ∧ pathfinish g = y"

lemma *valid_path_polynomial_function:*
  fixes *p::"real ⇒ 'a::euclidean_space"*
  shows *"polynomial_function p ⟹ valid_path p"*

This yields a valid path γ on some connected punctured set such that a holomorphic function has an integral along γ:

lemma *get_integrable_path:*
  fixes *s pts::"complex set"* **and** *a b::complex* **and** *f::"complex ⇒*
complex"
  assumes *"open s"* **and** *"connected (s - pts)"* **and** *"finite pts"*
    and *"f holomorphic_on (s - pts)"*
    and *"a ∈ s - pts"* **and** *"b ∈ s - pts"*
  obtains γ **where**
    "valid_path γ" **and** *"pathstart γ = a"* **and** *"pathfinish γ = b"*
    and *"path_image γ ⊆ s-pts"* **and** *"f contour_integrable_on γ"*

Finally, we obtain a lemma that reduces the integral along γ to a sum of integrals over small circles around singularities:

lemma *Cauchy_theorem_singularities:*
  fixes *s pts::"complex set"* **and** *f::"complex ⇒ complex"*
    and γ*::"real ⇒ complex"* **and** *h::"complex ⇒ real"*
  assumes *"open s"* **and** *"connected s"* **and** *"finite pts"*
    and *"f holomorphic_on (s - pts)"* **and** *"valid_path g"*
    and *"pathfinish g = pathstart g"* **and** *"path_image g ⊆ (s - pts)"*
    and *"∀z. (z ∉ s) ⟶ winding_number g z = 0"*
    and *"∀p∈s. h p>0 ∧ (∀w∈cball p (h p). w∈s ∧ (w≠p ⟶ w ∉ pts))"*
  shows *"contour_integral g f = (∑p∈pts. winding_number g p*
        * contour_integral (circlepath p (h p)) f)"

Proof. Since the number of singularities *pts* is finite, we do induction on them. Assuming the lemma holds when there are *pts* singularities, we aim to show the lemma for $\{q\} \cup pts$.

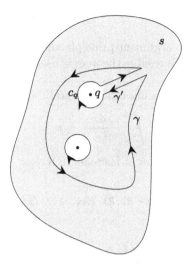

Fig. 2. Induction on the number of singularities

As illustrated in Fig. 2, suppose c_q is a (small) circle path around q, by the lemma `get_integrable_path`, we can obtain a valid path γ' from the end of γ to the start of c_q such that f has an integral along γ'.

Consider the path

$$\Gamma = \gamma + \gamma' + \underbrace{c_q + ... + c_q}_{n(\gamma,q)} + (-\gamma')$$

where $+$ is path concatenation, $n(\gamma, q)$ is the winding number of the path γ around q and $-\gamma'$ is the reverse path of γ'. We can show that Γ is a valid cycle path and the induction hypothesis applies to Γ, that is

$$\oint_\Gamma f = \sum_{p \in pts} n(\gamma, p) \oint_{c_p} f$$

hence

$$\oint_\gamma f + \oint_{\gamma'} f + n(\gamma, q) \oint_{c_q} f - \oint_{\gamma'} f = \sum_{p \in pts} n(\gamma, p) \oint_{c_p} f$$

and finally

$$\oint_\gamma f = \sum_{p \in \{q\} \cup pts} n(\gamma, p) \oint_{c_p} f$$

which concludes the proof.   □

By combining the lemma `Cauchy_theorem_singularities` and `base_residue`, we can finish the proof of Cauchy's residue theorem (i.e. the lemma `Residue_theorem`).

3.3   Applications

Besides corollaries like the argument principle and Rouché's theorem, which we will describe later, Cauchy's residue theorem is useful when evaluating improper integrals.

For example, evaluating an improper integral:

$$\int_{-\infty}^{\infty} \frac{dx}{x^2 + 1} = \pi$$

corresponds the following lemma in Isabelle/HOL:

lemma `improper_Ex`:
   `"Lim at_top (λR. integral {- R..R} (λx. 1 / (x² + 1))) = pi"`

Proof. Let

$$f(z) = \frac{1}{z^2 + 1}.$$

Now $f(z)$ is holomorphic on $\mathbb{C}$ except for two poles when $z = i$ or $z = -i$. We can then construct a semicircle path $\gamma_R + C_R$, where γ_R is a line path from $-R$ to R and C_R is an arc from R to $-R$, as illustrated in Fig. 3. From Cauchy's residue theorem, we obtain

$$\oint_{\gamma_R + C_R} f = \mathrm{Res}(f, i) = \pi$$

where $\mathrm{Res}(f, i)$ is the residue of f at i. Moreover, we have

$$\left| \oint_{C_R} f \right| \leq \frac{1}{R^2 - 1} \pi R$$

as $|f(z)|$ is bounded by $1/(R^2 - 1)$ when z is on C_R and R is large enough. Hence,

$$\oint_{C_R} f \to 0 \quad \text{when} \quad R \to \infty$$

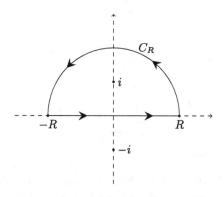

Fig. 3. A semicircle path centered at 0 with radius $R > 1$

and therefore

$$\int_{-\infty}^{\infty} \frac{dx}{x^2+1} = \oint_{\gamma_R} f = \oint_{\gamma_R+C_R} f = \pi \quad \text{when} \quad R \to \infty$$

which concludes the proof.   □

Evaluating such improper integrals was difficult for Avigad et al. [2] in their formalization of the Central Limit Theorem. We hope our development could facilitate such proofs in the future, though it may not be immediate as their proof is based on a different integration operator.

3.4 Remarks on the Formalization

It is surprising that we encountered difficulties when generalizing the lemma base_residue to the case of a finite number of poles. Several complex analysis textbooks [9,16] omit proofs for this part (giving the impression that the proof is trivial). Our statement of the lemma Cauchy_theorem_singularities follows the statement of Theorem 2.4, Chapter IV of Lang [15], but we were reluctant to follow his proof of generalizing paths to chains for fear of complicating existing theories. In the end, we devised proofs for this lemma on our own with inspiration from Stein and Shakarchi's concept of a *keyhole* [16].

Another tricky part we have encountered is in the proof of the lemma improper_Ex. When showing

$$\oint_{\gamma_R+C_R} f = \text{Res}(f,i) = \pi$$

it is necessary to show i ($-i$) is inside (outside) the semicircle path $\gamma_R + C_R$, that is,

$$n(i, \gamma_R + C_R) = 1 \wedge n(-i, \gamma_R + C_R) = 0$$

where n is the winding number operation. Such proof is straightforward for humans when looking at Fig. 3. However, to formally prove it in Isabelle/HOL, we ended up manually constructing some ad-hoc counter examples and employed proof by contradiction several times. Partially due to this reason, our proof of the lemma improper_Ex is around 300 lines of code, which we believe can be improved in the future.

4 The Argument Principle

In complex analysis, the *argument principle* is a lemma to describe the difference between the number of zeros and poles of a meromorphic[3] function.

---

[3] Holomorphic except for isolated poles.

lemma `argument_principle`:
  fixes `f h::"complex ⇒ complex"` **and** `poles s:: "complex set"`
  defines `"zeros≡{p. f p=0} - poles"`
  assumes `"open s"` **and** `"connected s"` **and**
          `"f holomorphic_on (s - poles)"` **and**
          `"h holomorphic_on s"` **and**
          `"valid_path γ"` **and**
          `"pathfinish γ = pathstart γ"` **and**
          `"path_image γ ⊆ s - (zeros ∪ poles)"` **and**
          `"∀z. (z ∉ s) ⟶ winding_number γ z = 0"` **and**
          `"finite (zeros ∪ poles)"` **and**
          `"∀p∈poles. is_pole f p"`
  shows `"contour_integral γ (λx. deriv f x * h x / f x) = 2 * pi * i *`
      `((∑p∈zeros. winding_number γ p * h p * zorder f p)`
      `- (∑p∈poles. winding_number γ p * h p * porder f p))"`

where

definition `is_pole :: "('a::topological_space ⇒ 'b::real_normed_vector)`
  `⇒ 'a ⇒ bool"` **where**
  `"is_pole f a = (LIM x (at a). f x :> at_infinity)"`

encodes the usual definition of poles (i.e. f approaches infinity as x approaches a). `zorder` and `porder` are the order of zeros and poles, which we will define in detail in the next subsection.

4.1 Zeros and Poles

A complex number z is referred as a *zero* of a holomorphic function f if $f(z) = 0$. And there is a local factorization property about $f(z)$:

lemma `holomorphic_factor_zero_Ex1`:
  fixes `s::"complex set"` **and** `f::"complex ⇒ complex"` **and** `z::complex`
  assumes `"open s"` **and** `"connected s"` **and** `"z ∈ s"` **and** `"f(z) = 0"`
    and `"f holomorphic_on s"` **and** `"∃w∈s. f w ≠ 0"`
  shows `"∃!n. ∃g r. 0 < n ∧ 0 < r ∧ ball z r ⊆ s ∧`
          `g holomorphic_on ball z r`
          `∧ (∀w∈ball z r. f w = (w-z)^n * g w ∧ g w≠0)"`

Here a *ball*, as usual, is an open neighborhood centred on a given point:

definition `ball :: "'a::metric_space ⇒ real ⇒ 'a set"`
  where `"ball x e = {y. dist x y < e}"`

Proof. [4]As f is holomorphic, f has a power expansion locally around z:

$$f(w) = \sum_{k=0}^{\infty} a_k(w - z)^k$$

and since f does not vanish identically, there exists a smallest n such that $a_n \neq 0$.

---

[4] The existence proof of such n, g and r is ported from HOL Light, while we have shown the uniqueness of n on our own.

Therefore

$$f(w) = \sum_{k=n}^{\infty} a_k(w-z)^k = (w-z)^n \sum_{k=0}^{\infty} a_{k+n}(w-z)^k = (w-z)^n g(w)$$

and the function $g(w)$ is holomorphic and non-vanishing near z due to $a_n \neq 0$.

Also, we can show that this n is unique, by assuming there exist m and another locally holomorphic function $h(w)$ such that

$$f(w) = (w-z)^n g(w) = (w-z)^m h(w)$$

and $h(w) \neq 0$. If $m > n$, then

$$g(w) = (w-z)^{m-n} h(w)$$

and this yields $g(w) \to 0$ when $w \to z$, which contradicts the fact that $g(w) \neq 0$. If $n > m$, then similarly $h(w) \to 0$ when $w \to z$, which contradicts $h(w) \neq 0$. Hence, $n = m$, and the proof is completed. □

The unique n in the lemma `holomorphic_factor_zero_Ex1` is usually referred as the *order/multiplicity of the zero* of f at z:

definition `zorder::"(complex ⇒ complex) ⇒ complex ⇒ nat"` **where**
`"zorder f z = (THE n. n>0 ∧ (∃g r. r>0 ∧ g holomorphic_on cball z r`
`∧ (∀w∈cball z r. f w = g w * (w-z)^n ∧ g w ≠0)))"`

We can also refer the complex function g in the lemma `holomorphic_factor_zero_Ex1` using Hilbert's epsilon operator in Isabelle/HOL:

definition `zer_poly::"[complex ⇒ complex, complex]⇒ complex ⇒ complex"`

where
`"zer_poly f z = (SOME g. ∃r . r>0 ∧ g holomorphic_on cball z r`
`∧ (∀w∈cball z r. f w = g w * (w-z)^(zorder f z) ∧ g w ≠0))"`

Given a complex function f that has a pole at z and is also holomorphic near (but not at) z, we know the function

$$\lambda x.\ \text{if } x = z \text{ then } 0 \text{ else } \frac{1}{f(x)}$$

has a zero at z and is holomorphic near (and at) z. On the top of the definition of the order of zeros, we can define the *order/multiplicity of the pole* of f at z:

definition `porder::"(complex ⇒ complex) ⇒ complex ⇒ nat"` **where**
`"porder f z = (let f'=(λx. if x=z then 0 else inverse (f x))`
`in zorder f' z)"`

definition `pol_poly::"[complex ⇒ complex,complex] ⇒ complex ⇒ complex"`
where
  `"pol_poly f z = (let f'=(λ x. if x=z then 0 else inverse (f x))`
                  `in inverse o zer_poly f' z)"`

and a lemma to describe a similar relationship among `f`, `porder` and `pol_poly`:

lemma `porder_exist:`
  fixes `f::"complex ⇒ complex"` **and** `s::"complex set"`
    and `z::complex`
  defines `"n≡porder f z"` **and** `"h≡pol_poly f z"`
  assumes `"open s"` **and** `"z ∈ s"`
    and `"f holomorphic_on (s - {z})"`
    and `"is_pole f z"`
  shows `"∃r. n>0 ∧ r>0 ∧ cball z r ⊆ s ∧ h holomorphic_on cball z r`
    `∧ (∀w∈cball z r. (w≠z ⟶ f w  = h w / (w-z)^n) ∧ h w ≠0)"`

Proof. With the lemma `holomorphic_factor_zero_Ex1`, we derive that there exist
n and g such that

$$\text{if } w = z \text{ then } 0 \text{ else } \frac{1}{f(w)} = (w - z)^n g(w)$$

and $g(w) \neq 0$ for w near z. Hence

$$f(w) = \frac{\frac{1}{g(w)}}{(w - z)^n} = \frac{h(w)}{(w - z)^n}$$

when $w \neq z$. Also, $h(w) \neq 0$ due to $g(w) \neq 0$. This concludes the proof. □

Moreover, `porder` and `pol_poly` can be used to construct an alternative definition of residue when the singularity is a pole.

lemma `residue_porder:`
  fixes `f::"complex ⇒ complex"` **and** `s::"complex set"`
    and `z::complex`
  defines `"n≡porder f z"` **and** `"h≡pol_poly f z"`
  assumes `"open s"` **and** `"z ∈ s"`
    and `"f holomorphic_on (s - {z})"`
    and `"is_pole f z"`
  shows `"residue f z = ((deriv ^^ (n - 1)) h z / fact (n - 1))"`

Proof. The idea behind the lemma `residue_porder` is to view $f(w)$ as $h(w)/(w - z)^n$, hence the conclusion becomes

$$\frac{1}{2\pi i} \oint_{c_\epsilon} \frac{h(w)}{(w - z)^n} dw = \frac{1}{(n - 1)!} \frac{d^{n-1}}{dw^{n-1}} h(z)$$

which can be then solved by Cauchy's integral formula. □

4.2   The Main Proof

The main idea behind the proof of the lemma `argument_principle` is to exploit the local factorization properties at zeros and poles, and then apply the Residue theorem.

Proof (the argument principle). Suppose f has a zero of order m when $w = z$. Then $f(w) = (w - z)^m g(w)$ and $g(w) \neq 0$. Hence,

$$\frac{f'(w)}{f(w)} = \frac{m}{w - z} + \frac{g'(w)}{g(w)}$$

which leads to

$$\oint_\gamma \frac{f'(w)h(w)}{f(w)} = \oint_\gamma \frac{mh(w)}{w - z} = mh(z) \tag{1}$$

since

$$\lambda w. \frac{g'(w)h(w)}{g(w)}$$

is holomorphic (g, g' and h are holomorphic and $g(w) \neq 0$).

Similarly, if f has a pole of order m when $w = z$, then $f(w) = g(w)/(w - z)^m$ and $g(w) \neq 0$. Hence,

$$\oint_\gamma \frac{f'(w)h(w)}{f(w)} = \oint_\gamma \frac{-mh(w)}{w - z} = -mh(z) \tag{2}$$

By combining (1), (2) and the lemma `Cauchy_theorem_singularities`[5], we can show

$$\oint_\gamma \frac{f'(w)h(w)}{f(w)} = 2\pi i \left(\sum_{p \in zeros} n(\gamma, p)h(p)\mathrm{zo}(f, p) - \sum_{p \in poles} n(\gamma, p)h(p)\mathrm{po}(f, p) \right)$$

where $\mathrm{zo}(f, p)$ (or $\mathrm{po}(f, p)$) is the order of zero (or pole) of f at p, and the proof is now complete.   □

4.3   Remarks

Our definitions and lemmas in Sect. 4.1 roughly follow Stein and Shakarchi [16], with one major exception. When f has a pole of order n at z, Stein and Shakarchi define residue as

$$\mathrm{Res}(f, z) = \lim_{w \to z} \frac{1}{(n - 1)!} \frac{d^{n-1}}{dw^{n-1}} [(w - z)^n f(w)]$$

while our lemma `residue_porder` states

$$\mathrm{Res}(f, z) = \frac{1}{(n - 1)!} \frac{d^{n-1}}{dw^{n-1}} h(z)$$

---

[5] Either the lemma `Cauchy_theorem_singularities` or the lemma `Residue_theorem` suffices in this place.

where $f(w) = \frac{h(w)}{(w-z)^n}$ and $h(w)$ is holomorphic and non-vanishing near z. Note, $h(w) = (w - z)^n f(w)$ only when $w \neq z$, since $f(w)$ is a pole (i.e. undefined) when $w = z$. Introducing the function h eliminates the technical difficulties of reasoning about limits formally.

5   Rouché's Theorem

Given two functions f and g holomorphic on an open set containing a path γ, if

$$|f(w)| > |g(w)|$$

for all $w \in \gamma$, then Rouché's Theorem states that f and $f + g$ have the same number of zeros counted with multiplicity and weighted with winding number:

lemma `Rouche_theorem`:
    fixes f g::"complex ⇒ complex" **and** s:: "complex set"
    defines "fg≡(λp. f p+ g p)"
    defines "zeros_fg≡{p. fg p =0}" **and** "zeros_f≡{p. f p=0}"
    assumes "open s" **and** "connected s" **and**
      "finite zeros_fg" **and** "finite zeros_f" **and**
      "f holomorphic_on s" **and** "g holomorphic_on s" **and**
      "valid_path γ" **and** "pathfinish γ = pathstart γ" **and**
      "path_image γ ⊆ s" **and**
      "∀z∈path_image γ. cmod(f z) > cmod(g z)" **and**
      "∀z. (z ∉ s) ⟶ winding_number γ z = 0"
    shows "(∑p∈zeros_fg. winding_number γ p * zorder fg p)
      = (∑p∈zeros_f. winding_number γ p * zorder f p)"

Proof. Let $\mathbb{Z}(f + g)$ and $\mathbb{Z}(f)$ be the number of zeros that $f + g$ and f has respectively (counted with multiplicity and weighted with winding number). By the argument principle, we have

$$\mathbb{Z}(f + g) = \frac{1}{2\pi i} \oint_\gamma \frac{(f+g)'}{f+g} = \frac{1}{2\pi i} \oint_\gamma \frac{f'}{f} + \frac{1}{2\pi i} \oint_\gamma \frac{(1+\frac{g}{f})'}{1+\frac{g}{f}}$$

and

$$\mathbb{Z}(f) = \frac{1}{2\pi i} \oint_\gamma \frac{f'}{f}$$

Hence, $\mathbb{Z}(f + g) = \mathbb{Z}(f)$ holds if we manage to show

$$\frac{1}{2\pi i} \oint_\gamma \frac{(1+\frac{g}{f})'}{1+\frac{g}{f}} = 0.$$

As illustrated in Fig. 4, let

$$h(w) = 1 + \frac{g(w)}{f(w)}.$$

Then the image of $h \circ \gamma$ is located within the disc of radius 1 centred at 1, since $|f(w)| > |g(w)|$ for all w on the image of γ. In this case, it can be observed that 0 lies outside $h \circ \gamma$, which leads to

$$\oint_{h \circ \gamma} \frac{dw}{w} = n(h \circ \gamma, 0) = 0$$

where $n(h \circ \gamma, 0)$ is the winding number of $h \circ \gamma$ at 0. Hence, we have

$$\oint_{\gamma} \frac{(1 + \frac{g}{f})'}{1 + \frac{g}{f}} = \int_0^1 \frac{h'(\gamma(t))}{h(\gamma(t))} \gamma'(t) dt = \int_0^1 \frac{(h \circ \gamma)'(t)}{(h \circ \gamma)(t)} dt = \oint_{h \circ \gamma} \frac{dw}{w} = 0$$

which concludes the proof.   □

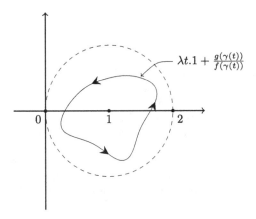

Fig. 4. The path image of $\lambda t.1 + \frac{g(\gamma(t))}{f(\gamma(t))}$ when $|f(w)| > |g(w)|$ for all w on the image of γ

Our proof of the lemma `Rouche_theorem` follows informal textbook proofs [3,15], but our formulation is more general: we do not require γ to be a regular closed path (i.e. where $n(\gamma, w) = 0 \vee n(\gamma, w) = 1$ for every complex number w that does not lie on the image of γ).

6   Related Work

HOL Light has a comprehensive library of complex analysis, on top of which the prime number theorem, the Kepler conjecture and other impressive results have been formalized [11–14]. A substantial portion of this library has been ported to Isabelle/HOL. It should be not hard to port our results to HOL Light.

Brunel [6] has described some non-constructive complex analysis in Coq, including a formalization of winding numbers. Also, there are other Coq libraries (mainly about real analysis), such as Coquelicot [5] and C-Corn [10]. However, as far as we know, Cauchy's integral theorem (which is the starting point of Cauchy's residue theorem) is not available in Coq yet.

7 Conclusion

We have described our formalization of Cauchy's residue theorem as well as two of its corollaries: the argument principle and Rouché's theorem. The proofs are drawn from multiple sources, but we were still obliged to devise some original proofs to fill the gaps. We hope our work will facilitate further work in formalizing complex analysis.

Acknowledgements. We are grateful to John Harrison for his insightful suggestions about mathematical formalization, and also to the anonymous reviewers for their useful comments on the first version of this paper. The first author was funded by the China Scholarship Council, via the CSC Cambridge Scholarship programme.

References

1. Ahlfors, L.V.: Complex Analysis: An Introduction to the Theory of Analytic Funtions of One Complex Variable. McGraw-Hill, New York (1966)
2. Avigad, J., Hölzl, J., Serafin, L.: A formally verified proof of the central limit theorem. CoRR abs/1405.7012 (2014)
3. Bak, J., Newman, D.: Complex Analysis. Springer, New York (2010)
4. Basu, S., Pollack, R., Roy, M.F.: Algorithms in Real Algebraic Geometry, vol. 10. Springer, Heidelberg (2006)
5. Boldo, S., Lelay, C., Melquiond, G.: Coquelicot: a user-friendly library of real analysis for Coq. Math. Comput. Sci. **9**(1), 41–62 (2014)
6. Brunel, A.: Non-constructive complex analysis in Coq. In: 18th International Workshop on Types for Proofs and Programs, TYPES 2011, Bergen, Norway, pp. 1–15, 8–11 September 2011
7. Bruno Brosowski, F.D.: An elementary proof of the Stone-Weierstrass theorem. Proc. Am. Math. Soc. **81**(1), 89–92 (1981)
8. Caviness, B., Johnson, J.: Quantifier Elimination and Cylindrical Algebraic Decomposition. Springer, New York (2012)
9. Conway, J.B.: Functions of One Complex Variable, vol. 11, 2nd edn. Springer, New York (1978)
10. Cruz-Filipe, L., Geuvers, H., Wiedijk, F.: C-CoRN, the constructive Coq repository at Nijmegen. In: Asperti, A., Bancerek, G., Trybulec, A. (eds.) MKM 2004. LNCS, vol. 3119, pp. 88–103. Springer, Heidelberg (2004)
11. Hales, T.C., Adams, M., Bauer, G., Dang, D.T., Harrison, J., Hoang, T.L., Kaliszyk, C., Magron, V., McLaughlin, S., Nguyen, T.T., Nguyen, T.Q., Nipkow, T., Obua, S., Pleso, J., Rute, J., Solovyev, A., Ta, A.H.T., Tran, T.N., Trieu, D.T., Urban, J., Vu, K.K., Zumkeller, R.: A formal proof of the Kepler conjecture. arXiv:1501.02155 (2015)
12. Harrison, J.: Formalizing basic complex analysis. In: Matuszewski, R., Zalewska, A. (eds.) From Insight to Proof: Festschrift in Honour of Andrzej Trybulec, vol. 10(23), pp. 151–165. University of Białystok (2007)
13. Harrison, J.: Formalizing an analytic proof of the Prime Number Theorem (dedicated to Mike Gordon on the occasion of his 60th birthday). J. Autom. Reasoning **43**, 243–261 (2009)

14. Harrison, J.: The HOL light theory of Euclidean space. J. Autom. Reasoning **50**, 173–190 (2013)
15. Lang, S.: Complex Analysis. Springer, New York (1993)
16. Stein, E.M., Shakarchi, R.: Complex Analysis, vol. 2. Princeton University Press, Princeton (2010)

Equational Reasoning with Applicative Functors

Andreas Lochbihler[(✉)] and Joshua Schneider[(✉)]

Institute of Information Security, Department of Computer Science,
ETH Zurich, Zürich, Switzerland
andreas.lochbihler@inf.ethz.ch, joshuas@student.ethz.ch

Abstract. In reasoning about effectful computations, it often suffices to focus on the effect-free parts. We present a package for automatically lifting equations to effects modelled by applicative functors. It exploits properties of the concrete functor thanks to a modular classification based on combinators. We formalise the meta theory and demonstrate the usability of our Isabelle/HOL package with two case studies. This is a first step towards practical reasoning with effectful computations.

1 Introduction

In functional languages, effectful computations are often captured by monads. Monadic effects also feature in many verification projects and formalisations (e.g., [4,8,20–22]). The reasoning support is typically tailored to the specific monad under consideration. Thus, the support must be designed and implemented anew for every monad. In contrast, reasoning about monadic effects in general has been largely neglected in the literature on both mechanised and pen-and-paper reasoning—one notable exception is [11]. One reason might be that the monadic operators can be used in too many different ways for one generic technique covering all of them.

Applicative functors (a.k.a. *idioms*) [25] are a less well-known alternative for modelling effects. Compared to monads, sequencing is restricted in idioms such that the effects of the second computation may not depend on the result of the first. In return, the structure of the computation becomes fixed. So, idiomatic expressions can be analysed statically and reasoned about. Every monad is an applicative functor and many real-world monadic programs can be expressed idiomatically [24].

In reasoning about effectful computations, only some steps involve reasoning about the effects themselves. Typically, many steps deal with the effect-free parts of the computations. In this case, one would like to get the effects out of the way, as they needlessly complicate the reasoning. *Lifting*, which transfers properties from the pure world to the effectful, can formally capture such an abstraction.

In this paper, we present a new package to automate the lifting for equational reasoning steps over effects modelled by applicative functors. We choose applicative functors (rather than monads) because they enjoy nicer properties: the computational structure is fixed and they compose. We focus on equational

© Springer International Publishing Switzerland 2016
J.C. Blanchette and S. Merz (Eds.): ITP 2016, LNCS 9807, pp. 252–273, 2016.
DOI: 10.1007/978-3-319-43144-4_16

reasoning as it is the fundamental reasoning principle in the verification of functional programs. The theory is inspired by Hinze's work on lifting [15] (see Sect. 5 for a comparison). We formalised, refined and implemented the theory in Isabelle/HOL. Our work is not specific to Isabelle; any HOL-based proof assistant could have been used.

Our package consists of two parts (see Sect. 1.2 for a usage example). First, the command **applicative** allows users to register HOL types as applicative functors. Second, two proof methods *applicative-nf* and *applicative-lifting* implement the lifting of equations as backwards-style reasoning steps over registered functors.

Crucially, lifting is generic in the applicative functor. That is, the implementation works uniformly for any applicative functor by relying only on the laws for applicative functors (Sect. 3). Yet, not all equations can be lifted in all idioms. If the functor provides additional laws like commutativity or idempotence of effects, then more equations can be lifted. So, it makes sense to specialise the reasoning infrastructure to some extent. To strike a balance between genericity and applicability, we identified classes of idioms for which the liftable equations can be characterised syntactically (Sect. 4). We achieve modularity in the implementation by using the same algorithm schema (borrowed from combinatory logic) for all classes.

Moreover, we have formalised a core idiomatic language and most of the metatheory in HOL itself (Sect. 2). In fact, we manually derived the implementation of the package from this formalisation. Thus, not only does the inference kernel check every step of our proof method, but we know that the algorithm is indeed correct.

Two small case studies on tree labelling (Sect. 1.2) and the Stern-Brocot tree (Sect. 4.4) demonstrate the reasoning power of the package and indicate directions for future extension (Sect. 6). The implementation and the examples are available online [9,23].

1.1 Background on Applicative Functors

McBride and Paterson [25] introduced the concept of applicative functors to abstract a recurring theme they observed in the programming language Haskell. An applicative functor (or idiom) is a unary type operator F (here written postfix) with two polymorphic operations $\mathsf{pure}_F :: \alpha \Rightarrow \alpha\ F$ and $(\diamond)_F :: (\alpha \Rightarrow \beta)\ F \Rightarrow \alpha\ F \Rightarrow \beta\ F$. The functor F models the effects of a computation with result type α, $\mathsf{pure}_F\ x$ represents a value x without effects, and $f \diamond_F x$ applies the function resulting from the computation f to the value of the computation x and combines their effects. That is, $(\diamond)_F$ lifts function application to effectful computations. When the functor F is clear from the context, we omit the subscript F. The infix operator $(\diamond)$ associates to the left like function application. Idioms must satisfy the following four laws called the applicative laws.

$$\text{pure}_F \ \text{id} \diamond_F x = x \qquad\qquad\qquad \text{(identity)}$$

$$\text{pure}_F \ (\circ) \diamond_F f \diamond_F g \diamond_F x = f \diamond_F (g \diamond_F x) \qquad\qquad \text{(composition)}$$

$$\text{pure}_F \ f \diamond_F \text{pure}_F \ x = \text{pure}_F \ (f \ x) \qquad\qquad \text{(homomorphism)}$$

$$f \diamond_F \text{pure}_F \ x = \text{pure}_F \ (\lambda f. \ f \ x) \diamond_F f \qquad\qquad \text{(interchange)}$$

Every monad is an applicative functor—take $\text{pure} = \text{return}$ and $f \diamond x = f \gg\!\!= (\lambda f'. \ x \gg\!\!= (\lambda x'. \ \text{return} \ (f' \ x')))$—but not vice versa. Thus, applicative functors are more general. For example, streams (**codatatype** α stream $= \alpha \prec \alpha$ stream) host an idiom (1) which cannot be extended to a monad [25]. More examples are given in Appendix A.

$$\text{pure} \ x = x \prec \text{pure} \ x \qquad\qquad (f \prec fs) \diamond (x \prec xs) = f \ x \prec (fs \diamond xs) \quad (1)$$

The more restrictive signature of ($\diamond$) imposes a fixed structure on the computation. In fact, any expression built from the applicative operators can be transformed into canonical form $\text{pure} \ f \diamond x_1 \diamond \ldots \diamond x_n$ using the applicative laws (see Sect. 3.1), namely "a single pure function [. . .] applied to the effectful parts in depth-first order" [25].

1.2   Motivating Example: Tree Labelling

To illustrate lifting and its benefits, we consider the problem of labelling a binary tree with distinct numbers. This example has been suggested by Hutton and Fulger [19]; Gibbons et al. [10, 11] explore it further. The classic solution shown below uses a state monad with an operation $\text{fresh} = \textbf{do} \ \{ \ x \leftarrow \text{get}; \ \text{put} \ (x + 1); \ \text{return} \ x \ \}$ to generate the labels, where we use Haskell-style **do** notation.

> **datatype** α tree $= \text{L} \ \alpha \ | \ \text{N} \ (\alpha \ \text{tree}) \ (\alpha \ \text{tree})$
> $\text{lbl} \ (\text{L} \ \_) \ \ = \textbf{do} \ \{ \ x \leftarrow \text{fresh}; \ \text{return} \ (\text{L} \ x) \ \}$
> $\text{lbl} \ (\text{N} \ l \ r) = \textbf{do} \ \{ \ l' \leftarrow \text{lbl} \ l; \ r' \leftarrow \text{lbl} \ r; \ \text{return} \ (\text{N} \ l' \ r') \ \}$

Hutton and Fulger expressed lbl concisely in the state idiom as follows.

$$\text{lbl} \ (\text{L} \ \_) = \text{pure} \ \text{L} \diamond \text{fresh} \qquad\qquad \text{lbl} \ (\text{N} \ l \ r) = \text{pure} \ \text{N} \diamond \text{lbl} \ l \diamond \text{lbl} \ r$$

The task is to prove that the labels in the resulting tree are distinct, i.e., $\text{pure} \ \text{lbls} \diamond \text{lbl} \ t$ returns only distinct lists where the function lbls given below extracts the labels in a tree and (++) concatenates two lists.

$$\text{lbls} \ (\text{L} \ x) = [x] \qquad\qquad \text{lbls} \ (\text{N} \ l \ r) = \text{lbls} \ l \ + \!\!+ \ \text{lbls} \ r \qquad (2)$$

As a warm-up, we prove that the list of labels in a relabelled tree equals a relabelling of the list of labels in the original tree. Formally, define relabelling for lists by $\text{lbl}' \ [] = \text{pure} \ []$ and $\text{lbl}' \ (x \cdot l) = \text{pure} \ (\cdot) \diamond \text{fresh} \diamond \text{lbl}' \ l$. We show $\text{pure} \ \text{lbls} \diamond \text{lbl} \ t = \text{lbl}' \ (\text{lbls} \ t)$ by induction on t. In each case, we first unfold the defining equations for lbl, lbl' and lbls, possibly the induction hypotheses and the auxiliary fact $\text{lbl}' \ (l \ + \!\!+ \ l') = \text{pure} \ (+ \!\!+) \diamond \text{lbl}' \ l \diamond \text{lbl}' \ l'$, which we prove similarly by induction on l and lifting the defining equations of $(++)$. Then, the two subgoals below remain.

$$\begin{aligned}
\textsf{pure lbls} \diamond (\textsf{pure L} \diamond \textsf{fresh}) &= \textsf{pure } (\cdot) \diamond \textsf{fresh} \diamond \textsf{pure } [\,] \\
\textsf{pure lbls} \diamond (\textsf{pure N} \diamond \textsf{lbl } l \diamond \textsf{lbl } r) &= \textsf{pure } (++) \diamond (\textsf{pure lbls} \diamond \textsf{lbl } l) \diamond (\textsf{pure lbls} \diamond \textsf{lbl } r)
\end{aligned} \quad (3)$$

Observe that they are precisely liftings of (2). We recover the latter equations if we remove all pures, replace $\diamond$ by function application and generalise fresh to a variable x.

Our new proof method *applicative-nf* performs this transition after the state idiom has been registered with the package using the command **applicative**. Registration takes the name of the idiom (here "state") and HOL terms for the applicative operations. Then, the applicative laws must be proven, which the proof method *auto* automates in this case.

applicative state **for** pure : $\textsf{pure}_{\textsf{state}}$ ap : $(\diamond)_{\textsf{state}}$ **by**(*auto simp:* $(\diamond)_{\textsf{state}}$-def)

After the registration, both subgoals in (3) are discharged automatically using the new proof method *applicative-nf* and term rewriting. The crucial point is that we have never unfolded the definitions of the state idiom or fresh. Thus, we do not break the abstraction.

Let us now return to the actual task. The main difficulty is stating distinctness of labels without looking into the state monad, as this would break the abstraction. Gibbons and Hinze [11] suggested to use an error monad; we adapt their idea to idioms. We consider the composition of the state idiom with the error idiom derived from the option monad (see Appendix A). Then, the correctness of fresh is expressed abstractly as

$$\forall n.\ \textsf{pure}_{\textsf{state}} \ (\textsf{assert distinct}) \diamond \textsf{nfresh } n = \textsf{nfresh } n \quad (4)$$

where $\textsf{pure}_{\textsf{state}}$ lifts the assertion from the error idiom to the state-error idiom. Further, the function $\textsf{nfresh } n = \textsf{pure}_{\textsf{state}} \ \textsf{pure}_{\textsf{option}} \diamond \textsf{repeat } n \ \textsf{fresh}$ produces n fresh symbols, where $\textsf{repeat } n \, x$ repeats the computation x for n times and collects the results in a list. Again, observe that $\textsf{pure}_{\textsf{state}} \textsf{pure}_{\textsf{option}}$ embeds the computation $\textsf{repeat } n \ \textsf{fresh}$ from the state idiom into the state-error idiom.

Moreover, in the error idiom, we can combine the extraction of labels from a tree and the test for disjointness in the subtrees of a node into a single function $\textsf{dlbls} :: \alpha \ \textsf{tree} \Rightarrow \alpha \ \textsf{list option}$. Here, $\textsf{disjoint } l \ l' \longleftrightarrow \textsf{set } l \cap \textsf{set } l' = \emptyset$ tests whether the lists l and l' are disjoint and $\lceil f \rceil$ uncurries the function f.

$$\begin{aligned}
\textsf{dlbls } (\textsf{L } x) &= \textsf{pure } [x] \\
\textsf{dlbls } (\textsf{N } l \ r) &= \textsf{pure } \lceil (++) \rceil \diamond (\textsf{assert } \lceil \textsf{disjoint} \rceil \diamond (\textsf{pure Pair} \diamond \textsf{dlbls } l \diamond \textsf{dlbls } r))
\end{aligned}$$

Finally, we can state correctness of lbl as follows ($\textsf{lvs } t$ counts the leaves in t).

Lemma 1. *If* (4) *holds, then* $\textsf{pure}_{\textsf{state}} \ \textsf{dlbls} \diamond_{\textsf{state}} \textsf{lbl } t = \textsf{nfresh } (\textsf{lvs } t)$.

Figure 1 shows the complete proof in Isar. The base case for L merely lifts the equation $\textsf{dlbls } (\textsf{L } x) = \textsf{pure } [x]$, which lives in the option idiom, to the state idiom.

lemma 1: **assumes** nfresh: $\forall n.$ pure$_{\text{state}}$ (assert distinct) $\diamond$ nfresh n = nfresh n
       shows pure$_{\text{state}}$ dlbls $\diamond_{\text{state}}$ lbl t = nfresh (lvs t)
proof (*induction t*)
  show pure dlbls $\diamond$ lbl (L x) = nfresh (lvs (L x)) **for** x
    unfolding lbl.simps lvs.simps repeat.simps **by** *applicative-nf* *simp*
next
  fix l r
  assume IH$_1$: pure dlbls $\diamond$ lbl l = nfresh (lvs l) **and** IH$_2$: pure dlbls $\diamond$ lbl r = nfresh (lvs r)
  let $?f = \lambda l\ r.$ pure $\lceil(+\!+)\rceil \diamond$ (assert $\lceil$disjoint$\rceil$ (pure Pair $\diamond$ l $\diamond$ r))
  have pure dlbls $\diamond$ lbl (N l r) = pure $?f$ $\diamond$ (pure dlbls $\diamond$ lbl l) $\diamond$ (pure dlbls $\diamond$ lbl r)
    unfolding lbl.simps **by** *applicative-nf* *simp*
  also have $\ldots$ = pure $?f$ $\diamond$ (pure (assert distinct) $\diamond$ nfresh (lvs l)) $\diamond$
                                     (pure (assert distinct) $\diamond$ nfresh (lvs r))
    unfolding IH$_1$ IH$_2$ nfresh ..
  also have $\ldots$ = pure (assert distinct) $\diamond$ nfresh (lvs (N l r))
    unfolding lvs.simps repeat-plus **by** *applicative-nf* *simp*
  also have $\ldots$ = nfresh (lvs (N l r)) **by** (*rule* nfresh)
  finally show pure dlbls $\diamond$ lbl (N l r) = nfresh (lvs (N l r)) .
qed

Fig. 1. Isar proof of Lemma 1. Our proof method is highlighted in grey. X.simps refers
to the defining equations of the function X, and repeat-plus to distributivity of repeat
over $(+)$.

As our package performs the lifting, the proof in Isabelle is automatic. The case
for N requires four reasoning steps, two of which involve lifting identities from
the error idiom to the state-error idiom; the other steps apply the induction
hypotheses and the assumption (4). This compares favourably with Gibbons'
and Hinze's proof for the monadic version [11], which requires one and a half
columns on paper and has not been checked mechanically.

2   Modelling Applicative Functors in HOL

We model applicative functors in HOL twice. In our first model, a functor F
appears as a family of HOL types α F with HOL terms for the applicative
operations. The package implementation rests on this basis. The second model
is used for the meta theory: we formalise a deep embedding of the idiomatic
language in order to establish the proof procedure and argue for its correctness.

Applicative functors in HOL. The general notion of an applicative functor cannot
be expressed in HOL for the same reasons as for monads [16]: there are no type
constructor variables in HOL, and the applicative operations occur with several
different type instances in the applicative laws. This implies that the first model
cannot be based on definitions and theorems that are generic in the functor.
Instead, we necessarily always work with a concrete applicative functor. The
corresponding terms and theorems can be expressed in HOL, as HOL constants
may be polymorphic. Our package keeps track of a set of applicative functors.

Thus, the user must register a functor using the command **applicative** before the package can use it. During the registration, the user must prove the applicative laws (and possibly additional properties, see Sect. 4).

The package follows the traditional LCF style of prover extensions. The proof procedures are written in ML, where they analyse the HOL terms syntactically and compose the inference rules accordingly. This approach shifts the problem of (functor) polymorphism to the program level, where it is easily solved. As usual, the logical kernel ensures that all the proofs are sound. Conversely, the proof procedures themselves should be correct, namely terminate and never attempt to create an invalid proof. Arguments to support this are in turn supplied by the meta theory studied in the second model.

Deep embedding of applicative functors. The second model serves two purposes: it formalises the meta theory and we derive our package implementation from it. The model separates the notion of idiomatic terms from the concrete applicative functor and represents them syntactically. Idiomatic terms consist of pure terms Pure t, opaque terms Opq x, and applications $t_1 \diamond t_2$ of two idiomatic terms.

$$\textbf{datatype } \alpha \text{ iterm} = \text{Pure term} \mid \text{Opq } \alpha \mid \alpha \text{ iterm} \diamond \alpha \text{ iterm}$$

Opaque terms represent impure (effectful) values of the functor, or variables in general; their representation is left abstract as it is irrelevant to most definitions in the model. In contrast, pure's argument needs some structure such that the applicative laws can be stated. To that end, we reuse Nipkow's formalisation of the untyped λ-calculus with de Bruijn indices [26]: **datatype** term = Var nat $\mid$ term \$ term $\mid$ Abs term. For readability, we write such terms as abstractions with named variables, e.g. $\overline{\lambda}x.\ x \equiv$ Abs (Var 0), where the notation $\overline{\lambda}$ distinguishes them from HOL terms. The relation $\simeq_{\beta\eta}$ on term denotes equivalence of λ-terms due to $\beta\eta$-conversion.

The model ignores types, as they are not needed for the meta theory. Thus, we cannot express type safety of our algorithms, either. However, we do not foresee any difficulties in extending our model with types, e.g., in the style of Berghofer [1].

Equational reasoning on the applicative functor is formalised by an equivalence relation $\simeq$ on α iterm. It is the least equivalence relation satisfying the rules in Fig. 2. They represent the applicative laws and the embedding of $\beta\eta$-equivalence on λ-terms. Clearly, if we interpret two idiomatic terms s and t in an applicative functor F in the obvious way as s' and t', and if s' and t' are type correct, then $s \simeq t$ implies $s' = t'$.

Connection between the two models. It is natural to ask how the verified meta model could be leveraged as part of the proofs in the shallow embedding. We decided to leave the connection informal and settled for the two-model approach for now. Formally bridging the gap is left as future work, for which two approaches appear promising.

$$\frac{}{\mathsf{Pure}\ \overline{\mathbf{B}}\ \overline{\diamond}\ f\ \overline{\diamond}\ g\ \overline{\diamond}\ x \simeq f\ \overline{\diamond}\ (g\ \overline{\diamond}\ x)}\ \text{(composition)} \qquad \frac{}{\mathsf{Pure}\ \overline{\mathbf{I}}\ \overline{\diamond}\ x \simeq x}\ \text{(identity)}$$

$$\frac{}{\mathsf{Pure}\ f\ \overline{\diamond}\ \mathsf{Pure}\ x \simeq \mathsf{Pure}\ (f\ \$\ x)}\ \text{(homomorphism)} \qquad \frac{t \simeq_{\beta\eta} t'}{\mathsf{Pure}\ t \simeq \mathsf{Pure}\ t'}\ \text{(cong-Pure)}$$

$$\frac{}{f\ \overline{\diamond}\ \mathsf{Pure}\ x \simeq \mathsf{Pure}\ (\overline{\lambda}f.\ f\ \$\ x)\ \overline{\diamond}\ f}\ \text{(interchange)} \qquad \frac{t_1 \simeq t_1' \qquad t_2 \simeq t_2'}{t_1\ \overline{\diamond}\ t_2 \simeq t_1'\ \overline{\diamond}\ t_2'}\ \text{(cong-}\overline{\diamond}\text{)}$$

Fig. 2. Equivalence of idiomatic terms, where $\overline{\mathbf{I}} \equiv \overline{\lambda}x.\ x$ and $\overline{\mathbf{B}} \equiv \overline{\lambda}f\ g\ x.\ f\ \$\ (g\ \$\ x)$.

Computational reflection makes the correspondence between objects of the logic and their representation explicit by an interpretation function with correctness theorems [3]. For idiomatic terms, interpretation cannot be defined directly in HOL, as a single term may refer to an arbitrary collection of types. Schropp and Popescu [29] circumvent this limitation by modelling the type universe as a single type parameter to the meta theory; additional machinery injects the actual types into this universe and transfers the obtained results. Similar injections could be crafted for idiomatic terms, but the connection would have to be built anew upon each usage. It is not clear that the overhead incurred is compensated by the savings in avoiding the replay of the lifting proof in the shallow embedding.

Alternatively, Tuong and Wolff [30] model the Isabelle API in HOL syntactically and can thus generate code for packages from the HOL formalisation. This could be used to express our proof tactics as HOL terms. Then, we could formally verify them and thus obtain a verified package. Before we can apply this technique in our setting, two challenges must be solved. First, their model merely defines the syntax, but lacks a semantics for the API. Hence, one would first have to model the semantics and validate it. Second, the additional code for usability aspects like preserving the names of bound variables would also have to be part of the HOL terms. This calls for some notion of refinement or abstraction, which is not yet available; otherwise, these parts would clutter the formalisation.

3   Lifting with Applicative Functors

The $\mathsf{pure_F}$ operation of an applicative functor F lifts values of type α to $\alpha\ \mathsf{F}$. If we view HOL terms as functions of their free variables, we can also lift terms via the following syntactic modification: free variables of type α are replaced by those of type $\alpha\ \mathsf{F}$, constants and abstractions[1] are embedded in $\mathsf{pure_F}$, and function application is replaced by $(\diamond)_\mathsf{F}$. Lifting extends to equations, where both sides are treated separately. (We assume that the free variables in an equation are implicitly

---

[1] As lifting wraps the types of *free* variables in F, it does not look into abstractions, but treats them like constants. For example, $\lambda x.\ x :: \alpha \Rightarrow \alpha$ is lifted to $\mathsf{pure}\ (\lambda x.\ x) :: (\alpha \Rightarrow \alpha)\ \mathsf{F}$ rather than $\lambda x.\ x :: \alpha\ \mathsf{F} \Rightarrow \alpha\ \mathsf{F}$. Thus, lifting effectively operates on first-order terms.

quantified universally, i.e., in the interpretation as functions, an equation denotes an equality of two functions.) Associativity $(x + y) + z = x + (y + z)$, e.g., gets lifted to pure $(+) \diamond ($pure $(+) \diamond x \diamond y) \diamond z = $ pure $(+) \diamond x \diamond ($pure $(+) \diamond y \diamond z)$. Conversely, unlifting removes the functor from an idiomatic expression or equation by dropping pures and $(\diamond)$ and replacing opaque terms with fresh variables. An equation is liftable in F iff the equation implies itself lifted to F. When we consider a term or equation and its lifted counterpart, we say that the former is at base level (relative to this lifting).

Hinze [15] characterised equations that are liftable in any idiom and showed that the proof of the lifting step follows a simple structure if both sides are in canonical form. In this section, we adapt his findings to our setting, formalise the lifting lemma in our deep model, and discuss its implementation in the package.

3.1 Conversion to Canonical Form

The first step of lifting converts an idiomatic expression into canonical form. Recall from Sect. 1.1 that an idiomatic term is in canonical form iff it consists of a single pure _ applied to the effectful (opaque) terms, i.e., pure $f \diamond x_1 \diamond \ldots \diamond x_n$. We formalise canonicity as the inductive set CF defined by (i) Pure $x \in$ CF, and (ii) $t \overline{\diamond}$ Opq $x \in$ CF if $t \in$ CF. Borrowing from Hinze's terminology, we say that n is a normal form of an idiomatic term t iff n is in canonical form and equivalent to t, i.e., $n \in$ CF and $t \simeq n$. If $t \in$ CF, we refer to the Pure x part as the single pure subterm of t.

Hinze [15, Lemma 1] gives an algorithm to compute a normal form in the monoidal representation of idioms, which is essentially an uncurried variant of the applicative representation from Sect. 1.1. Since HOL functions are typically curried, we want to retain the applicative style in lifted expressions (to which normalisation is applied). Therefore, we stick to curried functions and adapt the normalisation function accordingly. In the following, we first formalise the normalisation function $\downarrow$ in the deep model and then explain how a proof-producing function for the corresponding equation can be extracted. The latter step is a recurring theme in our implementation.

Figure 3 shows the specification for $\downarrow$. The cases of pure and opaque terms are easy. For applications, $\downarrow$ first normalises both arguments and combines the results using the auxiliary functions $\mathsf{norm_{nn}}$ and $\mathsf{norm_{pn}}$. The auxiliary function $\mathsf{norm_{pn}}$ handles the simplest case of applying a pure function f to a term in canonical form. By repeated application of the composition law, $\mathsf{norm_{pn}}$ splits the variables off until only two pure terms remain which can be combined by the

$$(\text{Pure } x)\!\downarrow = \text{Pure } x \qquad (\text{Opq } x)\!\downarrow = \text{Pure } \bar{\mathsf{I}} \,\overline{\diamond}\, \text{Opq } x \qquad (t \,\overline{\diamond}\, t')\!\downarrow = \mathsf{norm_{nn}} \ (t\!\downarrow) \ (t'\!\downarrow)$$

$$\mathsf{norm_{nn}} \ n \ (\text{Pure } x) = \mathsf{norm_{pn}} \ ((\overline{\lambda}a \ b. \ b \,\$\, a) \,\$\, x) \ n \qquad \mathsf{norm_{pn}} \ f \ (\text{Pure } x) = \text{Pure } (f \,\$\, x)$$

$$\mathsf{norm_{nn}} \ n \ (n' \,\overline{\diamond}\, x) = \mathsf{norm_{nn}} \ (\mathsf{norm_{pn}} \ \overline{\mathbf{B}} \ n) \ n' \,\overline{\diamond}\, x \qquad \mathsf{norm_{pn}} \ f \ (n \,\overline{\diamond}\, x) = \mathsf{norm_{pn}} \ (\overline{\mathbf{B}} \,\$\, f) \ n \,\overline{\diamond}\, x$$

Fig. 3. Specification of the normalisation function $t\!\downarrow$.

homomorphism law. The other function $\mathsf{norm_{nn}}$ assumes that both arguments are already in canonical form. The base case $n' = \mathsf{Pure}\ x$ reduces to the domain of $\mathsf{norm_{pn}}$ via the interchange law. In case of an application, $\mathsf{norm_{pn}}$ incorporates the added term $\overline{\mathsf{B}}$ into n before $\mathsf{norm_{nn}}$ recurses. Note that the equations for $\mathsf{norm_{nn}}$ and $\mathsf{norm_{pn}}$ are complete for terms in canonical form.

Lemma 2 (Correctness of $\downarrow$). *Let $t :: \alpha$ iterm, $f :: term$ and $n, n' \in CF$. Then,*

(a) $\mathsf{pure}\ f \diamond n \simeq \mathsf{norm_{pn}}\ f\ n$ *and* $\mathsf{norm_{pn}}\ f\ n \in CF$;
(b) $n \diamond n' \simeq \mathsf{norm_{nn}}\ n\ n'$ *and* $\mathsf{norm_{nn}}\ n\ n' \in CF$;
(c) $t \simeq t\downarrow$ *and* $t\downarrow \in CF$.

Proof. We prove each of (a)–(c) by structural induction. As a representative example, we focus on the three cases for (c): (i) The case $t = \mathsf{Pure}\ \_$ is trivial. (ii) For $t = \mathsf{Opq}\ x$, we justify $\mathsf{Opq}\ x \simeq \mathsf{Pure}\ \overline{\mathsf{I}} \,\overline{\diamond}\, \mathsf{Opq}\ x$ by the identity law (Fig. 2) and symmetry. (iii) For $(t\,\overline{\diamond}\,t')\downarrow$, the induction hypotheses are $t \simeq t\downarrow$ and $t\downarrow \in CF$, and analogously for t'. Thus, $t\,\overline{\diamond}\,t' \simeq t\downarrow\,\overline{\diamond}\,t'\downarrow \simeq \mathsf{norm_{nn}}\ (t\downarrow)\ (t'\downarrow) = (t\,\overline{\diamond}\,t')\downarrow$ by (b). □

In the shallow embedding, the proof method *applicative-nf* not only computes a normal form t' for an idiomatic term t. It also must prove them being equivalent, i.e., $t = t'$. Such a function from terms to equational theorems is known as a conversion. Paulson [27] designed a library of combinators for composing conversions, e.g., by transitivity. This way, each of (a)–(c) in Lemma 2 becomes one conversion which establishes the part about $\simeq$. (We ignore the part about $\_ \in CF$, as it is computationally irrelevant.) The inductive proofs yield the implementation of the conversions: the induction hypotheses are obtained by recursively calling the conversion on the subterms; case distinction is implemented by matching; and the concrete applicative laws are known to the package and instantiated directly. Thus, each proof step involving $\simeq$ indicates which conversion combinator has to be used.

3.2 Lifting

Hinze's condition for equations that can be lifted in all idioms is as follows: The list of variables, when reading from left to right, must be the same on both sides, and no variable may appear twice on either side. Then, the normal forms of the two lifted terms differ only in the the pure functions, which are just the base-level terms abstracted over all variables. The base equation implies that these functions are extensionally equal.

It is not entirely obvious that the normal form has this precise relationship with lifting, so we prove it formally. This gives us confidence that our proof procedure always succeeds if the conditions on the variables are met.

For practical reasons, our proof method performs unlifting rather than lifting. It takes as input an equality between idiomatic expressions, and reduces it to the weakest base-level equation that entails it—independent of the applicative functor. Unlifting has two advantages. First, the user can apply the method to instantiations of lifted equations, where the variables are replaced with concrete effects such as fresh. Thus, there is no need to manually generalise the lifted equation itself. Second, in the lifted equation, the pure terms distinguish constants (to be lifted) from opaque terms, but there are no such markers on the base level. Rather than lifting, we therefore formalise unlifting, which replaces each opaque term by a new bound variable ($|x|$ denotes the length of the list x).

$$\text{unlift } t = (\text{let } n = |\text{opq } t| \text{ in Abs}^n \text{ (unlift' } n \ 0 \ t))$$

$$\text{unlift' } n \ i \ (\text{Pure } x) = \text{shift } x \ n$$
$$\text{unlift' } n \ i \ (\text{Opq } x) = \text{Var } i$$
$$\text{unlift' } n \ i \ (t \bar{\diamond} t') = \text{unlift } n \ (i + |\text{opq } t'|) \ t \ \$ \ \text{unlift } n \ i \ t'$$

Here, the function opq t returns the list of all opaque terms from left to right, so $|\text{opq } t|$ counts the occurrences of Opq in t. Nipkow's function shift $x \ n$ increments all loose variables in x by n. For example, unlift (Opq $a \bar{\diamond}$ (Pure $f \bar{\diamond}$ Opq b)) $= \overline{\lambda}g \ x. \ g \ (f \ x)$, as expected. Note that this holds independent of a and b.

The benefit of the meta model is that we can characterise the normal form.

Lemma 3. *Let Pure f be the single pure subterm in $t{\downarrow}$. Then $f \simeq_{\beta\eta} \text{unlift } t$.*

Equality in a real theory generally has more axioms than those for term reductions. Let $\simeq'_{\beta\eta}$ be an extension of $\simeq_{\beta\eta}$, and $\simeq'$ the corresponding extension of $\simeq$. Then, we obtain the following lifting rule, which follows from Lemmas 2 and 3 and opq $(t{\downarrow}) = \text{opq } t$.

Lemma 4. *Let opq $s = \text{opq } t$. Then, unlift $s \simeq'_{\beta\eta} \text{unlift } t$ implies $s \simeq' t$.*

To implement its proof, we rewrite both idiomatic terms of the input equation with the normal form conversion, i.e., we are left with the subgoal of the form pure $f \diamond x_1 \ldots \diamond x_n = \text{pure } g \diamond x_1 \ldots \diamond x_n$. It suffices to prove $f = g$, which follows from the base-level equation $\forall x_1 \ldots x_n. \ f \ x_1 \ldots x_n = g \ x_1 \ldots x_n$ by extensionality.

Moreover, we get that the normal form is indeed unique.

Lemma 5. *If $s, t \in \text{CF}$ and $s \simeq' t$, then s and t have the same structure, and the pure terms are related by $\simeq'_{\beta\eta}$.*

4 Combinators

Lifting works for equations whose both sides contain the same list of variables without repetitions. Many equations, however, violate this conditions. Therefore, Hinze studied the class of idioms in which all equations can be lifted [15]. He proved that every equation can be lifted if the functor satisfies the two equations

$$\text{pure } \mathsf{S} \diamond f \diamond g \diamond x = f \diamond x \diamond (g \diamond x) \qquad \text{pure } \mathsf{K} \diamond x \diamond y = x \qquad (5)$$

for all f, g, x, and y, where $\mathsf{S} = (\lambda f \ g \ x. \ f \ x \ (g \ x))$ and $\mathsf{K} = (\lambda x \ y. \ x)$ denote the well-known combinators from combinatory logic. Similar to bracket abstraction for the λ-calculus, Hinze defines an abstraction algorithm $[x]t$ which removes an opaque term x from an idiomatic expression t such that $[x]t \bar{\diamond} x \simeq_E t$, where $\simeq_E$ extends $\simeq$ with the combinators' laws. For lifting, Hinze uses the abstraction algorithm to remove all variables from both sides of the equation in the same order (which may introduce S and K in the pure part), then applies the lifting technique and finally removes the combinators again.

However, only few applicative functors satisfy (5). In this section, we subject Hinze's idea to a finer analysis of equational lifting for various sets of combinators, present the implementation as a proof method in Isabelle and an application to the Stern-Brocot tree, and sketch the formalisation in the deep embedding.[2] The new proof method *applicative-lifting* subsumes the one from Sect. 3.

4.1   The Combinatorial Basis BCKW

While SK has become the canonical approach to combinatory logic, we argue that Curry's set of combinators BCKW works better for applicative lifting, where $\mathbf{B} = (\lambda f\ g\ x.\ f\ (g\ x))$ and $\mathbf{C} = (\lambda f\ x\ y.f\ y\ x)$ and $\mathbf{W} = (\lambda f\ x.\ f\ x\ x)$. We say that a functor has a combinator if the equation defining the combinator is liftable. For BICKW (where $\mathsf{I} = (\lambda x.\ x)$), the lifted equations are the following.

$$\text{pure } \mathbf{B} \diamond f \diamond g \diamond x = f \diamond (g \diamond x) \qquad \text{pure } \mathbf{C} \diamond f \diamond x \diamond y = f \diamond y \diamond x$$
$$\text{pure } \mathbf{K} \diamond x \diamond y = x \qquad \text{pure } \mathbf{W} \diamond f \diamond x = f \diamond x \diamond x \qquad \text{pure } \mathbf{I} \diamond x = x$$

Note that every applicative functor has combinators $\mathbf{B}$ and I as their equations are exactly the composition and identity law, respectively.

We focus on BCKW for two reasons. First, the combinators can be intuitively interpreted. A functor has $\mathbf{C}$ if effects can be swapped, it has $\mathbf{K}$ if effects may be omitted, and it has $\mathbf{W}$ if effects may be doubled. In contrast, Hinze's combinator S mixes doubling with a restricted form of swapping; full commutativity additionally requires $\mathbf{K}$. Second, our set of combinators yields a finer hierarchy of applicative functors. Thus, the proof method is more widely applicable because it can exploit more precisely the properties of the particular functor, although its implementation remains generic in the functor.

Table 1 lists for a number of applicative functors the combinators they possess. For reference, the functors are defined in Appendix A. The table is complete in the sense that there is a tick $\sqrt{}$ iff the functor has this combinator.

Most of the functors are standard, but some are worth mentioning. Hinze [15] proved that all functors which are isomorphic to the environment functor (a.k.a. the reader idiom, e.g., streams and infinite binary trees) have combinators S and $\mathbf{K}$. Thus, they also have the combinators $\mathbf{C}$ and $\mathbf{W}$, as the two can be expressed in terms of S and $\mathbf{K}$. However, some functors with combinators S and $\mathbf{K}$ are not isomorphic to the environment functor. One example is Huffman's construction of non-standard numbers in non-standard analysis [17].

---

[2] Hinze briefly considers functors with the combinators S and $\mathbf{C}$ and notes that the case with only the combinator $\mathbf{C}$ might be interesting, too, but omits the details.

Table 1. List of applicative functors, their combinators and the abstraction algorithm.

Applicative functor	B	I	C	K	W	S	Abstraction algorithm
environment, stream, non-standard numbers	√	√	√	√	√	√	(kibtcs)
option, zip list	√	√	√		√	√	(ibtcs)
probability, non-empty set	√	√	√	√			(kibtc)
subprobability, set, commutative monoid	√	√	√				(ibtc)
either, idempotent monoid	√	√			√		(ibtw)
distinct non-empty list	√	√		√			(kibt)
state, list, parser, monoid	√	√					(ibt)

Every monoid yields an applicative functor known as the writer idiom. Given a monoid on β with binary operation $+$ and neutral element 0, we turn the functor (β, α) monoid $= \beta \times \alpha$ into an applicative one via

$$\text{pure}_{\text{monoid}} \; x = (0, x) \qquad (a, f) \diamond_{\text{monoid}} (b, x) = (a + b, f \; x)$$

Commutative monoids have the combinator $\mathbf{C}$, idempotent ones have $\mathbf{W}$.

The idioms "probability" and "non-empty set" are derived from the monads for probabilities and non-determinism without failure. When the latter is implemented by distinct non-empty lists, commutativity is lost because lists respect the order of elements.

The attentive reader might have noticed that one combination of combinators is missing, namely $\mathbf{BIKW}$, i.e., only $\mathbf{C}$ is excluded. As $\mathbf{BCKW}$ is a minimal basis for combinatory logic, $\mathbf{C}$ cannot be expressed in terms of $\mathbf{BIKW}$. Surprisingly, an applicative functor always has $\mathbf{C}$ whenever it has $\mathbf{BIKW}$, as the following calculation shows, where Pair $x \; y = (x, y)$ and $\pi_1 \; (x, y) = x$ and $\pi_2 \; (x, y) = y$ and G abbreviates $\lambda f \; p \; q$. $\mathbf{C} \; f \; (\pi_2 \; p) \; (\pi_1 \; q)$. Steps (i) and (iii) are justified by the equations for $\mathbf{K}$ and $\mathbf{I}$ and $\mathbf{W}$; steps (ii) and (iv) hold by lifting of the identities $\mathbf{K} \; (\mathbf{C} \; f \; (\mathbf{K} \; \mathbf{I} \; z \; x) \; y) \; w = G \; f \; (z, x) \; (y, w)$ and $\mathbf{W} \; (G \; f) \; (y, x) = f \; y \; x$, respectively.

$$\text{pure } \mathbf{C} \diamond f \diamond x \diamond y \; \overset{(i)}{=} \; \text{pure } \mathbf{K} \diamond (\text{pure } \mathbf{C} \diamond f \diamond (\text{pure } \mathbf{K} \diamond \text{pure } \mathbf{I} \diamond y \diamond x) \diamond y) \diamond x$$
$$\overset{(ii)}{=} \; \text{pure } G \diamond f \diamond (\text{pure Pair} \diamond y \diamond x) \diamond (\text{pure Pair} \diamond y \diamond x)$$
$$\overset{(iii)}{=} \; \text{pure } \mathbf{W} \diamond (\text{pure } G \diamond f) \diamond (\text{pure Pair} \diamond y \diamond x)$$
$$\overset{(iv)}{=} \; f \diamond y \diamond x$$

The crucial difference between combinatory logic and idioms can be seen by looking at G, which is equivalent to $\mathbf{B} \; (\mathbf{B} \; (\mathbf{T} \; \pi_1)) \; (\mathbf{B} \; (\mathbf{B} \; \mathbf{B}) \; (\mathbf{B} \; (\mathbf{T} \; \pi_2) \; (\mathbf{B} \; \mathbf{B} \; \mathbf{C})))$ where $\mathbf{T} = (\lambda x \; f. \; f \; x)$. By the interchange and homomorphism laws, we have pure $(\mathbf{T} \; x) \diamond f = f \diamond \text{pure } x$ in every idiom. This is the very bit of reordering that $\mathbf{C}$ adds to $\mathbf{BKW}$. Note, however, that $\mathbf{T}$ is different from the other combinators: it may only occur applied to a term without $\mathbf{O}pq$ (as such terms are lifted to pure terms by the homomorphism law). In fact, if **T** was like the others, every applicative functor would have **C** thanks to **C** = **B** (**T** (**B B T**)) (**B B T**) [6].

4.2 Characterisation of Liftable Equations

The lifting technique from Sect. 3.2 requires that the list of opaque terms be the same on both sides and free of duplicates. With additional combinators, we can try to rewrite both sides such that the lists satisfy this condition. In this section, we derive for each set of combinators a simple criterion whether this can be achieved. Simplicity is important because users should be able to easily judge whether an equation can be lifted to a particular functor using our proof method. Our analysis heavily builds on the literature on representable λ-terms in various combinator bases [5]. Therefore, we refer to opaque terms as variables in the rest of this section.

By using normal forms (cf. Lemma 2), it suffices to consider only the list of variables on each side of the equation, say v_l and v_r. Our goal is to find a duplicate-free variable list v^* such that v_l and v_r can both be transformed into v^*. The permitted transformations are determined by the combinators:

- If **C** is available, we may reorder any two variables.
- If **K** is available, we may insert a variable anywhere.
- If **W** is available, we may duplicate any contiguous subsequence or drop a repetition of a contiguous subsequence (the repetition must be adjacent).

This yields the following characterisation of liftable equations. (The conditions for all the cases which include the combinator **C** are equal to the representation conditions for λ-terms with the given combinators [5].)

BI No transformation is possible. So we require $v^* = v_l = v_r$.

BIC v_l and v_r must be duplicate-free and permutations of each other. We choose for v^* any permutation of v_l.

BICK v_l and v_r must be duplicate-free. We choose for v^* any duplicate-free list of the union of the variables in v_l and v_r.

BICW v_l and v_r must contain the same variables, but need not be duplicate-free. We choose for v^* any duplicate-free list of the variables.

BICKW No constraints on v_l or v_r. We choose for v^* any duplicate-free list of the union of variables in v_l and v_r. (This is the case considered by Hinze [15].)

BIK v_l and v_r must be duplicate-free and the shared variables must occur in the same order. Take for v^* any merge of v_l and v_r, i.e., a duplicate-free sequence which contains v_l and v_r as subsequences.

BIW In this case, we work in the free idempotent monoid (FIM) whose letters are the variables in v_l and v_r. So, our task boils down to finding a duplicate-free word v^* such that $v_l \sim v^* \sim v_r$ where $\sim$ denotes equivalence in the FIM.

Green and Rees [13] characterised $\sim$ recursively: For a word x, let $\overrightarrow{x}$ and $\overleftarrow{x}$ denote the longest prefix or suffix of x that contains all but one letters of x. Then, $x \sim y$ iff x and y contain the same letters and $\overrightarrow{x} \sim \overrightarrow{y}$ and $\overleftarrow{x} \sim \overleftarrow{y}$.

This criterion yields the following conditions: (i) v_l and v_r contain the same variables; (ii) the orders in which the variables occur for the first or for the last time must be all the same in v_l and v_r (we choose v^* as the list of variables in this order); and (iii) recursively the same conditions hold for $\overrightarrow{v_l}$ and $\overrightarrow{v_r}$, and for $\overleftarrow{v_l}$ and $\overleftarrow{v_r}$. For example, the equation $\forall a\ b\ c.\ \mathsf{f}\ a\ b\ c = \mathsf{g}\ a\ b\ c\ a\ c\ b\ a\ b\ c$ satisfies this condition with $v^* = abc$.[3]

4.3   Implementation via Bracket Abstraction

Bracket abstraction converts a λ-calculus term into combinator form. The basic algorithm $[x]t$ abstracts the variable x from the term t (which must not contain any abstraction). Like λ-terms, applicative terms are built from constants (Pure _), variables (Opq _) and applications. So, bracket abstraction also makes sense for applicative terms. What is interesting about bracket abstraction is that the algorithm is modular in the combinators. That is, bracket abstraction allows us to deal with all the different combinator bases in a uniform way. In detail, we first abstract the variables on both sides of the equation in the order given by $v^* = v_1 \ldots v_n$. As $l \simeq_E ([v_1](\ldots ([v_n]l))) \otimes v_1 \otimes \ldots \otimes v_n$ and $r \simeq_E ([v_1](\ldots ([v_n]r))) \otimes v_1 \otimes \ldots \otimes v_n$ by the correctness of bracket abstraction, we thus obtain an equation whose two sides are in normal form. From there, our implementation proceeds as before (Sect. 3).

As usual, we specify a bracket abstraction algorithm by a list of rules, say (kibtcs). This means that the corresponding rules should be tried in that order and the first one matching should be taken. The algorithm for each set of combinators is listed in the last column of Table 1. The rules are shown in Table 2, where $\mathcal{V}(t)$ denotes the set of variables in t. All but (t) and (w) correspond to the standard abstraction rules for the λ-calculus [5]. The side condition of (t) reflects the restriction of the interchange law to pure computations (cf. Sect. 4.1). Rule (w) is used only if **C** is not available—otherwise (s) is used as **S** = **B** (**B W**) (**B B C**). It uses **T** to allow for pure computations between two occurrences of the same variable. This way, we avoid repeatedly converting the term to normal form, as otherwise **W** could only be used for terms of the form $t \otimes x \otimes x$ for some variable x.

Our bracket abstraction algorithm (ibtw) is not complete for **BIW**. A dedicated algorithm would be needed, as it seems not possible to construct equivalence proofs like in Footnote 3 using bracket abstraction, because the transformations are not local. Bersten's and Reutenauer's elementary proof [2, Theorem 2.4.1] of Green's and Rees' characterisation contains an algorithm, but we settle with (ibtw) nevertheless. Thus, our implementation imposes stronger conditions on v_l and v_r than those described in Sect. 4.2, namely v_l and v_r must

---

[3] The following shows the equivalence (bold face denotes doubling and underlining dropping of a repetition): $abcac\mathbf{babc} \sim abcacbabc\mathbf{abc} \sim abcacbabcabc\mathbf{cabc} \sim abcacbabc\mathbf{abc}acabc \sim abcacbabcabcac\mathbf{abc}cacabc \sim abcacbabcabcac\mathbf{bac}bcacabc \sim abcacbabcabcac\mathbf{bab}acbcacabc \sim \underline{abcacbabcabcacbabc}babcbacacabc \sim abc\underline{acbabc}babacbcacabc \sim abcac\underline{baba}cbcacabc \sim abc\underline{acbacb}cacabc \sim \underline{abcacbcacabc} \sim ab\underline{cacabc} \sim \underline{abcabc} \sim abc.$

$$[x]t \quad = \mathsf{Pure}\ \overline{\mathsf{K}}\ \overline{\diamond}\ t \qquad\qquad \text{if } x \notin \mathcal{V}(t) \quad \text{(k)}$$
$$[x]x \quad = \mathsf{Pure}\ \overline{\mathsf{I}} \qquad\qquad\qquad\qquad\qquad\qquad \text{(i)}$$
$$[x](s\ \overline{\diamond}\ t) = \mathsf{Pure}\ \overline{\mathsf{B}}\ \overline{\diamond}\ s\ \overline{\diamond}\ [x]t \qquad \text{if } x \notin \mathcal{V}(s) \quad \text{(b)}$$
$$[x](s\ \overline{\diamond}\ t) = \mathsf{Pure}\ \overline{\mathsf{T}}\ \overline{\diamond}\ t\ \overline{\diamond}\ [x]s \qquad \text{if } \mathcal{V}(t) = \emptyset \quad \text{(t)}$$
$$[x](s\ \overline{\diamond}\ t) = \mathsf{Pure}\ \overline{\mathsf{C}}\ \overline{\diamond}\ [x]s\ \overline{\diamond}\ t \qquad \text{if } x \notin \mathcal{V}(t) \quad \text{(c)}$$
$$[x](s\ \overline{\diamond}\ t) = \mathsf{Pure}\ \overline{\mathsf{S}}\ \overline{\diamond}\ [x]s\ \overline{\diamond}\ [x]t \qquad\qquad\qquad \text{(s)}$$
$$[x](s\ \overline{\diamond}\ t) = \mathsf{Pure}\ \overline{\mathsf{W}}\ \overline{\diamond}\ (\mathsf{Pure}\ \overline{\mathsf{B}}\ \overline{\diamond}\ (\mathsf{Pure}\ \overline{\mathsf{T}}\ \overline{\diamond}\ [x]t)\ \overline{\diamond}\ (\mathsf{Pure}\ (\overline{\mathsf{B}}\ \overline{\mathsf{B}})\ \overline{\diamond}\ [x]s))\ \text{if } \mathcal{V}([x]t) = \emptyset \quad \text{(w)}$$

Table 2. Bracket abstraction rules for applicative expressions.

use the same variables in the same order, but each variable may be repeated any number of times (with no other variable between the repetitions). In practice, we have not yet encountered a liftable sequence of variables that needs the full generality.

Again, we verify unlifting in the deep embedding. We show that the implementation with bracket abstraction yields the same equation (after reducing the combinators) as unlifting the lifted equation directly, where v^* determines the quantifier order. Thus, it suffices to rearrange the quantifiers according to v^*.

Unlike to Sect. 3.2, unlift must map identical opaque terms to the same variable. So, we assume that Opq's argument denotes the variable name. Then, the new function unlift* replaces $\mathsf{Opq}\ i$ with $\mathsf{Var}\ i$, $(\overline{\diamond})$ with $\$$ and $\mathsf{Pure}\ x$ with shift $x\ |v^*|$.

Further, we abstract from the concrete bracket abstraction algorithm. We model the algorithm as two partial functions $\overline{[\_]}$ and $[\_]$ on term and nat iterm and assume that they are correct ($\lfloor\_\rfloor$ denotes definedness): (i) if $\overline{[i]}t = \lfloor t' \rfloor$, then $t'\ \$\ \mathsf{Var}\ i \simeq_{\beta\eta} t$ and i is not free in t', (ii) if $[i]t = \lfloor t' \rfloor$, then $t' \diamond \mathsf{Opq}\ i \simeq_E t$ and $\mathsf{set}\ (\mathsf{opq}\ t') = \mathsf{set}\ (\mathsf{opq}\ t) - \{\mathsf{Opq}\ i\}$, and (iii) they commute with unlifting: $\overline{[i]}\ (\mathsf{unlift}^*\ n\ t) = \mathsf{unlift}^*\ n\ ([i]t)$ for $i < n$. Formalising and verifying bracket abstraction is left as future work. In the theorem below, the congruence relation $\simeq'_E$ combines $\simeq_E$ with the additional axioms from $\simeq'$.

Theorem 1. *Let $s, t :: \mathsf{nat\ iterm}$ and let v^* be a permutation of $\{0, \ldots, n-1\}$. Assume that $\mathsf{set}\ (\mathsf{opq}\ s) \cup \mathsf{set}\ (\mathsf{opq}\ t) \subseteq \mathsf{set}\ v^*$ and that $[\_]$ succeeds to abstract s and t over v^*. Then, $\mathsf{Abs}^n\ (\mathsf{unlift}^*\ n\ s) \simeq_{\beta\eta} \mathsf{Abs}^n\ (\mathsf{unlift}^*\ n\ t)$ implies $s \simeq'_E t$.*

4.4   Application: The Stern-Brocot Tree

Hinze uses his theory of lifting to reason about infinite trees of rational numbers [14]. In particular, he shows that a linearisation of the Stern-Brocot tree yields Dijsktra's fusc function [7]. We have formalised his reasoning in Isabelle as a benchmark for our package [9]. Here, we report on our findings.

The Stern-Brocot tree stern-brocot enumerates all the rationals in their lowest terms

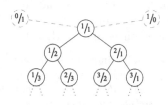

Fig. 4. The Stern-Brocot tree

(see [12]). It is an infinite binary tree of type frac cotree containing formal fractions (**type-synonym** frac = nat × nat). Each node is labelled with the mediant of its right-most and left-most ancestor (Fig. 4), where mediant (a, c) $(b, d) = (a + b, c + d)$. Formally, stern-brocot = sb-gen $(0, 1)$ $(1, 0)$ with

> **codatatype** α cotree = Node (root: α) (α cotree) (α cotree)
> **primcorec** sb-gen l u = (let m = mediant l u in Node m (sb-gen l m) (sb-gen m u))

The type constructor cotree forms an idiom analogous to stream, i.e., ($\diamond$) corresponds to zipping trees with function application. Combinators **C**, **K**, and **W** exist. The idiom homomorphism stream :: α cotree $\Rightarrow \alpha$ stream linearises a tree to a stream.

> **primcorec** chop (Node x l r) = Node (root l) r (chop l)
> **primcorec** stream t = root $t \prec$ stream (chop t)

Hinze shows that stream stern-brocot equals fusc $\circledast$ fusc' for Dijkstra's fusc and fusc' given by

$$\text{fusc} = 1 \prec \text{fusc}' \qquad \text{fusc}' = 1 \prec (\text{fusc} + \text{fusc}' - 2 * (\text{fusc mod fusc}'))$$

where all arithmetic operations are lifted to streams, e.g., $s + t$ denotes pure $(+) \diamond s \diamond t$, and $(\circledast)$ :: α stream $\Rightarrow \beta$ stream $\Rightarrow (\alpha \times \beta)$ stream zips two streams. The proof shows that stream stern-brocot satisfies the same recursion equation as fusc $\circledast$ fusc', so they must be equal. The crucial step is to show that chop den = num + den $- 2 * ($num mod den$)$ where num = pure $\pi_1 \diamond$ stern-brocot and den = pure $\pi_2 \diamond$ stern-brocot project the Stern-Brocot tree to numerators and denominators. Hinze proves this equality by lifting various arithmetic identities from integers to trees.

We instantiate Isabelle/HOL's fine-grained arithmetic type class hierarchy for cotree and stream up to the class for rings with characteristic 0. This way, we can use the algebraic operators and reason directly on trees and streams. Almost all algebraic laws are proven by our lifting package from the base equation. The only exception are the two cancellative laws in semigroups, namely $a = b$ whenever $a + c = b + c$ or $c + a = c + b$. Such conditional equations are not handled by our lifting machinery. So, we prove these two laws conventionally by coinduction.

Moreover, we discovered that Hinze's lifting framework cannot prove the identity for chop den, contrary to his claims. In detail, the proof relies on the identity $x \bmod (x + y) = x$ on natural numbers, but this holds only for $y > 0$. Hinze does not explain how to lift and handle such preconditions. As the combinators **K** and **W** exist, we express the lifted precondition as pure $(>) \diamond y \diamond 0 =$ pure True and split the proof into the three steps shown below: (i) and (iii) hold by lifting and (ii) by assumption.

$$x \bmod (x + y) \overset{(i)}{=} \text{pure} (\lambda b\ x.\ \text{if } b \text{ then } x \text{ else } 0) \diamond (\text{pure} (>) \diamond y \diamond 0) \diamond x \overset{(ii)}{=}$$
$$\text{pure} (\lambda b\ x.\ \text{if } b \text{ then } x \text{ else } 0) \diamond \text{pure True} \qquad \diamond x \overset{(iii)}{=} x$$

Overall, we found that the lifting package works well for algebraic reasoning and that we should extend lifting to handle arbitrary relations and preconditions.

5   Related Work

Most closely related to our work is Hinze's [15] on lifting. He focuses on the two extremes in the spectrum: the class of equations liftable in all idioms, and the idioms in which all equations are liftable. Our implementation for the former merely adapts his ideas to the HOL setting. For the latter, Hinze requires idioms to be strongly extensional in addition to them having **S** and **K**. This ensures that the idiom can emulate λ-abstraction, so lifting is defined for all λ-terms. Therefore, his proof of the Lifting Lemma does not carry over to weaker sets of combinators. As we focus on unlifting, we do not need such emulations and instead use bracket abstraction, which is hidden in Hinze's emulation of abstraction, uniformly for all sets of combinators. Hinze also models idiomatic expressions syntactically using GADTs, which ensures type correctness. He defines equivalence on idiomatic terms semantically. As the interpretation cannot be expressed in HOL, we use the syntactic relation $\simeq$ instead. This has the advantage that we can prove uniqueness of normal forms (Lemma 5) by induction.

Several other kinds of lifting are available as Isabelle/HOL packages. Huffman's transfer tactic [17] lifts properties to non-standard analysis (NSA) types like the hyperreals, which are formalised by the idiom star. The tactic can lift arbitrary first-order properties by exploiting the properties of star. To that end, the tactic first unlifts the property similar to our operation unlift and then proves equivalence by resolving with rules for logical and star operators. Our package subsumes Huffman's for equations, but it cannot lift first-order connectives yet.

The package Lifting [18] creates quotient types via partial equivalence relations. The companion package Transfer, which is different from aforementioned transfer tactic, exploits parametricity and representation independence to prove equivalences or implications between properties on the raw type and the quotient. Like for NSA, resolution guides the equivalence proof. Lifting and Transfer cannot handle lifting to applicative functors, as the functor's values are usually more complex than the base values, instead of more abstract. In comparison, our lifting step is much simpler, as it just considers pairs of extensionally equal functions; the whole automation is needed to extract these functions from idiomatic expressions. The other packages preserve the term structure and relate each component of the term as determined by the rules.

6   Conclusion and Future Work

This paper presents a first step towards a infrastructure for reasoning about effectful programs. Like applicative functors help in delimiting pure and effectful parts of the computation, our proof method supports separating the effectful and the pure aspects of the reasoning. The results from our case studies indicate that applicative functors are a suitable abstraction for reasoning. They seem to be better suited than monads, as applicative expressions can be analysed statically. Thus, one should prefer applicative functors over monads whenever possible.

There is much to be done before proof assistants support reasoning about effects smoothly. As a next step, we will investigate how to extend the scope of lifting. Going from equations to arbitrary relations should be easy: if the functor has a relator for which pure and ($\diamond$) are relationally parametric [28], then the lifting technique should work unchanged. The extension to preconditions and other first-order connectives seems to be harder. In any ring with $0 \neq 1$, e.g., $x = x + 1 \longrightarrow x = 2x$ holds, but it does not when interpreted in the set idiom over the same ring. We expect that combinators will help there, too. Moreover, we would like to study whether one should further refine the set of combinators. For example, the idiom "either" derived from the exception monad has the stronger combinator **H** with pure $\mathbf{H} \diamond f \diamond x \diamond y = f \diamond x \diamond y \diamond x$, which cannot be expressed by **BIW**. Experience will tell when specialisation is needed and when it goes too far.

The combinator laws can also be interpreted monadically. For example, **C** exists in commutative monads and **K** demands that $x \ggg (\lambda\_.y) = y$. Therefore, we experimented with lifting for monads, too. As ($\diamond$) and ($\ggg$) are related (cf. Sect. 1.1), one can express certain parts of a monadic term applicatively using ($\diamond$) and apply the lifting approach to those parts. In particular, the monadic laws for **C** and **K** can only be utilised if the affected part can be expressed applicatively. In a first attempt, we applied this idea to a security proof of the Elgamal encryption scheme [22], which uses the subprobability monad (which only has **C**). Our package successfully automates the arguments about commutativity in this proof, which previously were conducted by manual applications of the commutativity law. At present, we have to manually identify the right parts and rewrite them into applicative form. One reason is that monadic expressions in general contain several overlapping applicative subparts and consecutive applications of commutativity may require different parts for each application. Overall, the new Isar proof is more declarative, but also longer due to the manual rewrite steps. It will be an interesting problem to automate the identification of suitable parts and to combine the appropriate rewrites with lifting.

Acknowledgements. Peter Gammie triggered our interest in reasoning about applicative functors and helped with the Stern-Brocot tree. We thank Dmitriy Traytel, Joachim Breitner, and the anonymous reviewers for suggesting many textual improvements. The first author was supported by SNSF grant 153217 "Formalising Computational Soundness for Protocol Implementations".

A   Definitions of Applicative Functors

This appendix lists Isabelle/HOL definitions for the idioms mentioned in this paper. The definitional packages and their syntaxes are documented in the Isabelle/Isar reference manual. The proofs of the applicative laws and combinators are available online [23].

Environment (Reader)

> **type-synonym** (α, β) env $= (\beta \Rightarrow \alpha)$
> **definition** pure$_{env}$ $x = (\lambda\_.\ x)$
> **definition** $f \diamond_{env} x = (\lambda y.\ f\ y\ (x\ y))$

Stream

> **codatatype** α stream $= \alpha \prec \alpha$ stream
> **primcorec** pure$_{stream}$ $x = x \prec$ pure$_{stream}$ x
> **primcorec** $(f \prec fs) \diamond_{stream} (x \prec xs) = f\ x \prec (fs \diamond_{stream} xs)$

Infinite binary tree

> **codatatype** α cotree $=$ Node $\alpha\ (\alpha$ cotree$)\ (\alpha$ cotree$)$
> **primcorec** pure$_{cotree}$ $x =$ Node x (pure$_{cotree}$ x) (pure$_{cotree}$ x)
> **primcorec** (Node $f\ g\ h$) $\diamond_{cotree}$ (Node $x\ y\ z$) $=$ Node $(f\ x)\ (g \diamond_{cotree} y)\ (h \diamond_{cotree} z)$

Non-standard numbers as used in non-standard analysis in Isabelle/HOL [17]. The type α star is the quotient of the environment idiom $(\alpha,$ nat$)$ env over equality in some free ultrafilter $\mathcal{U}$ on nat.

> **quotient-type** α star $= (\alpha, nat)$ env $/ (\lambda X\ Y.\ (\lambda n.\ X n = Y n) \in \mathcal{U})$
> **lift-definition** pure$_{star}$ is $\lambda x\ \_.\ x$
> **lift-definition** $(\diamond)_{star}$ is $\lambda f\ x\ y.\ f\ y\ (x\ y)$

Option

> **datatype** α option $=$ None $|$ Some α
> **abbreviation** pure$_{option}$ $=$ Some
> **fun** $(\diamond)_{option}$ where (Some f) $\diamond_{option}$ (Some x) $=$ Some $(f\ x)$ $|$ $\_ \diamond_{option} \_ =$ None

Zip list

> **codatatype** α llist $= [\,]\ |\ \alpha \cdot \alpha$ llist
> **primcorec** pure$_{llist}$ $x = x \cdot$ pure$_{llist}$ x
> **primcorec** $(f \cdot fs) \diamond_{llist} (x \cdot xs) = f\ x \cdot (fs \diamond_{llist} xs)$ $|$ $\_ \diamond_{llist} \_ = [\,]$

Probability

> **typedef** α pmf $= \{\, f :: \alpha \Rightarrow$ real. $(\forall x.\ f\ x \geq 0) \wedge (\sum_x f\ x) = 1 \,\}$
> **lift-definition** pure$_{pmf}$ is $\lambda x\ y.$ if $x = y$ then 1 else 0
> **lift-definition** $(\diamond)_{pmf}$ is $\lambda F\ X\ y.\ \sum_{\{(f,x).\ f\ x=y\}} F\ f \cdot X\ x$

Subprobability

> **type-synonym** α spmf $= \alpha$ option pmf
> **definition** pure$_{spmf}$ $=$ pure$_{pmf}$ pure$_{option}$
> **definition** $f \diamond_{spmf} x =$ pure$_{pmf}$ $(\diamond)_{option} \diamond_{pmf} f \diamond_{pmf} x$

Set

> **definition** pure$_{set}$ $x = \{\, x \,\}$
> **definition** $F \diamond_{set} X = \{\, f\ x.\ f \in F \wedge x \in X \,\}$

Non-empty set

> **typedef** α neset $= \{\, A :: \alpha$ set. $A \neq \emptyset \,\}$
> **lift-definition** pure$_{neset}$ is pure$_{set}$
> **lift-definition** $(\diamond)_{neset}$ is $(\diamond)_{set}$

Monoid, commutative monoid, idempotent monoid

> **type-synonym** (α, β) monoid-ap $= \alpha \times \beta$
> **definition** pure$_{monoid}$ $x = (0, x)$
> **fun** $(\diamond)_{monoid}$ where $(a, f) \diamond_{monoid} (b, x) = (a + b, f\ x)$

The type variable α must have sort monoid-add. If α has sort comm-monoid-add, then monoid-ap has **C**. If α has sort idemp-monoid-add, then monoid-ap has **W**.

Either

> **datatype** (α, β) either $=$ Left α | Right β
> **definition** pure$_{either}$ $=$ Left
> **fun** $(\diamond)_{either}$ where
>   Left f $\diamond_{either}$ Left x $=$ Left $(f\ x)$
> | _ $\diamond_{either}$ Right y $=$ Right y
> | Right y $\diamond_{either}$ Left _ $=$ Right y

Distinct non-empty list. The function remdups removes duplicates from a list by retaining only the last occurrence of each element.

> **typedef** α dnelist $= \{\, xs :: \alpha$ list. distinct $xs \wedge xs \neq [\,] \,\}$
> **lift-definition** pure$_{dnelist}$ is pure$_{list}$
> **lift-definition** $(\diamond)_{dnelist}$ is $\lambda f\ x.$ remdups $(f \diamond_{list} x)$

State

> **type-synonym** (α, σ) state $= \sigma \Rightarrow \alpha \times \sigma$
> **definition** pure$_{state}$ $=$ Pair
> **definition** $f \diamond_{state} x = (\lambda s.$ case $f\ s$ of $(f', s') \Rightarrow$ case $x\ s'$ of $(x', s'') \Rightarrow (f'\ x', s''))$

List

> **datatype** α list $= [\,]$ | $\alpha \cdot \alpha$ list
> **definition** pure$_{list}$ $x = [x]$
> **definition** $f \diamond_{list} x =$ concat-map $(\lambda f'.$ map $f'\ x)\ f$

Parser. The function apfst applies a function to the first component of a tuple.

type-synonym (α, σ) parser $= \sigma \Rightarrow (\alpha \times \sigma)$ list
definition pure$_{\mathsf{parser}}$ $x = (\lambda s.\ [(x, s)])$
definition $f \diamond_{\mathsf{parser}} x = (\lambda s.\ \mathsf{concat\text{-}map}\ (\lambda(f', s').\ \mathsf{map}\ (\mathsf{apfst}\ f')\ (x\ s'))\ (f\ s))$

References

1. Berghofer, S.: Proofs, Programs and Executable Specifications in Higher Order Logic. Ph.D. thesis, Institut für Informatik, Technische Universität München (2003)
2. Berstel, J., Reutenauer, C.: Square-free words and idempotent semigroups. In: Lothaire, M. (ed.) Combinatorics on Words, 2nd edn., pp. 18–38. Cambridge University Press (1997)
3. Boutin, S.: Using reflection to build efficient and certified decision procedures. In: Abadi, M., Ito, T. (eds.) TACS 1997. LNCS, vol. 1281, pp. 515–529. Springer, Heidelberg (1997)
4. Bulwahn, L., Krauss, A., Haftmann, F., Erkök, L., Matthews, J.: Imperative functional programming with Isabelle/HOL. In: Mohamed, O.A., Muñoz, C., Tahar, S. (eds.) TPHOLs 2008. LNCS, vol. 5170, pp. 134–149. Springer, Heidelberg (2008)
5. Bunder, M.W.: Lambda terms definable as combinators. Theoret. Comput. Sci. **169**(1), 3–21 (1996)
6. Church, A.: The Calculi of Lambda-Conversion. Princeton University Press, Princeton (1941)
7. Dijkstra, E.W.: An exercise for Dr. R.M. Burstall. In: Selected Writings on Computing: A Personal Perspective. Texts and Monographs in Computer Science, pp. 215–216. Springer, New York (1982)
8. Eberl, M., Hölzl, J., Nipkow, T.: A verified compiler for probability density functions. In: Vitek, J. (ed.) ESOP 2015. LNCS, vol. 9032, pp. 80–104. Springer, Heidelberg (2015)
9. Gammie, P., Lochbihler, A.: The Stern-Brocot tree. Archive of Formal Proofs, (2015). http://isa-afp.org/entries/Stern_Brocot.shtml, Formal proof development
10. Gibbons, J., Bird, R.: Be kind, rewind: a modest proposal about traversal (2012). http://www.comlab.ox.ac.uk/jeremy.gibbons/publications/backwards.pdf
11. Gibbons, J., Hinze, R.: Just do it: simple monadic equational reasoning. In: ICFP 2011, pp. 2–14. ACM (2011)
12. Graham, R.L., Knuth, D.E., Patashnik, O.: Concrete Mathematics-A Foundation for Computer Science, 2nd edn. Addison-Wesley, Reading (1994)
13. Green, J.A., Rees, D.: On semi-groups in which $x^r = x$. Math. Proc. Camb. Philos. Soc. **48**, 35–40 (1952)
14. Hinze, R.: The Bird tree. J. Func. Programm. **19**(5), 491–508 (2009)
15. Hinze, R.: Lifting operators and laws (2010). http://www.cs.ox.ac.uk/ralf.hinze/Lifting.pdf
16. Homeier, P.V.: The HOL-omega logic. In: Berghofer, S., Nipkow, T., Urban, C., Wenzel, M. (eds.) TPHOLs 2009. LNCS, vol. 5674, pp. 244–259. Springer, Heidelberg (2009)
17. Huffman, B.: Transfer principle proof tactic for nonstandard analysis. In: Kanovich, M., White, G., Gottliebsen, H., Oliva, P. (eds.) NetCA 2005, pp. 18–26. Queen Mary, University of London, Dept. of Computer Science, Research report RR-05-06 (2005)

18. Huffman, B., Kunčar, O.: Lifting and Transfer: a modular design for quotients in Isabelle/HOL. In: Gonthier, G., Norrish, M. (eds.) CPP 2013. LNCS, vol. 8307, pp. 131–146. Springer, Heidelberg (2013)

19. Hutton, G., Fulger, D.: Reasoning about effects: seeing the wood through the trees. In: Trends in Functional Programming (TFP 2008) (2008)

20. Krebbers, R.: The C standard formalized in Coq. Ph.D. thesis, Radboud University (2015)

21. Lammich, P., Tuerk, T.: Applying data refinement for monadic programs to Hopcroft's algorithm. In: Beringer, L., Felty, A. (eds.) ITP 2012. LNCS, vol. 7406, pp. 166–182. Springer, Heidelberg (2012)

22. Lochbihler, A.: Probabilistic functions and cryptographic oracles in higher order logic. In: Thiemann, P. (ed.) ESOP 2016. LNCS, vol. 9632, pp. 503–531. Springer, Heidelberg (2016)

23. Lochbihler, A., Schneider, J.: Applicative lifting. Archive of Formal Proofs (2015). http://isa-afp.org/entries/Applicative_Lifting.shtml

24. Marlow, S., Peyton Jones, S., Kmett, E., Mokhov, A.: Desugaring Haskell's do-notation into applicative operations (2016). http://research.microsoft.com/en-us/um/people/simonpj/papers/list-comp/applicativedo.pdf

25. McBride, C., Paterson, R.: Applicative programming with effects. J. Funct. Program. **18**(1), 1–13 (2008)

26. Nipkow, T.: More Church-Rosser proofs (in Isabelle/HOL). J. Automat. Reason. **26**, 51–66 (2001)

27. Paulson, L.: A higher-order implementation of rewriting. Sci. Comput. Program **3**(2), 119–149 (1983)

28. Reynolds, J.C.: Types, abstraction and parametric polymorphism. In: IFIP 1983. Information Processing, vol. 83, pp. 513–523. North-Holland/IFIP (1983)

29. Schropp, A., Popescu, A.: Nonfree datatypes in Isabelle/HOL. In: Gonthier, G., Norrish, M. (eds.) CPP 2013. LNCS, vol. 8307, pp. 114–130. Springer, Heidelberg (2013)

30. Tuong, F., Wolff, B.: A meta-model for the Isabelle API. Archive of Formal Proofs (2015). http://isa-afp.org/entries/Isabelle_Meta_Model.shtml

Formally Verified Approximations
of Definite Integrals

Assia Mahboubi[1], Guillaume Melquiond[1,2](✉), and Thomas Sibut-Pinote[1]

[1] Inria, Palaiseau, France
[2] LRI, CNRS UMR 8623, Université Paris-Sud,
Orsay, France
guillaume.melquiond@inria.fr

Abstract. Finding an elementary form for an antiderivative is often a difficult task, so numerical integration has become a common tool when it comes to making sense of a definite integral. Some of the numerical integration methods can even be made rigorous: not only do they compute an approximation of the integral value but they also bound its inaccuracy. Yet numerical integration is still missing from the toolbox when performing formal proofs in analysis.

This paper presents an efficient method for automatically computing and proving bounds on some definite integrals inside the Coq formal system. Our approach is not based on traditional quadrature methods such as Newton-Cotes formulas. Instead, it relies on computing and evaluating antiderivatives of rigorous polynomial approximations, combined with an adaptive domain splitting. This work has been integrated to the CoqInterval library.

1 Introduction

Computing the value of definite integrals is the modern and generalized take on the ancient problem of computing the area of a figure. *Quadrature methods* hence refer to the numerical methods for estimating such integrals. Numerical integration is indeed often the preferred way of obtaining such estimations as symbolic approaches may be too difficult or even just impossible. Quadrature methods, as implemented in systems like Matlab, most often consist in interpolating the integrand function by a degree-n polynomial, integrating the polynomial and then bounding the error using a bound on the $n + 1$-th derivative of the integrand function. Estimating the value of integrals can be a crucial part of some mathematical proofs, making numerical integration an invaluable ally. Examples of such proofs occur in various areas of mathematics, such as number theory (*e.g.* Helfgott's proof of the ternary Goldbach conjecture [5]) or geometry (*e.g.* the first proof of the double bubble conjecture [4]). This motivates developing high-confidence methods for computing *reliable* yet accurate and fast estimations of integrals.

This work was supported in part by the project FastRelax ANR-14-CE25-0018-01.

© Springer International Publishing Switzerland 2016
J.C. Blanchette and S. Merz (Eds.): ITP 2016, LNCS 9807, pp. 274–289, 2016.
DOI: 10.1007/978-3-319-43144-4_17

The present paper describes a formal-proof producing procedure to obtain numerical enclosures of definite integrals $\int_u^v f(t)\,dt$, where f is a real-valued function that is Riemann-integrable on the bounded integration domain $[u, v]$. This procedure can deal with any function f for which we have an interval extension and/or a polynomial approximation. The enclosure is computed *inside* the Coq proof assistant and the computations are correct by construction. Interestingly, the formal proof that the integral exists comes as a by-product of these computations.

Our approach is based on interval methods, in the spirit of Moore et al. [10], and combines the computation of a numerical enclosure of the integrand with an adaptive dichotomy process. It is based on the CoqInterval library for computing interval extensions of elementary mathematical functions and is implemented as an improvement of the `interval` Coq tactic [8].

The paper is organized as follows: Sect. 2 introduces some definitions and notations used throughout the paper, and describes briefly the Coq libraries we build on. Section 3 describes the algorithms used to estimate integrals and Sect. 4 describes the design of the proof-producing Coq tactic. In Sect. 5 we provide cross-software benchmarks highlighting issues with both our and others' algorithms. In Sect. 6, we discuss the limitations and perspectives of this work.

2 Preliminaries

In this section we introduce some vocabulary and notations used throughout the paper and we summarize the existing Coq libraries the present work builds on.

2.1 Notations and First Definitions

An interval is a closed connected subset of the set of real numbers. We use $\mathbb{I}$ to denote the set of intervals: $\{[a, b] \mid a, b \in \mathbb{R} \cup \{\pm\infty\}\}$. A *point interval* is an interval of the shape $[a, a]$ where $a \in \mathbb{R}$. Any interval variable will be denoted using a bold font. For any interval $\mathbf{x} \in \mathbb{I}$, $\inf \mathbf{x}$ (resp. $\sup \mathbf{x}$) denotes its left (resp. right) bound, with $\inf \mathbf{x} \in \mathbb{R} \cup \{-\infty\}$ (resp. $\sup \mathbf{x} \in \mathbb{R} \cup \{+\infty\}$). An *enclosure* of $x \in \mathbb{R}$ is an interval $\mathbf{x} \in \mathbb{I}$ such that $x \in \mathbf{x}$.

In the following, we will not denote interval operators in any distinguishing way. In particular, whenever an arithmetic operator takes interval inputs, it should be understood as any interval extension of the corresponding operator on real numbers (see Sect. 2.3). Moreover, whenever a real number appears as an input of an interval operator, it should be understood as any interval that encloses this number. For instance, an expression like $(v - u) \cdot \mathbf{x}$ denotes the interval product of the interval $\mathbf{x}$ with any (hopefully tight) interval enclosing the real $v - u$.

2.2 Elementary Real Analysis in Coq

Coq's standard library `Reals`[1] axiomatizes real arithmetic, with a classical flavor [9]. It provides some notions of elementary real analysis, including the

---

[1] https://coq.inria.fr/distrib/current/stdlib/.

definition of continuity, differentiability and Riemann integrability. It also comes with a formalization of the properties of usual mathematical functions like sin, cos, exp, and so on.

The Coquelicot library is a conservative extension of this library [2]. It provides a *total* operator that outputs a real value from a function $f : \mathbb{R} \to \mathbb{R}$ and two bounds $u, v \in \mathbb{R}$:

```
Definition RInt (f : R -> R) (u : R) (v : R) : R := ...
```

When the function f is Riemann-integrable on $[u, v]$, the value (RInt f u v) is equal to $\int_u^v f(t)\, dt$. Otherwise it is left unspecified and thus most properties about the actual value of (RInt f u v) hold only if f is integrable on $[u, v]$.

The aim of this work is to provide a procedure that computes a numerical and formally proved enclosure of an expression (RInt f u v) –and justifies that this expression is well-defined. This procedure can then be used in an automated tactic that proves inequalities like $|\int_0^1 \sqrt{1 - x^2}\, dx - \frac{\pi}{4}| \le \frac{1}{100}$, stated as:

```
Goal Rabs (RInt (fun x => sqrt(1 - x * x)) 0 1 - PI / 4) <= 1/100.
```

Without Coquelicot's total operator RInt, the user would not be able to express such a statement as easily.

2.3   Numerical Computations in Coq

CoqInterval is a Coq library for computing numerical enclosures of real-valued expressions [8]. These expressions belong to a class $\mathcal{E}$ built from constants, variables, arithmetic operations, and some elementary functions. It also provides a tactic `interval` to automatically deduce certain goals from these enclosures.

The tactic typically takes a goal $A \le e \le B$ where e is such an expression, and A and B are constants. Using the paradigm of interval arithmetic, it builds a set **e** such that $e \in$ **e** holds by construction and such that **e** reduces to an interval $[\inf \mathbf{e}, \sup \mathbf{e}]$ by computation. Then it checks that $A \le \inf \mathbf{e}$ and $\sup \mathbf{e} \le B$, again by computation, from which it proves $A \le e \le B$. All the computations on interval bounds are performed using a rigorous yet efficient formalization of multi-precision floating-point arithmetic.

The library provides several ways to build the interval **e**: naive interval arithmetic, automatic differentiation, and rigorous polynomial approximations using Taylor models. Interval arithmetic is concerned with providing operators on intervals that respect the *inclusion property*. Given a binary operator $\diamond$ on real numbers, naive interval arithmetic provides a binary operator $\Diamond$ on intervals such that

$$\forall x, y \in \mathbb{R}, \ \forall \mathbf{x}, \mathbf{y} \in \mathbb{I}, \ x \in \mathbf{x} \wedge y \in \mathbf{y} \Rightarrow x \diamond y \in \mathbf{x} \Diamond \mathbf{y}.$$

This inclusion property is easily transported from operators to whole expressions by induction on these expressions. This ensures that the property $e \in$ **e** above can be easily proved when **e** is built using the operators from naive interval arithmetic. This approach, however, cannot keep track of correlations between subexpressions and might compute overestimated enclosures which are thus useless for proving some goals. For instance, assume that $x \in [3, 4]$, so $-x \in [-4, -3]$

using the interval extension of the negation, so $x + (-x) \in [3 + (-4), 4 + (-3)]$ using the interval extension of the addition. If the goal was to prove that $x - x$ is always 0, the interval $[-1, 1]$ obtained by naive interval arithmetic is useless. This is why the CoqInterval library also comes with refinements of naive interval arithmetic, such as rigorous polynomial approximations, so as to reduce this loss of correlations.

Our goal is to extend the class $\mathcal{E}$ of supported expressions with integrals whose bounds and bodies are in $\mathcal{E}$.

3   Interval Methods to Approximate an Integral

In this section, we describe how to compute a numerical enclosure of the real number $\int_u^v f(t)\, dt$ from enclosures of the finite bounds u and v and of the integrand function f. We describe two basic methods based respectively on the evaluation of a simple interval extension and on a polynomial approximation of f. They can be combined and improved by a dichotomy process.

3.1   Naive Integral Enclosure

Our first approach uses an *interval extension* of the integrand.

Definition 1. *For any function* $f : \mathbb{R}^n \to \mathbb{R}$, *a function* $F : \mathbb{I}^n \to \mathbb{I}$ *is an interval extension of* f *on* $\mathbb{R}$ *if*

$$\forall \mathbf{x}_1, \ldots, \mathbf{x}_n,\ \{f(x_1, \ldots, x_n) \mid \forall i, x_i \in \mathbf{x}_i\} \subseteq F(\mathbf{x}_1, \ldots, \mathbf{x}_n).$$

In the rest of the section we suppose that $F : \mathbb{I} \to \mathbb{I}$ is an interval extension of the univariate function f, and we want to compute an enclosure of $\int_u^v f(t)\, dt$, with $u, v \in \mathbb{R}$, and f integrable on $[u, v]$.

Definition 2. *The* convex hull *of a set* $A \subseteq \mathbb{R}$ *is the smallest convex set that contains* A, *denoted* $\mathrm{hull}(A)$. *Moreover, the interval* $\mathrm{hull}(\mathbf{a}, \mathbf{b})$ *denotes the convex hull of (the union of) two intervals* $\mathbf{a}$ *and* $\mathbf{b}$.

Lemma 1 (Naive integral enclosure)

$$\int_u^v f(t)\, dt \in (v - u) \cdot \mathrm{hull}\{f(t) \mid t \in [u, v] \vee t \in [v, u]\}. \tag{1}$$

Proof. Let us first suppose that $u \leq v$. Denote $f([u, v]) := \{f(t) \mid t \in [u, v]\}$. If $\mathrm{hull}(f([u, v])) = [m, M]$ (without loss of generality, m and M can be assumed to be finite) then for any $x \in [u, v]$, we have $m \leq f(x) \leq M$. So $(v - u)m \leq \int_u^v f(x) \leq (v - u)M$, hence (1). The case $v \leq u$ is symmetrical.

In practice we do not compute with f but only its interval extension F. Moreover, we want the computations to operate using only enclosures of the bounds. So we adapt Formula (1) accordingly.

Lemma 2 (Interval naive integral enclosure). *For any intervals* $\mathbf{u}, \mathbf{v}$ *such that* $u \in \mathbf{u}$ *and* $v \in \mathbf{v}$, *we have*

$$\int_u^v f(t)\, dt \in (\mathbf{v} - \mathbf{u}) \cdot F(\mathrm{hull}(\mathbf{u}, \mathbf{v})). \tag{2}$$

Note that if $\mathbf{u}$ *and* $\mathbf{v}$ *are point intervals and if* F *is the optimal interval extension of* f, *then* (2) *reduces to* (1).

Proof. If $u \in \mathbf{u}$ and $v \in \mathbf{v}$, then by (1) and reusing notations from the proof, we have $\int_u^v f(t)\, dt \in (v - u) \cdot \mathrm{hull}(f([u, v]))$. Since $(v - u) \in (\mathbf{v} - \mathbf{u})$, we only have to show that $\mathrm{hull}(f([u, v])) \subseteq F(\mathrm{hull}(\mathbf{u}, \mathbf{v}))$. If $y \in \mathrm{hull}(f([u, v]))$, then there exist $t_1, t_2 \in [u, v]$ such that $f(t_1) \leq y \leq f(t_2)$. Since $F(\mathrm{hull}(\mathbf{u}, \mathbf{v}))$ is an interval, we only need to show that $f(t_1), f(t_2) \in F(\mathrm{hull}(\mathbf{u}, \mathbf{v}))$. This holds because $t_1, t_2 \in \mathrm{hull}(\mathbf{u}, \mathbf{v})$, and F is an interval extension of f.

The `naive_integral` Coq function implements (2). Given $\mathbf{u}, \mathbf{v} \in \mathbb{I}$ and F a function of type $\mathbb{I} \to \mathbb{I}$, (`naive_integral prec F u v`) computes an interval $\mathbf{i}$ using floating-point arithmetic at precision `prec`. If F is an interval extension of f, if $u \in \mathbf{u}$ and $v \in \mathbf{v}$, and if f is integrable on $[u, v]$, then $\int_u^v f(t)\, dt \in \mathbf{i}$.

```
Definition naive_integral prec F u v :=
  I.mul prec (F (I.join u v)) (I.sub prec v u).
```

3.2   Polynomial Approximation

The enclosure method described in Sect. 3.1 is rather crude. Better knowledge of the integrated function allows for a more efficient approach.

The CoqInterval library defines a *rigorous polynomial approximation* (RPA) of $f : \mathbb{R} \to \mathbb{R}$ on the interval $\mathbf{x}$ as a pair $(\mathbf{p}, \mathbf{\Delta})$, with $\mathbf{p} \in \mathbb{I}[X]$, such that for some polynomial $p \in \mathbb{R}[X]$ enclosed[2] in $\mathbf{p}$ we have $f(x) - p(x) \in \mathbf{\Delta}$ for all $x \in \mathbf{x}$. CoqInterval computes these RPAs by composing and performing arithmetic operations on Taylor expansions of elementary functions [8]. Now that we have polynomial approximations, we can make use of the following lemma.

Lemma 3 (Polynomial approximation). *Suppose* f *is approximated on* $[u, v]$ *by* $p \in \mathbb{R}[X]$ *and* $\mathbf{\Delta} \in \mathbb{I}$ *in the sense that* $\forall x \in [u, v], f(x) - p(x) \in \mathbf{\Delta}$. *Then for any primitive* P *of* p *we have* $\int_u^v f(t)\, dt \in P(v) - P(u) + (v - u) \cdot \mathbf{\Delta}$.

Proof. We have $\int_u^v f(t)\, dt - (P(v) - P(u)) = \int_u^v (f(t) - p(t))\, dt$. By hypothesis, the constant function $\mathbf{\Delta}$ is an interval extension of $t \mapsto f(t) - p(t)$ on $[u, v]$, hence Lemma 1 applies (notice that $\mathrm{hull}(\mathbf{\Delta}) = \mathbf{\Delta}$).

Note that our method and proofs do not depend on the way RPAs are obtained.

---

[2] We say that $\mathbf{p} \in \mathbb{I}[X]$ is an enclosure of $p \in \mathbb{R}[X]$ if, for all $i \in \mathbb{N}$, the ith coefficient $\mathbf{p}_i$ of $\mathbf{p}$ is an enclosure of the ith coefficient p_i of p, where we take the convention that for $i > \deg \mathbf{p}$, $\mathbf{p}_i = \{0\}$ and for $i > \deg p$, $p_i = 0$.

3.3   Quality of the Integral Enclosures

Both methods described in Sects. 3.1 and 3.2 use a single approximation of
the integrand on the integration interval. A decomposition of this interval into
smaller pieces may increase the accuracy of the enclosure, if tighter approximations are obtained on each subinterval. In this section we give an intuition of how
the naive and polynomial approaches compare, from a time complexity point of
view. The naive (resp. polynomial) approach here consists in using a simple
interval approximation (resp. a valid polynomial approximation) to estimate the
integral on each subinterval. Let us suppose that we split the initial integration
interval, using Chasles' relation, before computing integral enclosures:

$$\int_u^v f = \int_{x_0}^{x_1} f + \ldots + \int_{x_{n-1}}^{x_n} f \quad \text{with } x_i = u + \frac{i}{n}(v - u).$$

Let $w(\mathbf{x}) = \sup \mathbf{x} - \inf \mathbf{x}$ denote the width of an interval. The smaller $w(\mathbf{x})$ is,
the more accurately any real $x \in \mathbf{x}$ is approximated by $\mathbf{x}$. Any sensible interval
arithmetic respects $w(\mathbf{x} + \mathbf{y}) \simeq w(\mathbf{x}) + w(\mathbf{y})$ and $w(k \cdot \mathbf{x}) \simeq k \cdot w(\mathbf{x})$.

We consider the case of the naive approach first. We assume that F is an optimal interval extension of f and that f has a Lipschitz-constant equal to k_0, that
is, $w(F(\mathbf{x})) \simeq k_0 \cdot w(\mathbf{x})$. Since $w(\text{naive}([x_i, x_{i+1}])) \simeq (x_{i+1} - x_i) \cdot w(F([x_i, x_{i+1}]))$,
we get the following accuracy when computing the integral:

$$w\left(\sum_i \text{naive}([x_i, x_{i+1}])\right) \simeq k_0 \cdot (v - u)^2/n.$$

To gain one bit of accuracy, we need to go from n to $2n$ integrals, which means
multiplying the computation time by two, hence an exponential complexity.

Now for the polynomial enclosure. Let us assume we can compute a poly-
nomial approximation of f on any interval $\mathbf{x}$ with an error $\boldsymbol{\Delta}(\mathbf{x})$. We can
expect this error to satisfy $w(\boldsymbol{\Delta}(\mathbf{x})) \simeq k_d \cdot w(\mathbf{x})^{d+1}$ with d the degree of the
polynomial approximation and k_d depending on f. Since $w(\text{poly}([x_i, x_{i+1}])) \simeq$
$(x_{i+1} - x_i) \cdot w(\boldsymbol{\Delta}([x_i, x_{i+1}]))$, the accuracy is now

$$w\left(\sum_i \text{poly}([x_i, x_{i+1}])\right) \simeq k_d \cdot (v - u)^{d+2}/n^{d+1}.$$

For a fixed d, one still has to increase n exponentially with respect to the target
accuracy. The power coefficient, however, is much smaller than for the naive
method. By doubling the computation time, one gets $d + 1$ additional bits of
accuracy.

In order to improve the accuracy of the result, one can increase d instead
of n. If f behaves similarly to exp or sin, Taylor-Lagrange formula tells us that
k_d decreases as fast as $(d!)^{-1}$. Moreover, the time complexity of computing a
polynomial approximation usually grows like d^3. So, if $n \simeq v - u$, doubling the
computation time by increasing d gives about 25 % more bits of accuracy.

As can be seen from the considerations above, striking the proper balance between n and d for reaching a target accuracy in a minimal amount of time is difficult, so we have made the decision of letting the user control d (see Sect. 4.3) while the implementation adaptively splits the integration interval.

3.4 Dichotomy and Adaptivity

Both methods presented in Sects. 3.1 and 3.2 can compute an interval enclosing $\int_u^v f(t)\,dt$. Polynomial approximations usually give tighter enclosures of the integral, but not always, so we combine both methods by taking the intersection of their result.

This may still not be sufficient for getting a tight enough enclosure, in which case we recursively split the integration domain in two parts, using Chasles' rule. The function `integral_float_absolute` performs this dichotomy and the integration on each subdomain. It takes an absolute error parameter ε; it stops splitting as soon as the width of the computed integral enclosure is smaller than ε. The function also takes a *depth* parameter, which means that the initial domain is split into at most $2^{depth+1}$ subdomains. Note that, because the depth is bounded, there is no guarantee that the target width will be reached.

Let us detail more precisely how the function behaves. It starts by splitting $[u, v]$ into $[u, m]$ and $[m, v]$ and computes some enclosures $\mathbf{i_1}$ of $\int_u^m f(t)\,dt$ and $\mathbf{i_2}$ of $\int_m^v f(t)\,dt$. If $depth = 0$, then the function returns $\mathbf{i_1} + \mathbf{i_2}$. Otherwise, several cases can occur:

- If $w(\mathbf{i_1}) \leq \frac{\varepsilon}{2}$ and $w(\mathbf{i_2}) \leq \frac{\varepsilon}{2}$, then the function simply returns $\mathbf{i_1} + \mathbf{i_2}$.
- If $w(\mathbf{i_1}) \leq \frac{\varepsilon}{2}$ and $w(\mathbf{i_2}) > \frac{\varepsilon}{2}$, then the first enclosure is sufficient but the second is not. So `integral_float_absolute` calls itself recursively on $[m, v]$ with $depth - 1$ as the new maximal depth and $\varepsilon - w(\mathbf{i_1})$ as the new target accuracy, yielding $\mathbf{i_2'}$. The function then returns $\mathbf{i_1} + \mathbf{i_2'}$.
- If $w(\mathbf{i_1}) > \frac{\varepsilon}{2}$ and $w(\mathbf{i_2}) \leq \frac{\varepsilon}{2}$, we proceed symmetrically.
- Otherwise, the function calls itself on $[u, m]$ and $[m, v]$ with $depth - 1$ as the new maximal depth and $\frac{\varepsilon}{2}$ as the new target accuracy, yielding $\mathbf{i_1'}$ and $\mathbf{i_2'}$. It then returns $\mathbf{i_1'} + \mathbf{i_2'}$.

4   Automating the Proof Process

In this section we explain how to compute the approximations of the integrand required by the theorems of Sect. 3, and how to automate the proof of its integrability. We conclude by describing how all the ingredients combine into the implementation of a parameterized Coq tactic.

4.1 Straight-Line Programs and Enclosures

As described in Sect. 2.3, enclosures and interval extensions are computed from expressions that appear as bounds or as the body of an integral, like for instance

$\ln 2$, 3, and $(t + \pi)\sqrt{t} - (t + \pi)$, in $\int_{\ln 2}^{3}((t + \pi)\sqrt{t} - (t + \pi))\,dt$. The tactic represents these expressions symbolically, as straight-line programs. This allows for explicit sharing of common subexpressions. Such a program is just a list of statements indicating what the operation is and where its inputs can be found. The place where the output is stored is left implicit: the result of an operation is always put at the top of the evaluation stack.[3] The stack is initially filled with values corresponding to the constants of the program. The result of evaluating a straight-line program is at the top of the stack.

Below is an example of a straight-line program corresponding to the expression $(t + \pi)\sqrt{t} - (t + \pi)$. It is a list containing the operations to be performed. Each list item first indicates the arity of the operation, then the operation itself, and finally the depth at which the inputs of the operation can be found in the evaluation stack. Note that, in this example, t and π are seen as constants, so the initial stack contains values that correspond to these subterms.[4] The comments in the term below indicate the content of the evaluation stack before evaluating each statement.

```
(* initial stack: [t, pi]                          *)      Binary Add 0 1
(* current stack: [t+pi, t, pi]                     *)   :: Unary Sqrt 1
(* current stack: [sqrt t, t+pi, t, pi]             *)   :: Binary Mul 1 0
(* current stack: [(t+pi)*sqrt t, sqrt t, ...]      *)   :: Binary Sub 0 2
(* current stack: [(t+pi)*sqrt t - (t+pi), ...]     *)   :: nil
```

The evaluation of a straight-line program depends on the interpretation of the arithmetic operations and on the values stored in the initial stack. For instance, if the arithmetic operations are the operations from the Reals library (*e.g.* Rplus) and if the stack contains the symbolic value of the constants, then the result is the actual expression over real numbers.

Let us denote $[\![p]\!]_\mathbb{R}(\vec{x})$ the result of evaluating the straight-line program p with operators from Reals over an initial stack $\vec{x}$ of real numbers. Similarly, $[\![p]\!]_\mathbb{I}(\vec{\mathbf{x}})$ denotes the result of evaluating p with interval operations over a stack of intervals. Then, thanks to the inclusion property of interval arithmetic, we can prove the following formula once and for all:

$$\forall p, \ \forall \vec{x} \in \mathbb{R}^n, \ \forall \vec{\mathbf{x}} \in \mathbb{I}^n, \ (\forall i \le n, \ x_i \in \mathbf{x}_i) \Rightarrow [\![p]\!]_\mathbb{R}(\vec{x}) \in [\![p]\!]_\mathbb{I}(\vec{\mathbf{x}}). \qquad (3)$$

Theorem (3) is the basic block used by the interval tactic for proving enclosures of expressions [8]. Given a goal $A \le e \le B$, the tactic first looks for a program p and a stack $\vec{x}$ of real numbers such that $[\![p]\!]_\mathbb{R}(\vec{x}) = e$. Note that this reification process is not proved to be correct, so Coq checks that both sides of the equality are convertible. More precisely, the goal $A \le e \le B$ is convertible to $[\![p]\!]_\mathbb{R}(\vec{x}) \in [A, B]$ if A and B are floating-point numbers and if the tactic successfully reified the term.

---

[3] Note that the evaluation model is quite simple: the stack grows linearly with the size of the expression since no element of the stack is ever removed.

[4] The only thing that will later distinguishes the integration variable t from an actual constant such as π is that its value is placed at the top of the initial evaluation stack.

The tactic then looks in the context for hypotheses of the form $A_i \leq x_i \leq B_i$ so that it can build a stack $\vec{\mathbf{x}}$ of intervals such that $\forall i, \ x_i \in \mathbf{x}_i$. If there is no such hypothesis, the tactic just uses $(-\infty, +\infty)$ for $\mathbf{x}_i$. The tactic can now apply Theorem (3) to replace the goal by $[\![p]\!]_{\mathbb{I}}(\vec{\mathbf{x}}) \subseteq [A, B]$. It then attempts to prove this new goal entirely by computation. Note that even if the original goal holds, this attempt may fail due to loss of correlation inherent to interval arithmetic.

Theorem (3) also implies that if a function f can be reified as $t \mapsto [\![p]\!]_{\mathbb{R}}(t, \vec{x})$, then $\mathbf{t} \mapsto [\![p]\!]_{\mathbb{I}}(\mathbf{t}, \vec{\mathbf{x}})$ is an interval extension of f if $\forall i, \ x_i \in \mathbf{x}_i$. This way, we obtain the interval extensions of the integrand that we need for Sect. 3.

There is also an evaluation scheme for computing RPAs for f. The program p is the same, but the initial evaluation stack now contains RPAs: a degree-1 polynomial for representing the domain of t, and constant polynomials for the constants. The result is an RPA of $t \mapsto [\![p]\!]_{\mathbb{R}}(t, \vec{x})$. By computing the image of this resulting polynomial approximation, one gets an enclosure of the expression that is usually better than the one computed by $\mathbf{t} \mapsto [\![p]\!]_{\mathbb{I}}(\mathbf{t}, \vec{\mathbf{x}})$.

4.2   Checking Integrability

When computing the enclosure of an integral, the tactic should first obtain a formal proof that the integrand is indeed integrable on the integration domain, as this is a prerequisite to all the theorems in Sect. 3. In fact we can be more clever: we prove that, if we succeed in numerically computing an *informative* enclosure of the integral, the function was actually integrable. This way, the tactic does not have to prove anything beforehand about the integrand.

This trick requires to explain the inner workings of the CoqInterval library in more detail. In particular, the library provides evaluation schemes that use bottom values. In all that follows $\overline{\mathbb{R}}$ denotes the set $\mathbb{R} \cup \{\perp_{\mathbb{R}}\}$ of *extended reals*, that is the set of real numbers completed with the extra point $\perp_{\mathbb{R}}$. The alternate scheme $[\![p]\!]_{\overline{\mathbb{R}}}$ produces the value $\perp_{\mathbb{R}}$ as soon as an operation is applied to inputs that are outside the usual definition domain of the operator. For instance, the resulting of dividing one by zero in $\overline{\mathbb{R}}$ is $\perp_{\mathbb{R}}$, while it is unspecified in $\mathbb{R}$. This $\perp_{\mathbb{R}}$ element is then propagated along the subsequent operations. Thus, the following equality holds, using the trivial embedding from $\mathbb{R}$ into $\overline{\mathbb{R}}$:

$$\forall p, \ \forall \vec{x} \in \mathbb{R}^n, \ [\![p]\!]_{\overline{\mathbb{R}}}(\vec{x}) \neq \perp_{\mathbb{R}} \Rightarrow [\![p]\!]_{\mathbb{R}}(\vec{x}) = [\![p]\!]_{\overline{\mathbb{R}}}(\vec{x}). \tag{4}$$

Moreover, the implementation of interval arithmetic uses not only pairs of floating-point numbers $[\inf \mathbf{x}, \sup \mathbf{x}]$ but also a special interval $\perp_{\mathbb{I}}$, which is propagated along computations. An interval operator produces the value $\perp_{\mathbb{I}}$ whenever the input intervals are not fully included in the definition domain of the corresponding real operator. In other words, an interval operator produces $\perp_{\mathbb{I}}$ whenever the corresponding operator on $\overline{\mathbb{R}}$ would have produced $\perp_{\mathbb{R}}$ for at least one value in one of the input intervals. Thus, by extending the definition of an enclosure so that $\perp_{\mathbb{R}} \in \perp_{\mathbb{I}}$ holds, we can prove a variant of Formula (3):

$$\forall p, \ \forall \vec{x} \in \overline{\mathbb{R}}^n, \ \forall \vec{\mathbf{x}} \in \mathbb{I}^n, \ (\forall i \leq n, \ x_i \in \mathbf{x}_i) \Rightarrow [\![p]\!]_{\overline{\mathbb{R}}}(\vec{x}) \in [\![p]\!]_{\mathbb{I}}(\vec{\mathbf{x}}). \tag{5}$$

In CoqInterval, Formula (3) is actually just a consequence of both Formulas (4) and (5). This is due to two other properties of $\perp_{\mathbb{I}}$. First, $(-\infty, +\infty) \subseteq \perp_{\mathbb{I}}$ holds, so the conclusion of Formula (5) trivially holds whenever $\llbracket p \rrbracket_{\mathbb{I}}(\vec{\mathbf{x}})$ evaluates to $\perp_{\mathbb{I}}$. Second, $\perp_{\mathbb{I}}$ is the only interval containing $\perp_{\mathbb{R}}$. As a consequence, whenever $\llbracket p \rrbracket_{\mathbb{I}}(\vec{\mathbf{x}})$ does not evaluate to $\perp_{\mathbb{I}}$ the premise of Formula (4) holds.

Let us go back to the issue of proving integrability. By definition, whenever $\llbracket p \rrbracket_{\overline{\mathbb{R}}}(\vec{x})$ does not evaluate to $\perp_{\mathbb{R}}$ the inputs $\vec{x}$ are part of the definition domain of the expression represented by p. But we can actually prove a stronger property: not only is $\vec{x}$ part of the definition domain, it is also part of the continuity domain. More precisely, we can prove the following property:

$$\forall p, \ \forall t_0 \in \mathbb{R}, \ \forall \vec{x} \in \mathbb{R}^n, \ \llbracket p \rrbracket_{\overline{\mathbb{R}}}(t_0, \vec{x}) \neq \perp_{\mathbb{R}} \Rightarrow$$
$$t \mapsto \llbracket p \rrbracket_{\mathbb{R}}(t, \vec{x}) \text{ is continuous at point } t_0. \quad (6)$$

Note that this property intrinsically depends on the operations that can appear inside p, *i.e.* the operations belonging to the class $\mathcal{E}$ of Sect. 2.3. Therefore, its proof has to be extended as soon as a new operator is supported in $\mathcal{E}$. In particular, it would become incorrect as such, if the integer part was ever supported.

By combining Formulas (3) and (6), we obtain a numerical method to prove that a function is continuous on a domain. Indeed, we just have to compute an enclosure of the function on that domain, and to check that it is not $\perp_{\mathbb{I}}$. A closer look at the way naive integral enclosures are computed provides the following corollary: whenever the enclosure of the integral is not $\perp_{\mathbb{I}}$, the function is actually continuous and thus integrable.

For the sake of completeness, we mention another scheme implemented in the CoqInterval library, which computes the enclosure of the derivative of a function through automatic differentiation. As before, the tactic does not have to prove beforehand that the function is actually differentiable, it is deduced from the computation not returning $\perp_{\mathbb{I}}$. This is of no use for computing integrals as done in this paper. It could however be used to implement a numeric quadrature such as the trapezoid method, since the latter requires bounding derivatives.

4.3   Integration into a Tactic

The interval tactic is primarily dedicated to computing/verifying the enclosure of an expression. For this purpose, the expression is first turned into a straight-line program, as described in Sect. 4.1. There is however no integral operator in the grammar $\mathcal{E}$ of programs: from the point of view of the reification process, integrals are just constants, and thus part of the initial stack used when evaluating the program.

The tactic supports constants for which it can get a formally-proved enclosure. In previous releases of CoqInterval, the only supported constants were floating-point numbers and π. Floating-point numbers are enclosed by the corresponding point interval, which is trivially correct. An interval function, and its correctness proof, provides enclosures of the constant π, at the required precision.

The tactic now supports constants expressed as integrals $\int_u^v e\,dt$. First, it reifies the bounds u and v into programs and it evaluates them over $\mathbb{I}$ to get hopefully tight enclosures of them. Second, it reifies e into a program p with t at the top of the initial evaluation stack. The tactic uses p to instantiate various evaluation methods, so that interval extensions and RPAs of e can be computed on all the integration subdomains, as described in Sect. 4.1. Third, using the formulas of Sect. 3, it creates a term of type $\mathbb{I}$ that, once reduced by Coq's kernel, has actual floating-point bounds. The tactic also proves that this term is an enclosure of the integral, using the theorems of Sects. 3 and 4.2.

4.4   Controlling the Tactic

The `interval` tactic now features three options that supply the user with some control over how it computes integral enclosures. First, the user can indicate the target accuracy for the integral, expressed as a relative error: the user indicates how many bits of the result should be significant (by default, 10 bits, so three decimal digits). It is an *a priori* error, that is, the implementation first computes a coarse magnitude of the integral value and uses it to turn the relative bound into an absolute one. It then performs computations using only this absolute bound.

The user can also indicate the degree of the RPAs used for approximating the integrand (default is 10). This value empirically provides a good trade-off between bisecting too deeply and computing costly RPAs when targeting the default accuracy of 10 bits. For poorly approximated integrands, choosing a smaller degree can improve timings significantly, while for highly regular integrands and a high target accuracy, choosing a larger degree might be worth a try.

Finally, the user can limit the maximal depth of bisection (default is 3). If the target absolute error is reached on each interval of the subdivision, then increasing the maximal depth does not affect timings. There might, however, be some points of the integration domain around which the target error is never reached. This setting prevents the computations from splitting the domain indefinitely, while the computed enclosure is already accurate enough to prove the goal.

Note that as in previous CoqInterval releases, the user can adjust the precision of floating-point computations used for interval computations, which has an impact on how integrals are computed. The default value is 30 bits, which is sufficient in practice for getting the default 10 bits of integral accuracy.

There are three reasons why the user-specified target accuracy might not be reached. If the computed magnitude during the initial estimate of the integral is too coarse, the absolute bound used by the adaptive algorithm will be too large and the final result might be less accurate than desired.[5] An insufficient bisection depth might also lead the result to be less accurate. This is also true with an insufficient precision of intermediate computations.

---

[5] The magnitude might be so coarse that it is computed as $+\infty$. In that case, the user setting is directly understood as an absolute bound.

The following script shows how to prove in Coq that the surface of a quarter unit disk is equal to $\pi/4$, at least up to 10^{-6}. The target accuracy is set to 20 bits, so that we can hope to reach the 10^{-6} bound. Since the integrand is poorly approximated near 1 (due to the square root), the integration domain has to be split into small pieces around 1. So we significantly increase the bisection depth to 15. Finally, since here the RPAs are poor, decreasing their degree to 5 shaves a few tenths of second off the time needed to check the result. In the end, it takes under a second for Coq to formally check the proof on a standard laptop.

```
Goal Rabs (RInt (fun t => sqrt (1 - t*t)) 0 1 - PI/4) <= 1/1000000.
interval with (i_integral_prec 20, i_integral_depth 15, i_integral_deg 5).
Qed.
```

5 Benchmarks

This section presents the behavior of the tactic on several integration problems, each given as a symbolic integral, its value (approximate if no closed form exists), and a set of absolute error bounds that must be reached by the tactic. Each problem is translated into a set of Coq scripts as follows, one for each bound:

```
Goal Rabs (RInt function domain - value) <= error.
interval with options.
Qed.
```

The tactic options have been set using the following experimental protocol. First, the target relative accuracy is computed from the error bound and the initial estimation of an integral. The floating-point precision is then set at about 10 more bits than the target accuracy, so that round-off errors do not make interval enclosures too large. The maximal depth is originally set to a large enough value. Then, various degrees of RPAs are tested and the one that leads to the fastest execution is kept. Finally, the maximal depth is reduced as long as the tactic succeeds in proving the bounds, so that we get an idea of how deep splitting has to be performed to compute an accurate enclosure of the integral. Note that reducing the maximal depth might improve timings in case the adaptive algorithm had been overly conservative and did too much domain splitting. Reducing the target relative accuracy could also improve timings (again by preventing some domain splitting), but this was not done. The tables below indicate, for each error bound, the time needed and the tactic settings. Timings are in seconds and are obtained on a standard-grade laptop.

For each integral, we also ran several quadrature methods from Octave [3]: quad, quadv, quadgk, quadl, quadcc. We also used IntLab [13]; it provides verifyquad, an interval arithmetic procedure that computes integral enclosures using a verified Romberg method. For each method, we ask for an absolute accuracy of 10^{-15}. We only comment when the answer is off, or when the execution time exceeds 1 s. Finally, we also tested VNODE-LP [11] on each example by representing the integral as the value of the solution of a differential equation.

The first problem is the integral of the derivative of arctan, a highly regular function. As expected, the tactic behaves well on it, since it takes about 3 s to compute 18 decimal digits of π by integration. Note that the time needed for reifying the goal and performing the initial computations is incompressible, so there is not much difference between 10^{-3} and 10^{-6}.

$$\int_0^1 \frac{dx}{1 + x^2} = \frac{\pi}{4}$$

Error	Time	Accuracy	Degree	Depth	Prec
10^{-3}	0.3	10	15	0	30
10^{-6}	0.3	20	6	2	30
10^{-9}	0.6	30	7	3	40
10^{-12}	1.0	40	7	4	50
10^{-15}	1.7	50	10	5	60
10^{-18}	2.9	60	12	5	70

The second problem is Ahmed's integral [1]. It is a bit less regular and uses more operators than the previous problem, but the tactic still behaves well enough: adding ten bits of accuracy doubles the computation time.

$$\int_0^1 \frac{\arctan \sqrt{x^2 + 2}}{\sqrt{x^2 + 2}\,(x^2 + 1)}\,dx = \frac{5\pi^2}{96}$$

Error	Time	Accuracy	Degree	Depth	Prec
10^{-3}	0.5	9	5	1	30
10^{-6}	1.2	19	7	3	30
10^{-9}	2.8	29	7	3	40
10^{-12}	5.5	39	10	3	50
10^{-15}	11.2	49	10	4	55

The third problem involves a function that is harder to approximate using RPAs, so the tactic performs more domain splitting, degrading performances.

$$\int_0^\pi \frac{x \sin x}{1 + \cos^2 x}\,dx = \frac{\pi^2}{4}$$

Error	Time	Accuracy	Degree	Depth	Prec
10^{-3}	1.1	11	9	2	30
10^{-6}	2.3	21	6	5	30
10^{-9}	5.0	31	9	5	40
10^{-12}	11.5	41	11	7	50
10^{-15}	27.2	51	11	7	65

The fourth problem is an example from Helfgott[6] in the spirit of [5]. The polynomial part crosses zero, so there is a point where the integrand is not differentiable because of the absolute value. Thus only degenerate Taylor models can be computed around that point. Although the tactic has to perform a lot of domain splitting to isolate that point, it still computes an enclosure of the integral quickly. Note that the approximate value of the integral was computed using the `interval_intro` tactic.

$$\int_0^1 \left|(x^4 + 10x^3 + 19x^2 - 6x - 6)\,\exp x\right|\,dx \simeq 11.14731055005714$$

On this example, quadrature methods have some troubles: `quad` gives only 10 correct digits; `verifyquad` gives a false answer (a tight interval not containing

---

[6] http://mathoverflow.net/questions/123677/rigorous-numerical-integration.

the value of the integral) without warning;[7] quadgk gives only 9 correct digits. VNODE-LP cannot be used because of the absolute value.

Error	Time	Accuracy	Degree	Depth	Prec
10^{-3}	0.7	14	5	8	30
10^{-6}	0.9	24	6	13	40
10^{-9}	1.3	34	8	18	50
10^{-12}	1.9	44	10	22	60
10^{-15}	2.7	54	12	28	70

The last two problems are inherently hard to numerically integrate. The first one is the 12-th coefficient of a Chebyshev expansion. Note that the initial estimation of the integral is completely off, which explains why the relative accuracy has to be set about 30 bits higher than one would expect. As with the previous problem, there are some points where no RPAs can be computed. The approximate value was again computed using the interval_intro tactic.

$$\int_{-1}^{1} \left(2048x^{12} - 6144x^{10} + 6912x^8 - 3584x^6 + 840x^4 - 72x^2 + 1\right)$$

$$\exp\left(-\left(x - \tfrac{3}{4}\right)^2\right) \sqrt{1 - x^2} \, dx \simeq -3.2555895745 \cdot 10^{-6}$$

The quad, quadl, and quadcc procedures give completely off but consistent answers without warning; quadv gives an answer which is off the mark as well, but it gives a warning "maximum iteration count reached"; verifyquad works only for functions that are four times differentiable, hence its failure here; quadgk gives yet another off answer with no warning. Finally, VNODE-LP fails here because of computational errors such as divisions by 0.

Error	Time	Accuracy	Degree	Depth	Prec
10^{-6}	10.7	32	8	17	40
10^{-9}	22.9	42	10	22	50
10^{-12}	48.3	52	13	28	60
10^{-15}	111.8	62	13	35	70

The last problem is an example taken from Tucker's book [14] and originally suggested by Rump in [13, p. 372]. This integral is often incorrectly approximated by computer algebra systems, because of the large number of oscillations (about 950 sign changes) and the large value of the n-th derivatives of the function. While the maximal depth is not too large, the tactic reaches it for numerous subdomains, hence the large computation time.

The quad, quadcc, and quadgk procedures give off values without any warning; quadv gives an off value with a warning; verifyquad takes 1.7 s to give a correct answer; quadl takes 9 s to return a correct answer.

---

[7] The bug lies in an incorrect implementation of Taylor models for absolute value.

$$\int_0^8 \sin(x + \exp x)\, dx \simeq 0.3474$$

Error	Time	Accuracy	Degree	Depth	Prec
10^{-1}	81.0	6	6	12	30
10^{-2}	123.6	9	8	12	30
10^{-3}	183.4	12	10	12	30
10^{-4}	277.6	15	12	12	30

6   Conclusion

We have presented a method for computing and formally verifying numerical enclosures of univariate definite integrals using the Coq proof assistant. It has been integrated into the `interval` tactic. The method just requires that there exist rigorous polynomial expressions of the elementary functions in the integrand, so it is only limited by the underlying library. At the time of writing, the supported functions are $\sqrt{\cdot}$, cos, sin, tan, exp, ln, arctan, and the integer power function. Any new function added to the CoqInterval library would be supported almost immediately by the integration module.

While our adaptive bisection algorithm and our rigorous quadrature based on primitives of polynomial might seem crude, they proved effective in practice: They produce accurate approximations of non-pathological integrals in a few seconds, and thus they are usable in an interactive setting. Moreover, they are able to handle functions with unbounded second derivatives in a rigorous way. Another contribution of this paper is the way we are able to infer that a function is integrable from a successful computation of its integral.

Nested integrals are not supported by our method. The naive enclosure approach could easily be adapted to support them, but performances would be even worse due to the curse of dimensionality. As there exists no general approach for integrating multivariate polynomials,[8] being able to compute rigorous multivariate polynomial approximations would presumably not help.

Improper integrals (infinite bounds) and definite integrals with poles are not supported either. This time, approximation methods are known (including rigorous ones), but we do not even have a good enough formalization of such integrals yet. Once we have it, improper integrals could be supported. Indeed, one would just split the integration domain into a bounded part (solvable using our current approach) and an infinite part on which the integrand is dominated by a function such as $t \mapsto \exp(-Ct)$ at $+\infty$. So the work would be mostly in automating the discovery of the dominating function.

For proper integrals, we could also have tried rigorous quadrature methods such as Newton-Cotes formulas. Indeed, rather than a degree-n polynomial approximation of the integrand, we could have integrated a degree-n polynomial interpolant, which would have given a much tighter enclosure of the integral at a fraction of the cost. The increased accuracy comes from the ability to compute a tight enclosure of the $n+1$-th derivative of the integrand. Unfortunately, we do not have any such tool yet. (CoqInterval only knows how to bound the

---

[8] Any 3-SAT instance can be reduced to approximating the integral of a multivariate polynomial.

first derivative.) Note that a very simplified version of this approach has already been implemented in Coq in the setting of exact real arithmetic by O'Connor and Spitters [12]. Since it does not involve a derivative, it is akin to our naive approach and thus the performances are dreadful.

We could also have tried a much more general method, that is, solving a differential equation built from the integrand, as we did with VNODE-LP. Again, there has been some work done for Coq in the setting of exact real arithmetic [7], but the performances are not good enough in practice. Much closer to actual numerical methods is Immler's work in Isabelle/HOL [6], which uses an arithmetic on affine forms. This approach is akin to computing with degree-1 RPAs.

References

1. Ahmed, Z.: Ahmed's integral: the maiden solution. Math. Spectr. **48**(1), 11–12 (2015)
2. Boldo, S., Lelay, C., Melquiond, G.: Coquelicot: a user-friendly library of real analysis for Coq. Math. Comput. Sci. **9**(1), 41–62 (2015)
3. Eaton, J.W., Bateman, D., Hauberg, S., Wehbring, R., Octave, G.N.U.: version 3.8.1 manual: a high-level interactive language for numerical computations (2014)
4. Hass, J., Schlafly, R.: Double bubbles minimize. Ann. Math. **151**(2), 459–515 (2000). Second Series
5. Helfgott, H.A.: Major arcs for Goldbach's problem (2013). http://arxiv.org/abs/1305.2897
6. Immler, F.: Formally verified computation of enclosures of solutions of ordinary differential equations. In: Badger, J.M., Rozier, K.Y. (eds.) NFM 2014. LNCS, vol. 8430, pp. 113–127. Springer, Heidelberg (2014)
7. Makarov, E., Spitters, B.: The picard algorithm for ordinary differential equations in Coq. In: Blazy, S., Paulin-Mohring, C., Pichardie, D. (eds.) ITP 2013. LNCS, vol. 7998, pp. 463–468. Springer, Heidelberg (2013)
8. Martin-Dorel, É., Melquiond, G.: Proving tight bounds on univariate expressions with elementary functions in Coq. J. Autom. Reason. 1–31 (2015)
9. Mayero, M.: Formalisation et automatisation de preuves en analyses réelle et numérique. Ph.D. thesis, Université Paris VI, December 2001
10. Moore, R.E., Kearfott, R.B., Cloud, M.J.: Introduction to Interval Analysis. SIAM, Philadelphia (2009)
11. Nedialkov, N.S.: Interval tools for ODEs and DAEs. In: Scientific Computing, Computer Arithmetic and Validated Numerics (SCAN) (2006). http://www.cas.mcmaster.ca/~nedialk/vnodelp/
12. O'Connor, R., Spitters, B.: A computer verified, monadic, functional implementation of the integral. Theor. Comput. Sci. **411**(37), 3386–3402 (2010)
13. Rump, S.M.: Verification methods: Rigorous results using floating-point arithmetic. Acta Numerica **19**, 287–449 (2010). http://www.ti3.tu-harburg.de/rump/intlab/
14. Tucker, W.: Validated Numerics: A Short Introduction to Rigorous Computations. Princeton University Press, Princeton (2011)

Certification of Classical Confluence Results for Left-Linear Term Rewrite Systems

Julian Nagele[(✉)] and Aart Middeldorp

Department of Computer Science, University of Innsbruck, Innsbruck, Austria
{julian.nagele,aart.middeldorp}@uibk.ac.at

Abstract. This paper presents the first formalization of three classic confluence criteria for first-order term rewrite systems by Huet and Toyama. We have formalized proofs, showing that (1) linear strongly closed systems, (2) left-linear parallel closed systems, and (3) left-linear almost parallel closed systems are confluent. The third result is extended to commutation. The proofs were carried out in the proof assistant Isabelle/HOL as part of the library IsaFoR and integrated into the certifier CeTA, significantly increasing the number of certifiable proofs produced by automatic confluence tools.

1 Introduction

Confluence of rewrite systems is an important property, which is intimately connected to uniqueness of normal forms, and hence to determinism of programs. In recent years there has been tremendous progress in establishing confluence or non-confluence of TRSs automatically, with a number of tools under active development, like ACP [2], Saigawa [8,11], CoLL [18], and our own tool, CSI [23].

The recent achievements in confluence research have enabled a competition[1] where such automated tools try to establish/refute confluence. As the proofs produced by these tools are often complicated and large, there is interest in checking them with trustable certifiers like CeTA [21]. (CeTA is a certifier for termination, confluence and complexity proofs for TRSs. Other certifiers exist for termination proofs, notably Rainbow [4] and CiME3 [5].) Given a certificate in CPF (certification problem format) [19], CeTA will either answer CERTIFIED or return a detailed error message why the proof was REJECTED. Its correctness is formally proven as part of IsaFoR, the Isabelle Formalization of Rewriting. IsaFoR contains executable "check"-functions for each formalized proof technique together with formal proofs that whenever such a check succeeds, the technique was indeed applied correctly. Isabelle's code-generation facility is used to obtain a trusted Haskell program from these check functions: the certifier CeTA.[2]

In the recent past, several confluence results have been formalized, starting from the fundamental result by Knuth and Bendix [12] that a terminating

---

This work is supported by FWF (Austrian Science Fund) project P27528.

[1] http://coco.nue.riec.tohoku.ac.jp/.

[2] IsaFoR/CeTA and CPF are available at http://cl-informatik.uibk.ac.at/software/ceta/.

© Springer International Publishing Switzerland 2016
J.C. Blanchette and S. Merz (Eds.): ITP 2016, LNCS 9807, pp. 290–306, 2016.
DOI: 10.1007/978-3-319-43144-4_18

rewrite system is confluent if and only if all its critical pairs are joinable. For non-terminating rewrite systems, weak orthogonality as well as sufficient conditions for non-joinability of critical pairs based on unification, discrimination pairs [1], interpretations, and tree automata [6] have been formalized. These results are described in [14]. More recently, redundant rules [13] and rule labeling [15] increased the number of certifiable confluence proofs significantly.

In this paper we report on the formalization of three classical confluence results. Two of these are due to Huet [10] and presented in full detail in the textbook of Baader and Nipkow [3, Lemma 6.3 and Sect. 6.4]. The third result is due to Toyama [22].

The remainder of this paper is organized as follows. After recalling basic notions of term rewriting in the next section, in Sect. 3 we report on the formalization of the result that linear strongly closed rewrite systems are confluent. Linearity is an important limitation, but the result does have its uses [7]. Section 4 is devoted to the formalization of the result of Huet that a left-linear rewrite system is confluent if its critical pairs are parallel closed. In Sect. 5 we consider Toyama's generalization of the previous result. Apart from a weaker joinability requirement on overlays, the result is extended to the commutation of two rewrite systems. Our formalization is an important first step for the certification of confluence proofs produced by CoLL [18], which is based on commutation. In Sect. 6 we explain what is needed for the automatic certification of confluence proofs that employ the formalized techniques and we present experimental results. In the final section we conclude with an outlook on future work, in particular the challenges that need to be overcome when extending the results from parallel closed rewrite systems to development closed higher-order rewrite systems [17]. The main Isabelle theories developed and integrated into IsaFoR are `Strongly_Closed.thy`, for the result on strongly closed rewrite systems, `Parallel_Closed.thy` for results on (almost) parallel closed systems (where we make heavy use of multihole contexts, cf. `Multihole_Context.thy`), and `Critical_Pair_Closure_Impl.thy` for the executable check functions.

2   Preliminaries

We assume familiarity with the basics of rewriting [3, 20]. Knowledge of Isabelle [16] is not essential but experience with an interactive theorem prover might be helpful.

Let $\mathcal{F}$ be a signature and $\mathcal{V}$ a set of variables disjoint from $\mathcal{F}$. By $\mathcal{T}(\mathcal{F}, \mathcal{V})$ we denote the set of terms over $\mathcal{F}$ and $\mathcal{V}$. Positions are strings of positive natural numbers, i.e., elements of $\mathbb{N}_+^*$. We write $q \leqslant p$ if $qq' = p$ for some position q', in which case $p \backslash q$ is defined to be q'. Furthermore $q < p$ if $q \leqslant p$ and $q \neq p$. Finally, positions q and p are parallel, written as $q \parallel p$, if neither $q \leqslant p$ nor $p < q$. Positions are used to address subterm occurrences. The set of positions of a term t is defined as $\mathcal{P}os(t) = \{\epsilon\}$ if t is a variable and as $\mathcal{P}os(t) = \{\epsilon\} \cup \{iq \mid 1 \leqslant i \leqslant n \text{ and } q \in \mathcal{P}os(t_i)\}$ if $t = f(t_1, \ldots, t_n)$. The subterm of t at position $p \in \mathcal{P}os(t)$ is defined as $t|_p = t$ if $p = \epsilon$ and as $t|_p = t_i|_q$ if $p = iq$ and $t = f(t_1, \ldots, t_n)$. We write $s[t]_p$ for the result of replacing the subterm at position

p of s with t. The size of a term t, i.e., the size of $\mathcal{P}os(t)$, is denoted by $|t|$. We write $\mathcal{V}ar(t)$ for the set of variables occurring in the term t. A term t is linear if every variable occurs at most once in it. A substitution is a mapping σ from $\mathcal{V}$ to $\mathcal{T}(\mathcal{F}, \mathcal{V})$ such that its domain $\{x \in \mathcal{V} \mid \sigma(x) \neq x\}$ is finite. We write $t\sigma$ for the result of applying σ to the term t.

Assume a fresh symbol $\square$, called hole. A *multihole* context is a term that may contain an arbitrary number of holes. Filling the holes in a multihole context C with terms $t_1, \ldots, t_n$ is written as $C[t_1, \ldots, t_n]$. (At this point we mention that in the formalization we of course have to make sure that the number of terms n matches the number of holes in C. To ease readability we usually do not make this explicit.) A term with exactly one hole is just called context and we also write $s[]_p$ for the context obtained by replacing position p in s by the hole. If $C[s] = t$ for some context C then s is called a subterm of t and we write $s \trianglelefteq t$. If additionally $C \neq \square$ then s is a proper subterm of t, which is denoted by $s \lhd t$.

A rewrite rule is a pair of terms (ℓ, r), written $\ell \to r$.[3] A rewrite rule $\ell \to r$ is left-linear if ℓ is linear, right-linear if r is linear, and linear if it is both left- and right-linear. A variant of a rewrite rule is obtained by renaming its variables. A term rewrite system (TRS) is set of rewrite rules over a signature. In the sequel, signatures are left implicit. A TRS is (left-)linear if all its rules are (left-)linear. A rewrite relation is a binary relation on terms that is closed under contexts and substitutions. For a TRS $\mathcal{R}$ we define $\to_\mathcal{R}$ (often written as $\to$) to be the smallest rewrite relation that contains $\mathcal{R}$. As usual $\to^=$ and $\to^*$ denote the reflexive, and reflexive and transitive closure of $\to$, respectively.

A relation $\to$ is said to have the diamond property if $\leftarrow \cdot \to \,\subseteq\, \to \cdot \leftarrow$ and is called confluent if its reflexive transitive closure has the diamond property. It is strongly confluent if $\leftarrow \cdot \to \,\subseteq\, \to^= \cdot \,{}^*\!\leftarrow$. The results in Sect. 5 will be proved in the more general setting of commutation. Two relations $\to_1$ and $\to_2$ commute if ${}^*_1\!\leftarrow \cdot \to^*_2 \,\subseteq\, \to^*_2 \cdot \,{}^*_1\!\leftarrow$, they strongly commute if ${}_1\!\leftarrow \cdot \to_2 \,\subseteq\, \to^=_2 \cdot \,{}^*_1\!\leftarrow$. The following lemma captures the well-known connections between the diamond property, (strong) confluence and (strong) commutation.

Lemma 1. *Let* $\to$, $\to_1$, $\to_2$, $\to_{1'}$, *and* $\to_{2'}$ *be binary relations.*

1. *If* $\to$ *has the diamond property then it is confluent.*
2. *If* $\to$ *is strongly confluent then it is confluent.*
3. *If* $\to_1$ *and* $\to_2$ *strongly commute then they commute.*
4. *If* $\to$ *commutes with itself then it is confluent.*
5. *If* $\to_1 \,\subseteq\, \to_{1'} \,\subseteq\, \to^*_1$ *and* $\to_{1'}$ *is confluent then* $\to_1$ *is confluent.*
6. *If* $\to_1 \,\subseteq\, \to_{1'} \,\subseteq\, \to^*_1$ *and* $\to_2 \,\subseteq\, \to_{2'} \,\subseteq\, \to^*_2$ *and* $\to_{1'}$ *and* $\to_{2'}$ *commute then* $\to_1$ *and* $\to_2$ *commute.*

Later, when applying the last two statements, the relations $\to_{1'}$ and $\to_{2'}$ between one and many step rewriting that we will use is parallel rewriting.

---

[3] We do not impose the common *variable conditions*, i.e., the restriction that ℓ is not a variable and all variables in r are contained in ℓ.

Definition 1. *For a TRS $\mathcal{R}$, the* parallel rewrite relation $\twoheadrightarrow_{\mathcal{R}}$ *is defined inductively by*

- *$x \twoheadrightarrow_{\mathcal{R}} x$ if x is a variable,*
- *$\ell\sigma \twoheadrightarrow_{\mathcal{R}} r\sigma$ if $\ell \to r \in \mathcal{R}$, and*
- *$f(s_1, \ldots, s_n) \twoheadrightarrow_{\mathcal{R}} f(t_1, \ldots, t_n)$ if f is a function symbol of arity n and $s_i \twoheadrightarrow_{\mathcal{R}} t_i$ for all $1 \leqslant i \leqslant n$.*

The following properties of parallel rewriting are well-known and follow by straight-forward induction proofs.

Lemma 2. *The following properties of $\twoheadrightarrow$ hold:*

- *$\to_{\mathcal{R}} \subseteq \twoheadrightarrow_{\mathcal{R}} \subseteq \to_{\mathcal{R}}^{*}$,*
- *$s \twoheadrightarrow_{\mathcal{R}} s$ for all terms s,*
- *if $x\sigma \twoheadrightarrow_{\mathcal{R}} x\tau$ for all $x \in \mathcal{V}ar(s)$ then $s\sigma \twoheadrightarrow_{\mathcal{R}} s\tau$.*

The confluence results formalized in this work are based on (left-)linearity and restricted joinability of critical pairs. Critical pairs arise from situations where two redexes overlap with each other. The definition we use here is slightly nonstandard in two regards. First we consider critical pairs for two rewrite systems to use them in a commutation setting later on. Second we do not exclude root overlaps of a rule with (a variant of) itself as is commonly done. This allows us to dispense with the variable condition that all variables in the right-hand side of a rule must also occur on the left. Moreover, if a TRS does satisfy the condition then all extra critical pairs that would normally be excluded are trivial.

A critical overlap $(\ell_1 \to r_1, C, \ell_2 \to r_2)_\mu$ of two TRSs $\mathcal{R}_1$ and $\mathcal{R}_2$ consists of variants $\ell_1 \to r_1$ and $\ell_2 \to r_2$ of rewrite rules in $\mathcal{R}_1$ and $\mathcal{R}_2$ without common variables, a context C, such that $\ell_2 = C[\ell']$ with $\ell' \notin \mathcal{V}$ and a most general unifier μ of ℓ_1 and ℓ'. From a critical overlap $(\ell_1 \to r_1, C, \ell_2 \to r_2)_\mu$ we obtain a critical peak $C\mu[r_1\mu] \,_{\mathcal{R}_1}\!\!\leftarrow C\mu[\ell_1\mu] \to_{\mathcal{R}_2} r_2\mu$ and the corresponding critical pair $C\mu[r_1\mu] \,_{\mathcal{R}_1}\!\!\leftarrow\!\!\bowtie\!\!\to_{\mathcal{R}_2} r_2\mu$. If $C = \Box$, the corresponding critical pair is called an overlay and written as $r_1\mu \,_{\mathcal{R}_1}\!\!\leftarrow\!\!\bowtie\!\!\to_{\mathcal{R}_2} r_2\mu$, otherwise it is called an inner critical pair, and denoted using $_{\mathcal{R}_1}\!\!\leftarrow\!\!\bowtie\!\!\to_{\mathcal{R}_2}$. When considering the critical pairs of a TRS $\mathcal{R}$ with itself we drop the subscripts and write $\leftarrow\!\!\bowtie\!\!\to$ instead of $_{\mathcal{R}}\!\!\leftarrow\!\!\bowtie\!\!\to_{\mathcal{R}}$.

3   Strongly Closed Critical Pairs

The first confluence criterion we consider is due to Huet [10] and based on the observation that in a linear rewrite system it suffices to have strong-confluence like joins for all critical pairs in order to guarantee strong confluence of the rewrite system. A preliminary version of the formalization described in this section was reported in [14].

Definition 2. *A TRS $\mathcal{R}$ is* strongly closed *if every critical pair $s \leftarrow\!\!\bowtie\!\!\to t$ of $\mathcal{R}$ satisfies both $s \to^{=} \cdot \,^{*}\!\!\leftarrow t$ and $s \to^{*} \cdot \,^{=}\!\!\leftarrow t$.*

The following folklore lemma tells us that in a linear term applying a substitution can be done by replacing the one subterm where the variable occurs and applying the remainder of the substitution.

Lemma 3. *Let t be a linear term and let $p \in \mathcal{P}os(t)$ be a position with $t|_p = x$. Then for substitutions σ and τ with $\sigma(y) = \tau(y)$ for all $y \in \mathcal{V}ar(t)$ different from x we have $t\tau = t\sigma[\tau(x)]_p$.*

The proof that linear strongly closed systems are strongly confluent is very similar to the one of the famous critical pair lemma, by analyzing the relative positions of the rewrite steps in a peak. The next lemma, which appears implicitly in Huet's proof of Corollary 1, takes care of the case where one position is above the other.

Lemma 4. *Let $\mathcal{R}$ be a linear, strongly closed TRS and assume $s \to_{\mathcal{R}} t$ with rule $\ell_1 \to r_1$ and substitution σ_1 at position p_1 and let $s \to_{\mathcal{R}} u$ with rule $\ell_2 \to r_2$ and substitution σ_2 at position p_2 with $p_1 \leqslant p_2$. Then there are terms v and w with $t \to_{\mathcal{R}}^* v \stackrel{=}{_{\mathcal{R}}}{\leftarrow} u$ and $t \to_{\mathcal{R}}^{\stackrel{=}{}} w \stackrel{*}{_{\mathcal{R}}}{\leftarrow} u$.*

Proof (Sketch). Since the proof is standard and the formalization closely follows the paper proof, we only sketch the idea and refer to the formalization for full details. We distinguish whether the step from s to u overlaps with the one from s to t or takes place in the substitution. If there is a critical pair, we can close it by the assumption that the system is strongly closed. If the step from s to u happens in the substitution we can join in the required shape due to linearity of $\mathcal{R}$, which avoids duplication of the redex, by using Lemma 3.

Now the main result of this section follows easily.

Corollary 1 (Huet [10]). *If a TRS $\mathcal{R}$ is linear and strongly closed then $\to_{\mathcal{R}}$ is strongly confluent.*

Proof. Assume $s \to_{\mathcal{R}} t$ and $s \to_{\mathcal{R}} u$. Then there are positions $p_1, p_2 \in \mathcal{P}os(s)$, substitutions σ_1, σ_2 and rules $\ell_1 \to r_1, \ell_2 \to r_2$ in $\mathcal{R}$ with $s|_{p_1} = \ell_1\sigma_1$, $s|_{p_2} = \ell_2\sigma_2$ and $t = s[r_1\sigma_1]_{p_1}$, $u = s[r_2\sigma_2]_{p_2}$. We show existence of a term v with $t \to^* v$ and $u \to^= v$ by analyzing the positions p_1 and p_2. If they are parallel then $t \to t[r_2\sigma_2]_{p_2} = u[r_1\sigma_1]_{p_1} \leftarrow u$. If they are not parallel then one is above the other. In both cases we conclude by Lemma 4.

Then by Lemma 1 $\mathcal{R}$ is also confluent.

Example 1. Consider the TRS $\mathcal{R}$ consisting of the two rewrite rules

$$\mathsf{f}(\mathsf{f}(x,y),z) \to \mathsf{f}(x,\mathsf{f}(y,z)) \qquad\qquad \mathsf{f}(x,y) \to \mathsf{f}(y,x)$$

There are four non-trivial critical pairs

$$\mathsf{f}(\mathsf{f}(x,\mathsf{f}(y,z)),v) \leftarrow\!\bowtie\!\to \mathsf{f}(\mathsf{f}(x,y),\mathsf{f}(z,v)) \qquad \mathsf{f}(x,\mathsf{f}(y,z)) \leftarrow\!\bowtie\!\to \mathsf{f}(z,\mathsf{f}(x,y))$$
$$\mathsf{f}(z,\mathsf{f}(x,y)) \leftarrow\!\bowtie\!\to \mathsf{f}(x,\mathsf{f}(y,z)) \qquad \mathsf{f}(\mathsf{f}(y,x),z) \leftarrow\!\bowtie\!\to \mathsf{f}(x,\mathsf{f}(y,z))$$

Since $\mathcal{R}$ is linear and all critical pairs are strongly closed, $\mathcal{R}$ is confluent.

The next example shows how to apply the criterion to a TRS that does not fulfill the variable conditions.

Example 2. Consider the linear TRS $\mathcal{R}$ consisting of the following three rules:

$$\mathsf{a} \rightarrow \mathsf{f}(x) \qquad\qquad \mathsf{f}(x) \rightarrow \mathsf{b} \qquad\qquad x \rightarrow \mathsf{f}(\mathsf{g}(x))$$

There are five critical pairs modulo symmetry:

$$\mathsf{f}(y) \leftarrow\!\!\times\!\!\rightarrow \mathsf{f}(x) \qquad\quad \mathsf{f}(\mathsf{g}(\mathsf{a})) \leftarrow\!\!\times\!\!\rightarrow \mathsf{f}(x) \qquad\quad \mathsf{b} \leftarrow\!\!\times\!\!\rightarrow \mathsf{b}$$
$$\mathsf{f}(\mathsf{g}(\mathsf{f}(x))) \leftarrow\!\!\times\!\!\rightarrow \mathsf{b} \qquad\quad \mathsf{f}(\mathsf{g}(x)) \leftarrow\!\!\times\!\!\rightarrow \mathsf{f}(\mathsf{g}(x))$$

Using the second rule it is easy to see that all of them are strongly closed. Hence $\mathcal{R}$ is confluent.

The next example shows that, if the variable condition is not satisfied, critical pairs that arise from overlapping a rule with itself at the root are essential.

Example 3. Consider the linear rewrite system $\mathcal{R}$ consisting of the rule $\mathsf{a} \rightarrow y$. Because of the peak $x \leftarrow \mathsf{a} \rightarrow y$, $\mathcal{R}$ is not confluent and indeed $x \leftarrow\!\!\times\!\!\rightarrow y$ is a non-joinable critical pair according to our definition.

In the next section we consider a criterion that drops the condition on $\mathcal{R}$ to be right-linear.

4   Parallel Closed Critical Pairs

The criterion from the previous section requires the TRS to be linear and while left-linearity is a common restriction, right-linearity is a rather unnatural one. Thus we turn our attention to criteria for left-linear systems that change the restriction on the joinability of critical pairs. The crucial observation is that in a non-right-linear system executing the upper step in variable overlap can duplicate the redex below. Thus to join such a situation multiple steps might be necessary, all of which take place at parallel positions. Consequently we consider parallel rewriting. The following definition describes the new joinability condition.

Definition 3. *A TRS $\mathcal{R}$ is* parallel closed *if every critical pair $s \leftarrow\!\!\times\!\!\rightarrow t$ of $\mathcal{R}$ satisfies $s \twoheadrightarrow_{\mathcal{R}} t$.*

Together with left-linearity this guarantees the diamond property of the parallel rewrite relation.

Theorem 1 (Huet [10]). *If a TRS $\mathcal{R}$ is left-linear and parallel closed then $\twoheadrightarrow_{\mathcal{R}}$ has the diamond property.*

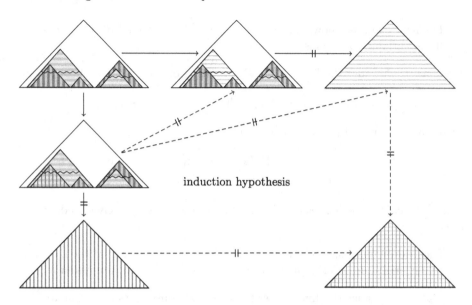

Fig. 1. Overview of the proof of Theorem 1.

The proof of this theorem is much more involved than the one for strongly closed systems. The first observation is that we will now have to consider a peak of parallel steps, in order to show the diamond property of ↠. In case the two parallel steps are orthogonal to each other, they simply commute by the well-known Parallel Moves Lemma. However, if they do interfere the assumption of the theorem only allows us to close a single critical pair to reduce the amount of interference. Thus we will have to use some form of induction on how much the patterns of the two parallel steps overlap. Figure 1 shows the setting for the overlapping case. The horizontal parallel step, described by the horizontally striped redexes, and the vertical step, described by the vertically striped redexes, overlap. Hence there is a critical pair, say the one obtained from overlapping the leftmost vertical redex with the leftmost horizontal redex. Then, by assumption there is a closing parallel step, which, since it takes place inside the critical pair, can be combined with the remaining horizontal redexes to obtain a new peak with less overlap, which can be closed by the induction hypothesis. When making this formal we identified two crucial choices. First the representation of the parallel rewrite relation and second the way to measure the amount of overlap between two parallel steps with the same source. Huet in his original proof heavily uses positions. That is, a parallel step is defined as multiple single steps that happen at parallel positions and for measuring overlap he takes the sum of the sizes of the subterms that are affected by both steps. More precisely, writing ↠$_P$ for a parallel step that takes place at positions in a set P, for a peak $t \;_{P_1}\!\!\twoheadleftarrow s \twoheadrightarrow_{P_2} u$ he uses

$$\sum_{q \in Q} |s|_q|$$

where $Q = \{p_1 \in P_1 \mid p_2 \leqslant p_1 \text{ for some } p_2 \in P_2\} \cup \{p_2 \in P_2 \mid p_1 \leqslant p_2 \text{ for some } p_1 \in P_1\}$. This formulation is also adopted in the text book by Baader and Nip-kow [3]. Consequently, when starting the present formalization, we also adopted this definition. However, the book keeping required by working with sets of positions as well as formally reasoning about this measure in Isabelle became so convoluted that it very much obscured the ingenuity and elegance of Huet's original idea while at the same time defeating our formalization efforts. Hence in the end we had to adopt a different approach.

Toyama [22], in the proof of his extension of Huet's result, does not use positions at all and instead relies on (multihole) contexts, which means a parallel step is then described by a context and a list of root steps that happen in the holes. To measure overlap he collects those redexes that are subterms of some redex in the other step, i.e., decorating the parallel rewrite relation with the redexes contracted in the step, for a peak $t \;{}_{t_1,\ldots,t_n}\!\!\leftsquigarrow\; s \;\rightsquigarrow_{u_1,\ldots,u_m}\; u$ Toyama's measure is

$$\sum_{s \in S} |s|$$

where $S = \{u_i \mid u_i \trianglelefteq t_j \text{ for some } t_j\} \cup \{t_j \mid t_j \trianglelefteq u_i \text{ for some } u_i\}$. However, this measure turns out to be problematic as shown in the following example.

Example 4. Consider the TRS consisting of the following five rewrite rules:

$$f(a, a, b, b) \to f(c, c, c, c) \qquad a \to b \qquad a \to c \qquad b \to a \qquad b \to c$$

Then we have the peak $f(b, b, a, a) \;\overset{a,a,b,b}{\leftsquigarrow}\; f(a, a, b, b) \;\xrightarrow{f(a,a,b,b)}\; f(c, c, c, c)$. The measure of this peak according to the definition above is 2, since $S = \{a, b\} \cup \varnothing$. Now after splitting of one of the four critical steps—it does not matter which one—and closing the corresponding critical pair, we arrive at

$$
\begin{array}{ccc}
f(a, a, b, b) & \longrightarrow & f(c, c, c, c) \\
\downarrow & \nearrow & \\
f(b, a, b, b) & & \\
\text{\Large $\ddagger$} & & \\
f(b, b, a, a) & &
\end{array}
$$

The measure of the new peak $f(b, b, a, a) \;\overset{a,b,b}{\leftsquigarrow}\; f(b, a, b, b) \;\xrightarrow{b,a,b,b}\; f(c, c, c, c)$ is still 2 since $S = \{a, b\} \cup \{a, b\}$.

Note that using multisets instead of sets does not help, since then the measure of the initial peak is 4 ($S = \{a, a, b, b\}$) and of the new peak, after closing the critical pair, it is 7 since $S = \{a, b, b\} \uplus \{b, a, b, b\}$ (and even if we take into account that three of the redexes are counted twice we still get 4). The problem is that in the new peak the redex at position 1 of the closing step is counted again, because b is a subterm of one the redexes of the other step. Hence it is crucial to only count redexes at overlapping positions.

To remedy this situation we will collect all *overlapping* redexes of a peak in a multiset. These multisets will then be compared by $\rhd_{\mathsf{mul}}$, the multiset extension of the proper superterm relation. We start by characterizing parallel rewrite steps using multihole contexts.

Definition 4. *We write* $s \xrightarrow{\;C,a_1,\ldots,a_n\;}_{\mathcal{R}} t$ *if* $s = C[a_1,\ldots,a_n]$ *and* $t = C[b_1,\ldots,b_n]$ *for some* $b_1,\ldots,b_n$ *with* $a_i \to^{\epsilon}_{\mathcal{R}} b_i$ *for all* $1 \leqslant i \leqslant n$.

To save space we sometimes abbreviate a list of terms $a_1,\ldots,a_n$ by $\bar{a}$ and write $s \xrightarrow{\;C,\bar{a}\;}_{\mathcal{R}} t$ leaving length implicit. The following expected correspondence is easily shown by induction.

Lemma 5. *We have* $s \mathbin{\mapstochar\rightarrow}_{\mathcal{R}} t$ *if and only if* $s \xrightarrow{\;C,\bar{s}\;}_{\mathcal{R}} t$ *for some* C *and* $\bar{s}$.

Now we can formally measure the overlap between two parallel rewrite steps by collecting those redexes that are below some redex in the other step.

Definition 5. *The overlap between two co-initial parallel rewrite steps is defined by the following equations*

$$\blacktriangle \left(\xleftarrow{\;\Box,a\;} s \xrightarrow{\;\Box,b\;} \right) = \{s\}$$

$$\blacktriangle \left(\xleftarrow{\;C,a_1,\ldots,a_c\;} s \xrightarrow{\;\Box,b\;} \right) = \{a_1,\ldots,a_c\}$$

$$\blacktriangle \left(\xleftarrow{\;\Box,a\;} s \xrightarrow{\;D,b_1,\ldots,b_d\;} \right) = \{b_1,\ldots,b_d\}$$

$$\blacktriangle \left(\xleftarrow{\;f(C_1,\ldots,C_n),\bar{a}\;} f(s_1,\ldots,s_n) \xrightarrow{\;f(D_1,\ldots,D_n),\bar{b}\;} \right) = \bigcup_{i=1}^{n} \blacktriangle \left(\xleftarrow{\;C_i,\bar{a}_i\;} s_i \xrightarrow{\;D_i,\bar{b}_i\;} \right)$$

where $\bar{a}_1,\ldots,\bar{a}_n = \bar{a}$ *and* $\bar{b}_1,\ldots,\bar{b}_n = \bar{b}$ *are partitions of* $\bar{a}$ *and* $\bar{b}$ *such that the length of* $\bar{a}_i$ *and* $\bar{b}_i$ *matches the number of holes in* C_i *and* D_i, *for all* $1 \leqslant i \leqslant n$.

Example 5. Applying this definition for the two peaks from Example 4 yields

$$\blacktriangle \left(\xleftarrow{\;f(\Box,\Box,\Box,\Box),a,a,b,b\;} f(a,a,b,b) \xrightarrow{\;\Box,f(a,a,b,b)\;} \right) = \{a,a,b,b\}$$

$$\blacktriangle \left(\xleftarrow{\;f(b,\Box,\Box,\Box),a,b,b\;} f(b,a,b,b) \xrightarrow{\;f(\Box,\Box,\Box,\Box),b,a,b,b\;} \right) = \{a,b,b\}$$

and $\{a,a,b,b\} \rhd_{\mathsf{mul}} \{a,b,b\}$ as desired.

Note that our definition of $\blacktriangle$ is in fact an over-approximation of the actual overlap between the steps. That is because we do not split redexes into the left-hand side of the applied rule and a substitution but take the redex as a whole. The following example illustrates the effect.

Example 6. Consider the rewrite system consisting of the two rules

$$\mathsf{f}(x) \to x \qquad\qquad \mathsf{a} \to \mathsf{b}$$

and the peak $\mathsf{a} \leftarrow \mathsf{f}(\mathsf{a}) \to \mathsf{f}(\mathsf{b})$. We have

$$\blacktriangle\left(\xleftarrow{\square,\mathsf{f}(\mathsf{a})}\!\!\!\Vert\!\!\!- \ \mathsf{f}(\mathsf{a}) \ \xrightarrow{\mathsf{f}(\square),\mathsf{a}}\!\!\!\Vert\!\!\!\to \right) = \{\mathsf{a}\}$$

although the two steps do not overlap—the step to the right takes place completely in the substitution of the one to the left (in fact the rewrite system in question is orthogonal).

However, since we are dealing with parallel rewriting, no problems arise from this over-approximation. This changes when extending the results to development steps, see Sect. 7 for further discussion.

The following properties of $\blacktriangle$ turned out to be crucial in our proof of Theorem 1.

Lemma 6. *For a peak* $\xleftarrow{C,\bar{a}}\!\!\!\Vert\!\!-\ s\ \xrightarrow{D,\bar{b}}\!\!\!\Vert\!\!\to$ *the following properties of* $\blacktriangle$ *hold.*

- *If* $s = f(s_1, \ldots, s_n)$ *with* $C = f(C_1, \ldots, C_n)$ *and* $D = f(D_1, \ldots, D_n)$ *then*

$$\blacktriangle\left(\xleftarrow{C_i,\bar{a}_i}\!\!\!\Vert\!\!-\ s_i\ \xrightarrow{D_i,\bar{b}_i}\!\!\!\Vert\!\!\to \right) \subseteq \blacktriangle\left(\xleftarrow{C,\bar{a}}\!\!\!\Vert\!\!-\ s\ \xrightarrow{D,\bar{b}}\!\!\!\Vert\!\!\to \right)$$

  for all $1 \leqslant i \leqslant n$.
- *The overlap is bounded by* $\bar{a}$, *i.e.,* $\{a_1, \ldots, a_c\} \trianglerighteq^=_{\mathsf{mul}} \blacktriangle\left(\xleftarrow{C,\bar{a}}\!\!\!\Vert\!\!-\ s\ \xrightarrow{D,\bar{b}}\!\!\!\Vert\!\!\to \right)$.
- *The overlap is symmetric, i.e.,* $\blacktriangle\left(\xleftarrow{C,\bar{a}}\!\!\!\Vert\!\!-\ s\ \xrightarrow{D,\bar{b}}\!\!\!\Vert\!\!\to \right) = \blacktriangle\left(\xleftarrow{D,\bar{b}}\!\!\!\Vert\!\!-\ s\ \xrightarrow{C,\bar{a}}\!\!\!\Vert\!\!\to \right)$.

There is one more high-level difference between the formalization and the paper proof. In the original proof one needs to combine the closing step for the critical pair with the remainder of the original step in order to obtain a new peak, to which the induction hypothesis can then be applied. This reasoning can be avoided, by using an additional induction on the source of the peak. Then the case where neither of the two parallel steps is a root step (and thus a single step) can be discharged by the induction hypothesis of that induction.

The following technical lemma tells us that a parallel rewrite step starting from $s\sigma$ is either inside s, i.e., we can split off a critical pair, or we can do the step completely inside σ.

Lemma 7. *Let* s *be a linear term. If* $s\sigma \xrightarrow{C,s_1,\ldots,s_n}\!\!\!\Vert\!\!\to_{\mathcal{R}} t$ *then either* $t = s\tau$ *for some substitution* τ *such that* $x\sigma \twoheadrightarrow x\tau$ *for all* $x \in \mathcal{V}ar(s)$ *or there exist a context* D, *a non-variable term* s', *a rule* $\ell \to r \in \mathcal{R}$, *a substitution* τ, *and a multihole context* C' *such that* $s = D[s']$, $s'\sigma = \ell\tau$, $D\sigma[r\tau] = C'[s_1, \ldots, s_{i-1}, s_{i+1}, \ldots, s_n]$ *and* $t = C'[t_1, \ldots, t_{i-1}, t_{i+1}, \ldots, t_n]$ *for some* $1 \leqslant i \leqslant n$.

We are now ready to prove the main result of this section. To ease presentation, the following proof does use the condition that the left-hand sides of rewrite rules are not variables. By employing additional technical case analyses this restriction can be easily dropped. We refer to the formalization for details.

Proof (of Theorem 1). Assume $t \overset{C,\bar{a}}{\longleftrightarrow\mkern-4mu\shortmid} s \overset{D,\bar{b}}{\shortmid\mkern-4mu\longleftrightarrow} u$. We show $t \twoheadrightarrow v \twoheadleftarrow u$ for some term v by well-founded induction on the overlap between the two parallel steps using the order $\rhd_{\mathsf{mul}}$ and continue by induction on s with respect to $\rhd$. If $s = x$ for some variable x then $t = u = x$. So let $s = f(s_1, \ldots, s_n)$. We distinguish four cases.

1. If $C = f(C_1, \ldots, C_n)$ and $D = f(D_1, \ldots, D_n)$ then $t = f(t_1, \ldots, t_n)$ and $u = f(u_1, \ldots, u_n)$ and we obtain partitions $\bar{a}_1, \ldots, \bar{a}_n = \bar{a}$ and $\bar{b}_1, \ldots, \bar{b}_n = \bar{b}$ of $\bar{a}$ and $\bar{b}$ with $t_i \overset{C_i, \bar{a}_i}{\longleftrightarrow\mkern-4mu\shortmid} s_i \overset{D_i, \bar{b}_i}{\shortmid\mkern-4mu\longleftrightarrow} u_i$ for all $1 \leqslant i \leqslant n$. Then, since we have

$$\blacktriangle \left(\overset{C_i, \bar{a}_i}{\longleftrightarrow\mkern-4mu\shortmid} s_i \overset{D_i, \bar{b}_i}{\shortmid\mkern-4mu\longleftrightarrow} \right) \subseteq \blacktriangle \left(\overset{C, \bar{a}}{\longleftrightarrow\mkern-4mu\shortmid} s \overset{D, \bar{b}}{\shortmid\mkern-4mu\longleftrightarrow} \right)$$

by Lemma 6 and thus also

$$\blacktriangle \left(\overset{C, \bar{a}}{\longleftrightarrow\mkern-4mu\shortmid} s \overset{D, \bar{b}}{\shortmid\mkern-4mu\longleftrightarrow} \right) \rhd^=_{\mathsf{mul}} \blacktriangle \left(\overset{C_i, \bar{a}_i}{\longleftrightarrow\mkern-4mu\shortmid} s_i \overset{D_i, \bar{b}_i}{\shortmid\mkern-4mu\longleftrightarrow} \right)$$

   we can apply the inner induction hypothesis and obtain terms v_i with $t_i \twoheadrightarrow v_i \twoheadleftarrow u_i$ for all $1 \leqslant i \leqslant n$ and thus we have $t \twoheadrightarrow f(v_1, \ldots, v_n) \twoheadleftarrow u$.
2. If $C = D = \square$ then both steps are root steps and thus single rewrite steps and we can write $t = r_1\sigma_1 \overset{\epsilon}{\leftarrow} \ell_1\sigma_1 = s = \ell_2\sigma_2 \overset{\epsilon}{\rightarrow} r_2\sigma_2 = u$. Hence, since $\ell_1\sigma_1 = \ell_2\sigma_2$, there is a critical pair $r'_1\mu \leftarrowtail\mkern-4mu\rightarrowtail r'_2\mu$ for variable disjoint variants $\ell'_1 \rightarrow r'_1, \ell'_2 \rightarrow r'_2$ of $\ell_1 \rightarrow r_1, \ell_2 \rightarrow r_2$ with μ a most general unifier of ℓ'_1 and ℓ'_2. Then by assumption $r'_1\mu \twoheadrightarrow r'_2\mu$ and by closure under substitution also $t = r_1\sigma_1 \twoheadrightarrow r_2\sigma_2 = u$.
3. If $C = f(C_1, \ldots, C_n)$ and $D = \square$ then the step to the right is a single root step and we write $t = f(t_1, \ldots, t_n) \overset{C, \bar{a}}{\longleftrightarrow\mkern-4mu\shortmid} s = \ell\sigma \overset{\epsilon}{\rightarrow} r\sigma = u$. Since ℓ is linear by assumption, we can apply Lemma 7 and either obtain τ with $t = \ell\tau$ and $x\sigma \twoheadrightarrow x\tau$ for all $x \in \mathcal{V}\mathsf{ar}(\ell)$ or a critical pair.
   - In the first case define

$$\delta(x) = \begin{cases} \tau(x) & \text{if } x \in \mathcal{V}\mathsf{ar}(\ell) \\ \sigma(x) & \text{otherwise} \end{cases}$$

   We have $t = \ell\tau = \ell\delta$ by definition of δ and hence $t \twoheadrightarrow r\delta$ by a single root step. Moreover we have $u = r\sigma \twoheadrightarrow r\delta$ since $x\sigma \twoheadrightarrow x\delta$ for all variables $x \in \mathcal{V}\mathsf{ar}(r)$. This holds because either $x \in \mathcal{V}\mathsf{ar}(\ell)$ and then $x\sigma \twoheadrightarrow x\tau = x\delta$ or $x \notin \mathcal{V}\mathsf{ar}(\ell)$ and then $x\sigma = x\delta$.
   - In the second case Lemma 7 yields a context E, a non-variable term ℓ'', a rule $\ell' \rightarrow r' \in \mathcal{R}$, a substitution τ, and a multihole context C' such

that $\ell = E[\ell'']$, $\ell''\sigma = \ell'\tau$, $E\sigma[r'\tau] = C'[a_1,\ldots,a_{i-1},a_{i+1},\ldots,a_c]$ and $t = C'[a_1',\ldots,a_{i-1}',a_{i+1}',\ldots,a_c']$ for some $1 \leqslant i \leqslant c$. Since $\ell''\sigma = \ell'\tau$ there is a critical pair $E\mu[r'\mu] \leftarrow\!\!\times\!\!\rightarrow r\mu$ and by assumption $E\mu[r'\mu] \twoheadrightarrow r\mu$ and thus also $E\sigma[r'\tau] \twoheadrightarrow r\sigma$. That is, we obtain a new peak

$$t \xleftarrow{C',\bar{a}'}_{\!+\!\!+} E\sigma[r'\tau] \twoheadrightarrow r\sigma$$

with $\bar{a}' = a_1,\ldots,a_{i-1},a_{i+1},\ldots,a_c$. Since

$$\blacktriangle\left(\xleftarrow{C,\bar{a}}_{\!+\!\!+} s \xrightarrow{\square,\ell\sigma}_{+\!\!+\!\!\!}\right) = \{a_1,\ldots,a_c\} \rhd_{\mathsf{mul}} \{a_1,\ldots,a_{i-1},a_{i+1},\ldots,a_c\}$$

$$\rhd^=_{\mathsf{mul}} \blacktriangle\left(\xleftarrow{C',\bar{a}'}_{\!+\!\!+} E\sigma[r'\tau] \twoheadrightarrow\right)$$

by Lemma 6, we can apply the induction hypothesis and obtain v with $t \twoheadrightarrow v \twoheadleftarrow r\sigma = u$.

4. The final case, $D = f(D_1,\ldots,D_n)$ and $C = \square$, is completely symmetric.

Finally, by Lemmas 1 and 2 we obtain confluence of $\to_{\mathcal{R}}$.

Example 7. Consider the TRS $\mathcal{R}$ consisting of the following three rewrite rules:

$$x + y \to y + x \quad (x+y)*z \to (x*z)+(y*z) \quad (y+x)*z \to (x*z)+(y*z)$$

Since the four critical pairs of $\mathcal{R}$

$$(y+x)*z \leftarrow\!\!\times\!\!\rightarrow (x*z)+(y*z) \quad (y*z)+(x*z) \leftarrow\!\!\times\!\!\rightarrow (x*z)+(y*z)$$
$$(x+y)*z \leftarrow\!\!\times\!\!\rightarrow (x*z)+(y*z) \quad (x*z)+(y*z) \leftarrow\!\!\times\!\!\rightarrow (y*z)+(x*z)$$

are parallel closed, $\mathcal{R}$ is confluent.

5   Almost Parallel Closed Critical Pairs and Commutation

In this section we consider two extensions to Huet's result due to Toyama [22]. The first one allows us to weaken the joining condition for some critical pairs.

When carefully examining the proof of Theorem 1 one realizes that in the case where both steps of the peak are single root steps, i.e., the case where $C = D = \square$, the induction hypothesis does not need to be applied, since closing the critical pair immediately closes the whole peak. This suggests that the joining condition can be weakened for overlays. A first idea could be to take $\leftarrow\!\!\times\!\!\rightarrow \subseteq \twoheadrightarrow \cdot \twoheadleftarrow$ since then we would still have the diamond property in the overlay case. However Toyama realized that one can do even better by weakening the diamond property to strong confluence. The following definition captures the new conditions.

Definition 6. *A TRS $\mathcal{R}$ is* almost parallel closed *if $s \twoheadrightarrow \cdot \xleftarrow{*} t$ for all overlays $s \leftarrow\!\!\times\!\!\rightarrow t$ and $s \twoheadrightarrow t$ for all inner critical pairs $s \leftarrow\!\!\times\!\!\rightarrow t$.*

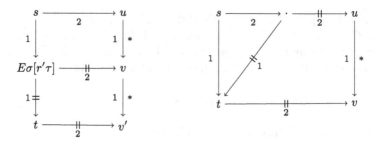

Fig. 2. Asymmetry in the proof of Theorem 2.

Using exactly the same proof structure as before we could now prove strong confluence of $\twoheadrightarrow$ for left-linear almost parallel closed systems. However, considering Toyama's second extension of Theorem 1, we will prove the theorem in the more general setting of commutation.

Theorem 2 (Toyama [22]). *Let $\mathcal{R}_1$ and $\mathcal{R}_2$ be left-linear TRSs. If $s \twoheadrightarrow_2 \cdot {}^*_1\!\leftarrow t$ for all critical pairs $s\ {}_1\!\leftarrow\!\Join\!\rightarrow_2 t$ and additionally $s \twoheadrightarrow_1 t$ for all inner critical pairs $s\ {}_2\!\leftarrow\!\Join\!\rightarrow_1 t$ then $\twoheadrightarrow_1$ and $\twoheadrightarrow_2$ strongly commute.*

Proof (Adaptations). We only highlight the differences to the proof of Theorem 1 and refer to the formalization for the full proof details. Assume

$$t\ {}_1\!\overset{C,\bar{a}}{\longleftarrow\!\shortmid\!\shortmid}\ s\ \overset{D,\bar{b}}{\shortmid\!\shortmid\!\longrightarrow}_2\ u$$

We show $t \twoheadrightarrow_2 v\ {}^*_1\!\leftarrow u$ for some term v. We apply the same inductions and case analyses as before. The cases $C = f(C_1, \ldots, C_n)$, $D = f(D_1, \ldots, D_n)$ and $C = D = \square$ require no noteworthy adaptation. The main difference is that now the cases $D = f(D_1, \ldots, D_n)$, $C = \square$ and $C = f(C_1, \ldots, C_n)$, $D = \square$ become asymmetric for the critical pair case—the corresponding diagrams are shown in Fig. 2.

First, suppose $C = f(C_1, \ldots, C_n)$ and $D = \square$, write $t = f(t_1, \ldots, t_n)\ {}_1\!\overset{C,\bar{a}}{\longleftarrow\!\shortmid\!\shortmid}$ $s = \ell\sigma \overset{\epsilon}{\rightarrow}_2 r\sigma = u$, and assume there is a critical pair according to Lemma 7. That is, we obtain $E\mu[r'\mu]\ {}_1\!\leftarrow\!\Join\!\rightarrow_2 r\mu$ with $E\sigma[r'\tau] \twoheadrightarrow_1 t$ and by assumption we obtain a v such that $E\sigma[r'\tau] \twoheadrightarrow_2 v\ {}^*_1\!\leftarrow r\sigma$. Then using the same reasoning as before, for the new peak

$$t\ {}_1\!\overset{C',\bar{a}'}{\longleftarrow\!\shortmid\!\shortmid}\ E\sigma[r'\tau] \twoheadrightarrow_2 v$$

we have

$$\blacktriangle\left(\overset{C,\bar{a}}{\longleftarrow\!\shortmid\!\shortmid}\ s\ \overset{\square,\ell\sigma}{\shortmid\!\shortmid\!\longrightarrow}\right) \rhd_{\mathsf{mul}} \blacktriangle\left(\overset{C',\bar{a}'}{\longleftarrow\!\shortmid\!\shortmid}\ E\sigma[r'\tau] \twoheadrightarrow\right)$$

and can apply the induction hypothesis to obtain a v' with $t \twoheadrightarrow_2 v'\ {}^*_1\!\leftarrow v$, which combined with $u = r\sigma \rightarrow^*_1 v$ concludes this case.

In the second case, i.e., when $D = f(D_1, \ldots, D_n)$ and $C = \square$, observe that the critical pair we obtain is an inner critical pair between $\mathcal{R}_2$ and $\mathcal{R}_1$, since

$D \neq \square$. Thus, after applying the assumption for critical pairs ${}_2\!\leftarrow\!\bowtie\!\rightarrow\!{}_1$, the proof is the same as for Theorem 1.

Instantiating $\mathcal{R}_1$ and $\mathcal{R}_2$ with the same TRS $\mathcal{R}$ yields the corresponding result for confluence.

Corollary 2 (Toyama [22]). *If the TRS $\mathcal{R}$ is left-linear and almost parallel closed then $\twoheadrightarrow_{\mathcal{R}}$ is strongly confluent.*

Proof. Immediate from the definition of almost parallel closed, Theorem 2 and the fact that $s \leftarrow\!\bowtie\!\rightarrow t$ if and only if $s \leftarrow\!\bowtie\!\rightarrow t$ or $s \leftarrow\!\bowtie\!\rightarrow t$.

Example 8. Recall the rewrite system from Example 4. One easily verifies that all its critical pairs are almost parallel closed, and hence it is confluent.

6   Certification and Experiments

To facilitate checking of confluence proofs generated by automatic tools based on Corollarys 1 and 2 we extended the CPF to represent such proofs. Since in order to check that a given TRS is strongly or almost parallel closed, CeTA has to compute all critical pairs anyway, in the certificate we just require the claim that the system is strongly or almost parallel closed, together with a bound on the length of the rewrite sequences to join the critical pairs.[4] Certificates for commutation are not yet supported, since currently no tool produces them, and CPF does not contain a specification for commutation proofs.

For experiments we considered all 277 TRSs in the Cops[5] database and used the confluence tool CSI to obtain certificates in CPF for confluence proofs. All generated certificates have been certified by CeTA. Table 1 shows the results of running CSI with different strategies. The first and second column show the results of applying just Corollarys 1 and 2 respectively, the third column is the combination. In the fourth column we show the result when additionally adding and removing redundant rewrite rules [13], which yields a considerable boost in power. The idea of that technique is to add and remove rules that can be simulated by other rules, which consequently does not change confluence of the system, but often makes other criteria, like the ones we consider here, applicable. Column "full" shows the results for the full certified strategy of CSI, which additionally includes Knuth and Bendix' criterion, weak orthogonality (which is subsumed by Corollary 2, however) and the rule labeling heuristic [15] as well as several criteria for non-confluence. The last column shows the difference to last year's version of CSI's certified strategy, which already included Corollary 1, but not Corollary 2. In addition to the new certifiable proofs, several existing proofs of CSI 2015 could be simplified and no longer require complicated reasoning via decreasing diagrams.

---

[4] This bound is necessary to ensure termination of the certifier.
[5] http://cops.uibk.ac.at.

Table 1. Experimental results.

	SC	PC	SC+PC	SC+PC+RR	full	2015
yes	38	21	41	92	110	104
no	0	0	0	0	48	48
maybe	239	256	236	185	119	125

7 Conclusion

In this paper we presented the first formalization of three classical criteria for confluence and commutation of (left-)linear rewrite systems. Building on top of IsaFoR—which provided invaluable support on the one hand, e.g. by its theories on critical pairs and multihole contexts, and on the other hand, as expected, was also extended with new basic facts about lists, multisets, multihole contexts etc.—we formalized proofs that linear strongly closed systems, and left-linear (almost) parallel closed systems are confluent (commute). The major difference to the paper proof is our definition of the overlap between two parallel steps that are represented via multihole contexts.

Concerning future work, another important extension of the results of Huet and Toyama due to van Oostrom [17] is using multisteps (also called development steps) ⊸ which allow nested non-overlapping redexes. This extension not only strengthens Huet's criterion in the first-order world but also makes it applicable to higher-order rewriting, where using parallel steps fails due to β-reduction.

However, although the paper proofs superficially look very similar, and do employ similar ideas, obtaining a formalized proof will require serious effort. In fact neither our representation of (parallel) rewrite steps, nor our definition of ▲, nor the idea of using an induction on the source of the peak to avoid reasoning about combining steps, carry over. To make the concepts that are hard to formalize in a proof assistant, e.g. measuring the amount of overlap between two multisteps or the descendants of a multistep, Hirokawa and Middeldorp [9] suggested to use proof terms to obtain a rigorous proof (and at the same time extended the result to commutation). This is a step forward but more is needed to obtain a formalized proof, also for the extension to higher-order systems. In particular, we anticipate the extensive use of sets of positions (in [9]) to be problematic without alternative notions. We plan to employ residual theory [20, Sect. 8.7] and to develop a notion of overlap for multisteps similar to Definition 5 to close the gap.

Acknowledgments. We thank Nao Hirokawa for suggesting Lemma 6 and Bertram Felgenhauer and Christian Sternagel for insightful discussion.

References

1. Aoto, T.: Disproving confluence of term rewriting systems by interpretation and ordering. In: Fontaine, P., Ringeissen, C., Schmidt, R.A. (eds.) FroCoS 2013. LNCS, vol. 8152, pp. 311–326. Springer, Heidelberg (2013). doi:10.1007/978-3-642-40885-4_22

2. Aoto, T., Yoshida, J., Toyama, Y.: Proving confluence of term rewriting systems automatically. In: Treinen, R. (ed.) RTA 2009. LNCS, vol. 5595, pp. 93–102. Springer, Heidelberg (2009). doi:10.1007/978-3-642-02348-4_7

3. Baader, F., Nipkow, T.: Term Rewriting and All That. Cambridge University Press, Cambridge (1998)

4. Blanqui, F., Koprowski, A.: CoLoR, a Coq library on well-founded rewrite relations and its application to the automated verification of termination certificates. Math. Struct. Comput. Sci. **21**(4), 827–859 (2011). doi:10.1017/S0960129511000120

5. Contejean, E., Courtieu, P., Forest, J., Pons, O., Urbain, X.: Automated certified proofs with CiME3. In: Schmidt-Schauß, M. (ed.) RTA 2011. LIPIcs, vol. 10. pp. 21–30. Schloss Dagstuhl–Leibniz-Zentrum für Informatik (2011). doi:10.4230/LIPIcs.RTA.2011.21

6. Felgenhauer, B., Thiemann, R.: Reachability analysis with state-compatible automata. In: Dediu, A.-H., Martín-Vide, C., Sierra-Rodríguez, J.-L., Truthe, B. (eds.) LATA 2014. LNCS, vol. 8370, pp. 347–359. Springer, Heidelberg (2014). doi:10.1007/978-3-319-04921-2_28

7. Geser, A., Middeldorp, A., Ohlebusch, E., Zantema, H.: Relative undecidability in the term rewriting, part 2: The confluence hierarchy. Inf. Comput. **178**(1), 132–148 (2002). doi:10.1006/inco.2002.3150

8. Hirokawa, N., Klein, D.: Saigawa: a confluence tool. In: Hirokawa, N., Middeldorp, A. (eds.) IWC 2012, p. 49 (2012). http://cl-informatik.uibk.ac.at/events/iwc-2012/

9. Hirokawa, N., Middeldorp, A.: Commutation via relative termination. In: Hirokawa, N., van Oostrom, V. (eds.) IWC 2013, pp. 29–33 (2013). http://www.jaist.ac.jp/~hirokawa/iwc2013/

10. Huet, G.: Confluent reductions: abstract properties and applications to term rewriting systems. J. ACM **27**(4), 797–821 (1980). doi:10.1145/322217.322230

11. Klein, D., Hirokawa, N.: Confluence of non-left-linear TRSs via relative termination. In: Bjørner, N., Voronkov, A. (eds.) LPAR-18 2012. LNCS, vol. 7180, pp. 258–273. Springer, Heidelberg (2012). doi:10.1007/978-3-642-28717-6_21

12. Knuth, D., Bendix, P.: Simple word problems in universal algebras. In: Leech, J. (ed.) Computational Problems in Abstract Algebra, pp. 263–297. Pergamon Press (1970)

13. Nagele, J., Felgenhauer, B., Middeldorp, A.: Improving automatic confluence analysis of rewrite systems by redundant rules. In: Fernández, M. (ed.) RTA 2015. LIPIcs, vol. 36, pp. 257–268. Schloss Dagstuhl–Leibniz-Zentrum für Informatik (2015). doi:10.4230/LIPIcs.RTA.2015.257

14. Nagele, J., Thiemann, R.: Certification of confluence proofs using CeTA. In: Aoto, T., Kesner, D. (eds.) IWC 2014, pp. 19–23 (2014). http://www.nue.riec.tohoku.ac.jp/iwc2014/

15. Nagele, J., Zankl, H.: Certified rule labeling. In: Fernández, M. (ed.) RTA 2015. LIPIcs, vol. 36, pp. 269–284. Schloss Dagstuhl–Leibniz-Zentrum für Informatik (2015). doi:10.4230/LIPIcs.RTA.2015.269

16. Nipkow, T., Paulson, L., Wenzel, M.: Isabelle/HOL - A Proof Assistant for Higher-Order Logic. LNCS, vol. 2283. Springer, Heidelberg (2002)

17. van Oostrom, V.: Developing developments. Theoret. Comput. Sci. **175**(1), 159–181 (1997). doi:10.1016/S0304-3975(96)00173-9

18. Shintani, K., Hirokawa, N.: CoLL: A confluence tool for left-linear term rewrite systems. In: Felty, A., Middeldorp, A. (eds.) CADE 2015. LNCS, vol. 9195, pp. 127–136. Springer, Heidelberg (2015). doi:10.1007/978-3-319-21401-6_8

19. Sternagel, C., Thiemann, R.: The certification problem format. In: Benzmüller, C., Woltzenlogel Paleo, B. (eds.) UITP 2014. EPTCS, vol. 167, pp. 61–72. Open Publishing Association (2014). doi:10.4204/EPTCS.167.8

20. Terese: Term Rewriting Systems. Cambridge Tracts in Theoretical Computer-Science, vol. 55. Cambridge University Press, Cambridge (2003)

21. Thiemann, R., Sternagel, C.: Certification of termination proofs using CeTA. In: Berghofer, S., Nipkow, T., Urban, C., Wenzel, M. (eds.) TPHOLs 2009. LNCS, vol. 5674, pp. 452–468. Springer, Heidelberg (2009). doi:10.1007/978-3-642-03359-9_31

22. Toyama, Y.: Commutativity of term rewriting systems. In: Fuchi, K., Kott, L. (eds.) Programming of Future Generation Computers II, pp. 393–407. North-Holland (1988)

23. Zankl, H., Felgenhauer, B., Middeldorp, A.: CSI – a confluence tool. In: Bjørner, N., Sofronie-Stokkermans, V. (eds.) CADE 2011. LNCS, vol. 6803, pp. 499–505. Springer, Heidelberg (2011). doi:10.1007/978-3-642-22438-6_38

Automatic Functional Correctness Proofs for Functional Search Trees

Tobias Nipkow$^{(\boxtimes)}$

Technische Universität München, Munich, Germany
nipkow@in.tum.de
http://www.in.tum.de/~nipkow

Abstract. In a new approach, functional correctness specifications of *insert/update* and *delete* operations on search trees are expressed on the level of lists by means of an inorder traversal function that projects trees to lists. With the help of a small lemma library, functional correctness and preservation of the search tree property are proved automatically (in Isabelle/HOL) for a range of data structures: unbalanced binary trees, AVL trees, red-black trees, 2-3 and 2-3-4 trees, 1-2 brother trees, AA trees and splay trees.

1 Introduction

Most books and articles on search tree data structures do not discuss functional correctness, which is taken to be obvious, but concentrate on non-obvious structural invariants like balancedness. This paper confirms that this is the right attitude by providing a framework for proving the functional correctness of eight different search tree data structures automatically (in Isabelle/HOL [19,21]).

What is proved automatically? Functional correctness of *insert*, *delete* and *isin* together with the preservation of the search tree invariant, i.e. sortedness, by *insert* and *delete*. Structural invariants like balancedness are proved manually, depend on the specific data structure, and are not discussed here.

Which data structures are covered? Unbalanced binary trees, AVL trees, red-black trees, 2-3 and 2-3-4 trees, 1-2 brother trees, AA trees and splay trees.[1] As far as we know, these are the first formal proofs for 2-3 and 2-3-4 trees, 1-2 brother trees and AA trees, and the first automatic proofs for most of the eight data structures.

What does automatic mean? It means that all the required theorems are proved by induction followed by a single invocation of Isabelle's `auto` proof method, parameterized with a fixed set of basic lemmas plus further lemmas about auxiliary functions. The lemmas to be proved about *insert*, *delete* and *isin* are fixed; lemmas about auxiliary functions need to be invented but (mostly) follow a simple pattern.

T. Nipkow—Supported by DFG Koselleck grant NI 491/16-1.

[1] See http://isabelle.in.tum.de/library/HOL/HOL-Data_Structures/ or the source directory `src/HOL/Data_Structures/` in the Isabelle distribution.

© Springer International Publishing Switzerland 2016
J.C. Blanchette and S. Merz (Eds.): ITP 2016, LNCS 9807, pp. 307–322, 2016.
DOI: 10.1007/978-3-319-43144-4_19

The paper is structured as follows. Section 3 presents two approaches to the specification and verification of set implementations: the standard approach and our new approach. Section 4 details the verification framework behind the new approach. In Sect. 5 eight different search tree implementations and their correctness proofs are discussed. In a final section it is shown how the framework can be generalized from sets to maps.

Related work is discussed in the body of the paper. With one exception, the proofs in previous work are not automatic. We refrain from stating this each time and we do not describe how far from automatic they are, although this varies significantly (from a few to more than a hundred lines).

2   Lists and Trees

Lists (type $'a$ *list*) are constructed from the empty list [] via the infix cons-operator ".". The notation $[x,y,z]$ is short for $x \cdot y \cdot z \cdot []$. The infix @ concatenates two lists.

Binary trees are defined as the data type $'a$ *tree* with two constructors: the empty tree or leaf $\langle\rangle$ and the node $\langle l, a, r \rangle$ with subtrees $l, r :: 'a$ *tree* and contents $a :: 'a$.

There is also a type $'a$ *set* of sets with their usual operations.

3   Set Implementations

We require that an implementation of sets provides some type $'a$ t (where $'a$ is the element type) and the operations

empty :: $'a$ t
insert :: $'a \Rightarrow 'a$ $t \Rightarrow 'a$ t
delete :: $'a \Rightarrow 'a$ $t \Rightarrow 'a$ t
isin :: $'a$ $t \Rightarrow 'a \Rightarrow bool$

In the rest of the paper we ignore *empty* because it is trivial.

In order to specify these operations we assume that there is also an *abstraction function set* :: $'a$ $t \Rightarrow 'a$ *set* and a data type invariant *invar* :: $'a$ $t \Rightarrow bool$. These are not part of the interface and need not be executable but have to be provided in order to prove an implementation correct w.r.t. the specification in Fig. 1. Specifications phrased in terms of abstraction functions that are required to be homomorphisms go back to Hoare [11] and became an integral part of the model-oriented specification language VDM [13]. In the first-order context of universal algebra it was shown that there are always fully abstract models such that any concrete implementation can be shown correct with a homomorphism [16]. In Isabelle, implementations that satisfy such specifications can automatically be plugged in for the abstract type by regarding the abstraction function as a constructor [9]. For example, turning the equation *isin* s $x = (x \in set\ s)$ around tells us how to evaluate $x \in set\ s$ with the help of *isin*.

From now on we assume that the element type $'a$ is linearly ordered.

$$invar\ s \implies set\ (insert\ x\ s) = \{x\} \cup set\ s$$
$$invar\ s \implies set\ (delete\ x\ s) = set\ s - \{x\}$$
$$invar\ s \implies isin\ s\ x = (x \in set\ s)$$
$$invar\ s \implies invar\ (insert\ x\ s)$$
$$invar\ s \implies invar\ (delete\ x\ s)$$

Fig. 1. Specification of set implementations

3.1   The Standard Approach

The most compact form of the standard approach to the verification of search tree implementations of sets consists of the following items:

- An abstraction function *set* that extract the set of elements in a tree.
- A recursively defined (binary) search tree invariant
  $bst\ \langle l,a,r \rangle = (bst\ l \wedge bst\ r \wedge (\forall x \in set\ l.\ x < a) \wedge (\forall x \in set\ r.\ a < x)$
- Proof of the correctness conditions in Fig. 1 where *invar* is *bst*, possibly conjoined with additional structural invariants.

There are many variations of the above setup, some of which address two complications that arise when automating the proofs, the quantifiers and the non-free data type of sets:

- In the definition of *bst*, quantifiers are replaced by auxiliary functions that check if all elements in a tree are less/greater than a given element.
- Instead of extensional equality of sets, e.g. $set\ (insert\ x\ s) = \{x\} \cup set\ s$, pointwise equality is proved, e.g. $isin\ (insert\ x\ s)\ y = (x = y \vee isin\ s\ y)$.
- The predicate *bst* is defined inductively rather than recursively.

We subsume all of these variations under the "standard approach". Unless stated otherwise, all related work follows the standard approach.

3.2   The *inorder* Approach

Why is it perfectly obvious that the following equation (where R/B construct red/black nodes) preserves sortedness and the set of elements of a tree?

$$balance\ (R\ (R\ t_1\ a\ t_2)\ b\ t_3)\ c\ t_4 = R\ (B\ t_1\ a\ t_2)\ b\ (B\ t_3\ c\ t_4)$$

Because the sequence of subtrees and elements is the same on both sides! We merely need to make the machine see this as well as we do.

The key idea of our approach is to base it on the inorder traversal of trees. That is, we use lists as an intermediate data type between sets and trees. To this end we need four auxiliary functions on lists:

- $\uparrow :: {}'a\ list \Rightarrow bool$
  $\uparrow$ means that the list is sorted in ascending order w.r.t. $<$.

$$invar\ t \implies \lfloor insert\ x\ t \rfloor = ins_{list}\ x\ \lfloor t \rfloor \tag{1}$$
$$invar\ t \implies \lfloor delete\ x\ t \rfloor = del_{list}\ x\ \lfloor t \rfloor \tag{2}$$
$$invar\ t \implies isin\ t\ x = (x \in elems\ \lfloor t \rfloor) \tag{3}$$
$$invar\ t \implies inv\ (insert\ x\ t) \tag{4}$$
$$invar\ t \implies inv\ (delete\ x\ t) \tag{5}$$

Fig. 2. Specification of set implementations over ordered types

$$\uparrow [] = True$$
$$\uparrow [x] = True$$
$$\uparrow (x \cdot y \cdot zs) = (x < y \land \uparrow (y \cdot zs))$$

– $ins_{list} :: {'}a \Rightarrow {'}a\ list \Rightarrow {'}a\ list$

ins_{list} inserts an element at the correct position into a sorted list if the element is not present in the list yet.

$$ins_{list}\ x\ [] = [x]$$
$$ins_{list}\ x\ (a \cdot xs) =$$
$$(if\ x < a\ then\ x \cdot a \cdot xs\ else\ if\ x = a\ then\ a \cdot xs\ else\ a \cdot ins_{list}\ x\ xs)$$

– $del_{list} :: {'}a \Rightarrow {'}a\ list \Rightarrow {'}a\ list$

del_{list} deletes the first occurrence of an element from a list.

$$del_{list}\ x\ [] = []$$
$$del_{list}\ x\ (a \cdot xs) = (if\ x = a\ then\ xs\ else\ a \cdot del_{list}\ x\ xs)$$

– $elems :: {'}a\ list \Rightarrow {'}a\ set$

$elems$ turns a list into the set of its elements.

$$elems\ [] = \emptyset$$
$$elems\ (x \cdot xs) = \{x\} \cup elems\ xs$$

A new specification of sets (over a linearly ordered type ${'}a$) is shown in Fig. 2. The crucial ingredient is an additional specification function

$$inorder :: {'}a\ t \Rightarrow {'}a\ list$$

As the name *inorder* suggests, you should now think of ${'}a\ t$ as a type of trees. We abbreviate *inorder* t by $\lfloor t \rfloor$.

The fact that some t is a search tree, i.e. sorted, can now be expressed as $\uparrow \lfloor t \rfloor$. This invariant will be dealt with automatically. Of course search trees frequently have additional structural invariants. These can be supplied via yet another specification function $inv :: {'}a\ t \Rightarrow bool$. Both kinds of invariants are combined into *invar*:

$$invar\ t = (inv\ t \land \uparrow \lfloor t \rfloor)$$

The first three propositions in Fig. 2 demand the functional correctness of *insert*, *delete* and *isin* w.r.t. ins_{list}, del_{list} and *elems*. The next two propositions demand that *inv* is invariant. If we interpret function *set* as *elems* ∘ *inorder* it is easy to show that Fig. 2 implies Fig. 1 with the help of the following simple inductive lemmas:

$$elems \ (ins_{list} \ x \ xs) = \{x\} \cup elems \ xs$$
$$\uparrow xs \implies distinct \ xs$$
$$distinct \ xs \implies elems \ (del_{list} \ x \ xs) = elems \ xs - \{x\}$$
$$\uparrow xs \implies \uparrow (ins_{list} \ x \ xs)$$
$$\uparrow xs \implies \uparrow (del_{list} \ x \ xs)$$

In summary: the functional correctness of an implementation of *insert*, *delete* and *isin* on some data structure can be verified by proving the properties in Fig. 2 for some suitable definition of *inorder* and *inv*. In the following section we introduce a library of lemmas about $\uparrow$, ins_{list} and del_{list} that allows us to automate the proofs of (1)–(3).

Note that although we equate (1)–(3) with "functional correctness", it is more: (1)–(3) also imply that sortedness is an invariant.

4   The Verification Framework

We do not claim to provide a framework that can prove any implementation of sets by search trees automatically correct. Instead we provide lemmas that work in practice (they automate the correctness proofs for a list of benchmark implementations presented in Sect. 5) and are well motivated by general considerations concerning the shape of formulas that arise in the verification.

As a motivating example we consider ordinary unbalanced binary trees $'a \ tree$. The textbook definitions of *insert*, *delete* and *isin* are omitted. Let us examine how to prove

$$\uparrow \lfloor t \rfloor \implies \lfloor insert \ x \ t \rfloor = ins_{list} \ x \ \lfloor t \rfloor$$

The proof is by induction on t and we consider the case $t = \langle l, a, r \rangle$ such that $x < a$. Ideally the proof looks like this:

$$\lfloor insert \ x \ t \rfloor = \lfloor insert \ x \ l \rfloor @ a \cdot \lfloor r \rfloor = ins_{list} \ x \ \lfloor l \rfloor @ a \cdot \lfloor r \rfloor$$
$$= ins_{list} \ x \ (\lfloor l \rfloor @ a \cdot \lfloor r \rfloor) = ins_{list} \ x \ t$$

The first and last step are by definition, the second step by induction hypothesis, but the third step requires two lemmas:

$$\uparrow (xs @ y \cdot ys) = (\uparrow (xs @ [y]) \wedge \uparrow (y \cdot ys))$$
$$\uparrow (xs @ [a]) \wedge x < a \implies ins_{list} \ x \ (xs @ a \cdot ys) = ins_{list} \ x \ xs @ a \cdot ys$$

The first lemma rewrites the assumption $\uparrow \lfloor t \rfloor$ to $\uparrow (\lfloor l \rfloor @ [a]) \wedge \uparrow (a \cdot \lfloor r \rfloor)$, thus allowing the second lemma to rewrite the term $ins_{list} \ x \ (\lfloor l \rfloor @ a \cdot \lfloor r \rfloor)$ to $ins_{list} \ x \ \lfloor l \rfloor @ a \cdot \lfloor r \rfloor$.

It may seem that the two lemmas just shown are rather arbitrary, but we will see that in the context of trees, where each node is a tuple $\langle s_0, a_1, s_1, \ldots, s_n \rangle$ of subtrees s_i alternating with elements a_i, there is an underlying principle. In the properties in Fig. 2 the following three terms are crucial: $\uparrow \lfloor t \rfloor$, $ins_{list} \ x \ \lfloor t \rfloor$ and $del_{list} \ x \ \lfloor t \rfloor$. Assuming that the properties are proved by induction, t will be some (possibly complicated) tree constructor term. Evaluating $\lfloor t \rfloor$ will thus lead to a list of the following form where sublists and individual elements alternate:

$$\uparrow (xs \text{ @ } y \cdot ys) = (\uparrow (xs \text{ @ } [y]) \land \uparrow (y \cdot ys)) \tag{6}$$

$$\uparrow (x \cdot xs \text{ @ } y \cdot ys) = (\uparrow (x \cdot xs) \land x < y \land \uparrow (xs \text{ @ } [y]) \land \uparrow (y \cdot ys)) \tag{7}$$

$$\uparrow (x \cdot xs) \Longrightarrow \uparrow xs \tag{8}$$

$$\uparrow (xs \text{ @ } [y]) \Longrightarrow \uparrow xs \tag{9}$$

$$\uparrow (xs \text{ @ } [a]) \Longrightarrow ins_{list} \; x \; (xs \text{ @ } a \cdot ys) = \tag{10}$$
$$(\textit{if } x < a \textit{ then } ins_{list} \; x \; xs \text{ @ } a \cdot ys \textit{ else } xs \text{ @ } ins_{list} \; x \; (a \cdot ys))$$

$$\uparrow (xs \text{ @ } a \cdot ys) \Longrightarrow del_{list} \; x \; (xs \text{ @ } a \cdot ys) = \tag{11}$$
$$(\textit{if } x < a \textit{ then } del_{list} \; x \; xs \text{ @ } a \cdot ys \textit{ else } xs \text{ @ } del_{list} \; x \; (a \cdot ys))$$

$$elems \; (xs \text{ @ } ys) = elems \; xs \cup elems \; ys \tag{12}$$

$$\uparrow (y \cdot xs) \land x \le y \Longrightarrow x \notin elems \; xs \tag{13}$$

$$\uparrow (xs \text{ @ } [y]) \land y \le x \Longrightarrow x \notin elems \; xs \tag{14}$$

Fig. 3. Lemmas for $\uparrow$, ins_{list}, del_{list} and $elems$

$$\lfloor t_1 \rfloor \text{ @ } a_1 \cdot \lfloor t_2 \rfloor \text{ @ } a_2 \cdot \ldots \cdot \lfloor t_n \rfloor$$

Now we discuss a set of lemmas (see Fig. 3) that allow us to simplify the application of $\uparrow$, ins_{list} and del_{list} to such terms.

Terms of the form $\uparrow (xs_1 \text{ @ } a_1 \cdot xs_2 \text{ @ } a_2 \cdot \ldots \cdot xs_n)$ are decomposed into the following *basic* formulas

$$\uparrow (xs \text{ @ } [a]) \qquad (\text{simulating } \forall x \in set \; xs. \; x < a)$$
$$\uparrow (a \cdot xs) \qquad (\text{simulating } \forall x \in set \; xs. \; a < x)$$
$$a < b$$

by the rewrite rules (6)–(7). Lemmas (8)–(9) enable deductions from basic formulas.

Terms of the form $ins_{list} \; x \; (xs_1 \text{ @ } a_1 \cdot xs_2 \text{ @ } a_2 \cdot \ldots \cdot xs_n)$ are rewritten with equation (10) (and the defining equations for ins_{list}) to push ins_{list} inwards. Terms of the form $del_{list} \; x \; (xs_1 \text{ @ } a_1 \cdot xs_2 \text{ @ } a_2 \cdot \ldots \cdot xs_n)$ are rewritten with Eq. (11) (and the defining equations for del_{list}) to push del_{list} inwards.

Finally we need lemmas (12)–(14) about *elems* on sorted lists.

The lemmas in Fig. 3 form the complete set of basic lemmas on which the automatic proofs of almost all search trees in the paper rest; only splay trees need additional lemmas.

4.1   Proof Automation by Rewriting

The automatic proofs rely on conditional, contextual term rewriting with the following bells and whistles (which Isabelle's simplifier provides):

– Conjunctions in the context are split up into their conjuncts.
– Conditionals and case-expressions can be split automatically.
– A decision procedure for linear orders that can decide if some literal (a possibly negated atom $a < b$ or $a \le b$) follows from a set of literals in the context.

- Implications (8)–(9) lead to nontermination when used as conditional rewrite rules. It must be possible to direct the simplifier to solve the preconditions of those rules by assumptions in the context rather than a recursive simplifier invocation. In Isabelle there is a constant $ASSUMPTION = (\lambda x.\ x)$ that can be wrapped around a precondition of a rewrite rule and prevents recursive applications of the simplifier to that precondition.

5 An Arboretum

In the rest of this section we focus on (1) and (2) when discussing the proofs of the properties in Fig. 2. This is because requirement (3) can always (except for splay trees) be proved automatically without further lemmas and (4) and (5) are specific to the individual data structures and not part of functional correctness.

Because there is not enough space to present all definitions and proofs, Table 1 gives an overview in terms of lines of code and numbers of functions needed for each data structure. Because *isin* is (almost) the same for all of them (except splay trees), it is excluded. The table shows that there is at most one lemma per function, except for splay trees.

Table 1. Code and proof statistics for *insert* + *delete* (l.o. = lines of)

	Unbal	AVL	Red-Black	2-3	2-3-4	Brother	AA	Splay
l.o. code	17	45	61	88	143	66	55	46
functions	3	8	11	12	16	10	8	4
lemmas	3	6	11	10	14	10	6	5

The majority of lemmas about auxiliary functions follow a simple pattern. Typical examples are balancing functions, e.g. $\lfloor bal\ t \rfloor = \lfloor t \rfloor$, or smart constructors, e.g. $\lfloor node\ l\ a\ r \rfloor = \lfloor l \rfloor\ @\ a \cdot \lfloor r \rfloor$. We call these *trivial lemmas*. More complicated lemmas are discussed explicitly in the text; we call them *non-trivial*.

All our implementations compare elements with a comparison operator *cmp* that returns an element of the **datatype** $cmp = LT \mid EQ \mid GT$.

5.1 Unbalanced Trees

Function *insert* is trivial and (1) is proved directly. Function *delete* is more interesting because it is defined with the help of an auxiliary function:

$delete\ x\ \langle\rangle = \langle\rangle$
$delete\ x\ \langle l,\ a,\ r \rangle =$
(**case** $cmp\ x\ a$ **of** $LT\ \Rightarrow\ \langle delete\ x\ l,\ a,\ r \rangle$
$\mid\ EQ\ \Rightarrow$ **if** $r = \langle\rangle$ **then** l **else let** $(x,\ y) = del\_min\ r$ **in** $\langle l,\ x,\ y \rangle$
$\mid\ GT\ \Rightarrow\ \langle l,\ a,\ delete\ x\ r \rangle)$

$del\_min \ \langle l, \ a, \ r \rangle =$
$(if \ l = \langle\rangle \ then \ (a, \ r) \ else \ let \ (x, \ l') = del\_min \ l \ in \ (x, \ \langle l', \ a, \ r \rangle))$

The proof of (2) requires the following lemma about del_min that the user has to formulate himself; the proof is again automatic.

$del\_min \ t = (x, \ t') \land t \neq \langle\rangle \implies x \cdot \lfloor t' \rfloor = \lfloor t \rfloor$

This is one of the more "difficult" lemmas to invent.

5.2 AVL Trees

Our starting point was an existing formalization [20] which follows the standard approach. Functional correctness of AVL trees can be proved without assuming any structural (height) invariants. The only non-trivial lemma we require is

$del\_max \ t = (t', \ a) \land t \neq \langle\rangle \implies \lfloor t' \rfloor \ @ \ [a] = \lfloor t \rfloor$

Related Work. Filliâtre and Letouzey [6] report on a verification of AVL trees in Coq. They follow the standard approach, except that the executable functions are extracted from constructive proofs. An updated version of their proofs in the Coq distribution gives the functions explicitly. Ralston [26] reports a proof with ACL2. The verification by Clochard [4] in Why3 is interesting because he also abstracts trees to their inorder traversal and reports that the proofs for AVL trees are automatic.

5.3 Red-Black Trees

Red-black trees were invented by Bayer [3]. Guibas and Sedgewick [8] introduced the red/black color convention. Red-black trees can be seen as an encoding of 2-3-4 trees as binary trees.

Our starting point was an existing formalization in the Isabelle distribution (in `HOL/Library/RBT_Impl.thy`, by Reiter and Krauss) which in turn is based on the code by Okasaki [22] (for *insert*) and Stefan Kahrs [14] (for *delete* see the URL given in the article). The original verification has a certain similarity to ours because it also involves an inorder listing of the tree (function *entries*), but a number of the proofs are distinctly long and manual. In contrast, the only non-trivial lemmas we require are the following ones that need to be proved simultaneously about three auxiliary functions:

$\uparrow \lfloor t \rfloor \implies \lfloor del \ x \ t \rfloor = del_{list} \ x \ \lfloor t \rfloor$
$\uparrow \lfloor l \rfloor \implies \lfloor delL \ x \ l \ a \ r \rfloor = del_{list} \ x \ \lfloor l \rfloor \ @ \ a \cdot \lfloor r \rfloor$
$\uparrow \lfloor r \rfloor \implies \lfloor delR \ x \ l \ a \ r \rfloor = \lfloor l \rfloor \ @ \ a \cdot del_{list} \ x \ \lfloor r \rfloor$

Of course the proof is automatic, as usual.

Functional correctness of red-black trees can be proved without assuming any structural (red-black) invariants.

Related Work. Filliâtre and Letouzey [6] and Appel [2] verified red-black trees in Coq.

5.4   2-3 Trees

In a 2-3 tree (invented by Hopcroft in 1970 [5]), every non-leaf node has either two or three children: $\langle l, a, r \rangle$ or $\langle l, a, m, b, r \rangle$ where l, m, r are trees and a, b are elements. One can view $\langle l, a, m, b, r \rangle$ as a more compact representation of $\langle l, a, \langle m, b, r \rangle \rangle$ (see AA trees). Their structural invariant is that they are balanced, i.e. all leaves occur at the same depth.

Our code is based on the lecture notes by Turbak [29], who presents the key transformations in a graphical format. We present the more complex *delete* function in Fig. 4. Function *del* descends into the tree until the element (or a leaf) is found. Modified subtrees are recombined with smart constructors *nodeij* that combines i subtrees where subtree j has been modified and is wrapped up in either T_d (if the height of the subtree is unchanged) or $U p_d$ (if the height of the subtree has decreased). We only show the functions *nodei*1 because the other *nodeij* are symmetric.

The lemmas required for the correctness proof are similar to what we have seen already, with one new complication: the balancedness invariant *bal* is frequently required as a precondition, e.g. here:

$$del\_min \ t = (x, t') \wedge bal \ t \wedge 0 < height \ t \implies x \cdot \lfloor tree_d \ t' \rfloor = \lfloor t \rfloor$$

Our automatic framework can cope because *bal* and *height* are defined in a straightforward manner by primitive recursion.

Related Work. The existing formalization of 2-3 trees in the Isabelle distribution (in `HOL/ex/Tree23.thy`, by Huffman and Nipkow) proves invariants but not functional correctness. Hoffmann and O'Donnell [12] give an equational definition of insertion. Reade [27] gives a similar equational definition of insertion and adds deletion; Turbak's version of deletion appears a bit simpler. Reade sketches (because there are too many cases) a pen-and-paper correctness proof and writes: "Mechanical support for such reasoning and the potential for partial automation of similar proofs are topics currently being investigated by the author".

5.5   2-3-4 Trees

2-3-4 trees are an extension of 2-3 trees where nodes may also have 4 children: $\langle t_1, a, t_2, b, t_3, c, t_4 \rangle$. Their structural invariant is that they are balanced, i.e. all leaves occur at the same depth. The code for 2-3-4 trees can also be viewed as an extension of that for 2-3 trees with additional cases. There are also new smart constructors *node4j*, e.g. *node4*1:

$$node41 \ (T_d \ t_1) \ a \ t_2 \ b \ t_3 \ c \ t_4 = T_d \ \langle t_1, a, t_2, b, t_3, c, t_4 \rangle$$
$$node41 \ (U p_d \ t_1) \ a \ \langle t_2, b, t_3 \rangle \ c \ t_4 \ d \ t_5 = T_d \ \langle\langle t_1, a, t_2, b, t_3 \rangle, c, t_4, d, t_5 \rangle$$

datatype $'a\ up_d = T_d\ ('a\ tree23)\ |\ Up_d\ ('a\ tree23)$

$tree_d\ (T_d\ t) = t$
$tree_d\ (Up_d\ t) = t$

$node21\ (T_d\ t_1)\ a\ t_2 = T_d\ \langle t_1,\ a,\ t_2 \rangle$
$node21\ (Up_d\ t_1)\ a\ \langle t_2,\ b,\ t_3 \rangle = Up_d\ \langle t_1,\ a,\ t_2,\ b,\ t_3 \rangle$
$node21\ (Up_d\ t_1)\ a\ \langle t_2,\ b,\ t_3,\ c,\ t_4 \rangle = T_d\ \langle\langle t_1,\ a,\ t_2 \rangle,\ b,\ \langle t_3,\ c,\ t_4 \rangle\rangle$

$node31\ (T_d\ t_1)\ a\ t_2\ b\ t_3 = T_d\ \langle t_1,\ a,\ t_2,\ b,\ t_3 \rangle$
$node31\ (Up_d\ t_1)\ a\ \langle t_2,\ b,\ t_3 \rangle\ c\ t_4 = T_d\ \langle\langle t_1,\ a,\ t_2,\ b,\ t_3 \rangle,\ c,\ t_4 \rangle$
$node31\ (Up_d\ t_1)\ a\ \langle t_2,\ b,\ t_3,\ c,\ t_4 \rangle\ d\ t_5 = T_d\ \langle\langle t_1,\ a,\ t_2 \rangle,\ b,\ \langle t_3,\ c,\ t_4 \rangle,\ d,\ t_5 \rangle$

$del\_min\ \langle\langle\rangle,\ a,\ \langle\rangle\rangle = (a,\ Up_d\ \langle\rangle)$
$del\_min\ \langle\langle\rangle,\ a,\ \langle\rangle,\ b,\ \langle\rangle\rangle = (a,\ T_d\ \langle\langle\rangle,\ b,\ \langle\rangle\rangle)$
$del\_min\ \langle l,\ a,\ r \rangle = (\textbf{let}\ (x,\ l') = del\_min\ l\ \textbf{in}\ (x,\ node21\ l'\ a\ r))$
$del\_min\ \langle l,\ a,\ m,\ b,\ r \rangle = (\textbf{let}\ (x,\ l') = del\_min\ l\ \textbf{in}\ (x,\ node31\ l'\ a\ m\ b\ r))$

$del\ x\ \langle\rangle = T_d\ \langle\rangle$
$del\ x\ \langle\langle\rangle,\ a,\ \langle\rangle\rangle = (\textbf{if}\ x = a\ \textbf{then}\ Up_d\ \langle\rangle\ \textbf{else}\ T_d\ \langle\langle\rangle,\ a,\ \langle\rangle\rangle)$
$del\ x\ \langle\langle\rangle,\ a,\ \langle\rangle,\ b,\ \langle\rangle\rangle =$
$T_d\ (\textbf{if}\ x = a\ \textbf{then}\ \langle\langle\rangle,\ b,\ \langle\rangle\rangle\ \textbf{else if}\ x = b\ \textbf{then}\ \langle\langle\rangle,\ a,\ \langle\rangle\rangle\ \textbf{else}\ \langle\langle\rangle,\ a,\ \langle\rangle,\ b,\ \langle\rangle\rangle)$
$del\ x\ \langle l,\ a,\ r \rangle =$
$(\textbf{case}\ cmp\ x\ a\ \textbf{of}\ LT \Rightarrow node21\ (del\ x\ l)\ a\ r$
$\quad |\ EQ \Rightarrow \textbf{let}\ (a',\ t) = del\_min\ r\ \textbf{in}\ node22\ l\ a'\ t\ |\ GT \Rightarrow node22\ l\ a\ (del\ x\ r))$
$del\ x\ \langle l,\ a,\ m,\ b,\ r \rangle =$
$(\textbf{case}\ cmp\ x\ a\ \textbf{of}\ LT \Rightarrow node31\ (del\ x\ l)\ a\ m\ b\ r$
$\quad |\ EQ \Rightarrow \textbf{let}\ (a',\ m') = del\_min\ m\ \textbf{in}\ node32\ l\ a'\ m'\ b\ r$
$\quad |\ GT \Rightarrow \textbf{case}\ cmp\ x\ b\ \textbf{of}\ LT \Rightarrow node32\ l\ a\ (del\ x\ m)\ b\ r$
$\qquad\qquad |\ EQ \Rightarrow \textbf{let}\ (b',\ r') = del\_min\ r\ \textbf{in}\ node33\ l\ a\ m\ b'\ r'$
$\qquad\qquad |\ GT \Rightarrow node33\ l\ a\ m\ b\ (del\ x\ r))$

$delete\ x\ t = tree_d\ (del\ x\ t)$

Fig. 4. Deletion in 2-3 trees

$node41\ (Up_d\ t_1)\ a\ \langle t_2,\ b,\ t_3,\ c,\ t_4 \rangle\ d\ t_5\ e\ t_6 =$
$T_d\ \langle\langle t_1,\ a,\ t_2 \rangle,\ b,\ \langle t_3,\ c,\ t_4 \rangle,\ d,\ t_5,\ e,\ t_6 \rangle$
$node41\ (Up_d\ t_1)\ a\ \langle t_2,\ b,\ t_3,\ c,\ t_4,\ d,\ t_5 \rangle\ e\ t_6\ f\ t_7 =$
$T_d\ \langle\langle t_1,\ a,\ t_2 \rangle,\ b,\ \langle t_3,\ c,\ t_4,\ d,\ t_5 \rangle,\ e,\ t_6,\ f,\ t_7 \rangle$

Related Work. It appears that the only (partially) published functional implementations of 2-3-4 trees is one in Maude [15] where the full code is available online. No formal proofs are reported.

5.6   1-2 Brother Trees

A 1-2 brother tree [23,24] is a binary tree with one further constructor $N1$ from trees to trees for unary nodes. The structural invariant is that the tree is balanced (all leaves at the same depth) and that every unary node has a binary brother. Unary nodes allow us to balance any tree. There is a bijection between 1-2 brother trees and AVL trees: remove the unary nodes from a 1-2 brother tree and you obtain an AVL tree. Our formalization is based on the article by Hinze [10] where all code and invariants can be found. Hinze captures the invariant by two sets $B\ h$ and $U\ h$, the sets of brother trees of height h that have a binary (or nullary) respectively unary root node. The actual brother trees are captured by B; U is an auxiliary notion. The correctness lemmas (1)–(3) for *insert, delete* and *isin* employ the abbreviation $T\ h = B\ h \cup U\ h$:

$$t \in T\ h \wedge \uparrow \lfloor t \rfloor \implies \lfloor insert\ a\ t \rfloor = ins_{list}\ a\ \lfloor t \rfloor$$
$$t \in T\ h \wedge \uparrow \lfloor t \rfloor \implies \lfloor delete\ x\ t \rfloor = del_{list}\ x\ \lfloor t \rfloor$$
$$t \in T\ h \wedge \uparrow \lfloor t \rfloor \implies isin\ t\ x = (x \in elems\ \lfloor t \rfloor)$$

The non-trivial but automatic auxiliary lemmas are

$$t \in T\ h \wedge \uparrow \lfloor t \rfloor \implies \lfloor ins\ a\ t \rfloor = ins_{list}\ a\ \lfloor t \rfloor$$
$$t \in T\ h \wedge \uparrow \lfloor t \rfloor \implies \lfloor del\ x\ t \rfloor = del_{list}\ x\ \lfloor t \rfloor$$

$$t \in T\ h \implies$$
$$(del\_min\ t = None) = (\lfloor t \rfloor = []) \wedge$$
$$(del\_min\ t = Some\ (a,\ t') \longrightarrow \lfloor t \rfloor = a \cdot \lfloor t' \rfloor)$$

5.7   AA Trees

Arne Anderson [1] invented a particularly simple form of balanced trees, named AA trees by Weiss [30]. They encode 2-3 trees as binary trees (with the help of an additional height field, although a single bit would suffice). Their main selling point is simplicity and compactness of the code. Our verification started from the functional version of AA trees published by Ragde [25] without proofs. The proofs for insertion were automatic as usual, but deletion posed problems.

The use of non-linear patterns in the Haskell code for **delete** was easily fixed. Then a failed correctness proof revealed that function **dellrg** goes down the wrong branch in the recursive case. After this bug was corrected the next complication was the fact that the definition of function **adjust** (which is supposed to restore the invariant after deletion) does not cover certain trees that cannot arise. Therefore I needed to introduce the following invariant corresponding to the textual invariants AA1–AA3 in [25]; function *lvl* returns the height field of a node:

$$invar\ \langle\rangle = True$$
$$invar\ \langle h,\ l,\ a,\ r \rangle =$$
$$(invar\ l \wedge invar\ r \wedge h = lvl\ l + 1 \wedge$$
$$(h = lvl\ r + 1 \vee (\exists\ lr\ b\ rr.\ r = \langle h,\ lr,\ b,\ rr \rangle \wedge h = lvl\ rr + 1)))$$

Proving that insertion and deletion preserve the invariant was non-trivial, in particular because there were two more bugs:

- Function dellrg fails to call adjust to restore the invariant. This is the correct code (we call dellrg *del_max*):

$del\_max\ \langle lv,\ l,\ a,\ \langle\rangle\rangle = (l,\ a)$
$del\_max\ \langle lv,\ l,\ a,\ r\rangle = (\textit{let}\ (r',\ b) = del\_max\ r\ \textit{in}\ (adjust\ \langle lv,\ l,\ a,\ r'\rangle,\ b))$

- The auxiliary function nlvl is incorrect. The correct version is as follows:
$nlvl\ t = (\textit{if}\ sngl\ t\ \textit{then}\ lvl\ t\ \textit{else}\ lvl\ t + 1)$

For the verification of functional correctness of deletion the domain of the partial adjust had to be characterized by a predicate *pre_adjust* (not in [25]). With its help we can formulate and prove the trivial *inorder*-lemma for *adjust*:

$$t \neq \langle\rangle \wedge pre\_adjust\ t \Longrightarrow \lfloor adjust\ t\rfloor = \lfloor t\rfloor$$

The main correctness theorem (2) requires a number of further lemmas:

$$del\_max\ t = (t',\ x) \wedge t \neq \langle\rangle \wedge invar\ t \Longrightarrow \lfloor t'\rfloor\ @\ [x] = \lfloor t\rfloor$$
$$invar\ \langle lv,\ l,\ a,\ r\rangle \wedge post\_del\ l\ l' \Longrightarrow pre\_adjust\ \langle lv,\ l',\ b,\ r\rangle$$
$$invar\ \langle lv,\ l,\ a,\ r\rangle \wedge post\_del\ r\ r' \Longrightarrow pre\_adjust\ \langle lv,\ l,\ a,\ r'\rangle$$
$$invar\ t \wedge (t',\ x) = del\_max\ t \wedge t \neq \langle\rangle \Longrightarrow post\_del\ t\ t'$$
$$invar\ t \Longrightarrow post\_del\ t\ (delete\ x\ t)$$

As usual, the proofs of the *inorder*-lemmas and theorems are automatic. The last four lemmas and the pre- and post-conditions involved are part of the invariant proofs and are merely reused. Hence they are not included in Table 1.

5.8   Splay Trees

Splay trees [28] are self-adjusting binary search trees where query and update operations modify the tree by rotating the accessed element to the root of the tree. The logarithmic amortized complexity of splay trees has been verified before [18]. The functional correctness proofs [17] followed the standard approach. Starting from the same code we automated those proofs.

Splay trees are different from the other trees we cover. All operations are based on a function $splay :: 'a \Rightarrow 'a\ tree \Rightarrow 'a\ tree$ that rotates the given element (or an element close to it) to the root of the tree. For example, this is *isin*:

$$isin\ t\ x = (\textit{case}\ splay\ x\ t\ \textit{of}\ \langle\rangle \Rightarrow False\ |\ \langle l,\ a,\ r\rangle \Rightarrow x = a)$$

See elsewhere [17,18] for *insert* and *delete*. Note that *isin* should return the new tree as well to achieve amortized logarithmic complexity. This is awkward in a functional language and gives the data structure an imperative flavour.

The verification is more demanding than before and we present all the required lemmas in Fig. 5. Lemmas (15)–(20) extend our lemma library in Fig. 3 but are only required for splay trees. With the help of these lemmas, the proofs of (1)–(3) are automatic.

$$\uparrow (x \cdot xs) \wedge y \leq x \Longrightarrow \uparrow (y \cdot xs) \tag{15}$$
$$\uparrow (xs \text{ @ } [x]) \wedge x \leq y \Longrightarrow \uparrow (xs \text{ @ } [y]) \tag{16}$$
$$\uparrow (x \cdot xs) \Longrightarrow ins_{list} \ x \ xs = x \cdot xs \tag{17}$$
$$\uparrow (xs \text{ @ } [x]) \Longrightarrow ins_{list} \ x \ xs = xs \text{ @ } [x] \tag{18}$$
$$\uparrow (x \cdot xs) \Longrightarrow del_{list} \ x \ xs = xs \tag{19}$$
$$\uparrow (xs \text{ @ } [x]) \Longrightarrow del_{list} \ x \ (xs \text{ @ } ys) = xs \text{ @ } del_{list} \ x \ ys \tag{20}$$

$$(splay \ a \ t = \langle\rangle) = (t = \langle\rangle)$$
$$(splay\_max \ t = \langle\rangle) = (t = \langle\rangle)$$
$$splay \ x \ t = \langle l, \ a, \ r\rangle \wedge \uparrow \lfloor t \rfloor \Longrightarrow (x \in elems \lfloor t \rfloor) = (x = a)$$
$$\lfloor splay \ x \ t \rfloor = \lfloor t \rfloor$$
$$\uparrow \lfloor t \rfloor \wedge splay \ x \ t = \langle l, \ a, \ r\rangle \Longrightarrow \uparrow (\lfloor l \rfloor \text{ @ } x \cdot \lfloor r \rfloor)$$
$$splay\_max \ t = \langle l, \ a, \ r\rangle \wedge \uparrow \lfloor t \rfloor \Longrightarrow \lfloor l \rfloor \text{ @ } [a] = \lfloor t \rfloor \wedge r = \langle\rangle$$

Fig. 5. Lemmas for splay tree verification

6 Maps

6.1 Specifications

Search trees can implement maps as well as sets. Although sets are a special case of maps, we presented sets first because their simplicity facilitates the explanation of the basic concepts. Now we present the modifications required for maps. An implementation of maps must provides a type $('a, 'b) \ t$ (where $'a$ are the keys and $'b$ the values) with the operations

$$empty :: ('a, 'b) \ t$$
$$update :: 'a \Rightarrow 'b \Rightarrow ('a, 'b) \ t \Rightarrow ('a, 'b) \ t$$
$$delete :: 'a \Rightarrow ('a, 'b) \ t \Rightarrow ('a, 'b) \ t$$
$$lookup :: ('a, 'b) \ t \Rightarrow 'a \Rightarrow 'b \ option$$

where **datatype** $'a \ option = None \mid Some \ 'a$ is predefined. Function *lookup* also plays the role of the abstraction function. In addition there is a data type invariant $invar :: ('a, 'b) \ t \Rightarrow bool$. The specification of maps is shown in Fig. 6 (corresponding to Fig. 1). It uses the function update notation

$$f(a := b) = (\lambda x. \ \text{if } x = a \text{ then } b \text{ else } f \ x)$$

$$invar \ m \Longrightarrow lookup \ (update \ a \ b \ m) = (lookup \ m)(a := Some \ b)$$
$$invar \ m \Longrightarrow lookup \ (delete \ a \ m) = (lookup \ m)(a := None)$$
$$invar \ m \Longrightarrow invar \ (update \ a \ b \ m)$$
$$invar \ m \Longrightarrow invar \ (delete \ a \ m)$$

Fig. 6. Specification of map implementations

Now we assume that the keys are linearly ordered. Search trees are abstracted to a list of key-value pairs sorted by their keys. The auxiliary functions $\uparrow$, ins_{list} and del_{list} are replaced by

$\uparrow_1 :: ('a \times 'b) \; list \Rightarrow bool$
$upd_{list} :: 'a \Rightarrow 'b \Rightarrow ('a \times 'b) \; list \Rightarrow ('a \times 'b) \; list$
$del_{list} :: 'a \Rightarrow ('a \times 'b) \; list \Rightarrow ('a \times 'b) \; list$

- $\uparrow_1 xs = \uparrow (map \; fst \; xs)$ where $fst \; (a, b) = a$.
- $upd_{list} \; a \; b$ updates a sorted (w.r.t. $\uparrow_1$) list by either inserting (a, b) at the correct position (w.r.t. $<$) if no $(a, \_)$ is in the list, or replacing the first $(a, \_)$ by (a, b) otherwise.
- $del_{list} \; a$ deletes the first occurrence of a pair $(a, \_)$ from a list.

$$invar \; t \Longrightarrow \lfloor update \; a \; b \; t \rfloor = upd_{list} \; a \; b \; \lfloor t \rfloor$$
$$invar \; t \Longrightarrow \lfloor delete \; a \; t \rfloor = del_{list} \; a \; \lfloor t \rfloor$$
$$invar \; t \Longrightarrow lookup \; t \; a = map\_of \; \lfloor t \rfloor \; a$$
$$invar \; t \Longrightarrow inv \; (update \; a \; b \; t)$$
$$invar \; t \Longrightarrow inv \; (delete \; a \; t)$$

Fig. 7. Specification of map implementations over ordered types

Our second specification of maps (over a linearly ordered type $'a$) is shown in Fig. 7 (corresponding to Fig. 2). It is again based on an inorder function:

$inorder :: ('a, 'b) \; t \Rightarrow ('a \times 'b) \; list$

Again, we abbreviate $inorder \; t$ by $\lfloor t \rfloor$.

The search tree invariant is now expressed as $\uparrow_1 \lfloor t \rfloor$. Structural invariants can be added via the specification function $inv :: ('a, 'b) \; t \Rightarrow bool$ and we define

$invar \; t = (inv \; t \wedge \uparrow_1 \lfloor t \rfloor)$

The first three propositions in Fig. 7 express functional correctness of *update*, *delete* and *lookup* w.r.t. upd_{list}, del_{list} and map_of. The latter is a predefined function on key-value lists:

$map\_of \; [] = (\lambda\_. \; None)$
$map\_of \; ((a, b) \cdot ps) = (map\_of \; ps)(a := b)$

The next two propositions demand that inv is invariant. It is easy to show that Fig. 7 implies Fig. 6.

6.2   Proof Automation

Figure 8 (corresponding to Fig. 3) shows the set of lemmas used to automate the correctness proofs of implementations of maps. There are no lemmas about $\uparrow_1$ because its definition is simply unfolded and the lemmas (6)–(9) about $\uparrow$ apply.

The litmus tests for the lemma collection are the correctness proofs for the map-variants of all the search trees discussed in Sect. 5. The code of the map-variants is structurally the same as their set-counterparts. The same is true for the lemmas required in the verification. In the end, the proofs of the map-variants are just as automatic as the ones of their set-counterparts.

$\uparrow_1 (ps \ @ \ [(a, \ b)]) \Longrightarrow upd_{list} \ x \ y \ (ps \ @ \ (a, \ b) \cdot qs) =$
$(if \ x < a \ then \ upd_{list} \ x \ y \ ps \ @ \ (a, \ b) \cdot qs \ else \ ps \ @ \ upd_{list} \ x \ y \ ((a, \ b) \cdot qs))$

$\uparrow_1 (ps \ @ \ (a, \ b) \cdot qs) \Longrightarrow del_{list} \ x \ (ps \ @ \ (a, \ b) \cdot qs) =$
$(if \ x < a \ then \ del_{list} \ x \ ps \ @ \ (a, \ b) \cdot qs \ else \ ps \ @ \ del_{list} \ x \ ((a, \ b) \cdot qs))$

$map\_of \ (ps \ @ \ qs) \ x =$
$(case \ map\_of \ ps \ x \ of \ None \Rightarrow map \ of \ qs \ x \ | \ Some \ y \Rightarrow Some \ y)$
$\uparrow (a \cdot map \ fst \ ps) \wedge x < a \Longrightarrow map\_of \ ps \ x = None$
$\uparrow (map \ fst \ ps \ @ \ [a]) \wedge a \leq x \Longrightarrow map\_of \ ps \ x = None$

Fig. 8. Lemmas for upd_{list}, del_{list} and map_of

7   Conclusion

Our proof method works well because all the trees we considered follow the same
ordering principle: inorder traversal yields a sorted list. Two referees suspected
that for Trie-like trees [7] it would not work so well. I formalized binary trees
where nodes are addressed by bit lists indicating the path to the node. A direct
correctness proof is easy. The methods of this paper can also be applied (the list
of addresses of the nodes in a tree, in prefix order, is lexicographically ordered)
but the proof is more complicated and less automatic. Our approach seems
overkill and awkward for such search trees.

Acknowledgement. Daniel Stüwe found and corrected the two invariant-related bugs
in AA trees and proved preservation of the invariant under deletion for AA trees and
1-2 Brother trees.

References

1. Andersson, A.: Balanced search trees made simple. In: Dehne, F., Sack, J.-R., Santoro, N. (eds.) WADS 1993. LNCS, vol. 709, pp. 60–71. Springer, Heidelberg (1993)
2. Appel, A.: Efficient verified red-black trees (2011)
3. Bayer, R.: Symmetric binary B-trees: Data structure and maintenance algorithms. Acta Inform. **1**, 290–306 (1972)
4. Clochard, M.: Automatically verified implementation of data structures based on AVL trees. In: Giannakopoulou, D., Kroening, D. (eds.) VSTTE 2014. LNCS, vol. 8471, pp. 167–180. Springer, Heidelberg (2014)
5. Cormen, T.H., Leiserson, C.E., Rivest, R.L.: Introduction to Algorithms. MIT Press, Cambridge (1990)
6. Filliâtre, J.-C., Letouzey, P.: Functors for proofs and programs. In: Schmidt, D. (ed.) ESOP 2004. LNCS, vol. 2986, pp. 370–384. Springer, Heidelberg (2004)
7. Fredkin, E.: Trie memory. Commun. ACM **3**(9), 490–499 (1960)
8. Guibas, L.J., Sedgewick, R.: A dichromatic framework for balanced trees. In: 19th Annual Symposium on Foundations of Computer Science, pp. 8–21. IEEE Computer Society (1978)

9. Haftmann, F., Krauss, A., Kunčar, O., Nipkow, T.: Data refinement in Isabelle/HOL. In: Blazy, S., Paulin-Mohring, C., Pichardie, D. (eds.) ITP 2013. LNCS, vol. 7998, pp. 100–115. Springer, Heidelberg (2013)
10. Hinze, R.: Purely functional 1-2 brother trees. J. Funct. Program. **19**(6), 633–644 (2009)
11. Hoare, C.: Proof of correctness of data representations. Acta Inform. **1**, 271–281 (1972)
12. Hoffmann, C.M., O'Donnell, M.J.: Programming with equations. ACM Trans. Program. Lang. Syst. **4**(1), 83–112 (1982)
13. Jones, C.B.: Software Development. A Rigourous Approach. Prentice Hall, London (1980)
14. Kahrs, S.: Red black trees with types. J. Funct. Program. **11**(4), 425–432 (2001)
15. Martí-Oliet, N., Palomino, M., Verdejo, A.: A tutorial on specifying data structures in Maude. Electr. Notes Theor. Comput. Sci. **137**(1), 105–132 (2005)
16. Nipkow, T.: Are homomorphisms sufficient for behavioural implementations of deterministic and nondeterministic data types? In: Brandenburg, F.J., Wirsing, M., Vidal-Naquet, G. (eds.) STACS 1987. LNCS, vol. 247, pp. 260–271. Springer, Heidelberg (1987)
17. Nipkow, T.: Splay tree. Archive of Formal Proofs, Formal proof development, August 2014. http://isa-afp.org/entries/Splay_Tree.shtml
18. Nipkow, T.: Amortized complexity verified. In: Urban, C., Zhang, X. (eds.) ITP 2015. LNCS, vol. 9236, pp. 310–324. Springer, Heidelberg (2015)
19. Nipkow, T., Klein, G.: Concrete Semantics with Isabelle/HOL. Springer (2014). http://concrete-semantics.org
20. Nipkow, T., Kunčar, O., Pusch, C.: AVL trees. Archive of Formal Proofs, Formal proof development, March 2004. http://isa-afp.org/entries/AVL-Trees.shtml
21. Nipkow, T., Paulson, L.C., Wenzel, M. (eds.): Isabelle/HOL. A Proof Assistant for Higher-Order Logic. LNCS, vol. 2283. Springer, Heidelberg (2002)
22. Okasaki, C.: Purely Functional Data Structures. Cambridge University Press, Cambridge (1998)
23. Ottmann, T., Six, H.W.: Eine neue Klasse von ausgeglichenen Binärbäumen. Angewandte Informatik **18**(9), 395–400 (1976)
24. Ottmann, T., Wood, D.: 1-2 brother trees or AVL trees revisited. Comput. J. **23**(3), 248–255 (1980)
25. Ragde, P.: Simple balanced binary search trees. In: Caldwell, J., Hölzenspies, P., Achten, P. (eds.) Trends in Functional Programming in Education, EPTCS, vol. 170, pp. 78–87 (2014)
26. Ralston, R.: ACL2-certified AVL trees. In: Proceedings of 8th International Workshop ACL2 Theorem Prover and its Applications, pp. 71–74. ACM (2009)
27. Reade, C.: Balanced trees with removals: an exercise in rewriting and proof. Sci. Comput. Program. **18**(2), 181–204 (1992)
28. Sleator, D.D., Tarjan, R.E.: Self-adjusting binary search trees. J. ACM **32**(3), 652–686 (1985)
29. Turbak, F.: CS230 Handouts – Spring 2007 (2007). http://cs.wellesley.edu/cs230/spring07/handouts.html
30. Weiss, M.A.: Data Structures and Algorithm Analysis, 2nd edn. Benjamin/Cummings, Redwood City (1994)

A Framework for the Automatic Formal Verification of Refinement from COGENT to C

Christine Rizkallah[1,2]([⊠]), Japheth Lim[1,2], Yutaka Nagashima[1],
Thomas Sewell[1,2], Zilin Chen[1,2], Liam O'Connor[1,2], Toby Murray[1,2],
Gabriele Keller[1,2], and Gerwin Klein[1,2]

[1] Data61 (formerly NICTA), Sydney, Australia
[2] UNSW, Sydney, Australia
christine.rizkallah@data61.csiro.au

Abstract. Our language COGENT simplifies verification of systems software using a certifying compiler, which produces a proof that the generated C code is a refinement of the original COGENT program. Despite the fact that COGENT itself contains a number of refinement layers, the semantic gap between even the lowest level of COGENT semantics and the generated C code remains large.

In this paper we close this gap with an automated refinement framework which validates the compiler's code generation phase. This framework makes use of existing C verification tools and introduces a new technique to relate the type systems of COGENT and C.

1 Introduction

In previous work, we designed a new language called COGENT [9] for easing the verification of certain classes of systems code such as file systems. COGENT is a linearly-typed, pure, polymorphic, functional language with a *certifying* compiler. We used it in separate work to write two Linux filesystems, ext2 and BilbyFs, and achieved performance comparable to their native C implementations [2].

From a COGENT program the COGENT compiler produces three artefacts: C code, a shallow embedding of the COGENT program in Isabelle/HOL [8], and an Isabelle/HOL proof relating the two. The compiler certificate is a series of

NICTA is funded by the Australian Government through the Department of Communications and the Australian Research Council through the ICT Centre of Excellence Program.

This material is based on research sponsored by Air Force Research Laboratory and the Defense Advanced Research Projects Agency (DARPA) under agreement number FA8750-12-9-0179. The U.S. Government is authorised to reproduce and distribute reprints for Governmental purposes notwithstanding any copyright notation thereon. The views and conclusions contained herein are those of the authors and should not be interpreted as necessarily representing the official policies or endorsements, either expressed or implied, of Air Force Research Laboratory, the Defense Advanced Research Projects Agency or the U.S. Government.

© Springer International Publishing Switzerland 2016
J.C. Blanchette and S. Merz (Eds.): ITP 2016, LNCS 9807, pp. 323–340, 2016.
DOI: 10.1007/978-3-319-43144-4_20

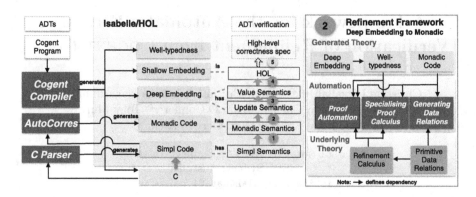

Fig. 1. An overview of the verification chain and our refinement framework.

language-level proofs and per-program translation validation phases that are combined into one top-level theorem in Isabelle/HOL. The most involved phase, and the phase we discuss in this paper, is the translation validation phase relating COGENT's imperative semantics to the generated C.

We present a refinement framework that enables the full automation of this phase of COGENT's certifying compilation. This framework has several components that relate COGENT values, states, types, and statements to their C counterparts. We put significant proof engineering work into enabling the framework to bridge the gap between the COGENT store and the C heap semantics. Moreover, we introduced the idea of *partial type erasure* to eliminate linearity information from a COGENT type in order to relate it to the corresponding C type. Furthermore, to relate COGENT and C statements, we developed a refinement calculus which contains a set of compositional proof rules. Given a program, our framework then customises the proof rules based on the values, types, and states that are used in this program. Finally, our refinement tactic applies the customised rules in a syntax-directed manner, certifying the refinement for this phase.

The method scales to significant COGENT code size, as demonstrated in the two Linux filesystems [2] mentioned above. A snapshot of our work is available online [1].

2 Overview and Background

This section explains the contribution of this paper within the broader COGENT project. The heart of the COGENT project is its certifying compiler. The certificate the compiler produces is a refinement theorem relating the generated shallow embedding and the generated C code. To ensure the C code is run correctly on the binary level, it can be compiled by CompCert [7].[1] It also falls into

---

[1] Mind the potential logical gap between our C parser's C semantics [13] and that of CompCert.

the subset of Sewell *et al.*'s gcc translation validator [12], which can be made to compose directly with our compiler certificate.[2]

The shallow Isabelle/HOL embedding is convenient for manual reasoning; however, the compiler additionally produces a deep embedding of each COGENT program, for the sake of structuring the generated certificate theorem and proof. There are two formal semantics for this deep embedding: (1) a functional *value semantics* where programs evaluate to values and (2) an imperative *update semantics* where programs manipulate references to mutable global state.

The left side of Fig. 1 summarises the generated program representations and the breakdown of the compiler certificate. The program representations are (from the bottom of Fig. 1): the C code, the semantics of the C code [13] expressed in Simpl [11], which is a generic imperative language inside Isabelle/HOL, the same expressed as a monadic program [4], an A-normal [10] deep embedding of the COGENT program, and a shallow embedding. Several theorems rely on the COGENT program being well-typed, which we prove automatically using type inference information from the compiler.

The labelled arrows and the arrow from C to Simpl represent refinement proofs and the arrow labels correspond to the numbers in the following description. The only arrow that is not verified is the one crossing from C code into Isabelle/HOL at the bottom of Fig. 1 — this is the C parser [13], which is a mature tool used in a number of large-scale verifications [5]. It could additionally be checked by Sewell *et al.*'s gcc translation validation tool.

We briefly describe each intermediate theorem, starting with Simpl at the bottom. For well-typed COGENT programs, we automatically prove the following four theorems, which together form the compiler certificate:

① The C parser's Simpl code corresponds to a monadic representation of the C code.

② The monadic code terminates and is a refinement of the update semantics of the COGENT deep embedding. To relate COGENT's linear type system to the monadic one, we introduce the reusable idea of *partial type erasure*.

③ If a COGENT deep embedding evaluates in the update semantics, it evaluates to the same result in the value semantics.

④ If the COGENT deep embedding evaluates in the value semantics then the COGENT shallow embedding evaluates to a corresponding shallow Isabelle/HOL value.

In order to prove high-level functional correctness, an additional step is necessary:

Arrow ⑤ indicates verification of user-supplied abstract data types (ADTs) implemented in C and manual high-level proofs on top of the shallow embedding. We demonstrated that this step is enabled by the previous steps for two real-world filesystems [2].

Step ③ is a consequence of linear types. It is a general property about the language and has been proven manually once and for all [9]. Steps ①, ②, and ④,

---

[2] COGENT's occasionally larger stack frames lead to `memcpy()` calls that, while conceptually straightforward, the translation validator does not yet cover.

as well as their respective proofs, are generated by our compiler for every program. The proof for step ① is generated by an adjusted version of the AutoCorres tool [4]. For steps ② and ④ we define compositional refinement calculi which enable the automation of the proofs. The most involved refinement proof is the one for step ② which we present in this paper. It took about three person years to develop tools for automating this proof. The calculus for step ④ is similar but much simpler, as at this stage one does not reason about the state. In comparison, its development only took a few person weeks.

The right side of Fig. 1 expands on the refinement framework used for proving step ②. The bottom layer represents the underlying theory we developed for defining primitive value and type relations which we use to create a refinement calculus between COGENT deeply embedded expressions and corresponding monadic statements. The middle layer represents the proof tools that automate the refinement proof on a per-program basis. These proof tools rely on the underlying theories about the language in general, and on compiler generated theories specific to the program. In particular, we have a tool for generating non-primitive data relations, one for specialising complex rules in the calculus to support automation, and finally a proof automation tactic which composes the proof rules to provide a fully-automatic refinement proof.

2.1 COGENT

COGENT is a restricted, polymorphic, higher-order, and purely functional language with *linear types*. The linear types ensure that resources such as memory are disposed of correctly without run-time support like garbage collection. Crucially for us, they also allow COGENT to be compiled into efficient C, including *destructive updates* to values rather than the repeated copying common in purely functional styles.

Variables of linear type must be used *exactly once*. This means each active mutable heap object has exactly one active pointer in scope at any point in the program. Hence, the difference between a destructive update and a pure copy-update is unobservable.

The COGENT compiler generates C code, a shallow embedding, and a collection of "hints" used by the proof tactic to certify the compilation. Importantly, the performance of the generated C is comparable to carefully handwritten C.

COGENT's certifying compilation makes the verification of filesystems more cost-effective, fully automating a significant part of the low-level proofs. We demonstrate this on two real-world COGENT filesystems, with a minimal TCB [2].

This paper focuses on the lower-level generated refinement proofs, which connect COGENT's *update semantics* to C. Figure 2 introduces a relevant fragment of COGENT. Many features of the full language are omitted here and described in detail elsewhere [9], including polymorphism, sum types, and the foreign function interface. The following gives a brief summary.

Much of the syntax presented in our fragment is standard for a functional language, such as handling control flow (**if**) and local bindings (**let**). The main point of difference is COGENT's record system: Some care is needed to reconcile

$$\boxed{\text{Statics}}$$

primops	o	$\in \{+, *, /, <=, ==, ||, <<, \ldots\}$
literals	ℓ	$\in \{123, \text{True}, \text{'a'}, \ldots\}$
expressions	e	$::= x \mid () \mid f \mid o(\overline{e}) \mid e_1\, e_2 \mid \ell$
		$\mid \text{let } x = e_1 \text{ in } e_2 \mid \text{if } e_1 \text{ then } e_2 \text{ else } e_3$
		$\mid \{\overline{f = e}\} \mid \text{put } e_1.f := e_2 \mid \text{take } x\, \{f = y\} = e_1 \text{ in } e_2 \mid \cdots$
fn. names	$\ni$	f, g
variables	$\ni$	x, y
record fields	$\ni$	f, g
prim. types	t	$::= \text{U8} \mid \text{U16} \mid \text{U32} \mid \text{U64} \mid \text{Bool}$
types	τ	$::= \alpha \mid t \mid \tau_1 \rightarrow \tau_2 \mid () \mid \{\overline{f :: \tau^?}\}\, m \mid \cdots$
field types	$\tau^?$	$::= \tau \mid \mathsf{T}$
modes	m	$::= \mathtt{r} \mid \mathtt{w} \mid \mathtt{u}$

$$\boxed{\text{Dynamics}}$$

update semantics values	u	$::= \ell\, r \mid () \mid \langle \lambda x.\, e \rangle \mid \{\overline{f = u}\}\, r \mid p\, r \mid \cdots$
type representations	r	$::= t \mid () \mid \text{Fun} \mid \{\overline{f :: r}\} \mid \text{Ptr}\, r \mid \cdots$
environments	U	$::= \overline{x \mapsto u}$ pointers p stores $\mu : p \rightharpoonup u$

Fig. 2. Definitions for COGENT fragment

record types and linear types. If a record contains at least one linear field, the whole record is of linear type. Otherwise, the linear field could be shared by sharing the record.

Accessing records becomes more complex as well. For instance, assume that Object is a type synonym for a record type containing an integer and two (linear) buffers, where Object $= \{\text{size} :: \text{U32}, b_1 :: \text{Buf}, b_2 :: \text{Buf}\}\, \mathtt{u}$. Let us say we want to extract the field b_1 from an Object. If we extract just a single Buf, we have implicitly discarded the other buffer b_2. However, we cannot return the entire Object along with Buf, as this would introduce aliasing. Our solution is to return along with Buf an Object where the field b_1 cannot be extracted again, and reflect this in the field's type, written as $b_1 :: \overline{\text{Buf}}$. This field extractor, whose general form is **take** $x\,\{f = y\} = e_1$ **in** e_2, operates as follows: given a record e_1, it binds the field f of e_1 to the variable y, and the new record to the variable x in e_2. If that field is linear, it will then be marked as unavailable, or *taken*, in the type of the new record x.

Conversely, we also introduce a **put** operation, which, given a record with a taken field, allows a new value to be supplied in its place. The expression **put** $e_1.f := e_2$ returns the record in e_1 where the field f has been replaced with the result of e_2. Unless the type of the field f allows it to be discarded, it must already be taken, to avoid accidentally destroying our only reference to a linear resource.

```
1 flip :: {f :: U8} w → {f :: U8} w
2 flip x =
3    take x' {f = y} = x
4    in if y == 0
5       then put x'.f := 1
6       else put x'.f := 0
```

Fig. 3. Example function in COGENT. *flip* updates a record on the heap in place.

We distinguish boxed records stored on the heap from unboxed records that are passed by value. Unboxed records can be created using a simple struct literal $\{\overline{f_i = e_i}\}$. Boxed records are created by invoking an externally-defined C allocator function. For these allocation functions, it is often convenient to allocate a record with all fields already taken, to indicate that they are uninitialised. That is, a function for allocating Object-like records might return values of type: $\{\text{size} :: \overline{\text{U32}}, b_1 :: \overline{\text{Buf}}, b_2 :: \overline{\text{Buf}}\}$ w.

Also included in a record type is the *storage mode* of the type. A record is stored on the heap when its associated mode m is not unboxed. For boxed records, the storage mode distinguishes between those that are writable vs. read-only.

Example 1. Figure 3 defines a simple function in COGENT which, given a mutable record x, first **take**s the field f and, depending on its value, destructively updates the field with a new value, returning the updated record.

The details of COGENT's type system, semantics, and this proof are presented in [9], we only repeat the top-level concepts here.

The dynamic big step *update* semantics maps a triple of environment U, expression e, and mutable store μ to a result value u and a new mutable environment μ', written $U \vdash e \mid \mu \Downarrow_u u \mid \mu'$. The rules [9] for variables and **let** are straightforward. Functions are top-level functions in COGENT, and a function name simply evaluates to the lambda-expression it represents. The **take** and **put** rules evaluate as described above.

The static semantics include the standard typing judgement $\Gamma \vdash e : \tau$. Unlike conventional type systems, linear type systems are *substructural*, which means that the context Γ cannot be treated merely as a set of assumptions that always grows as one descends into the syntax tree. Instead, assumptions may also be removed from the context. This complication requires us to occasionally generalise the **corres** rules presented in Sect. 3.4 with multiple typing assumptions with different contexts.

To state *type preservation* for COGENT, we define the corresponding typing judgement for dynamic *values*, written $u \mid \mu : \tau$ and a generalisation of it to *environments* and *contexts*, written $U \mid \mu : \Gamma$. With this, we can prove the following (see also [9]).

Theorem 1 (type preservation). *For a program e, if $\Gamma \vdash e : \tau$ and $U \mid \mu : \Gamma$ and $U \vdash e \mid \mu \Downarrow_u u \mid \mu'$, then $u \mid \mu' : \tau$*

$$\boxed{\textbf{repr}(\cdot) : \tau \to r}$$

$\textbf{repr}(()) = ()$ $\quad$ $\textbf{repr}(\{\overline{\texttt{f} :: \tau^?}\}\ \texttt{u}) = \{\overline{\texttt{f} :: \textbf{repr}(\tau)}\}$ $\qquad$ $\textbf{repr}(\tau \to \rho)$ $\quad = \texttt{Fun}$

$\textbf{repr}(t)\ \ = t$ $\quad$ $\textbf{repr}(\{\overline{\texttt{f} :: \tau^?}\}\ \texttt{r}) = \texttt{Ptr}\ \{\overline{\texttt{f} :: \textbf{repr}(\tau)}\}$ $\qquad$ $\textbf{repr}(\{\overline{\texttt{f} :: \tau^?}\}\ \texttt{w}) = \texttt{Ptr}\ \{\overline{\texttt{f} :: \textbf{repr}(\tau)}\}$

Fig. 4. Partial type erasure of dynamic typing relation for update semantics

For a COGENT value to be well-typed, all accessible pointers in this value, e.g. a record, must be valid. This is important for proving safety, but becomes cumbersome when showing refinement to C as there exist values in the C code, such as those for taken fields, which may include temporarily invalid pointers. We therefore include additional information in each COGENT value, called its *representation*, which provides enough type information to determine the corresponding C type, without requiring recursive descent into the heap. In other words, the representation shown in Fig. 4 contains only the type information which is pertinent to C, with the linearity information erased. We call this technique *partial type erasure*. The value typing relation ensures that the representation information agrees with the value's type.

2.2   AutoCorres and C Monads in Isabelle/HOL

We use the C-to-Simpl [13] parser to provide a formal semantics for the generated C code. In principle, we could work from the C parser's output directly; however, this would mean dealing with the details of its low-level memory model. Instead, we opt to work with a *typed* heap model, provided by AutoCorres [4]. Specifically, the *state* of the AutoCorres monadic representation contains a set of *typed heaps*, each of type τ ptr $\Rightarrow \tau$, one for each type τ used on the heap in the C input program.

As AutoCorres was designed for human-guided verification, it uses many context-sensitive rules to simplify the generated code. As we aim to verify code automatically, we switch off most of these simplification stages in order to obtain predictable output.

AutoCorres generates shallow embeddings of code in the *nondeterministic state monad* of Cock *et al.* [3]. In this monad, computation is represented by functions of type *state* $\Rightarrow (\alpha \times$ *state*$)$ *set* $\times$ *bool*. Here *state* is the global state of the monadic program, including global variables, while α is the return-type of the computation. A computation takes as input the global state and returns a set, *results*, of pairs with new state and result value. Additionally the computation returns a boolean, *failed*, indicating whether there potentially was undefined behaviour.

As C does not guarantee that all pointer locations are valid, AutoCorres emits *is-valid* guards before each memory access. When proving refinement between COGENT and monadic code, we need to discharge those guards using a state invariant (Sect. 3.2).

Figure 5 shows an example AutoCorres specification, using the following keywords:

do ...; ... **od**   sequence of statements
condition *cond* e_1 e_2 run e_1 if *cond* is true, otherwise run e_2
return v monadic return
gets f access part of monadic state given by f
modify h update part of monadic state given by h
guard G program fails if monadic state does not satisfy G

3 Refinement Framework

Recall that for a well-typed COGENT program, the compiler emits C code, a deep embedding of the program's semantics, and a proof that the C code correctly refines this embedding. We choose C as a compilation target because most existing systems code is written in C, and thanks to tools like CompCert and gcc translation validation, our C subset has formalised semantics and an existing formal verification infrastructure.

The right side of Fig. 1 provides an overview of the generation of our refinement proof. To phrase the refinement statement, we first define how deeply-embedded COGENT values relate to values in the monadic embedding (Sect. 3.2).

The C code generation is straightforward and this step itself does not perform global optimisations or transformations. Such transformations, for instance A-normalisation, are performed in earlier compiler phases. A-normalisation in particular is performed to simplify code generation, but it also simplifies our C refinement. Since it is performed early (and verified early on top of the shallow embedding [9]), it is sufficient for us to only consider COGENT expressions in A-normal form here, where nested subexpressions are replaced with explicit

```
1 flip :: {f :: U8} → {f :: U8}
2 flip x =
3   take x' {f = y} = x
.
4   in let tmp₁ = 0
5     and tmp₂ = (y == tmp₁)
6   in if tmp₂
7     then let tmp₃ = 1
8       and x'' = put x'.f := tmp₃
.
9     in x''
10    else let tmp₄ = 0
11      and x'' = put x'.f := tmp₄
.
12    in x''
.
.
```

```
1 flip_C :: rec₁ ptr ⇒ (rec₁ ptr, σ) nondet_monad
2 flip_C x = do
3   guard (λσ. is-valid σ x);
4   y ← gets (λσ. σ[r].f);
5   tmp₁ ← return 0;
6   tmp₂ ← return bool (y = tmp₁);
7   tmp_result ← condition (bool tmp₂ ≠ 0)
8     (do tmp₃ ← return 1;
9       guard (λσ. is-valid σ x);
10      modify (λσ. σ[x].f := tmp₃);
11      return x od)
12    (do tmp₄ ← return 0;
13      guard (λσ. is-valid σ x);
14      modify (λσ. σ[x].f := tmp₄);
15      return x od);
16  return tmp_result
17 od
```

Fig. 5. Intermediate representations of COGENT function from Fig. 3. Left: A-normalised source code, embedded into Isabelle/HOL. Right: AutoCorres monadic semantics for generated C code.

variable bindings. With this, the refinement calculus contains a set of compositional **corres** proof rules, typically one for each A-normal COGENT construct, which are applied automatically in a syntax-directed manner (Sect. 3.4).

The **corres** proof rules depend on preconditions about the expected state of the program, such as preconditions about the type and validity of pointers in the heap. We propagate the conditions similarly to the proof calculus of Cock *et al.* [3]. Our refinement theorem does not need an explicit assumption of well-typedness for the whole COGENT program — The proof tactic will simply fail for programs that are ill-typed.

Since our **corres** proof rules are specialised to COGENT and to the operation of the compiler, we can predict the form of their preconditions and design proof rules to combine them. This forms the basis for automation.

3.1 Refinement Statement

We define refinement generically between a monadic computation p_m and a COGENT expression e, evaluated under the update semantics. We denote the refinement predicate **corres**. The state relation R changes for each COGENT program, so we parametrise **corres** by an arbitrary state relation R. It is additionally parametrised by the typing context Γ and the environment U, as well as by the initial update semantics store μ and monadic shallow embedding state σ.

Definition 1 (correspondence)

$$\textbf{corres } R\, e\, p_m\, U\, \Gamma\, \mu\, \sigma \overset{def}{=}$$
$$U \mid \mu : \Gamma \longrightarrow (\mu, \sigma) \in R \longrightarrow$$
$$(\neg\, failed\, (p_m\, \sigma) \wedge$$
$$(\forall v_m\, \sigma'.\, (v_m, \sigma') \in results\, (p_m\, \sigma) \longrightarrow$$
$$(\exists \mu'\, u.\, U \vdash e \mid \mu \Downarrow_u u \mid \mu' \wedge (\mu', \sigma') \in R \wedge val\text{-}rel\, u\, v_m)))$$

Definition 1 states for well-typed stores μ that if the state relation R holds initially, then the monadic computation p_m cannot fail and, moreover, for all executions of p_m there must exist a corresponding execution under the update semantics of the expression e such that the final states are related by a state relation R and a value relation $val\text{-}rel$ holds between the results of e and p_m.[3] We present the state and value relations in Sect. 3.2.

AutoCorres proves that if the monadic code never fails, then the C code is type- and memory-safe, and is free of undefined behaviour [4]. We prove non-failure as a side-condition of the refinement statement, essentially using COGENT's type system to guarantee C memory safety during execution. The **corres** predicate can compose with itself sequentially: it both assumes and shows the relation R, and the additional typing assumptions are preserved thanks to type preservation (Theorem 1).

---

[3] Although **corres** technically permits the monadic code to return no results, the code that we generate will additionally always return $results \neq \emptyset$ as long as it has not *failed*.

3.2   Data Relations

For each program, based on a library for primitive types, we generate a set of relations between the values, types and heaps of the COGENT and monadic code. We denote these as *val-rel*, *type-rel* and $\mathcal{R}$ respectively.

We must give these relations separate definitions for each COGENT type, because each C struct type is embedded as a distinct Isabelle/HOL record. We use Isabelle's ad-hoc overloading mechanism for this.

Recall that AutoCorres generates different typed heaps for each C type. The type relation *type-rel* is used by the state relation $\mathcal{R}$ to select the corresponding typed heap for each COGENT type. It is defined using the **repr** function (Fig. 4) which performs *partial type erasure*, unifying COGENT types that differ only in linear annotations in order to relate them to the same C type.

Given *val-rel* and *type-rel* for a particular COGENT program, the *state relation* $\mathcal{R}$ defines the correspondence between the store μ over which the COGENT update semantics operates, and the state σ of the monadic shallow embedding. This relation is made into an invariant in **corres** (Sect. 3.1); it allows us to show that all C pointer accesses satisfy *is-valid*, whenever there are corresponding objects in the COGENT store μ.

Definition 2 (state relation). $(\mu, \sigma) \in \mathcal{R}$ *if and only if for all pointers p in the domain of μ, there exists a value v in the appropriate heap of σ (as defined by type-rel) at location p, such that val-rel $(\mu\, p)\ v$ holds.*

Generating Data Relations. We generate $\mathcal{R}$, *val-rel* and *type-rel* after obtaining the monadic program and its typed heaps from AutoCorres. Our COGENT compiler outputs a list of (COGENT, C) type pairs, which is used by an Isabelle/ML procedure to generate the needed relations.

Example 2. The program in Fig. 5 uses the types U8, Bool and {f :: U8}, which correspond to the C types word8, bool and rec_1, respectively. For *val-rel* and *type-rel*, the U8–word8 relation can be defined a priori, but bool and rec_1 are generated with the monadic program and their data relations are generated dynamically:

$$\text{(pre-defined)} \quad val\text{-}rel\ (a :: \text{U8})\ (a_C :: \text{word8}) \stackrel{\text{def}}{=} (a = a_C)$$
$$val\text{-}rel\ (a :: \text{Bool})\ (a_C :: \text{bool}) \stackrel{\text{def}}{=} (a = (bool\ a_C \neq 0))$$
$$val\text{-}rel\ (a :: \{\text{f} :: \text{U8}\})\ (a_C :: rec_1) \stackrel{\text{def}}{=} val\text{-}rel\ (a.\text{f})\ (a_C.\text{f})$$

Note that the *val-rel* definition for {f :: U8} depends on the definition for its field of type U8. The COGENT compiler always outputs the type list in dependency order, so this does not pose a problem.

The state relation $\mathcal{R}$ cannot be overloaded in the same way as *val-rel* and *type-rel*, because it relates the heaps for every type simultaneously. We introduce an intermediate state relation, *heap-rel*, which relates a particular typed heap with a portion of the COGENT store. Like the other relations, this intermediate relation can make use of type-based overloading. Following Definition 2, we define *heap-rel* for each type τ that appears on the heap as follows:

$$\textit{heap-rel } \sigma_\tau \ \mu \overset{\text{def}}{=} \forall p.\ \mu(p) \mapsto v \wedge \textit{type-rel } (\mathbf{vrepr}(v)) \ \tau \longrightarrow$$
$$\textit{is-valid } \sigma_\tau \ p \wedge \textit{val-rel } v \ \sigma_\tau[p]$$

where **vrepr** gives the partially-erased type for a value, similar to **repr**. The state relation over all typed heaps σ_{τ_k} is $\mathcal{R} \ \sigma \ \mu \overset{\text{def}}{=} (\textit{heap-rel } \sigma_{\tau_1} \ \mu \wedge \textit{heap-rel } \sigma_{\tau_2} \ \mu \wedge \ldots)$.

3.3 Refinement Theorem

We state the overall top-level C refinement theorem below. In addition to the assumptions listed here, it also assumes that **corres** holds for all the foreign functions used in the program.

Theorem 2. *Let f be a* COGENT *function, with type τ and body e. Let p_m be the monadic embedding of its generated C code. Let u and v_m be arguments of appropriate type for f and p_m respectively. Then:*

$$\forall \mu \ \sigma. \ \textit{val-rel } u \ v_m \longrightarrow \mathbf{corres} \ \mathcal{R} \ e \ (p_m \ v_m) \ (x \mapsto u) \ (x : \tau) \ \mu \ \sigma$$

Example 3. In Fig. 5, $f = flip$, $p_m = flip_C$, and $\tau = \tau' = \{\mathsf{f} :: \mathsf{U8}\}$.

3.4 Refinement Proof

This section describes the main components of the refinement proof automation, as shown in Fig. 1: the proof calculus used to relate COGENT and C programs, the generation of well-typedness theorems for COGENT, and the automated tactic that combines these two components to perform the overall refinement proof.

Refinement Calculus. Figure 6 depicts the **corres** rules in our calculus for variables, **let**, **if**, and for **take** and **put** expressions for boxed records. The full calculus is available online [1] under c − refinement/COGENT_Corres.thy. The proofs of the **corres** rules for compound expressions rely on Theorem 1 to infer value well-typedness.

The assumptions for these rules fall under three main groups:

1. Well-typedness assumptions; we generate typing theorems to discharge these.
2. Assumptions relating the values and mutable heaps of COGENT and C. Once a C program is read and concrete data relations (Sect. 3.2) are defined, we *specialise* the **corres** rules to simplify these assumptions.
3. **corres** assumptions on sub-expressions, discharged through our proof automation.

The rules VAR and LET correspond respectively to the two basic monadic operations **return**, which yields values, and **do** ...; ... **od**, for sequencing computations.

$$\frac{(x \mapsto v_u) \in U \qquad \textit{val-rel } v_u \, v_m}{\textbf{corres } R \; x \; (\textbf{return } v_m) \; U \; \Gamma \; \mu \; \sigma} \text{Var}$$

$$\frac{\begin{array}{c} \Gamma_1 \vdash e_1 : \tau \quad \textbf{corres } R \, e_1 \, e_1' \; U \, \Gamma_1 \, \mu \, \sigma \\ (\forall v_u \, v_m \, \mu' \, \sigma'. \; \textit{val-rel } v_u v_m \longrightarrow \textbf{corres } R \, e_2 \, (e_2' \, v_m) \, (x \mapsto v_u, U) \, (x : \tau, \Gamma_2) \, \mu' \sigma') \end{array}}{\textbf{corres } R \, (\textbf{let } x = e_1 \textbf{ in } e_2) \, (\textbf{do } e_1'; \, e_2' \textbf{ od}) \; U \, (\Gamma_1 \Gamma_2) \, \mu \, \sigma} \text{Let}$$

$$\frac{\begin{array}{c} \Gamma_1 \vdash c : \textbf{Bool} \quad (\textit{bool } c' = 0 \lor \textit{bool } c' = 1) \quad c \text{ is a Cogent boolean equal to } (\textit{bool } c' \ne 0) \\ \textbf{corres } R \, e_1 \, e_1' \; U \, \Gamma_2 \, \mu \, \sigma \qquad \textbf{corres } R \, e_2 \, e_2' \; U \, \Gamma_2 \, \mu \, \sigma \end{array}}{\begin{array}{c} \textbf{corres } R \, (\textbf{if } c \textbf{ then } e_1 \textbf{ else } e_2) \\ (\textbf{do } x \leftarrow \textbf{condition } (\textit{bool } c' \ne 0) \, e_1' \, e_2'; \textbf{return } x \textbf{ od}) \; U \, (\Gamma_1 \Gamma_2) \, \mu \, \sigma \end{array}} \text{If}$$

$$\frac{\begin{array}{c} \exists \tau. \, (\Gamma_1 \Gamma_2) \vdash (\textbf{let } x = \textbf{put } e_1.\mathbf{f}_k := e_2 \textbf{ in } e_3) : \tau \qquad (\Gamma_1 \Gamma_2) \text{ and } \Gamma_1 \vdash e_1 : \{\overline{\mathbf{f}_i :: \tau_i}, \mathbf{f}_k :: \tau_k^?\} \, \mathbf{w} \\ (\Gamma_1 \Gamma_2) \vdash (\textbf{put } e_1.\mathbf{f}_k := e_2) : \{\overline{\mathbf{f}_i :: \tau_i}, \mathbf{f}_k :: \tau_k\} \, \mathbf{w} \qquad (e_1 \mapsto p \, (\textbf{Ptr } r)) \in U \qquad (e_2 \mapsto u_k') \in U \\ (\forall \overline{u_i} \, u_k. \, (\mu, \sigma) \in R \longrightarrow \mu(p) = \{\overline{\mathbf{f}_i} = u_i, \mathbf{f}_k = u_k\} \, r \longrightarrow p \, (\textbf{Ptr } r) \, | \, \mu : \{\overline{\mathbf{f}_i :: \tau_i}, \mathbf{f}_k :: \tau_k^?\} \, \mathbf{w} \longrightarrow \\ p \, (\textbf{Ptr } r) \, | \, \mu(p := \{\overline{\mathbf{f}_i} = u_i, \mathbf{f}_k = u_k'\} \, r) : \{\overline{\mathbf{f}_i :: \tau_i}, \mathbf{f}_k :: \tau_k^?\} \, \mathbf{w} \longrightarrow \\ (\textit{is-valid } \sigma \, p' \land (\mu(p := \{\overline{\mathbf{f}_i} = u_i, \mathbf{f}_k = u_k'\} \, r), \, h \, \sigma) \in R)) \\ (\forall \mu', \sigma'. \, \textbf{corres } R \, e_3 \, e_3' \, (e_1 \mapsto p \, (\textbf{Ptr } r), U) \, (e_1 \mapsto \{\overline{\mathbf{f}_i :: \tau_i}, \mathbf{f}_k :: \tau_k\} \, \mathbf{w}, \Gamma_2) \, \mu' \, \sigma') \end{array}}{\begin{array}{c} \textbf{corres } R \, (\textbf{let } x = \textbf{put } e_1.\mathbf{f}_k := e_2 \textbf{ in } e_3) \\ (\textbf{do } \_ \leftarrow \textbf{guard } (\lambda \sigma. \, \textit{is-valid } \sigma \, p'); \, \_ \leftarrow \textbf{modify } h; \, e_3' \textbf{ od}) \; U \, (\Gamma_1 \Gamma_2) \, \mu \, \sigma \end{array}} \text{Put}$$

$$\frac{\begin{array}{c} (\Gamma_1 \Gamma_2) \vdash (\textbf{take } x \, \{\mathbf{f}_k = y\} = e_1 \textbf{ in } e_2) : \tau' \\ (\Gamma_1 \Gamma_2) \text{ and } \Gamma_1 \vdash e_1 : \{\overline{\mathbf{f}_i :: \tau_i}, \mathbf{f}_k :: \tau_k\} \, \mathbf{w} \qquad (e_1.\mathbf{f}_k \mapsto \tau_k, e_1 \mapsto \{\overline{\mathbf{f}_i :: \tau_i}, \mathbf{f}_k :: \tau_k^?\} \, \mathbf{w}, \Gamma_2) \vdash e_2 : \tau' \\ p' \text{ has a C pointer type} \qquad (e_1 \mapsto p \, (\textbf{Ptr } r)) \in U \qquad \textit{val-rel } (p \, (\textbf{Ptr } r)) \, p' \\ (\forall \overline{u_i}, u_k. \, (\mu, \sigma) \in R \longrightarrow \mu(p) = \{\overline{\mathbf{f}_i} = u_i, \mathbf{f}_k = u_k\} \, r \longrightarrow \\ p \, (\textbf{Ptr } r) \, | \, \mu : \{\overline{\mathbf{f}_i :: \tau_i}, \mathbf{f}_k :: \tau_k\} \, \mathbf{w} \longrightarrow \textit{is-valid } \sigma \, p' \land \textit{val-rel } u_k \, (f' \, \sigma)) \\ (\forall v_u \, v_m. \, \textit{val-rel } v_u \, v_m \longrightarrow \textbf{corres } R \, e_2 \, (e_2' \, v_m) \, (e_1.\mathbf{f}_k \mapsto v_u, e_1 \mapsto (p \, (\textbf{Ptr } r)), U) \\ (\mathbf{f}_k \mapsto \tau_k, e_1 \mapsto \{\overline{\mathbf{f}_i :: \tau_i}, \mathbf{f}_k :: \tau_k^?\} \, \mathbf{w}, \Gamma_2) \, \mu \, \sigma) \end{array}}{\begin{array}{c} \textbf{corres } R \, (\textbf{take } x \, \{\mathbf{f}_k = y\} = e_1 \textbf{ in } e_2) \\ (\textbf{do } \_ \leftarrow \textbf{guard } (\lambda \sigma. \, \textit{is-valid } \sigma \, p'); \, y' \leftarrow \textbf{gets } f'; \, e_2' \, y' \textbf{ od}) \; U \, (\Gamma_1 \Gamma_2) \, \mu \, \sigma \end{array}} \text{Take}$$

Fig. 6. Some of the important **corres** rules

Observe that Let is *compositional*: to prove that **let** $x = e_1$ **in** e_2 corresponds to **do** e_1'; e_2' **od**, we must prove that (1) e_1 corresponds to e_1' and (2) e_2 corresponds to e_2' when each are executed over corresponding results v_u and v_m (e.g. as yielded by e_1 and e_1' respectively). This compositionality, which is present in our whole calculus, significantly simplifies the automation of the refinement proof.

The If rule relates **if** c **then** e_1 **else** e_2 expressions to monadic **condition** $(\textit{bool } c' \ne 0) \, e_1' \, e_2'$ statements. It works similarly to Let, requiring an equivalence between c and $(\textit{bool } c' \ne 0)$, and correspondences between e_1 and e_1', and between e_2 and e_2'. Note that we represent booleans in C using a struct **bool** with an integer field named *bool*; we avoid C's builtin type _Bool because it may be an alias for an existing integer type like U8 and therefore indistinguishable from that integer type.

The more intricate rules in Fig. 6 are PUT and TAKE, which apply to **put** and **take** on boxed records (additional rules exist for unboxed records). Recall that boxed records are stored on the heap and are subject to the linear typing rules. These two rules are involved and contain many assumptions. They are mainly presented here to illustrate to the reader why we have a separate phase later on dedicated to simplifying them.

The PUT rule handles the correspondence between (**let** $x = $ **put** $e_1.f_k :=$ e_2 **in** e_3) expressions and (**do** _ $\leftarrow$ **guard** $(\lambda\sigma.\ \textit{is-valid}\ \sigma\ p')$; _ $\leftarrow$ **modify** h; e'_3 **od**) statements. Note that unlike **let**, **if**, and **take**, **put** does not contain a continuation. Therefore, the compiler ensures that **put** expressions always appear within **let** expressions, which allows us to have a compositional rule for **put** in the same style as the other operators.

Recall that if e_1 is a pointer p, **put** updates the field f_k, of the record pointed to by p to the value of e_2. Similarly, the monadic code asserts that the corresponding p' is a valid pointer, then modifies the record at p' in h. At this stage h and *is-valid* are left unspecified, as these rules are defined generically regardless of type. Therefore, our PUT rule additionally includes a number of assumptions describing the expected properties of h and *is-valid*. In the next subsection, we specialise this rule to eliminate these assumptions.

TAKE is similar, it relates (**take** $x\ \{f_k = y\} = e_1$ **in** e_2) expressions and

$$(\mathbf{do}\ \_ \leftarrow \mathbf{guard}\ (\lambda\sigma.\ \textit{is-valid}\ \sigma\ p');\ y' \leftarrow \mathbf{gets}\ f';\ e'\ y'\ \mathbf{od})$$

statements. Recall that **take** removes the field f_k from e_1, binds it to a new variable y and runs e_2. The **corres** assumptions of TAKE are that (1) p' and e_1's value are related, and (2) given related values v_u and v_m, e_2 corresponds to $e'_2\ v_m$ under the extended value environment $(f_k \mapsto v_u, e_1 \mapsto p\,(\mathtt{Ptr}\ r), U)$. We need to re-add e_1 to U because it is linear and cannot be reused.

Generating Specialised Rules. As mentioned earlier, we generate program-specific proof rules for operators involving specific C types, such as **take** and **put**. This is because the set of C types, different for each program, is shallowly embedded into Isabelle/HOL. Thus, the assumptions for rules involving those types can only be discharged once the C code has been parsed into Isabelle/HOL.

We could prove these assumptions while applying the **corres** rules, but this would be inefficient for rules that are applied many times. Thus, we generate specialised rules in a separate preprocessing phase. Implemented as an Isabelle/ML program, this phase reads the (COGENT, C) type list used for generating data relations to produce rules for the appropriate C and COGENT types.

Example 4. For the COGENT record $\{f :: \mathtt{U8}\}$ in Fig. 5, we generate the following specialised rules for **take** and **put**:

$$\frac{\begin{array}{c}(\Gamma_1\Gamma_2) \vdash (\textbf{take } x \ \{f \ = y\} = e_1 \textbf{ in } e_2) : \tau' \\ (\Gamma_1\Gamma_2) \vdash e_1 : \{f :: \text{U8}\} \ \texttt{w} \qquad (y \mapsto \text{U8}, x \mapsto \{f :: \cancel{\text{U8}}\} \ \texttt{w}, \Gamma_2) \vdash e_2 : \tau' \\ p' \text{ has type rec}_1 \text{ ptr} \qquad (e_1 \mapsto p \ (\texttt{Ptr } r)) \in U \qquad \textit{val-rel } (p \ (\texttt{Ptr } r)) \ p' \\ \textit{type-rel } (\textbf{repr}(\text{U8})) \ \texttt{word8} \qquad \textit{type-rel } (\textbf{repr}(\{f :: \cancel{\text{U8}}\} \ \texttt{w})) \ (\text{rec}_1 \text{ ptr}) \\ (\forall v_u \ v_m. \ \textit{val-rel } v_u \ v_m \longrightarrow \textbf{corres } \mathcal{R} \ e_2 \ (e_2' \ v_m) \ (y \mapsto v_u, \ x \mapsto (p \ (\texttt{Ptr } r)), \ U) \\ (y \mapsto \text{U8}, x \mapsto \{f :: \cancel{\text{U8}}\} \ \texttt{w}, \Gamma_2) \ \mu \ \sigma)\end{array}}{\begin{array}{c}\textbf{corres } \mathcal{R} \ (\textbf{take } x \ \{f \ = y\} = e_1 \textbf{ in } e_2) \\ (\textbf{do} \ \_ \leftarrow \textbf{guard } (\lambda\sigma. \ \textit{is-valid } \sigma \ p'); \ y' \leftarrow \textbf{gets } (\lambda\sigma. \ \sigma[p'].f); \ e' \ y' \textbf{ od}) \\ U \ (\Gamma_1\Gamma_2) \ \mu \ \sigma\end{array}} \text{ TAKE}$$

$$\frac{\begin{array}{c}\exists \tau. \ (\Gamma_1\Gamma_2) \vdash (\textbf{let } x = \textbf{put } e_1.f := e_2 \textbf{ in } e_3) : \tau \qquad (\Gamma_1\Gamma_2) \vdash e_1 : \{f :: \cancel{\text{U8}}\} \ \texttt{w} \\ \Gamma_1 \vdash (\textbf{put } e_1.f := e_2) : \{f :: \text{U8}\} \ \texttt{w} \qquad (e_1 \mapsto p \ (\texttt{Ptr } r)) \in U \qquad (e_2 \mapsto v) \in U \\ \textit{val-rel } (p \ (\texttt{Ptr } r)) \ p' \qquad \textit{type-rel } (\textbf{repr}(\{f :: \cancel{\text{U8}}\} \ \texttt{w})) \ (\text{rec}_1 \text{ ptr}) \qquad \textit{val-rel } v \ v' \\ (\forall \mu', \sigma'. \ \textbf{corres } \mathcal{R} \ e_3 \ e_3' \ (e_1 \mapsto p \ (\texttt{Ptr } r), U) \ (e_1 \mapsto \{f :: \text{U8}\} \ \texttt{w}, \Gamma_2) \ \mu' \ \sigma')\end{array}}{\begin{array}{c}\textbf{corres } \mathcal{R} \ (\textbf{let } x = \textbf{put } e_1.f := e_2 \textbf{ in } e_3) \\ (\textbf{do} \ \_ \leftarrow \textbf{guard } (\lambda\sigma. \ \textit{is-valid } \sigma \ p'); \ \_ \leftarrow \textbf{modify } (\lambda\sigma. \ \sigma[p'].f := v'); \ e_3' \textbf{ od}) \\ U \ (\Gamma_1\Gamma_2) \ \mu \ \sigma\end{array}} \text{ PUT}$$

Note that the cumbersome record-update assumptions from Fig. 6 have been reduced to *val-rel* and *type-rel* statements. This is only possible after we obtain the concrete program and its data relations. We also instantiate the state relation $\mathcal{R}$ and show that **take** and **put** preserve it, allowing us to simplify the heap-update assumptions.

Well-Typedness. The COGENT compiler proves, via an automated Isabelle tactic, that the deep embedding of the input program is well-typed. Specifically, it shows for each function f with argument x, body e, and type $\tau_1 \rightarrow \tau_2$, that $x \mapsto \tau_1 \vdash e : \tau_2$.

Recall that the type system is substructural, and that proving refinement requires access to the typing judgements for each sub-expression of the program. To solve this, the COGENT compiler instructs Isabelle to store all intermediate typing judgements established during type checking. These theorems are stored in a tree structure, isomorphic to the COGENT program's type derivation tree. Each node is a typing theorem for a program sub-expression, and can be retrieved by the refinement proof tactic as it descends into the program.

Proof Automation. The core of our refinement prover is an Isabelle/ML tactic that proves the **corres** refinement theorem (Sect. 3.3) for each COGENT function in the program, by applying the **corres** rules previously proven, both generic and specialised (Sect. 3.4). This algorithm is straightforward as our rules are syntax-directed.

The tactic also expands definitions of *val-rel* and *type-rel* (Sect. 3.2) in order to discharge data relation assumptions in those **corres** rules, and retrieves the type derivation tree for the given COGENT function to discharge all well-typedness assumptions.

Example 5. For *flip* in Fig. 5, we wish to prove the refinement theorem

corres $\mathcal{R}$ *flip* $(flip_C v_m)$ $(x \mapsto u)$ $(x : \{\mathtt{f} :: \mathtt{U8}\}\ )\ \mu\ \sigma$

or after unfolding

corres $\mathcal{R}$ (**take** x' $\{\mathtt{f} = y\} = x$ **in** ...)
        (**do guard** $(\lambda\sigma.\ is\text{-}valid\ \sigma\ x)$; $y \leftarrow$ **gets** $(\lambda\sigma.\ \sigma[r].\mathtt{f})$; ... **od**)
        $(x \mapsto u)$ $(x : \{\mathtt{f} :: \mathtt{U8}\}\ )\ \mu\ \sigma$

The first step of the proof applies the specialised **take** rule for $\{\mathtt{f} :: \mathtt{U8}\}$ (Sect. 3.4). After discharging its typing and *val-rel* assumptions, we are left with a **corres** obligation on the remainder of the function, which can in turn be solved using the other proof rules.

Our tactic can be used easily for single functions, but extending it to whole programs required significant proof engineering effort, as we must handle function calls both to externally-defined C functions and to (potentially higher-order) COGENT functions.

Foreign functions. COGENT code depends on calls to foreign C functions to perform loops and I/O. Our framework requires these functions to be well-behaved, i.e. they respect COGENT's termination order and do not break the COGENT type system (e.g. by modifying variables they do not have access to).

Foreign functions are user-supplied and not verified automatically. Thus, when proving refinement theorems for COGENT code that calls these functions, we automatically insert assumptions that they are well-behaved. These assumptions remain until they are resolved by manual verification.

Whole-program refinement. COGENT is a total language and does not permit recursion, so we have, in principle, a well-ordering on function calls in any program. However, for higher-order functions, this well-ordering is non-obvious and difficult to work with.

In practice, most function calls in systems code are direct calls to first-order functions. For such functions, we can simply prove the **corres** theorems in bottom-up fashion, starting from the leaf functions and ending at the top-level functions.

There is one major exception: COGENT code cannot express loops using only first-order functions. Our COGENT programs use iteration *combinators*, which are second-order foreign functions that take a COGENT function pointer as the loop body (similar to the *map* or *fold* combinators in functional programming).

Therefore, our framework also supports second-order calls to foreign functions. Before assuming **corres** for these functions, we first prove **corres** for the argument function (i.e. the loop body).

This technique allows us to automate refinement for code with first- and second-order calls. While this restriction means that not all COGENT programs can be verified in our framework, we developed COGENT code for two file system drivers [2] in this fragment, demonstrating that substantial programs can be written in this subset.

4   Related Work

To date, the largest trustworthy compilation projects are the CompCert [7] C compiler and the CakeML [6] ML environment. In contrast to COGENT, they compile general-purpose programming languages and rely more heavily on verified compilation passes.

CompCert translates (a subset of) C to binary while our compiler translates the functional COGENT language to C. CompCert's core compilation process is verified and its optimisation passes are validated; the compiler executable itself is extracted from Coq into Caml. There is ongoing work to validate the Coq code extraction process and the Caml compiler for CompCert.

We chose to use certificates for most of COGENT's compiler passes, because our proof tools for C run in Isabelle directly, and our COGENT compiler is written in Haskell, which does not have a formal semantics nor a verified runtime at present. On one hand, processing the certificates is time-intensive. On the other hand, we do not need to trust the code extractor, nor the runtime for the extracted language. We do need to either trust the C compiler or use a verified one.

COGENT is closer to CakeML in that it is a high-level source language. However, COGENT targets a different application area. CakeML is a Turing-complete dialect of ML with complex semantics, and is suited for application code. On the other hand, COGENT is a restricted language of total functions with simple semantics that facilitate equational reasoning. COGENT avoids the need for a large runtime and a garbage collector so it can be used for embedded systems code, especially layered systems code with minimal sharing such as the control code of filesystems or network protocol stacks.

5   Take Away Lessons and Future Work

When designing the certifying compiler, we made a trade-off by writing the COGENT compiler tool-chain in Haskell, while the proof component was written in Isabelle's Standard ML environment. This divide allows the COGENT tool-chain to be used outside the theorem prover, and allows the proof tools to build on the existing C parser and AutoCorres framework.

On the other hand, this choice leads to complexity in designing the interface between these components. This is illustrated by our well-typedness proof of Sect. 3.4, where the COGENT compiler generates a certificate with the necessary type derivation hints. Initially, we used a naïve format consisting of the entire derivation tree, resulting in gigabyte-sized certificates. We implemented various compression techniques to reduce the certificates to a reasonable size (a few megabytes). It is possible to avoid these certificates entirely by duplicating the type inference algorithm in Isabelle/ML, but this would increase the code maintenance burden.

Even though reusing the C parser and AutoCorres is desirable, they take a long time to process our verbose generated C code. They take a total of 12 CPU hours to translate the `ext2` filesystem into a monadic embedding and they take 32 CPU hours when applied to `BilbyFs`. Further proof optimisation is needed.

Optimisation of the generated code is another topic for future work. High-level COGENT-to-COGENT optimisations will be easy, as they can be verified over the shallow embedding of COGENT using equational rewriting. For instance, we verified A-normalisation using rewriting; while it is not an optimisation, it is an example of a code transformation that does not affect the COGENT-to-C proof. For low-level optimisations, we rely on the C compiler so as not to complicate our syntax-directed proof approach.

6 Conclusions

We developed a compositional refinement calculus and proof tools to create a fully automatic refinement certificate from COGENT's update semantics to C, including the use of *partial type erasure* to relate COGENT's expressive types to simpler C types. This refinement certificate is the most involved step in the full automation of the overall compiler certificate. Through the co-generation of code and proofs, our framework significantly reduces the cost of reasoning about efficient C code, by automatically discharging cumbersome safety obligations, and providing an embedding more amenable to verification. Our framework has been applied successfully to two real-world file-systems.

References

1. COGENT material (2016). https://github.com/NICTA/cogent/tree/itp_2016
2. Amani, S., Hixon, A., Chen, Z., Rizkallah, C., Chubb, P., O'Connor, L., Beeren, J., Nagashima, Y., Lim, J., Sewell, T., Tuong, J., Keller, G., Murray, T., Klein, G., Heiser, G.: Cogent: Verifying high-assurance file system implementations. In: ASPLOS, pp. 175–188, April 2016
3. Cock, D., Klein, G., Sewell, T.: Secure microkernels, state monads and scalable refinement. In: Mohamed, O.A., Muñoz, C., Tahar, S. (eds.) TPHOLs 2008. LNCS, vol. 5170, pp. 167–182. Springer, Heidelberg (2008)
4. Greenaway, D., Lim, J., Andronick, J., Klein, G.: Don't sweat the small stuff: formal verification of C code without the pain. In: PLDI, pp. 429–439 (June 2014)
5. Klein, G., Elphinstone, K., Heiser, G., Andronick, J., Cock, D., Derrin, P., Elkaduwe, D., Engelhardt, K., Kolanski, R., Norrish, M., Sewell, T., Tuch, H., Winwood, S.: seL4: Formal verification of an OS kernel. In: SOSP, pp. 207–220 (October 2009)
6. Kumar, R., Myreen, M., Norrish, M., Owens, S.: CakeML: a verified implementation of ML. In: POPL, pp. 179–191, January 2014
7. Leroy, X.: Formal verification of a realistic compiler. CACM **52**(7), 107–115 (2009)
8. Nipkow, T., Paulson, L.C., Wenzel, M. (eds.): Isabelle/HOL — A Proof Assistant for Higher-Order Logic. LNCS, vol. 2283. Springer, Heidelberg (2002)
9. O'Connor, L., Chen, Z., Rizkallah, C., Amani, S., Lim, J., Murray, T., Nagashima, Y., Sewell, T., Klein, G.: Refinement through restraint: bringing down the cost of verification. In: ICFP (to appear, 2016)

10. Sabry, A., Felleisen, M.: Reasoning about programs in continuation-passing style. SIGPLAN Lisp Pointers **V**(1), 288–298 (January 1992)
11. Schirmer, N.: Verification of Sequential Imperative Programs in Isabelle/HOL. Ph.D. thesis, Technische Universität München (2006)
12. Sewell, T., Myreen, M., Klein, G.: Translation validation for a verified OS kernel. In: PLDI, pp. 471–481 (June 2013)
13. Tuch, H., Klein, G., Norrish, M.: Types, bytes, and separation logic. In: POPL, pp. 97–108 (January 2007)

Formalization of the Resolution Calculus for First-Order Logic

Anders Schlichtkrull[(✉)]

DTU Compute, Technical University of Denmark, 2800 Kongens Lyngby, Denmark
andschl@dtu.dk

Abstract. A formalization in Isabelle/HOL of the resolution calculus for first-order logic is presented. Its soundness and completeness are formally proven using the substitution lemma, semantic trees, Herbrand's theorem, and the lifting lemma. In contrast to previous formalizations of resolution, it considers first-order logic with full first-order terms, instead of the propositional case.

Keywords: First-order logic · Resolution · Isabelle/HOL · Herbrand's theorem · Soundness · Completeness

1 Introduction

The resolution calculus plays an important role in automatic theorem proving for first-order logic as many of the most efficient automatic theorem provers, e.g. E [23], SPASS [25], and Vampire [18], are based on resolution and an extension called superposition. Studying the resolution calculus is furthermore an integral part of many university courses on logic in computer science. The resolution calculus was introduced by Robinson in his groundbreaking paper which also introduced most general unifiers (MGUs) [20].

The calculus reasons about first-order literals, i.e. atoms and their negations. Since the literals are first-order, they may contain full first-order terms. Literals are collected in clauses, i.e. disjunctions of literals. The calculus is refutationally complete, which means that if a set of clauses is unsatisfiable, then the resolution calculus can derive a contradiction (the empty clause) from it. One can also use the calculus to prove any valid formula by first negating it, then transforming it to an equisatisfiable set of clauses, and lastly refuting this set with the resolution calculus. Resolution is a calculus for first-order logic, but it does not have any machinery to handle equality or any other theories.

We mostly follow textbooks by Ben-Ari [1], Chang and Lee [8], and Leitsch [15]. The idea of Chang and Lee's completeness proof is to consider semantic trees, which are binary trees that represent interpretations. Such a tree is cut smaller and smaller, and for each cut, a derivation is done towards the empty clause. The theorem that cuts the tree down to finite size is Herbrand's theorem, which we also formalize. We prove the completeness theorem for Herbrand universes only, but e.g. Chang and Lee's Theorem 4.2 states that this is sufficient to prove it complete for any universe. That theorem is, however, not formalized.

J.C. Blanchette and S. Merz (Eds.): ITP 2016, LNCS 9807, pp. 341–357, 2016.
DOI: 10.1007/978-3-319-43144-4_21

The formalization is included in the IsaFoL project [3], which formalizes several logical calculi in Isabelle/HOL. IsaFoL is part of a larger effort to formally prove theorems about logics and logical calculi. This also includes formalizations of ground resolution, which is propositional by nature. The formalization in this paper stands out from these by formalizing resolution for first-order logic. The theory needed to do this is very different from that of ground resolution since first-order logic involves a richer syntax and semantics. To the best of my knowledge, I present the first formalized completeness proof of the resolution calculus for first-order logic.

Harrison formalizes Herbrand's theorem in a model theoretic formulation [10]. It says that if a purely existential formula is valid, then some disjunction of instances of the body is propositionally valid. In automatic theorem proving, the theorem is viewed in a different, equivalent way: A finite set of clauses is unsatisfiable if some finite set of ground, i.e. variable free, instances of its clauses is as well. This is what SAT solvers take advantage of when refuting first-order formulas. Essentially, they enumerate ground instances and try to refute them. We formalize a third equivalent view stating exactly what the completeness proof needs: If a set of clauses is unsatisfiable, then it has a finite closed semantic tree. This bridges first-order unsatisfiability with decisions made in a semantic tree.

Since this paper is a case study in formalizing mathematics, it is also worthwhile to consider which tools were helpful in this regard:

- The Isabelle/jEdit Prover IDE has many useful features to navigate proof documents. This was advantageous when the theory grew larger.
- The structured proof language Isar was beneficial because it allows formal proofs to be written as sequences of claims that follow from the previous claims. This clearly mirrors mathematical paper proof, which is what we are formalizing. Furthermore, it makes the proofs easy to read, and this is important when a formalization is to help in the understanding of a theory.
- The proof methods of Isabelle such as auto, blast, and metis were effective in discharging proof goals.
- The Sledgehammer tool finds proofs by picking important facts from the theory and then employing top-of-the-line automatic theorem provers and satisfiability modulo solvers. It often helps proving claims that we know are true, but where finding the necessary facts from the theory and libraries as well as choosing and instructing a proof method would be tedious.

Understanding proofs of logical systems can be challenging since one must keep separate which parts of the proofs are about the syntactic level, and which are about the semantic level. It can be tempting to mix intuition about semantics and syntax. Fortunately, a formalization makes the distinction very clear, and hopefully this can aid in understanding the proofs.

2 Overview

A *literal* l is either an atom or its negation. The *sign* of an atom is *True*, while that of its negation is *False*. The *complement* p^c of an atom p is $\neg p$, and the complement $(\neg p)^c$ of its negation is p. The complement L^C of a set of literals L is $\{l^c \mid l \in L\}$. The set of variables in a clause is $vars_{ls}\ C$. A clause with an empty set of variables is called *ground*. A *clause* is a set of literals representing the universal quantification of the disjunction of the literals in the clause. The empty clause represents a contradiction since it is an empty disjunction. A *substitution* σ is a function from variables to terms, and is applied to a clause C by applying it to all variables in C. The result is written $C \mathbin{\gamma_{ls}} \sigma$ and is called an instance of C. We can likewise apply a substitution to a single literal: $l \mathbin{\gamma} \sigma$.

We will consider the following formulation of the resolution calculus:

$$\frac{C_1 \qquad C_2}{((C_1 - L_1) \cup (C_2 - L_2)) \mathbin{\gamma_{ls}} \sigma} \quad \begin{array}{l} vars_{ls}\ C_1 \cap vars_{ls}\ C_2 = \{\} \\ L_1 \subseteq C_1,\ L_2 \subseteq C_2 \\ \sigma \text{ is a substitution and an MGU of } L_1 \cup L_2^C \end{array}$$

The conclusion of the rule is called a *resolvent* of C_1 and C_2. L_1 and L_2 are called *clashing* sets of literals. Additionally, the calculus allows us to apply variable renaming to clauses before we apply the resolution rule. Renaming variables in two clauses C_1 and C_2 such that $vars_{ls}\ C_1 \cap vars_{ls}\ C_2 = \{\}$ is called *standardizing apart*. Notice that L_1 and L_2 are sets of literals. Some other resolution calculi instead let L_1 and L_2 be single literals. These calculi then have an additional rule called factoring, which allows unification of subsets of clauses.

The completeness proof we consider is very much inspired by that of Chang and Lee [8], and the proof of the lifting lemma by that of Leitsch [15].

Semantic trees are defined from an enumeration of Herbrand, i.e. ground, atoms. A semantic tree is essentially a binary decision tree in which the decision of going left in a node on level i corresponds to mapping the ith atom of the enumeration to *True*, and in which going right corresponds to mapping it to *False*. See Fig. 1. Therefore, a finite path in a semantic tree can be seen as a *partial interpretation*. This differs from the usual interpretations in first-order logic in two ways. Firstly, it does not consist of a function denotation and a predicate denotation, but instead assigns *True* and *False* to ground atoms directly. Secondly, it is finite, which means that some ground literals are assigned neither *True* nor *False*. A partial interpretation is said to *falsify a ground clause* if it, to all literals in the clause, assigns the opposite of their signs. A *branch* is a path from the root of a tree to one of its leaves. A *closed branch* is a branch whose corresponding partial interpretation falsifies some ground instance of a clause in the set of clauses. A *closed semantic tree* for a set of clauses is a minimal tree in which all branches are closed.

Herbrand's theorem is proven in the following formulation: If a set of clauses is unsatisfiable, then there is a finite and closed semantic tree for that set. We prove it in its contrapositive formulation and therefore assume that all finite semantic trees of a set of clauses have an open (non-closed) branch. Obtaining longer and longer branches of larger and larger finite semantic trees, we can, using König's lemma, obtain an infinite path all of whose prefixes are open branches of finite semantic trees. Thus these branches satisfy, that is, do not falsify, the set of clauses. We can then prove that this infinite path, when seen as

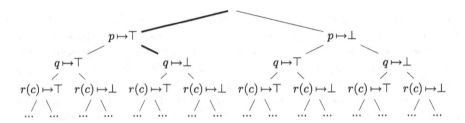

Fig. 1. Semantic tree with partial interpretation $[p \mapsto \mathit{True}, q \mapsto \mathit{False}]$

an Herbrand interpretation, also satisfies the set of clauses, and this concludes the proof. Converting the infinite path to a full interpretation can be seen as the step that goes from syntax to semantics.

The lifting lemma lifts resolution derivation steps done on the ground level up to the first-order world. The lemma considers two instances, C_1' and C_2', of two first-order clauses, C_1 and C_2. It states that if C_1' and C_2' can be resolved to a clause C' then also C_1 and C_2 can be resolved to a clause C. And not only that, but it can even be done in such a way that C' is an instance of this C. See Fig. 2. To prove the theorem we look at the clashing sets of literals $L_1' \subseteq C_1'$ and $L_2' \subseteq C_2'$. We partition C_1' in L_1' and the rest, $R_1' = C_1' - L_1'$. Then we lift this up to C_1 by partitioning it in L_1, the part that instantiates to L_1', and the rest R_1 which instantiates to R_1'. We do the same for C_2. Since L_1' and $L_2'^C$ can be unified, so can L_1 and L_2^C, and therefore they have an MGU. Thus C_1 and C_2 can be resolved to a resolvent C. With some bookkeeping of the substitutions and unifiers, we can also show that C has the ground resolvent C' as an instance.

Lastly, *completeness* itself is proven. It states that the empty clause can be derived from any unsatisfiable set of clauses. We start by obtaining a finite closed semantic tree for the set of clauses. Then we cut off two sibling leaves. The branches ending in these leaves falsify a ground clause each, and these clauses can be resolved. We lift this up to the first-order world by the lifting lemma and resolve the first-order clauses. Repeating this procedure, we obtain a derivation

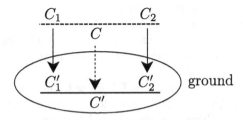

Fig. 2. The lifting lemma. An arrow from C to C' indicates that C' is an instance of C. The bars are derivations. Full bars or arrows are relations we know, and the stippled ones are established by the lemma.

which ends when we have cut the tree down to the root. Only the empty clause can be falsified here, and so we have a derivation of the empty clause.

3   Clausal First-Order Logic

We briefly explain the formalization of first-order clausal logic. A first-order term is either a variable consisting of a variable symbol (a string) or it is a function application consisting of a function symbol (a string) and a list of subterms:

> **datatype** *fterm = Var var-sym | Fun fun-sym (fterm list)*

A literal is either positive or negative, and it contains a predicate symbol (a string) and a list of terms. The datatype is parametrized with the type of terms $'t$ since it will both represent first-order literals (*fterm literal*) and Herbrand literals. A clause is a set of literals.

> **datatype** $'t$ *literal = Pos pred-sym ($'t$ list) | Neg pred-sym ($'t$ list)*

> **type-synonym** $'t$ *clause = $'t$ literal set*

We formalize the ground *fterm literals* using a predicate $ground_l$ which holds for l if it contains no variables. Likewise, we formalize ground *fterm clauses* using a predicate $ground_{ls}$.

A substitution is a function from variable symbols into terms:

> **type-synonym** *substitution = var-sym $\Rightarrow$ fterm*

This is very different from Chang and Lee where they are represented by finite sets [8]. The advantage of functions is that they make it much easier to apply and compose substitutions. If C' is an instance of C we write *instance-of$_{ls}$*. The composition of two substitutions, σ_1 and σ_2, is also defined, and written $\sigma_1 \cdot \sigma_2$. We also define unifiers and most-general unifiers of literals (and similarly of terms):

> **definition** $unifier_{ls}$ σ L $\longleftrightarrow$ ($\exists l'.$ $\forall l \in L.$ l $_\cdot$ $\sigma = l'$)

> **definition** mgu_{ls} σ L $\longleftrightarrow$ $unifier_{ls}$ σ $L \wedge$ ($\forall u.$ $unifier_{ls}$ u L $\longrightarrow$ $\exists i.$ $u = \sigma \cdot i$)

One important theorem is that if a finite set of literals has a unifier, then it also has an MGU. This theorem is formalized in the IsaFoR project [24] by means of a unification algorithm, and we obtain it by proving the literals, unifiers, and MGUs of IsaFoR equivalent to ours.

> **lemma** *unification*:
> **assumes** *finite L*
> **assumes** $unifier_{ls}$ σ L
> **shows** $\exists\theta.$ mgu_{ls} θ L

We also formalize a semantics of terms and literals. A variable denotation, $var\text{-}denot$, maps variable symbols to values of the domain. The domain is represented by the type variable $'u$:

type-synonym $'u\ var\text{-}denot = var\text{-}sym \Rightarrow 'u$

Interpretations consist of denotations of functions and predicates. A function denotation maps function symbols and lists of values to values:

type-synonym $'u\ fun\text{-}denot = fun\text{-}sym \Rightarrow 'u\ list \Rightarrow 'u$

Likewise, a predicate denotation maps predicate symbols and lists of values to the two boolean values:

type-synonym $'u\ pred\text{-}denot = pred\text{-}sym \Rightarrow 'u\ list \Rightarrow bool.$

Similar to other formalizations of first-order logic, the predicate and function symbols do not have fixed arities. The semantics of a term is then defined by the recursive function $eval_t$.

fun $eval_t :: 'u\ var\text{-}denot \Rightarrow 'u\ fun\text{-}denot \Rightarrow fterm \Rightarrow 'u$ **where**
$eval_t\ E\ F\ (Var\ x) = E\ x$
$\mid eval_t\ E\ F\ (Fun\ f\ ts) = F\ f\ (map\ (eval_t\ E\ F)\ ts)$

Here, $map\ (eval_t\ E\ F)\ [e_1, \ldots, e_n] = [eval_t\ E\ F\ e_1, \ldots, eval_t\ E\ F\ e_n]$, and from now on we abbreviate $map\ (eval_t\ E\ F)\ ts$ as $eval_{ts}\ E\ F\ ts$.

If an expression evaluates to $True$ in an interpretation, we say that it is satisfied by the interpretation. If it evaluates to $False$, we say that it is falsified. The semantics of literals is a function $eval_l$ that evaluates literals.

fun $eval_l :: 'u\ var\text{-}denot \Rightarrow 'u\ fun\text{-}denot \Rightarrow 'u\ pred\text{-}denot$
$\Rightarrow fterm\ literal \Rightarrow bool$ **where**
$eval_l\ E\ F\ G\ (Pos\ p\ ts) \longleftrightarrow G\ p\ (eval_{ts}\ E\ F\ ts)$
$\mid eval_l\ E\ F\ G\ (Neg\ p\ ts) \longleftrightarrow \neg G\ p\ (eval_{ts}\ E\ F\ ts)$

We extend the semantics to clauses.

definition $eval_c :: 'u\ fun\text{-}denot \Rightarrow 'u\ pred\text{-}denot$
$\Rightarrow fterm\ clause \Rightarrow bool$ **where**
$eval_c\ F\ G\ C \longleftrightarrow (\forall E.\ \exists l \in C.\ eval_l\ E\ F\ G\ l)$

A set of clauses Cs is satisfied, written $eval_{cs}\ F\ G\ Cs$, if all its clauses are satisfied.

4   The Resolution Calculus

We first formalize resolvents, i.e. the conclusion of the resolution rule.

definition $resolution\ C_1\ C_2\ L_1\ L_2\ \sigma = ((C_1 - L_1) \cup (C_2 - L_2)) \cdot_{ls} \sigma$

In Sect. 2 we saw that the resolution rule had three side-conditions. We additionally restrict the rule to require that L_1 and L_2 are non-empty. When these side-conditions are fulfilled, the rule is applicable.

definition *applicable* C_1 C_2 L_1 L_2 σ $\longleftrightarrow$
$$C_1 \neq \{\} \wedge C_2 \neq \{\} \wedge L_1 \neq \{\} \wedge L_2 \neq \{\}$$
$$\wedge\ vars_{ls}\ C_1 \cap vars_{ls}\ C_2 = \{\}$$
$$\wedge\ L_1 \subseteq C_1 \wedge L_2 \subseteq C_2$$
$$\wedge\ mgu_{ls}\ \sigma\ (L_1 \cup L_2^C)$$

A step in the resolution calculus either inserts a resolvent of two clauses in a set of clauses, or it inserts a variable renaming of one of the clauses. Two clauses are variable renamings of each other if they can be instantiated to each other. Alternatively we could say that we apply a substitution which is a bijection between the variables in the clause and another set of variables.

definition *var-renaming-of* :: *fterm clause* $\Rightarrow$ *fterm clause* $\Rightarrow$ *bool* **where**
var-renaming-of C_1 C_2 $\longleftrightarrow$ *instance-of*$_{ls}$ C_1 C_2 $\wedge$ *instance-of*$_{ls}$ C_2 C_1

The rule for variable renaming allows us to standardize clauses apart.

inductive *resolution-step*
:: *fterm clause set* $\Rightarrow$ *fterm clause set* $\Rightarrow$ *bool* **where**
resolution-rule:
$C_1 \in Cs \Longrightarrow C_2 \in Cs \Longrightarrow applicable\ C_1\ C_2\ L_1\ L_2\ \sigma \Longrightarrow$
resolution-step Cs $(Cs \cup \{resolution\ C_1\ C_2\ L_1\ L_2\ \sigma\})$
| standardize-apart:
$C \in Cs \Longrightarrow var\text{-}renaming\text{-}of\ C\ C' \Longrightarrow resolution\text{-}step\ Cs\ (Cs \cup \{C'\})$

Derivation steps are extended to derivations by taking the reflexive transitive closure of *resolution-step*, which is given by *rtranclp*.

definition *resolution-deriv* = *rtranclp resolution-step*

We will prove the resolution rule sound by combining several simpler rules. The first we need looks as follows:

$$\frac{C}{C \downarrow_{ls} \sigma}$$

It is not entirely trivial to prove, but the needed insight is that given a function denotation and a variable denotation, any substitution can be converted to a variable denotation by evaluating the terms of its domain. We do this using function composition $\circ$:

definition *evalsub* E F σ = *eval*$_t$ E $F \circ \sigma$

We can then prove the substitution lemma:

lemma *substitution*: *eval*$_l$ E F G $(l \downarrow_l \sigma)$ $\longleftrightarrow$ *eval*$_l$ (*evalsub* E F σ) F G l

Next, we prove a special version of the resolution rule sound. The rule is special since it is only allowed to remove two literals instead of two sets of literals:

$$\frac{C_1 \qquad C_2}{(C_1 - \{l_1\}) \cup (C_2 - \{l_2\})} \begin{array}{l} l_1 \in C_1 \\ l_2 \in C_2 \\ l_1 = l_2^c \end{array}$$

Lastly, we prove that from a clause follows any superset of the clause:

$$\frac{C_1}{C_1 \cup C_2}$$

The proofs of all four rules are made as short structured Isar-proofs.

These four sound rules are combined to give the resolution rule, which must consequently be sound. We are of course allowed to use the assumptions of the resolution rule, so we know that when σ is applied to L_1 and L_2, they turn in to a complementary pair of literals, which we denote $l_1 \,_{\mathsf{ls}} \sigma$ and $l_2 \,_{\mathsf{ls}} \sigma$. This justifies the book keeping inference below. It also means that we can apply the special resolution rule. The bottommost rule application uses the superset rule.

$$\frac{\dfrac{C_1}{C_1 \,_{\mathsf{ls}} \sigma} \qquad \dfrac{C_2}{C_2 \,_{\mathsf{ls}} \sigma}}{\dfrac{\dfrac{(C_1 \,_{\mathsf{ls}} \sigma - \{l_1 \,_{\mathsf{ls}} \sigma\}) \cup (C_2 \,_{\mathsf{ls}} \sigma - \{l_2 \,_{\mathsf{ls}} \sigma\})}{(C_1 \,_{\mathsf{ls}} \sigma - L_1 \,_{\mathsf{ls}} \sigma) \cup (C_2 \,_{\mathsf{ls}} \sigma - L_2 \,_{\mathsf{ls}} \sigma)}}{((C_1 - L_1) \cup (C_2 - L_2)) \,_{\mathsf{ls}} \sigma}} \begin{array}{l} \text{special resolution} \\ \\ \text{book keeping} \end{array}$$

All this reasoning is made as a structured Isar-proofs.

> **lemma** *resolution-sound*:
> **assumes** $eval_c \ F \ G \ C_1 \wedge eval_c \ F \ G \ C_2$
> **assumes** $applicable \ C_1 \ C_2 \ L_1 \ L_2 \ \sigma$
> **shows** $eval_c \ F \ G \ (resolution \ C_1 \ C_2 \ L_1 \ L_2 \ \sigma)$

5   Herbrand Interpretations

Herbrand interpretations are a special kind of interpretations, which are characterized by two properties. The first is that their universe is the set of Herbrand terms. Since we chose that the universe should be a type, we need to represent the universe of Herbrand terms by a type. We do it by introducing a new type *hterm* which is similar to *fterm*, but does not have a constructor for variables.

> **datatype** *hterm* = *HFun fun-sym* (*hterm list*)

This is the same datatype as in Berghofer's formalization of natural deduction [2]. Had we chosen to represent the universes by sets like Ridge and Margetson [19], then we could have represented the Herbrand universe by the set of ground *fterms*. Unfortunately, we would then need wellformedness predicates for variable and function denotations. We introduce functions *fterm-of-hterm* and *hterm-of-fterm*, converting between *hterms* and ground *fterms*.

The second characteristic property is that the function denotation of an Herbrand interpretation is *HFun*, and thus, evaluating a ground term under such an interpretation corresponds to replacing all applications of *Fun* with *HFun*, that is, the ground term is interpreted as itself.

As we saw in Sect. 2, we need an enumeration of Herbrand atoms, such that we can construct our semantic trees. So we define the type of atoms:

type-synonym *'t atom = pred-sym * 't list*

Isabelle/HOL provides the proof method *countable-datatype* that can automatically prove that a given datatype, in our case *hterm*, is countable. Since also the predicate symbols are countable, then so must *hterm atom* be. Furthermore, it is easy to prove that there are infinitely many *hterm atoms*. Using these facts and Hilbert's choice operator, we specify a bijection *hatom-from-nat* between the natural numbers and the *hterm atoms*. We call its inverse *nat-from-hatom*. Additionally, we write functions, *nat-from-fatom* and *fatom-from-nat*, enumerating the ground *fterm* atoms in the same order. We also introduce a function *get-atom* which returns the atom corresponding to a literal.

5.1 Semantic Trees

We need to formalize semantic trees. In paper-proofs the trees are often labeled with the atoms which we add to or remove from our partial interpretations. In this formalization the trees are unlabeled, because for a given level we can always calculate the corresponding atom.

datatype *tree = Leaf | Branching tree tree*

Our formalization contains a quite substantial, approximately 700 lines, theory on these unlabeled binary trees, paths within them, and their branches. The details are not particularly interesting, but a theory of binary trees is necessary because we, in contrast to paper proofs, cannot rely on intuition about trees.

In our formalization, *bool lists* represent both paths in trees and partial interpretations, denoted by the type *partial-pred-denot*. E.g., if we consider the path [*True, True, False*], then it is the path from the root of a semantic tree that goes first left, then left again, and lastly right. On the other hand, it is also the partial interpretation which considers *hatom-from-nat 0* to be *True*, *hatom-from-nat 1* to be *True* and *hatom-from-nat 2* to be *False*. Our formalization illustrates the correspondence between partial interpretations and paths clearly by identifying their types.

Infinite trees and paths can not be represented by datatypes. We, thus, model possibly infinite trees as sets of paths with a wellformedness property:

abbreviation *wf-tree* :: *dir list set ⇒ bool* **where**
wf-tree T ≡ (∀ds d. (ds @ d) ∈ T ⟶ ds ∈ T)

Similarly, we model infinite paths as functions from natural numbers into finite paths. Applying the function to number i gives us the prefix of length i. We call such functions infinite paths, and their characteristic property is:

abbreviation *wf-infpath* :: $(nat \Rightarrow {'a\ list}) \Rightarrow bool$ **where**
wf-infpath $f \equiv (f\ 0 = []) \wedge (\forall n.\ \exists a.\ f\ (Suc\ n) = (f\ n)\ @\ [a])$

We must make formal, what it means for a partial interpretation to falsify an expression. A partial interpretation G falsifies, written *falsifies$_l$ G l*, a ground literal l, if the opposite of its sign occurs on index *nat-from-fatom (get-atom l)* of the interpretation.

definition *falsifies$_l$* :: *partial-pred-denot* $\Rightarrow$ *fterm literal* $\Rightarrow$ *bool* **where**
falsifies$_l$ G l $\longleftrightarrow$ *ground$_l$ l*
$\wedge$ (*let* $i = $ *nat-from-fatom (get-atom l) in*
$i < $ *length* $G \wedge G\ !\ i = (\neg sign\ l))$

A ground clause C is falsified, written *falsifies$_g$ G C*, if all its literals are falsified. A first-order clause C is falsified, written *falsifies$_c$ G C*, if it has a falsified ground instance. A partial interpretation satisfies an expression if it does not falsify it. Lastly, a semantic tree T is closed, written *closed-tree T Cs*, for a set of clauses Cs if it is a minimal tree that falsifies all the clauses in Cs.

5.2 Herbrand's Theorem

The formalization of Herbrand's theorem is mostly straightforward and is done as an Isar-proof that follows the sketch from Sect. 2. The challenging part is to take an infinite path, all of whose prefixes satisfy a set of clauses Cs and then prove that its conversion to an interpretation also satisfies Cs. Chang and Lee [8] do not elaborate much on this, but it takes up a large part of the formalization and illustrates the interplay of syntax and semantics.

First we must define how to convert the infinite path to an Herbrand interpretation. We know that the function denotation must be *HFun*, so we just need to convert the infinite path to a predicate denotation. We do it as follows:

abbreviation *extend*
:: $(nat \Rightarrow partial\text{-}pred\text{-}denot) \Rightarrow hterm\ pred\text{-}denot$ **where**
extend f P ts $\equiv$
let $n = $ *nat-from-hatom* (P, ts) *in*
$f\ (Suc\ n)\ !\ n$

We use currying, so P and ts can be thought of as the predicate symbol and list of values which we wish to evaluate in our semantics. We do it by collecting them to an Herbrand atom, and finding its index. Then we look up a prefix of our infinite path that is long enough to have decided whether the atom is considered *True* or *False*.

We now prove that if the prefixes collected in the infinite path f satisfy a set of clauses Cs then so does its extension to a full predicate denotation $extend\ f$.

Since we want to prove that the clauses in Cs are satisfied, we fix one C and prove that it has the same property.

> **lemma** *extend-infpath*:
> **assumes** *wf-infpath* $(f :: nat \Rightarrow partial\text{-}pred\text{-}denot)$
> **assumes** $\forall n.\ \neg falsifies_c\ (f\ n)\ C$
> **assumes** *finite* C
> **shows** $eval_c\ HFun\ (extend\ f)\ C$

We will consider four ways in which clauses can be satisfied:

1. A *first-order clause* can be satisfied by a *partial interpretation*.
2. A *ground clause* can be satisfied by a *partial interpretation*.
3. A *ground clause* can be satisfied by an *interpretation*.
4. A *first-order clause* can be satisfied by an *interpretation*.

The *extend-infpath* lemma relates 1 and 4, and does so by using lemmas that relate 1 to 2 to 3 to 4. The four ways seem similar, but they are in fact very different. That a ground clause is satisfied is very different from a first-order clause being satisfied since we do not need to worry about any ground instances or variables. Likewise, a ground clause being satisfied by a partial interpretation is clearly different from being satisfied by an interpretation since the two types are vastly different: a partial interpretation is a *bool list* while an interpretation consists of a *fun-sym* $\Rightarrow$ *hterm list* $\Rightarrow$ *hterm* and a *pred-sym* $\Rightarrow$ *hterm list* $\Rightarrow$ *bool*.

We relate 1 and 2: If a first-order clause is satisfied by all prefixes of an infinite path, then so is any, in particular ground, instance. This follows from the definition of being satisfied by a partial interpretation.

We relate 2 and 3: If a ground clause is satisfied by all prefixes of an infinite path f, then it is also satisfied by $extend\ f$. This follows almost directly from the definition of *extend*.

We relate 3 and 4: Ideally we would prove that if a ground clause is satisfied by an Herbrand interpretation, then so is a first-order clause of which it is an instance. That is, however, too general. Fortunately, we notice a similarity that ties first-order clauses and ground clauses together by considering a variable denotation in the Herbrand universe, i.e. of type *var-sym* $\Rightarrow$ *hterm*. We can create a function that converts its domain to *fterms*, and thus get a substitution.

> **fun** *sub-of-denot* :: *hterm var-denot* $\Rightarrow$ *substitution*
> *sub-of-denot* $E = fterm\text{-}of\text{-}hterm \circ E$

Now we have the machinery to state the needed lemma: If the ground clause $C \cdot_{ls} sub\text{-}of\text{-}denot\ E$ is satisfied by an Herbrand interpretation under E, then so is the first-order clause C. The reason is simply that if we look at a variable in C, then it is replaced by a ground term in *sub-of-denot* E. This term evaluates to the same as the Herbrand term that it is interpreted as in E.

The final step is to chain 1, 2, 3, and 4 together to relate 1 and 4.

1. Assume that C is satisfied by all prefixes of f.
2. Then the ground instance C $_\text{ls}$ *sub-of-denot* E is satisfied by all f's prefixes.
3. Then the ground instance C $_\text{ls}$ *sub-of-denot* E is satisfied by *extend* f under E in particular.
4. Then C is satisfied by *extend* f under E.

With this, we can formalize Herbrand's theorem:

> **theorem** *herbrand*:
> **assumes** $\forall G.\ \neg eval_\text{cs}\ HFun\ G\ Cs$
> **assumes** *finite* $Cs \wedge (\forall C \in Cs.\ finite\ C)$
> **shows** $\exists T.\ closed\text{-}tree\ T\ Cs$

6   Completeness

The completeness proof combines Herbrand's theorem, the lifting lemma, and reasoning about semantic trees and derivations. We will take a look at the most challenging parts of the formalization of the proof.

6.1   Lifting Lemma

Our formalization of the resolution rule removes literals from clauses before it applies the MGU. This is similar to several presentations from the literature [15,20]. Another approach, which our formalization used in an earlier version, is to apply the MGU before the literals are removed:

$$\frac{C_1 \qquad C_2}{(C_1 \ _\text{ls}\ \sigma - L_1 \ _\text{ls}\ \sigma) \cup (C_2 \ _\text{ls}\ \sigma - L_2 \ _\text{ls}\ \sigma)} \begin{array}{l} vars_\text{ls}\ C_1 \cap vars_\text{ls}\ C_2 = \{\} \\ L_1 \subseteq C_1,\ L_2 \subseteq C_2 \\ \sigma \text{ is an MGU of } L_1 \cup L_2^C \end{array}$$

This is exactly the rule used by Ben-Ari [1]. Chang and Lee use a similar approach [8]. However, we were not able to formalize their proofs of the lifting lemma because they had some flaws. The flaws are described in my MSc thesis [21]. The most critical flaw is that the proofs seem to use that $B \subseteq A \implies (A - B) \ _\text{ls}\ \sigma = A \ _\text{ls}\ \sigma - B \ _\text{ls}\ \sigma$, which does not hold in general. Leitsch [14, Proposition 4.1] noticed flaws in Chang and Lee's proof already, and presented a counter-example to it.

With our current approach, however, the lifting lemma is straightforward to formalize as an Isar-proof using the proof by Leitsch [15]. The lemma uses the *unification* lemma from Sect. 3 to obtain MGUs.

> **lemma** *lifting*:
> **assumes** *finite* $C \wedge finite\ D$
> **assumes** $vars_\text{ls}\ C \cap vars_\text{ls}\ D = \{\}$
> **assumes** *instance-of*$_\text{ls}$ $C'\ C \wedge instance\text{-}of_\text{ls}\ D'\ D$
> **assumes** *applicable* $C'\ D'\ L'\ M'\ \sigma$
> **shows** $\exists L\ M\ \tau.\ applicable\ C\ D\ L\ M\ \tau\ \wedge$
> $\qquad\qquad instance\text{-}of_\text{ls}\ (resolution\ C'\ D'\ L'\ M'\ \sigma)\ (resolution\ C\ D\ L\ M\ \tau)$

6.2   The Formal Completeness Proof

Like Herbrand's theorem, we formalize completeness as an Isar-proof following Chang and Lee [8]. This time, however, the proof is much longer than its informal counterpart. The paper proof is about 30 lines while the formal proof is approximately 150 lines. There are several reasons for this:

- We explicitly have to standardize our clauses apart.
- We need to reason very precisely about the numbers of the ground atoms.
- We need to cut the tree twice.
  - First to remove two leaves.
  - Next to minimize it.
- In both cases we must prove that all branches are closed.
- We must tie our derivation-steps together.

Our completeness proof consists of two steps. First we apply Herbrand's theorem to obtain a finite tree. Next we take a finite tree and cut it smaller while making a derivation. Then we repeat the process on that tree. To prove that this works, we formalize the process using induction on the size of the tree. Our formalization uses the induction rule *measure_induct_rule* instantiated with the size of a tree. This gives us the following induction principle.

$$(\bigwedge x.\ (\bigwedge y.\ treesize\ y < treesize\ x \Longrightarrow ?P\ y) \Longrightarrow ?P\ x) \Longrightarrow ?P\ ?a$$

Here, the induction hypothesis holds for any tree of a smaller size, and we need this since we will cut off several nodes in each step.

6.3   Standardizing Apart

In each step we need to make sure that the clauses we resolve are standardized apart. We create functions to do this.

abbreviation $std_1\ C \equiv C\ {}_{ls}(\lambda x.\ Var\ (''1''\ @\ x))$

abbreviation $std_2\ C \equiv C\ {}_{ls}(\lambda x.\ Var\ (''2''\ @\ x))$

They take clauses C_1 and C_2 and create the clauses $std_1\ C_1$ and $std_2\ C_2$ which have added respectively 1 and 2 to the beginning of all variables. The most important property is that the clauses actually have distinct variables after we apply it. We need this such that we can apply the resolution rule, and so we can use the lifting lemma.

lemma std-$apart$-$apart$: $vars_{ls}\ (std_1\ C_1) \cap vars_{ls}\ (std_2\ C_2) = \{\}$"

We also need to prove that it actually renames the variables. This was a prerequisite for the standardize apart rule of the calculus.

lemma std_1-$renames$: var-$renaming$-$of\ C_1\ (std_1\ C_1)$

In the completeness proof C_1 is falsified by B_1, but not by B. The same holds for $std_1\ C_1$ since it is falsified by the same partial interpretations as C_1.

lemma std_1-$falsifies$: $falsifies_c\ G\ C_1 \longleftrightarrow falsifies_c\ G\ (std_1\ C_1)$

6.4 Branches and Ground Clauses

In each step, the completeness proof removes two sibling leaves and resolves the clauses, C_1 and C_2, that were falsified by the branches, $B_1 = B \, @ \, [\textit{True}]$ and $B_2 = B \, @ \, [\textit{False}]$, ending in the leaves. The resolvent is falsified by B. This is first proven on the ground level and then lifted to the first-order level using the lifting lemma. Thus, on the ground level we must prove two properties.

1. The two ground clauses C_1' and C_2' falsified by B_1 and B_2 can be resolved.
2. Their ground resolvent C' is falsified by B.

We prove 1 first. We do it by proving that C_1' contains the negative literal of number $\textit{length} \ B$ and that C_2' contains its complement. Here, the case for C_1' is presented. C_1' is falsified by B_1, but not B, since the closed semantic tree is minimal. Thus, it must be the decision of going left that was necessary to falsify C_1'. Going left falsified the negative literal l with number $\textit{length} \ B$ in the enumeration, and hence it must be in C_1'.

We prove 2 next. To prove it we must show that the ground resolvent $C' = (C_1' - \{l\}) \cup (C_2' - \{l^c\})$ is falsified by B. We do it by proving that the literals in both $C_1' - \{l\}$ and $C_2' - \{l^c\}$ are falsified. The case for $C_1' - \{l\}$ is presented here. The overall idea is that l is falsified by B_1, but not by B. The decision of going left falsified l, and then all of C_1' was falsified. Therefore, the other literals must have been falsified before we made the decision, in other words, they must have been falsified already by B.

To formalize this we must prove that all the literals in $C_1' - \{l\}$ are indeed falsified by B. We do it by a lemma showing that any other literal $lo \in C_1'$ than l is falsified by B. Its proof first shows that lo has another number than l has, i.e. other than $\textit{length} \ B$. It seems obvious since $lo \neq l$, but we also need to ensure that $lo \neq l^c$. We do this by proving another lemma which says that a clause only can be falsified by a partial interpretation if it does not contain two complementary literals. Then we show that lo has a number smaller than $\textit{length} \ B \, @ \, [\textit{True}]$, since lo is falsified by $B \, @ \, [\textit{True}]$. This concludes the proof. We abstracts from $\textit{True}$ to d such that the lemma also works for $B \, @ \, [\textit{False}]$.

> **lemma** *other-falsified*:
> **assumes** $\textit{ground}_\text{ls} \ C_1' \wedge \textit{falsifies}_\text{g} \ (B \, @ \, [d]) \ C_1'$
> **assumes** $l \in C_1' \wedge \textit{nat-from-fatom} \ (\textit{get-atom} \ l) = \textit{length} \ B$
> **assumes** $lo \in C_1' \wedge lo \neq l$
> **shows** $\textit{falsifies}_\text{l} \ B \ lo$

6.5 The Derivation

At the end of the proof we must tie the derivations together:

$$\frac{\dfrac{C_1}{\textit{std}_1 \ C_1} \qquad \dfrac{C_2}{\textit{std}_2 \ C_2}}{\textit{resolution} \ C_1 \ C_2 \ L_1 \ L_2 \ \sigma}$$

$$\vdots$$

$$\{\}$$

The dots represent the derivation we obtain from the induction hypothesis. It is done using the definitions of *resolution-step* and *resolution-deriv*. The completeness lemma is formalized as follows:

> **theorem** *completeness*:
> **assumes** *finite Cs* $\wedge$ ($\forall C \in Cs.$ *finite C*)
> **assumes**
> $\forall (F :: htcrm\ fun\text{-}denot)\ (G :: hterm\ pred\text{-}denot).\ \neg eval_{cs}\ F\ G\ Cs$
> **shows** $\exists Cs'.\ resolution\text{-}deriv\ Cs\ Cs' \wedge \{\} \in Cs'$

7 Related Work

The literature contains several formalizations of first-order logic. Harrison proves model theoretic results about first-order logic, including the compactness theorem, the Löwenheim-Skolem theorem, and Herbrand's theorem [10]. There are also formalizations of the completeness of several logical calculi for first-order logic. Margetson and Ridge [16] prove, in Isabelle/HOL, a sequent calculus sound and complete, and they formalize a verified prover based on the calculus [19]. Braselmann and Koepke prove, in Mizar, a sequent calculus sound and complete [6,7]. Schlöder and Koepke prove it complete even for uncountable languages [22]. Berghofer proves, in Isabelle/HOL, a natural deduction calculus sound and complete [2]. Illik formalizes constructive versions of completeness proofs for classical logic and full intuitionistic predicate logic [12]. Blanchette, Popescu, and Traytel formalize, in Isabelle/HOL, an abstract completeness proof that is independent of any specific proof system and syntax for first-order logic [5]. Other important formalizations of logic are Paulson's formalization of Gödel's incompleteness theorems [17], and Harrison's soundness proof of HOL Light [11] which is extended upon by Kumar, Arthan, Myreen and Owens [13].

There are also formalizations of sound and complete propositional resolution calculi. Blanchette and Traytel formalize, in Isabelle/HOL, propositional resolution [4]. Fleury formalizes, in Isabelle/HOL, many ground calculi including SAT solvers and propositional resolution [9].

8 Conclusion

This paper describes a formalization of the resolution calculus for first-order logic as well as its soundness and completeness. This includes formalizations of the substitution lemma, Herbrand's theorem, and the lifting lemma. As far as I know, this is the first formalized soundness and completeness proof of the resolution calculus for first-order logic.

The paper emphasizes how the formalization illustrates details glanced over in the paper proofs, which are necessary in a formalization. For instance it shows the jump from satisfiability by an infinite path in a semantic tree to satisfiability by an interpretation. It likewise illustrates how and when to standardize clauses apart in the completeness proof, and the lemmas necessary to allow this. Furthermore, the formalization combines theory from different sources. The proofs of Herbrand's theorem and completeness are based mainly on those by Chang and Lee [8], while the proof of the lifting lemma is based on that by Leitsch [15]. The existence proof of MGUs for unifiable clauses comes from IsaFoR [24].

Proof assistants take advantage of automatic theorem provers by using them to dispense of subgoals. This formalization could be a step towards mutual benefit between the two. Perhaps formalizations in proof assistants can help automatic theorem provers by contributing a highly rigorous understanding of their meta-theory.

Acknowledgement. Jørgen Villadsen, Jasmin Blanchette, and Dmitriy Traytel supervised me in making the formalization. Jørgen and Jasmin provided valuable feedback on the paper.

References

1. Ben-Ari, M.: Mathematical Logic for Computer Science, 3rd edn. Springer (2012)
2. Berghofer, S.: First-order logic according to Fitting. Archive of Formal Proofs, Formal proof development. http://isa-afp.org/entries/FOL-Fitting.shtml
3. Blanchette, J.C., Fleury, M., Schlichtkrull, A., Traytel, D.: IsaFoL: Isabelle Formalization of Logic. https://bitbucket.org/jasmin_blanchette/isafol
4. Blanchette, J.C., Traytel, D.: Formalization of Bachmair and Ganzinger's "Resolution Theorem Proving". https://bitbucket.org/jasmin_blanchette/isafol/src/master/Bachmair_Ganzinger/
5. Blanchette, J.C., Popescu, A., Traytel, D.: Unified classical logic completeness – A coinductive pearl. In: Demri, S., Kapur, D., Weidenbach, C. (eds.) IJCAR 2014. LNCS, vol. 8562, pp. 46–60. Springer, Heidelberg (2014)
6. Braselmann, P., Koepke, P.: Gödel completeness theorem. Formalized Math. **13**(1), 49–53 (2005)
7. Braselmann, P., Koepke, P.: A sequent calculus for first-order logic. Formalized Math. **13**(1), 33–39 (2005)
8. Chang, C.L., Lee, R.C.T.: Symbolic Logic and Mechanical Theorem Proving, 1st edn. Academic Press Inc., Orlando (1973)
9. Fleury, M.: Formalisation of ground inference systems in a proof assistant. Master's thesis, École normale supérieure Rennes (2015). http://www.mpi-inf.mpg.de/fileadmin/inf/rg1/Documents/fleury_master_thesis.pdf
10. Harrison, J.V.: Formalizing basic first order model theory. In: Grundy, J., Newey, M. (eds.) TPHOLs 1998. LNCS, vol. 1479, pp. 153–170. Springer, Heidelberg (1998)
11. Harrison, J.: Towards self-verification of HOL Light. In: Furbach, U., Shankar, N. (eds.) IJCAR 2006. LNCS (LNAI), vol. 4130, pp. 177–191. Springer, Heidelberg (2006)
12. Illik, D.: Constructive completeness proofs and delimited control. Ph.D. thesis, École Polytechnique (2010)
13. Kumar, R., Arthan, R., Myreen, M.O., Owens, S.: Self-formalisation of higher-order logic – Semantics, soundness, and a verified implementation. J. Autom. Reason **56**(3), 221–259 (2016)
14. Leitsch, A.: On different concepts of resolution. Math. Logic Q. **35**(1), 71–77 (1989)

15. Leitsch, A.: The Resolution Calculus. Springer, Texts in theoretical computer science (1997)
16. Margetson, J., Ridge, T.: Completeness theorem. Archive of Formal Proofs, Formal proof development. http://isa-afp.org/entries/Completeness.shtml
17. Paulson, L.C.: A mechanised proof of Gödel's incompleteness theorems using Nominal Isabelle. J. Autom. Reason. **55**(1), 1–37 (2015)
18. Riazanov, A., Voronkov, A.: Vampire. In: Ganzinger, H. (ed.) CADE 1999. LNCS (LNAI), vol. 1632, pp. 292–296. Springer, Heidelberg (1999)
19. Ridge, T., Margetson, J.: A mechanically verified, sound and complete theorem prover for first order logic. In: Hurd, J., Melham, T. (eds.) TPHOLs 2005. LNCS, vol. 3603, pp. 294–309. Springer, Heidelberg (2005)
20. Robinson, J.A.: A machine-oriented logic based on the resolution principle. J. ACM **12**(1), 23–41 (1965)
21. Schlichtkrull, A.: Formalization of resolution calculus in Isabelle. Master's thesis, Technical University of Denmark (2015). https://people.compute.dtu.dk/andschl/Thesis.pdf
22. Schlöder, J.J., Koepke, P.: The Gödel completeness theorem for uncountable languages. Formalized Math. **20**(3), 199–203 (2012)
23. Schulz, S.: System description: E 1.8. In: McMillan, K., Middeldorp, A., Voronkov, A. (eds.) LPAR-19 2013. LNCS, vol. 8312, pp. 735–743. Springer, Heidelberg (2013)
24. Sternagel, C., Thiemann, R.: An Isabelle/HOL formalization of rewriting for certified termination analysis. http://cl-informatik.uibk.ac.at/software/ceta/
25. Weidenbach, C., Dimova, D., Fietzke, A., Kumar, R., Suda, M., Wischnewski, P.: SPASS version 3.5. In: Schmidt, R.A. (ed.) CADE-22. LNCS, vol. 5663, pp. 140–145. Springer, Heidelberg (2009)

Verified Operational Transformation for Trees

Sergey Sinchuk[1], Pavel Chuprikov[1(✉)], and Konstantin Solomatov[2]

[1] JetBrains, St. Petersburg, Russia
sinchukss@gmail.com, pschuprikov@gmail.com
[2] JetBrains, Boston, USA
konstantin.solomatov@gmail.com

Abstract. Operational transformation (OT) is an approach to concurrency control in groupware editors first proposed by C. Ellis and S. Gibbs in 1989. Google Wave and Google Docs are examples of better known OT-based systems and there are many other experimental ones described in the literature. In their recent articles A. Imine et al. have shown that many OT implementations contain mistakes and do not possess claimed consistency properties.

The present work describes an experimental library which is based on SSReflect/Coq and contains several operational transformation algorithms and proofs of their correctness.

1 Introduction

A collaborative groupware editor is an application that allows multiple users to edit shared data objects (e.g., a text document or a spreadsheet). We will be mainly concerned with *synchronous* groupware editors (or real-time collaborative editors), i.e. editors which allow simultaneous editing of the shared data and provide automatic real-time synchronization between users. Moreover, such editors usually do not use locks in the implementation of their synchronization algorithm. Instead, every user is provided with his own replica of the data and is allowed to modify it freely.

Due to the network latency and the lock-free nature of the editor, a naive synchronization algorithm applying remote operations to a local replica *unchanged* will not be consistent. The replicas' states may diverge significantly from each other and a remote operation may not have its intended effect when applied to the local replica. Let us consider the simplest scenario in which this problem occurs. Alice removes symbol "b" and Bob inserts character "c" at the second position. After these commands are processed the state of the network becomes invalid (see Fig. 1a).

Operational transformation was conceived to overcome this problem. In the simplest case of two communicating clients its idea can be roughly stated as follows. Instead of applying Bob's operation o_B directly, Alice first "transforms" o_B through the history of her recent local changes o_A thereby calculating the operation o'_B which is applicable to the actual version of Alice's data and has the same effect as o_B. Similarly, Bob computes o'_A from known o_A and o_B. Then,

J.C. Blanchette and S. Merz (Eds.): ITP 2016, LNCS 9807, pp. 358–373, 2016.
DOI: 10.1007/978-3-319-43144-4_22

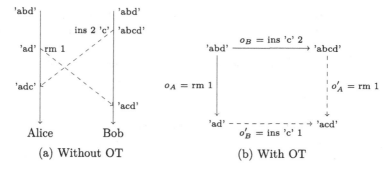

Fig. 1. An example illustrating the need of an OT. On (a) unmodified operations lead to the invalid state. On (b) the state remains consistent if we use an OT

after the operations o'_A and o'_B are executed, the replicas are again in the same state. Figure 1b illustrates the relationship between o_A, o_B, o'_A and o'_B.

A typical implementation of an OT algorithm consists of two separated components: a *transformation function* which carries out transformation of operations (i.e. computes o'_B from specified o_B and o_A) and an *integration algorithm* which is responsible for storing local histories and maintaining communication between the clients. Only the former component typically depends on the semantics of the data.

An OT algorithm is said to possess the *convergence property* if the replicas of a shared document become identical at all sites after all user operations have been executed. This property is essential for the correct operation of an OT algorithm because the existence of a counterexample to it necessarily means that the loss of user data is possible. To guarantee that an OT algorithm satisfies this convergence property one needs to ensure that the integration algorithm is correct and, moreover, that the transformation function satisfies correctness properties C_1 and C_2 (see, Sect. 2.1 for more details).

- Roughly speaking, the property C_1 requires that the effect of executing operations $o_B \circ o_{A'}$ and $o_A \circ o_{B'}$ is the same (compare with Fig. 1b). The property C_1 suffices to ensure the convergence of an OT algorithm in the case where the network has a tree topology (e.g., in the case of a network with a central server, see [2]),
- The more complicated property C_2 becomes necessary in a more general setting when one considers less restrictive network topologies which may have loops (e.g., peer-to-peer networks).

Several generic integration algorithms together with proofs of their correctness have been proposed in the literature (e.g., in [8,11]). On the other hand, for each datatype the transformation function must be implemented separately and such an implementation is known to be a hard and error-prone task even in the simplest case when the shared data object in question is a string buffer. Indeed, in the recent works of Imine and others it was shown that many implementations of transformation functions for strings do not possess the above correctness properties (see [7,9,10]).

Furthermore, even less is known about the correctness of OT algorithms for more general data types such as trees, despite the fact that such datatypes are useful in practice. Indeed, strings can be used only as a data model for a text editor, while trees can represent a variety of different structured documents (e.g., a spreadsheet or an XML document). Besides that, earlier efforts mainly concentrated on algorithms with a very small set of user operations (e.g., consisting only of operations of insertion and deletion of a single character). While formal verfications of such algorithms is easier, their behavior is less satisfactory from the semantic viewpoint as compared to algorithms with a more complicated set of operations. For example, if the last-mentioned reduced set of operations is used then it becomes impossible to take into account any higher-level semantic entities such as words or sentences in the implementation of the transformation function and, as a result, the loss of character order in the document may occur under a certain scenario, see [6, p. 325].

The main goal of this paper is to describe our attempt of formalizing several transformation functions for two different kinds of a tree datatype, namely the datatype of ordered trees (which may represent, e.g., an XML-like document) and the datatype of unordered trees (which may represent the directory structure of a filesystem). The choice of an OT for trees as a subject for verification was motivated by the fact that Jetpad platform[1] stores shared data in a tree-like structure.

Our definition of the transformation function for ordered trees is a generalized and corrected variant of algorithms of Ressel and Sun (cf. [10]). Using Coq we verify that our transformation functions satisfy convergence property C_1 and inversion property IP_1 (in the terminology of [12]). The choice of Coq as our verification tool was made due to the following considerations:

- many complex algorithms including compilers and static analyzers have been verified in Coq (e.g., CompCert, Verasco, etc.);
- Coq includes a comprehensive standard library and allows tactic-based proofs, whose power is greatly increased by the SSReflect library (cf. [5]);

The rest of the article is organized as follows. In Sect. 2 we formalize the basic terminology related to the operational transformation approach. Then, in Sect. 3 we present the main results of the paper, namely the precise definitions of the transformation functions whose correctness has been verified in Coq. The library source code in Coq can be found at github.com/JetBrains/ot-coq/.

---

[1] Jetpad is a closed-source proprietary collaborative platform of JetBrains upon which several products such as CoachingSpaces or CensusAnalyzer are based.

2 Formalization of OT Algorithms and Their Correctness Properties

2.1 Operation Model

In this section we formally define the notion of a transformation function and formulate its correctness properties. Our definitions generally follow [7], however, there are certain differences which are explained in detail below.

First of all, we need to give names to variables corresponding to a set of possible states of a shared data object and a set of atomic user operations which can modify it. Let us denote these two variables by X and cmd, respectively.

Now, we define the type class abstracting the minimal functionality of an OT algorithm. This class should encapsulate the following three entities:

1. an *interpretation* (or *transition*) function `interp` specifying how user operations are applied to the data;
2. a *transformation* function `it` which performs the transformation of operations;
3. a formula expressing the convergence property C_1 of the function `it`.

First of all, notice that we do not require the function `interp` to be total, i.e. we allow certain operations to be inapplicable to certain states of the data. For example, an operation of file deletion is only applicable if the file exists. Thus, we choose the following signature for `interp`:

$$\texttt{interp}: cmd \to X \to \texttt{option } X.$$

The equality `interp` $op\ x = \texttt{None}$, thus, should be interpreted as "op is inapplicable to x".

Now we are going to specify the signature of the transformation function. In the literature (e.g., in [4,7,10]) it is typically defined as `it`: $cmd \to cmd \to cmd$. The operation $op'_1 = \texttt{it } op_1\ op_2$ is interpreted as the result of a transformation of op_1 through op_2.

In this context, convergence property C_1 can be stated as follows: any pair of atomic operations op_1, op_2 applicable to s can be completed to a square by means of operations `it` $op_1\ op_2$, `it` $op_2\ op_1$ (see Fig. 2a).

For the sake of completeness, we also give the precise statement of the convergence property C_2. The property C_2 requires that for any op_1, op_2, op_3 one has the following equality of operations (cf. [7, Definition 2.12]):

$$\texttt{it }(\texttt{it }\ op_1\ op_2)\ (\texttt{it }\ op_3\ op_2) = \texttt{it }(\texttt{it }\ op_1\ op_3)\ (\texttt{it }\ op_2\ op_3).$$

This property is rather restrictive and difficult to prove in practice. Also, it has been suggested in [10] that it is not possible to implement a transformation function for text buffers satisfying C_2. Furthermore, as we said before, the property C_2 is not necessary to achieve convergence for networks with dedicated servers which is the main case of interest for us. For these reasons, we do not include

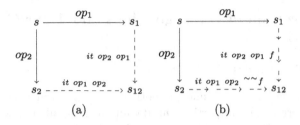

Fig. 2. Diagrams expressing different variants of convergence property C_1

the statement of C_2 into our formal definition of a convergent OT algorithm presented below.

There are two things that we change in the signature of the function `it`. The first is that we make the definition of `it` asymmetric by adding a special boolean flag which allows `it` to take into account possible difference in the priority of clients. We give an example illustrating why such a prioritization may be necessary.

Consider the following situation: Alice and Bob simultaneously insert two different characters into the same position of a shared text (say, op_1 = ins "a" 0, op_2 = ins "b" 0). The problem with the signature `it`: cmd $\rightarrow$ cmd $\rightarrow$ cmd is that there appears to be no way to implement `it` so that the resulting conflict is resolved in a semantically satisfactory manner and the property C_1 is also preserved.

A possible way to resolve this problem is to assign different priorities to different clients and then take them into account in the implementation of the function `it`. Thus, in the last example, we could agree to always insert the string typed by Alice before Bob's if both are inserted into the same position.

Thus, we are either forced to store the information about client priorities inside *cmd*, despite the fact that it is not used in the implementation of an interpretation function, or we should make the signature of `it` asymmetric. In our algorithm we use the latter approach, while the former was used, for example, in the algorithms of Ellis–Gibbs and Ressel (see [10, Sect. 2.3]). Notice that in these algorithms the priority of operations is encoded with natural numbers since they assume the presence of multiple clients with different priorities.

In contrast, in our integration algorithm we do not assign different priorities to clients. Instead, all users are connected to the central server and each client-server connection is considered as a network with two collaborating users of which the central server has the higher priority. Notice that the server does not modify the shared data by itself and is only used to propagate user-made changes. With this approach a boolean flag is sufficient to distinguish between the client and the server on each client-server connection.

The second modification is that we allow the result of a transformation of two atomic operations to be a composite operation, i.e. a list of atomic operations.

This not only makes our definition more flexible and general, but also simplifies the definition of the operation type *cmd* in practice. For example, it is very common that the result of a transformation of two nonempty primitive operations is empty (e.g., if two users simultaneously delete the same file). The fact that we allow it to return composite operations eliminates the need to define a dedicated empty operation constructor for *cmd*. Instead, it can simply return an empty composite operation. We conclude with the following two definitions which will be used throughout the rest of the article.

Definition 1. *A transformation function is a function with the following signature:*

$$\mathtt{it}: cmd \to cmd \to bool \to list \ cmd.$$

Definition 2. *We say that the transformation function* it *satisfies the property* C_1 *if for any boolean flag* f, *any pair of primitive operations* op_1, op_2, *applicable to a state* s, *can be completed to a square Fig. 2b.*

We can put together all our formal definitions stated above into the following Coq class.

```
Class OTBase (X cmd: Type) := {
interp:cmd → X → option X;
it     :cmd → cmd → bool → list cmd;
it_c1 :forall (op₁ op₂: cmd)(f: bool) m (s₁ s₂: X),
interp op₁ s = Some s₁ → interp op₂ s = Some s₂ →
let s₂₁:= exec_all interp (Some s₂) (it op₁ op₂ f) in
let s₁₂:= exec_all interp (Some s₁) (it op₂ op₁ ˜˜f)in
s₂₁ = s₁₂ /\ s₂₁ <> None}.
```

In the above code example exec_all interp is the function extending the interpretation function to composite operations in an obvious way.

In our model we also want to have a special type class formalizing the notion of an OT algorithm with invertible user operations. We can define it as a descendant class of OTBase by adding the following two members: the inversion function inv and the formula expressing the property IP_1 (see [12]). The latter asserts that the effect of any operation op applicable to a state s can be undone by means of inv op.

```
Class OTInv (X cmd : Type) (M : OTBase X cmd) := {
  inv  : cmd → cmd;
  ip1  : forall op s s₁, interp op s = Some s₁ → interp (inv op) s₁
        = Some s}.
```

Other, more subtle inversion properties have also been described in the literature (e.g., properties IP_2, IP_3, see [12]). However, these properties are not satisfied by the transformation algorithms described in Sect. 3. For this reason we do not include them into the definition of OTInv.

Apart from the property C_1, we do not impose any semantical constraints on the behavior of the transformation function. In particular, C_1 is satisfied by the trivial transformation function, which always cancels operations of both clients (e.g., it op_1 op_2 $f = [:: \text{inv } op_1]$). Another trivial example of the function satisfying C_1 is the function that always rolls back an operation of the client with a lower priority:

$$\text{it } op_1 \ op_2 \ true = [::], \quad \text{it } op_1 \ op_2 \ false = [:: \text{ inv } op_2; op_1].$$

2.2   Transformation of Composite Operations (file Comp.v)[2]

In the previous subsection we defined the signature of a transformation function in such a way that the result of transforming two atomic operations could be a composite operation. While this approach has multiple advantages mentioned above it also creates difficulties associated with the transformation of composite operations.

Imagine that we want to write a function transforming a composite operation op_1 through another composite operation op_2 provided that we already know how atomic operations are transformed. Of course, there is only one way to do this: first cut atomic pieces off op_1 and op_2, then transform these pieces using the transformation function it for atomic operations and, finally, run the transformation recursively on the remaining chunks. The following piece of code implements this behavior (cf. Fig. 3a):

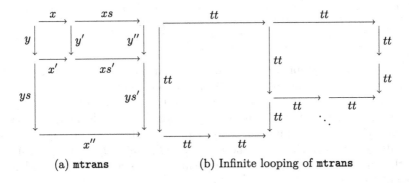

(a) mtrans   (b) Infinite looping of mtrans

Fig. 3. Illustration of the transformation of a composite operation

```
Fixpoint mtrans (it: cmd → cmd → bool → list cmd) (op1 op2:
     list cmd) (nSteps: nat): option ((list cmd) * (list cmd)):=
match nSteps with
| 0 ⇒ None
| S nSteps' ⇒
match op1, op2 with
| nil, _ | _, nil ⇒ Some (op2, op1)
| x :: xs, y :: ys ⇒
  match mtrans it xs (it y x false) nSteps' with
  | Some (y'', xs') ⇒
    match mtrans it ((it x y true) ++ xs') ys nSteps' with
```

---

[2] https://github.com/JetBrains/ot-coq/blob/master/Comp.v.

```
  | Some (ys', x'') ⇒ Some (y'' ++ ys', x'')
  | _ ⇒ None
  end
  | _ ⇒ None
  end
 end
end.
```

Notice that we had to add a parameter *nSteps* to the definition of `mtrans` to limit the recursion depth, otherwise, Coq would reject the definition as potentially nonterminating. It is easy to see that such nontermination, indeed, may occur. For example, consider the following trivial implementation of OT (with the property C_1 also trivially satisfied).

```
Definition bad_it := (fun (_ _ : unit) (_ : bool) ⇒ [::tt;
    tt]).
Instance nonterm : OTBase unit unit :=
 {interp := (fun _ _ ⇒ Some tt); it := bad_it}.
```

Although this function works well for elementary operations, we will get an infinite loop if we try to transform composite operations. Indeed, `mtrans` loops on the following simple example after the first two iterations and the result of transformation can not be computed (cf. Fig. 3b).

```
Eval compute in mtrans bad_it [::tt] [::tt] 2.
    = Some ([:: tt; tt], [:: tt; tt])
Eval compute in mtrans bad_it [::tt] [::tt; tt] 100.
    = None
```

Although the above example looks somewhat artificial it is, in fact, similar to the following situation often encountered in practice. Imagine that two concurrent user operations are semantically inconsistent with each other, i.e. there is no sensible way to transform them so that the effect of both operations is preserved. For example, this may happen if the operation model of OT becomes sufficiently complex. In this case the only way to enforce property C_1 is to rollback operation of one of the users and execute the operation of the other. Such implementation forces the transformation function to return a composite operation when invoked on a pair of elementary ones.

In the remainder of this section we describe a condition sufficient for practical purposes which guarantees that the result of transformation of two composite operations can be computed in a finite number of steps. We will further refer to this condition as the *computability property*.

The idea is to define two natural-valued functionals which should be interpreted as assignments of "size" and "cost" to an atomic operation:

$$\texttt{sz0}: cmd \rightarrow nat, \quad \texttt{si0}: cmd \rightarrow nat.$$

The rationale here is the following: if the total "size" of a composite operation increases in the process of transformation then the value of the "cost" functional should at the same time decrease. Since the latter is a natural number, it is guaranteed that the total "size" of the operation will stop increasing at some point and that the transformation function will terminate.

Formally speaking, we extend sz0 and si0 by additivity to composite operations, denoting them by sz and si. Every atomic operation is required to have nonzero "size" and both functionals sz and si are required not to increase on each transformation square from Definition 1 (i.e. the sum of values on the "transformed" right and bottom arrows should not exceed the sum of values on the "original" left and top arrows). We also assume that for each transformation square at least one of the following two statements holds:

- The operations are transformed without rollbacks. In this situation "size" functional sz does not increase for both operations, i.e. $sz\ o'_i \leq sz\ o_i$, $i = 1, 2$.
- If one of the operations is rolled back, the "size" of one of the transformed arrows may increase. In this situation we require that "cost" functional strictly decreases on such transformation square.

It is easy to see that under these assumptions the result of transforming two composite operations o_1 and o_2 can be computed in less than $(sz\ o_1 + sz\ o_2)^2 +$ $si\ o_1 + si\ o_2$ steps (i.e. atomic transformations). We formulate this as a theorem in Coq.

```
Context {X cmd: Type} (ot: OTBase X cmd) (comp: OTComp X cmd
    ot).
Theorem ot_computable: forall (op₁ op₂: list cmd),
    exists nSteps, mtrans it op₁ op₂ nSteps <> None.
```

3 Examples of Verified Transformation Functions

In the current section we present two concrete implementations of abstract classes defined in Sect. 2. In each of the two cases we describe how exactly the abstract classes and signatures are instantiated and also outline the idea of the proof of corresponding convergence and computability properties. The algorithms described below are contained in the following modules of our library: TreeOt.v[3], Fs.v[4], RichText.v[5].

3.1 The Case of Ordered Trees with Labels

In this subsection we describe an OT algorithm for concurrent editing of ordered trees. The algorithm in question is a modification of the OT algorithm of Ressel for text buffers (cf. [10, 2.3.2]).

We are working with ordered trees labeled by elements of some fixed type T (i.e. a label of type T is assigned to every vertex of a tree). Of course, we will need the two most basic operations of tree editing: insertion and removal of a tree branch. Similarly to Sun's algorithm for strings (see [10, 2.3.3], [13]) we allow sequential insertion and deletion of multiple tree branches in a single

---

[3] https://github.com/JetBrains/ot-coq/blob/master/TreeOt.v

[4] https://github.com/JetBrains/ot-coq/blob/master/Fs.v

[5] https://github.com/JetBrains/ot-coq/blob/master/RichText.v

operation. We also include one more operation `EditLabel` which modifies the label of a single tree node via some user-defined set of commands TC and, otherwise, leaves the tree structure unchanged.

Since we want our OT algorithm for trees to be C_1-consistent we should first assume the OT algorithm for labels to be C_1-consistent (see Sect. 2). We implement this in Coq by adding several parameter variables.

```
Context {T: eqType} (TC: Type) {otT: OTBase T TC}.
```

Now, we can give the following definition for the type of elementary operations:

```
Inductive tree_cmd : Type :=
 | EditLabel of TC
 | TreeInsert of nat & list (tree T)
 | TreeRemove of nat & list (tree T)
 | OpenRoot of nat & tree_cmd.
```

The first three constructors of this type correspond to the operations modifying the root node of a tree, while a sequence of `OpenRoot` constructors can be used to specify a position in the tree to which the first three operations are to be applied.

The semantics of the interpretation function `interp` for this operation set is as follows.

- Case `EditLabel` tc. The operation executes command tc on the label of the tree's root node using the function `interp` specified in the "parameter" class otT.
- Case `TreeInsert` $n\ l$. The operation inserts the list l into n-th position of the children list of the root node. `None` is returned if a range check error occurs during this process.
- Case `TreeRemove` $n\ l$. The operation compares the list l with the sublist of branches of the root node starting at n-th position. If these lists are the same then the corresponding sublist of children is removed from the root node. Otherwise, or if a range check error occurs, the function returns `None`.
- Case `OpenRoot` $n\ c$. The operation applies operation c to the n-th child of the root. The operation returns `None` if there is no child with such index.

The behavior of the interpretation function is illustrated in Fig. 4.

It is easy to see that the above set of operations satisfies property IP_1 provided so does the algorithm for labels. The inversion function can be defined by swapping `TreeInsert` and `TreeRemove` constructors:

```
Context (ipT : OTInv _ _ otT).
Fixpoint tree_inv (c : tree_cmd) :=
match c with
 | EditLabel c'  ⇒ EditLabel (@inv _ _ _ ipT c')
 | TreeInsert n l ⇒ TreeRemove n l
 | TreeRemove n l ⇒
  if l is [::] then TreeInsert 0 [::] else TreeInsert n l
 | OpenRoot n c'  ⇒ OpenRoot n (tree_inv c')
```

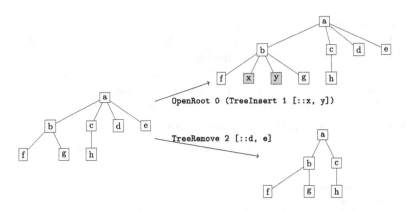

Fig. 4. An example illustrating the behavior of operations `TreeInsert` and `TreeRemove`

```
end.
Instance treeInv: (OTInv (tree T) tree_cmd treeOT) := {inv :=
    tree_inv}.
```

Notice that the fact that we compare the list of nodes in the actual model with the list of nodes specified in `TreeRemove` command before executing the latter is essential for checking property IP_1.

The next step is to define the transformation function for tree operations. First, we define some auxiliary functions.

```
Definition tr_ins (len: nat) (n₁ n₂: nat): nat :=
  if (n₁ < n₂) then n₁ else n₁ + len.
Definition tr_rem (len: nat) (n₁ n₂: nat): option nat :=
  if (n₁ < n₂) then Some n₁ else
    (if (n₁ >= n₂ + len) then Some (n₁ - len) else None).
Fixpoint cut {X} (l : list X) (sc rc : nat) :=
match sc, rc, l with
| S sc', _, x :: xs ⇒ x :: (cut xs sc' rc)
| 0, S rc', x :: xs ⇒ cut xs sc rc'
| _, _, _ ⇒ l
end.
```

The function `cut` $l\ n\ m$ removes from l the sublist of length m starting from n-th position. Now we are all set to define the transformation function for trees. The idea behind its definition is that we attempt to preserve the intuitive "effect" of both operations op_1 and op_2. To accomplish this we have to perform a large case-by-case analysis.

```
Fixpoint tree_it (op₁ op₂: tree_cmd) (f: bool): list tree_cmd :=
let triv := [:: op₁] in
match op₁, op₂ with
| EditLabel c₁, EditLabel c₂ ⇒
  map EditLabel (@it _ _ otT c₁ c₂ f)
| EditLabel _, _ | _, EditLabel _ ⇒ triv
| OpenRoot n₁ tc₁, OpenRoot n₂ tc₂ ⇒
  if n₁==n₂ then map(OpenRoot n₁) (tree_it tc₁ tc₂ f) else triv
```

```
| TreeRemove n₁ l₁, OpenRoot n₂ tc₂ ⇒
match tr_rem (size l₁) n₂ n₁ with
| None ⇒ let i := n₂ - n₁ in
  match replace i (tree_interp tc₂ (nth i l₁)) l₁ with
  | Some l₁′ ⇒ [:: TreeRemove n₁ l₁′]
  | _ ⇒ triv
  end
| _ ⇒ triv
end
| _, OpenRoot _ _ ⇒ triv
| OpenRoot n₁ tc₁, TreeInsert n₂ l₂ ⇒
[:: OpenRoot (tr_ins (size l₂) n₁ n₂) tc₁]
| OpenRoot n₁ tc₁, TreeRemove n₂ l₂ ⇒
match tr_rem (size l₂) n₁ n₂ with
| Some n₁′ ⇒ [:: OpenRoot n₁′ tc₁]
| None ⇒ nil
end
| TreeInsert n₁ l₁, TreeInsert n₂ l₂ ⇒
if (n₁==n₂) then
(if f then triv else [::TreeInsert (n₁+size l₂) l₁])
else [:: TreeInsert (tr_ins (size l₂) n₁ n₂) l₁]
| TreeInsert n₁ l₁, TreeRemove n₂ l₂ ⇒
let len := size l₂ in
if n₁≤n₂ then triv else
if n₁≥n₂+len then[::TreeInsert (n₁-len) l₁] else nil
| TreeRemove n₁ l₁, TreeRemove n₂ l₂ ⇒
let (len₁, len₂) := (size l₁, size l₂) in
(if n₂ + len₂≤n₁ then [::TreeRemove (n₁-len₂) l₁] else
  (if n₂ ≤ n₁ then
    (if n₂ + len₂ < n₁ + len₁
      then [:: TreeRemove n₂ (cut l₁ 0 (len₂+n₂-n₁))]
      else nil)
    else [:: TreeRemove n₁ (cut l₁ (n₂-n₁) len₂)])))
| TreeRemove n₁ l₁, TreeInsert n₂ l₂ ⇒
let (len₁, len₂) := (size l₁, size l₂) in
if n₁ + len₁ ≤ n₂ then triv else
(if n₂ ≤ n₁ then [:: TreeRemove (n₁+len₂) l₁] else
  match insert (n₂ - n₁) l₂ l₁ with
  | Some l₁′ ⇒ [:: TreeRemove n₁ l₁′]
  | None ⇒ triv
  end)
end.

Instance treeOT: (OTBase (tree T) tree_cmd) :=
{interp := tree_interp; it := tree_it}.
```

The above transformation function always cancels **TreeInsert** whenever it conflicts with a concurrent **TreeRemove**. In order to transform two concurrent **EditLabel** operations the transformation function for the label type is invoked. Notice that the priority flag f is used in this piece of code to resolve the conflicting situation when two lists of trees are concurrently inserted into the same position.

The proof of the fact that tree_it satisfies property C_1 is rather bulky and technical. After applying induction over op_1 it essentially comes down to proving a number of commutation lemmas about list operations. The proof of C_1 itself takes 250 lines of Coq code, in addition, further 700 lines are occupied by commutation lemmata for list operations.

Notice that the problem of transforming composite operations mentioned in Sect. 2.2 does not arise for tree_it provided it does not arise for the "parameter" algorithm otT. The reason for this is that tree_it can only return a proper composite operation as a result of transformation of two EditLabel operations (which is clear from the examination of the definition).

3.2   The Case of Unordered Trees

Now we describe a transformation algorithm for unordered trees i.e., trees for which the order among siblings is unimportant. Informally speaking, the difference between the datatypes of ordered and unordered trees is the same as between the datatypes of ordered lists and sets. The directory structure of a filesystem may serve as an example of an unordered tree.

From the implementation viewpoint it will be convenient for us to consider unordered trees as usual ordered trees whose branches are sorted with respect to some total ordering defined on the type of labels. In particular, such implementation simplifies the test for equality and also allows us to implement unordered trees in Coq as a *subset type*, i.e. as an ordered pair consisting of a tree and the evidence (proof object) of its sortedness (see, e.g., [1, Sect. 6]).

The main difference between the operation set for unordered trees and the operation set from the previous subsection is that now we refer to nodes of an unordered tree using their labels rather than indices. We support 3 different atomic tree operations: modification of the node's label (file renaming), insertion and deletion of a subtree.

```
Inductive raw_fs_cmd T :=
 | Edit of T & T
 | Create of tree T
 | Remove of tree T
 | Open of T & raw_fs_cmd.
```

More formally, semantics of these operations can be described as follows.

- Case Edit l_1 l_2. The operation seeks a child with label l_1 among the children of the root node. If such a child is found, its label is changed to l_2. None is returned if there is no such child, or if there is another child with label l_2 among the children of the root node.
- Case Create t. The operation adds t to the set of children of the root node. None is returned if there is already another node with the same label.
- Case Remove t. The operation looks for a child of the root node which coincides with t and then removes it. None is returned if such a child is not found.
- Case Open l c. The operation looks for a child with label l and applies operation c to it. As before, None is returned if there is no such child.

The transformation function presented below is even simpler as compared to the function **tree_it** from Sect. 3.1 because in the context of unordered trees there is no need to compare indices of operations and only node labels are to be taken into account (**value** t denotes the label of the root node of t).

```
Fixpoint fs_it (op₁ op₂ : raw_fs_cmd) (f : bool) : seq fs_cmd :=
match op₁, op₂ with
| Edit l₁ l'₁, Edit l₂ l'₂ ⇒
 match l₁ == l₂, l'₁ == l'₂ with
 |false, false ⇒ [:: op₁]
 |true, true ⇒ [::]
 |true, false ⇒ (if f then [::] else [:: Edit l'₂ l'₁])
 |false, true ⇒ [:: Edit l'₂ l₂]
 end
| Edit l₁ l'₁, Create t₂ ⇒
 if l'₁ == value t₂ then [:: Remove t₂; op₁] else [:: op₁]
| Edit l₁ _, Remove t₂ ⇒
 if l₁ == value t₂ then [::] else [:: op₁]
| Create t₁, Edit l₂ l'₂ ⇒
 if value t₁ == l'₂ then [::] else [:: op₁]
| Create t₁, Create t₂ ⇒
 if value t₁ == value t₂ then merge_trees t₁ t₂ else [:: op₁]
| Remove t₁, Edit l₂ l'₂ ⇒
 if l₂ == value t₁ then [:: Remove (Node l'₂ (children t₁)) ] else [::
  op₁]
| Remove t₁, Remove t₂ ⇒
 if value t₁ == value t₂ then [::] else [:: op₁]
| Remove t₁, Open l₂ yc ⇒
 if value t₁ == l₂ then
 (if fs_interp yc t₁ is Some t then [:: Remove t]
  else [::])
 else [:: op₁]
| Open l₁ c₁, Edit l₂ l'₂ ⇒
 if l₁ == l₂ then [:: Open l'₂ c₁ ] else [:: op₁]
| Open l₁ _, Remove t₂ ⇒
 if value t₂ == l₁ then [::] else [:: op₁]
| Open l₁ c₁, Open l₂ c₂ ⇒
 if l₁ == l₂ then map (Open l₁) (fs_it c₁ c₂ f)
 else [:: op₁]
| _, _ ⇒ [:: op₁]
end.
```

In the above piece of code the priority flag f is used to resolve the conflicting situation when two users concurrently attempt to replace a node's label with two different labels. There is a notable conflicting situation which did not occur in the previous section, namely when two different trees with identical labels are concurrently inserted into the same node. We handle such a conflict by recursively merging the contents of these trees (**merge_trees** t_1 t_2 generates a sequence of create operations for all descendants of t_1 that are not descendants of t_2).

Notice that the above transformation function can return a composite operation as a result of transforming two atomic **Create** operations. In order to verify that the result of transformation can always be computed in a finite number of steps we use the sufficient condition for computability i.e. we define certains functionals `fs_sz0, fs_si0: raw_fs_cmd → nat` and check that they satisfy the inequalities of Sect. 2.2.

4   Related Work

In [3] A.H. Davis et al. have proposed an OT algorithm for editing structured data (e.g., trees). However, their protocol has not been formally verified and, moreover, assumed the presence of a garbage collector.

In [7,10] SPIKE theorem prover and UPPAAL TIGA model checker were used to find counterexamples violating the correctness of several published OT algorithms for strings. As a result, it was shown that all tested algorithms violate property C_2 while some of them violate even C_1. In [9] Coq was applied to the same problem of finding counterexamples to property C_2.

However, to the best of our knowledge, our work is the first to obtain a formal proof of property C_1 by means of a proof assistant.

5   Conclusions and Future Work

In our work we attempted to formalize OT algorithms for two different datatypes commonly met in practice and obtained the following results:

- a library containing commutation lemmas for lists and trees has been developed (file `ListTools.v`)[6];
- the correctness of several inclusive transformation functions has been verified by means of Coq (Sects. 3.1, 3.2).

Our algorithms may be easily extracted from the library (either automatically or manually) and be used in software applications.

Along the way, we propose a generalized signature of the transformation function `it` which allows the transformation result to be a composite operation. The latter allows the implementation of `it` to be more flexible and, moreover, makes possible to define semantically correct transformation functions without polluting the set of commands with domain-irrelevant entities such as *nop* operation. We also present a sufficient condition ensuring computability of transformed operations in this setting, which is an essential requirement for any practical implementation. We believe that the ideas behind our approach are general enough to be successfully applied to other data models as well.

There is a number of different directions one could take for future research within this topic. For example, one could try to figure out whether property C_2 holds for the algorithm formulated in Sect. 3.2. One could also attempt to formalize OT algorithms for different datatypes and try different sets of operations for the datatypes used in this article. For example, our library already contains an attempt to extend the operation set of Sect. 3.1 with two more operations

---

[6] https://github.com/JetBrains/ot-coq/blob/master/ListTools.v

`TreeUnite` and `TreeFlatten` which may be useful for the implementation of rich-text editors (see `RichText.v`) https://github.com/JetBrains/ot-coq/blob/master/RichText.v.

References

1. Chlipala, A.: Certified Programming with Dependent Types. The MIT press, Cambridge (2013)
2. Cormack, G.: A counterexample to the distributed operational transform and a corrected algorithm for point-to-point communication. Technical report, CS-95-06, Univ. Waterloo (1995). http://cs.uwaterloo.ca/research/tr/1995/08/dopt.pdf
3. Davis, A.H., Sun, C., Lu, J.: Generalizing operational transformation to the standard general markup language. In: Proceedings of Computer Supported Cooperative Work, pp. 58–67 (2002). http://dx.doi.org/10.1145/587078.587088
4. Ellis, C., Gibbs, S.: Concurrency control in groupware systems. In: Proceedings of 1989 ACM SIGMOD International Conference on Management of Data, vol. 18, pp. 399–407 (1989). http://dx.doi.org/10.1145/66926.66963
5. Gonthier, G., Mahboubi, A., Tassi, E.: A Small Scale Reflection Extension for the Coq system. Research Report RR-6455, Inria Saclay Ile de France (2015). https://hal.inria.fr/inria-00258384
6. Ignat, C.L., Norrie, M.C.: Customizable collaborative editor relying on treeOPT algorithm. In: Proceedings of 8th European Conference Computer Supported Cooperative Work, ECSCW 2003, pp. 315–334 (2003). http://citeseer.ist.psu.edu/viewdoc/download?doi=10.1.1.70.9273&rep=rep1&type=pdf
7. Imine, A., Rusinowitch, M., Molli, P., Oster, G.: Formal design and verification of operational transformation for copies convergence. Theor. Comput. Sci. **351**(2), 167–183 (2006). http://dx.doi.org/10.1016/j.tcs.2005.09.066
8. Lushman, B., Cormack, G.V.: Proof of correctness of Ressel's adOPTed algorithm. Inform. Process. Lett. **86**(6), 303–310 (2003). http://www.sciencedirect.com/science/article/pii/S0020019003002278
9. Lushman, B., Roegist, A.: An automated verication of property tp2 for concurrent collaborative text buffers. Technical report, CS-2012-25, Univ. Waterloo, December 2012. http://cs.uwaterloo.ca/sites/ca.computer-science/files/uploads/files/CS-2012-25.pdf
10. Randolph, A., Boucheneb, H., Imine, A., Quintero, A.: On consistency of operational transformation approach. Elec. Proc. Theor. Comput. Sci. **107**, 45–59 (2013). http://dx.doi.org/10.4204/EPTCS.107.5
11. Suleiman, M., Cart, M., Ferrié, J.: Concurrent operations in a distributed and mobile collaborative environment. In: Proceedings of 14th International Conference Data Engineering, Orlando, FL, USA, 23–27 February, pp. 36–45 (1998). http://dx.doi.org/10.1109/ICDE.1998.655755
12. Sun, C.: Operational transformation frequently asked questions and answers (2010). http://cooffice.ntu.edu.sg/otfaq/
13. Sun, C., Jia, X., Zhang, Y., Yang, Y., Chen, D.: Achieving convergence, causality preservation, and intention preservation in real-time cooperative editing systems. ACM Trans. Comput. Hum. Interact. **5**(1), 63–108 (1998). http://dx.doi.org/10.1145/274444.274447

Hereditarily Finite Sets
in Constructive Type Theory

Gert Smolka[✉] and Kathrin Stark[✉]

Saarland University, Saarbrücken, Germany
{smolka,kstark}@ps.uni-saarland.de

Abstract. We axiomatize hereditarily finite sets in constructive type
theory and show that all models of the axiomatization are isomorphic.
The axiomatization takes the empty set and adjunction as primitives and
comes with a strong induction principle. Based on the axiomatization, we
construct the set operations of ZF and develop the basic theory of finite
ordinals and cardinality. We construct a model of the axiomatization as a
quotient of an inductive type of binary trees. The development is carried
out in Coq.

1 Introduction

An HF set (hereditarily finite set) is a finite and well-founded set whose elements
are HF sets. The class of HF sets may be defined inductively:

– The empty set is an HF set.
– If x and y are HF sets, then $\{x\} \cup y$ is an HF set.

We call the operation $x.y := \{x\} \cup y$ adjunction. Set membership can be
expressed with adjunction and equality: $x \in y \leftrightarrow x.y = y$. Ackermann [1]
discovered that the natural numbers are in one-to-one correspondence with the
HF sets, and that the class of HF sets satisfies all axioms of ZF set theory but
infinity.

We present an axiomatization of HF sets in a constructive type theory without
inductive types and obtain the following results:

– All models of the axiomatization are isomorphic.
– The usual set operations, including separation, replacement, union, power,
  and transitive closure, can be constructed.
– A cardinality operation mapping sets to equipotent ordinals can be constructed.
– A model of the axiomatization can be constructed as a quotient of an inductive
  type of binary trees.

Our axiomatization of HF sets assumes a type X of sets and constants for
the empty set and adjunction. There are four basic axioms

– $x.(x.y) = x.y$

© Springer International Publishing Switzerland 2016
J.C. Blanchette and S. Merz (Eds.): ITP 2016, LNCS 9807, pp. 374–390, 2016.
DOI: 10.1007/978-3-319-43144-4_23

- $x.(y.z) = y.(x.z)$
- $x.y \neq \emptyset$
- $x.(y.z) = y.z \;\to\; x = y \;\vee\; x.z = z$

and a strong induction principle:

- $\forall p : X \to \mathsf{Type}.\; p\emptyset \to (\forall xy.\; px \to py \to p(x.y)) \to \forall x.\, px$

We speak of a strong induction principle since it applies to functions into Type rather than just functions into Prop (i.e., predicates). The strong induction principle provides for the recursive definition of functions and ensures that all sets are finite and well-founded. In contrast to recursors for inductive types, the induction principle does not come with equations.

Related work. Several axiomatizations of hereditarily finite sets appear in the literature: Takahashi 1977 [8], Givant and Tarski 1977 [2], Previale 1994 [6], Świerczkowski 2003 [7], and Kirby 2009 [3]. All of them are formulated as first-order theories, and all of them employ the empty set, adjunction, and an adjunction-based induction principle as ingredients. Except for Kirby's axiomatization, which uses no additional constant, the existing axiomatizations are formulated with an additional constant for set membership. Except for Previale's axiomatization, which is studied in an intuitionistic setting, the existing axiomatizations are studied in a classical setting. Previale's axiomatization employs both membership and its transitive closure as additional constants. Previale derives the decidability of equality and membership. Our axiomatization extends Kirby's axiomatization by strengthening the induction principle to types.

A type of HF sets is available in Isabelle/HOL. It is realized with Ackermann's [1] encoding. Paulson [4,5] makes essential use of the type of HF sets in his formalizations of finite automata and Godel's incompleteness theorems in Isabelle/HOL. Interestingly, Isabelle/HOL can also define a type of HF sets by recursion through a type constructor for finite sets over a given base type.

Contribution of the paper. The paper explores for the first time an axiomatization of HF sets in constructive type theory. We show that the axiomatization is categorical, a result that has not been shown before for any of the existing axiomatizations. Our axiomatization extends Kirby's axiomatization by strengthening the induction principle from predicates to general functions.

We construct a model of the axiomatization as a quotient of an inductive type of binary trees. This natural model construction (from the perspective of constructive type theory) does not appear in the literature. It complements a conventional model construction based on numbers and Ackermann's encoding. To obtain the quotient with minimal assumptions, we base the construction on a normalizing sorting function for the lexical tree ordering.

Organization of the paper. We start with a section recalling the underlying type theory and basic notions like decidability. We also recall how quotients can be obtained as subtypes based on normalizers. We then introduce HF structures and establish basic results including extensionality, decidability, and strong epsilon induction. In Sect. 4 we show that all HF structures are isomorphic. We then construct the basic set operations using the strong induction principle and membership-based specifications. In Sects. 6 and 7 we consider ordinals and define equipotence of sets. Based on an inductively defined cardinality relation, we show that every equipotence class contains exactly one ordinal, and that equipotence is a decidable equivalence relation. We use the strong induction principle to construct a cardinality operator. In Sect. 8 we extend the underlying type theory with an inductive type of binary trees and show that every HF structure is a quotient of the tree type. In Sect. 9 we define tree equivalence and construct a normalizing sorting function for the lexical tree order. Based on the sorting function, we obtain an HF structure, thus showing consistency of the axiomatization of HF sets.

Accompanying Coq development. The development of the paper is formalized in Coq. The Coq development is available at http://www.ps.uni-saarland.de/extras/hfs and contains additional results that for space reasons could not be included in the paper.

2   Preliminaries

We assume a constructive type theory with dependent function types, dependent pair types, sum types, and an impredicative universe Prop of propositions. We do not use inductive types except for the model construction in Sect. 9, which requires an inductive type of binary trees and an inductive proposition $\top$ with exactly one proof.

We will frequently use inductively defined predicates, which are always obtained as impredicatively defined intersection predicates. The logical operations and the equality predicate are also defined impredicatively.

We write $P \mathbin{\underline{\vee}} Q$ for strong disjunctions (sums $P + Q$ in Coq) and $\underline{\exists}x.px$ for strong existentials (sigT p in Coq). A proposition P is *decidable* if $P \mathbin{\underline{\vee}} \neg P$.

A *decidable predicate* on a type X is a pair consisting of a predicate $p : X \to$ Prop and a function $\forall x.\ px \mathbin{\underline{\vee}} \neg px$. Decidable binary predicates are defined analogously.

A *discrete type* is a pair of a type X and a function $\forall xy.\ x = y \mathbin{\underline{\vee}} x \neq y$.

In Sect. 9 we will construct a quotient of an inductive tree type. The quotient will be obtained as a subtype of the tree type consisting of the fixed points of a normalizer for the underlying equivalence relation. The quotient construction will be an instance of the abstract subtype construction described in the following.

We assume a proposition $\top$ such that $\forall AB : \top.\ A = B$.
A predicate $p : X \to$ Prop is *pure* if $\forall x \forall AB : px.\ A = B$.

Fact 1. *For every decidable predicate there is an equivalent pure predicate.*

Proof. Let p be a decidable predicate. Then $\lambda x.$ if px then $\top$ else $\bot$ is an equivalent pure predicate ($\bot := \forall P :$ Prop. P is falsity). $\square$

Fact 2 (Subtype). *Let f be an idempotent function on a discrete type A. Then there are a discrete type X and functions $S : A \to X$ and $I : X \to A$ such that $S(Ix) = x$ and $I(Sa) = fa$ for all x and all a.*

Proof. The assumptions suffice to construct a pure predicate p such that $pa \leftrightarrow fa = a$ and a function $F : \forall a.\ p(fa)$. We define $X := \exists a.pa$, $Sa := (fa, Fa)$, and $I(a, \phi) := a$. $\square$

We may see the type X established by Fact 2 in several ways:

1. X is the subtype of A consisting of all fixed points of f.
2. X is the subtype of A consisting of all points in the range of f.
3. X is the quotient of A under the equivalence relation $\lambda ab.\ fa = fb$ induced by f.

A *normalizer* for a relation $\sim$ on a type X is an idempotent function $f : X \to X$ such that $x \sim y \leftrightarrow fx = fy$ for all x and y. Obviously, a relation is an equivalence relation if it has a normalizer. Moreover, a relation on a discrete type is decidable if it has a normalizer.

3   HF Structures

We axiomatize HF sets with a type of sets, a constant $\emptyset$ for the empty set, and a binary operation $x.y$ on sets we call adjunction. Informally, $x.y$ is the set $\{x\} \cup y$.

Formally, an *HF structure* consists of the following:

- A type X. The elements of X are called *sets*.
- A set $\emptyset$ called *empty set*.
- A function $X \to X \to X$ called *adjunction*. We write $x.y$ for the adjunction of two sets x and y.
- A function $\forall p : X \to$ Type. $p\emptyset \to (\forall xy.\ px \to py \to p(x.y)) \to \forall x.px$ called *strong induction principle*.
- The following laws:
  - $x.(x.y) = x.y$ $\hfill$ *cancellation law*
  - $x.(y.z) = y.(x.z)$ $\hfill$ *swap law*
  - $x.y \neq \emptyset$ $\hfill$ *discrimination law*
  - $x.(y.z) = y.z \ \to \ x = y \ \vee \ x.z = z$ $\hfill$ *membership law*

We write $x.y.z$ for $x.(y.z)$.

Given an HF structure, we define *membership* and *inclusion*:

$$x \in y := (x.y = y)$$
$$x \subseteq y := \forall z.\ z \in x \to z \in y$$

Using the notation for membership, we may write the membership law more suggestively as $x \in y.z \to x = y \vee x \in z$.

Example 3. We prove $(\emptyset.\emptyset).\emptyset \neq \emptyset.\emptyset$. Suppose $A : (\emptyset.\emptyset).\emptyset = \emptyset.\emptyset$. Cancellation gives us $(\emptyset.\emptyset).(\emptyset.\emptyset).\emptyset = \emptyset.\emptyset$. Thus $\emptyset.\emptyset \in \emptyset.\emptyset$ using A. Thus either $\emptyset.\emptyset = \emptyset$ or $\emptyset.\emptyset \in \emptyset$ with the membership law. In either case we have a contradiction by the discrimination law.

We assume an HF structure X and use the letters x, y, z, a, and b to denote sets in X.

Fact 4 (Decomposition). $x = \emptyset \;\bar{\vee}\; \bar{\exists}a\bar{\exists}y.\; x = a.y.$

Proof. Immediate consequence of the strong induction principle.

Fact 5.

1. $z \notin \emptyset$.
2. $z \in x.y \;\leftrightarrow\; z = x \vee z \in y$.
3. $x.y \subseteq z \;\leftrightarrow\; x \in z \wedge y \subseteq z$.
4. $x \subseteq \emptyset \leftrightarrow x = \emptyset$.
5. $a \notin x \rightarrow x \subseteq a.y \rightarrow x \subseteq y$.

Proof. Straightforward.

The sets of an HF structure are extensional in that two sets are equal if they have the same elements. Proving this basic fact constructively is not straightforward. We employ a nested HF induction and interleave the extensionality proof with proofs for the decidability of membership, inclusion, and equality of HF sets. The proof is organized in three lemmas.

Lemma 6.

1. $\emptyset \subseteq x$ and $x \subseteq \emptyset$ and $x \in \emptyset$ and $x = \emptyset$ are decidable.
2. If $x = a$ and $x \in y$ are decidable, $x \in a.y$ is decidable.
3. If $a \in y$ and $x \subseteq y$ are decidable, $a.x \subseteq y$ is decidable.
4. $\emptyset \in x$ is decidable.

Proof. Claim (1) follows with Fact 4. Claims (2) and (3) follow with Claims (2) and (3) of Fact 5. Claim (4) follows by induction on x using Claims (2) and (1).

Lemma 7 (Partition). *Let $a \in x$. Then there strongly exists a set u such that $x = a.u$ and $a \notin u$, provided the propositions $a \in z$ and $a = z$ are decidable for all sets z.*

Proof. By induction on x. The case $x = \emptyset$ is contradictory. Let $x = b.x$. By assumption, $a \in x$ is decidable. If $a \notin x$, the claim follows with $u = x$ and Fact 5 (2). Otherwise, let $a \in x$. By the inductive hypothesis we have a set u such that $x = a.u$ and $a \notin u$. By assumption, $a = b$ is decidable. If $a = b$, the claim follows with $u = x$. If $a \neq b$, the claim follows with $u = b.x$.

Lemma 8. *For all sets x and y:*

1. *$x \subseteq y$ and $y \subseteq x$ are decidable.*
2. *$x \in y$ and $y \in x$ are decidable.*
3. *$x \subseteq y \to y \subseteq x \to x = y$.*
4. *$x = y$ is decidable.*

Proof. We prove the claims simultaneously by nested induction on x and y. If $x = \emptyset$ or $y = \emptyset$, the claims follow with Lemma 6. Otherwise, we have $x = a.x$ and $y = b.y$ and inductive hypotheses for a, x, b, and y.

1. $a.x \subseteq b.y$ and $b.y \subseteq a.x$ are decidable. Follows by Lemma 6 (3) and the inductive hypotheses for a, x, b, and y.
2. $a.x \in b.y$ and $b.y \in a.x$ are decidable. Follows by Lemma 6 (2) and the inductive hypotheses for b, y, a, and x.
3. $a.x \subseteq b.y \to b.y \subseteq a.x \to a.x = b.y$. Let $a.x \subseteq b.y$ and $b.y \subseteq a.x$. We show $a.x = b.y$. By the inductive hypothesis for x we know that $a \in x$ is decidable. Case analysis.
   (a) $a \in x$. Then $a.x = x$ and the claim follows by Claim (3) of the inductive hypothesis for x.
   (b) $a \notin x$. We have $a \in b.y$. By Lemma 7 we have a set u such that $b.y = a.u$ and $a \notin u$. Thus it suffices to show $x = u$, which follows by Claim (3) of the inductive hypothesis for x provided we have $x \subseteq u$ and $u \subseteq x$. The two inclusions hold since $a.x \subseteq a.u$, $a \notin x$, $a.u \subseteq a.x$, and $a \notin u$.
4. $a.x = b.y$ is decidable. Case analysis based on (1).
   (a) $a.x \subseteq b.y$ and $b.y \subseteq a.x$. Then $a.x = b.y$ by (3).
   (b) $a.x \not\subseteq b.y$ or $b.y \not\subseteq a.x$. Then $a.x \neq b.y$. □

Theorem 9 (Extensionality). $(\forall z.\ z \in x \leftrightarrow z \in y) \to x = y$.

Proof. Follows with Lemma 8 (3).

Corollary 10. *Set inclusion is a partial ordering on sets.*

Theorem 11 (Decidability). *Equality, membership, and inclusion of HF sets are decidable.*

Proof. Follows with Lemma 8.

Fact 12 (Decidability). *The propositions $\exists z.\ z \in x \land pz$ and $\forall z.\ z \in x \to pz$ are decidable if p is a decidable predicate.*

Proof. Follows by induction on x.

Fact 13 (Partition). $a \in x \to \bar{\exists} u.\ x = a.u \land a \notin u$.

Proof. Follows with Lemmas 7 and 8.

Fact 14 (Strong Epsilon Induction).
$\forall p : X \to \mathsf{Type}.\ (\forall x.\ (\forall z \in x.\ pz) \to px) \to \forall x.\ px.$

Proof. Assume $p : X \to$ Type and $A : \forall x. (\forall z \in x. \ pz) \to px$. By A it suffices to prove $\forall z \in x. \ pz$. We prove this claim by induction on x. The case for $x = \emptyset$ is obvious. Let $x = a.x'$ and $z \in a.x'$. It suffices to show pz. If $z = a$, then pa follows by A and the inductive hypothesis for a. Otherwise, $z \in x'$ and the claim follows by the inductive hypothesis for x'.

Fact 15. $x \notin x$.

Proof. By ϵ-induction we have $z \notin z$ for all $z \in x$. The claim follows.

Corollary 16. *There is no set that contains all sets.*

Fact 17. $x \in y \to y \notin x$.

Proof. By ϵ-induction.

The operation $\lambda x.x.x$ of self-adjunction is known as *successor operation*. The successor of x contains the elements of x plus one additional element, which is x itself. That x is in fact a new element is asserted by Fact 15.

The successor operation is injective.

Fact 18 (Successor Injectivity). $x.x = y.y \to x = y$.

Proof. Let $x.x = y.y$. Then $x \in y.y$ and $y \in x.x$. By the membership law, we have either $x = y$ or $y = x$ or $x \in y \in x$. The third case is impossible by Fact 17.

4   Categoricity

We show that all HF structures are isomorphic. In fact, given two HF structures X and Y, there is exactly one homomorphism from X to Y. We obtain the homomorphism with the recursion principle from an inductively defined relational version of the homomorphism.

We start with the definition of an inductive predicate[1]

$$R : \forall X \, Y : \text{HF.} \ X \to Y \to \text{Prop}$$

homomorphically relating the sets of two HF structures:

$$\frac{}{R\emptyset\emptyset} \qquad \frac{Rab \qquad Rxy}{R(a.x)(b.y)}$$

Given HF structures X and Y, we will show that R_{XY} is a bijection.

Fact 19 (Symmetry). $R_{XY} \, xy \to R_{YX} \, yx$.

Proof. By induction on Rxy.

---

[1] An impredicative definition of R looks as follows: $\lambda(XY : \text{HF})(x : X)(y : Y)$. $\forall S : X \to Y \to \text{Prop.} \ S\emptyset\emptyset \to (\forall axby. \ Sab \to Sxy \to S(a.x)(b.y)) \to Sxy$.

Fact 20 (Strong Totality). $\forall x \bar{\exists} y.\ Rxy.$

Proof. By induction on x.

Proving that R is functional requires some effort. The key ingredients are extensionality and a simulation lemma for membership.

Lemma 21 (Simulation). $Rxy \to a \in x \to \exists b.\ b \in y \wedge Rab.$

Proof By induction on Rxy. The case for the first rule is trivial. For the second rule, we have $Ra'b'$, $Rx'y'$, and $a \in a'.x'$ and need a set $b \in b'.y'$ such that Rab. If $a = a'$, $b := b'$ does the job. Otherwise, we have $a \in x'$. By the inductive hypothesis we obtain $b \in y'$ with Rab. The claim follows since $y' \subseteq b'.y'$.

Fact 22 (Functionality). $Rxy \to Rxy' \to y = y'.$

Proof. By ϵ-induction on x. We show $y = y'$ using extensionality (Theorem 9). Let $b \in y$. We show $b \in y'$. By the facts for symmetry and simulation we obtain an $a \in x$ such that Rab. By the simulation lemma we obtain $b' \in y'$ such that Rab'. By the inductive hypothesis we have $b = b'$. Thus $b \in y'$. We now have $y \subseteq y'$. The other direction $y' \subseteq y$ follows analogously.

A *homomorphism* from an HF structure X to an HF structure Y is a function $f : X \to Y$ such that $f\emptyset = \emptyset$, and $f(a.x) = fa.fx$ for all a and x. Two HF structures X and Y are *isomorphic* if there are homomorphisms $f : X \to Y$ and $g : Y \to X$ such that $g(fx) = x$ and $f(gy) = y$ for all x and y.

Fact 23. *Let f be a homomorphism from an HF structure X to an HF structure Y. Then $Rx(fx)$ for all x.*

Proof. By induction on x.

Fact 24. *All homomorphisms between two HF structures are equivalent.*

Proof. Follows with Facts 23 and 22.

Theorem 25 (Categoricity). *All HF structures are isomorphic.*

Proof. Follows with Facts 19, 20, and 22.

5 Set Operations

We now construct basic set operations known from ZF for HF structures using the strong induction principle and preexisting specifications of the desired operations. The specifications are needed since the induction principle does not come with equations (in contrast to a full recursor). Suitable specifications for the basic set operations are easily obtained using to the extensionality property of sets.

Fact 26 (Binary Union). *There is a function $x \cup y$ from sets to sets as follows:*

1. $z \in x \cup y \leftrightarrow z \in x \vee z \in y$.
2. $\emptyset \cup y = y$.
3. $(a.x) \cup y = a.(x \cup y)$.

Proof. We fix y and define $Uxu := \forall z.\ z \in u \leftrightarrow z \in x \vee z \in y$. We construct a function $F : \forall x \exists u.\ Uxu$ using the strong induction principle. The base case follows with $U\emptyset y$. For the adjunction case it suffices to prove $\forall a x u.\ Uxu \rightarrow U(a.x)(a.u)$, which is straightforward. We define $x \cup y := \pi_1(Fx)$. We have $Ux(\pi_1(Fx))$ and thus Claim 1. Claims 2 and 3 follow with extensionality from Claim 1.

Note that the two cases of the inductive construction of F choose their witnesses according to Claims 2 and 3. This is by design. Claims 2 and 3 explicate the ideas behind the construction of F.

The following constructions all follow the scheme used for binary union.

Fact 27 (Big Union). *There is a function $\bigcup x$ from sets to sets as follows:*

1. $z \in \bigcup x \leftrightarrow \exists y \in x.\ z \in y$.
2. $\bigcup \emptyset = \emptyset$.
3. $\bigcup(a.x) = a \cup \bigcup x$.

A set x is *transitive* if every element of x is a subset of x. For transitive sets, big union undoes the successor operation.

Fact 28 (Predecessor). *Let x be transitive. Then $\bigcup(x.x) = x$.*

Fact 29 (Separation). *For every decidable predicate p on sets there is a function $x|p$ from sets to sets as follows:*

1. $z \in x|p \leftrightarrow z \in x \wedge pz$.
2. $\emptyset|p = \emptyset$.
3. $(a.x)|p = $ if pa then $a.(x|p)$ else $x|p$.

Constructions and correctness proofs for the remaining set operations of ZF can be found in the accompanying Coq development, which also covers a transitive closure operation.

6   Ordinals

We define the class of *ordinals* as an inductive predicate on sets:

$$\frac{}{\mathcal{O}\,\emptyset} \qquad \frac{\mathcal{O}x}{\mathcal{O}(x.x)}$$

The ordinals represent the natural numbers as HF sets, where a number n is represented as the unique ordinal having n elements. The ordinal for n can be obtained by applying the successor function n-times to the empty set.

We use the letters α and β for sets that should be thought of as ordinals.

Fact 30 (Transitivity). *Ordinals are transitive sets whose elements are ordinals.*

Proof. By induction on $\mathcal{O}\alpha$.

Fact 31 (Empty Ordinal). *Let α be an ordinal. Then $\alpha = \emptyset \;\bar{\vee}\; \emptyset \in \alpha$.*

Proof. Show $\alpha \neq \emptyset \rightarrow \emptyset \in \alpha$ by induction on $\mathcal{O}\alpha$.

Fact 32 (Predecessor Ordinal). *Let α be an ordinal. Then $\bigcup\alpha$ is an ordinal. Moreover, $\alpha = (\bigcup\alpha).(\bigcup\alpha)$ if $\alpha \neq \emptyset$.*

Proof. Both claims follow by induction on $\mathcal{O}\alpha$ using Facts 28 and 30.

Fact 33 (Inversion). *Let α be an ordinal. Then $\alpha = \emptyset \;\bar{\vee}\; \bar{\exists}\gamma.\; \mathcal{O}\gamma \wedge \alpha = \gamma.\gamma$.*

Proof. Follows with Fact 32.

Fact 34 (Strong Ordinal Induction).
$\forall p : X \rightarrow \mathsf{Type}.\; p\emptyset \rightarrow (\forall\alpha.\; \mathcal{O}\alpha \rightarrow p\alpha \rightarrow p(\alpha.\alpha)) \rightarrow \forall\alpha.\; \mathcal{O}\alpha \rightarrow p\alpha$.

Proof. Follows by strong epsilon induction (Fact 14) using Fact 33.

7   Cardinality

Given an HF structure, we would expect that we can construct a model of the natural numbers by taking the subtype of the ordinals as type for the numbers. In constructive type theory, however, the subtype of ordinals does not come for free. We need an idempotent function on sets whose fixed points are the ordinals. A natural choice for this function is a cardinality function mapping every set to the unique equipotent ordinal. Two sets are equipotent if they have same number of elements.

We define *equipotence of sets* with an inductive predicate $x \sim y$:

$$\frac{}{\emptyset \sim \emptyset} \qquad\qquad \frac{a \notin x \qquad b \notin y \qquad x \sim y}{a.x \sim b.y}$$

Our definition of equipotence is tuned for finite sets. From the definition of equipotence it is not obvious that equipotence is an equivalence relation.

We will construct a function Γ from sets to sets such that Γx is the unique ordinal equipotent to x. Similar to what we did in the section on categoricity, we obtain Γ with the strong induction principle from an inductively defined predicate $Cx\alpha$:

$$\frac{}{C\emptyset\emptyset} \qquad\qquad \frac{a \notin x \qquad Cx\alpha}{C(a.x)(\alpha.\alpha)}$$

We will show that the relation C is strongly total and functional. Γ will be defined as the function accompanying C.

Fact 35 (Soundness). *Let Cxy. Then $x \sim y$ and y is an ordinal.*

Proof. By induction on Cxy using Fact 15.

Fact 36 (Strong Totality). $\forall x \exists \alpha. \ C x \alpha.$

Proof. By induction on x.

Fact 37 (Idempotence). *Let α be an ordinal. Then $C\alpha\alpha$.*

Proof. By induction on $\mathcal{O}\alpha$ using Fact 15.

Sets related to the same ordinal by C are equipotent.

Fact 38 (Injectivity). $Cx\alpha \to Cy\alpha \to x \sim y.$

Proof. By induction on $Cx\alpha$. The case for the first rule is straightforward. For the second rule we have $x = a.x'$, $a \notin x'$, $Cx'\alpha'$, and $\alpha = \alpha'.\alpha'$. By inversion of $Cy\alpha$ we obtain b, y', and β such that $y = b.y'$, $b \notin y'$, $\alpha = \beta.\beta$, and $Cy'\beta$. By Fact 18 we have $\alpha' = \beta$. Thus $x' \sim y'$ by the inductive hypothesis for $Cx'\alpha'$. Hence $x \sim y$.

Proving that C is functional takes effort. We need an inversion lemma whose proof requires a further lemma involving an instance of separation (Fact 29). We use the notation $x \div y := x | (\lambda z. z \neq y)$ (read x without y).

Lemma 39. $Cx\alpha \to a \in x \to \exists \beta. \ \alpha = \beta.\beta \wedge C(x \div a)\beta.$

Proof. By induction on $Cx\alpha$. The case for the first rule is straightforward. For the second rule we have $x = a'.x'$, $a' \notin x'$, $Cx'\alpha'$, and either $a = a'$ or $a \in x'$. It suffices to show that $C((a'.x') \div a)\alpha'$.

Let $a = a'$. Then the claim follows with $(a'.x') \div a = x'$.

Let $a \neq a'$ and $a \in x'$. The inductive hypothesis for $Cx'\alpha'$ gives us some β such that $\alpha' = \beta.\beta$ and $C(x' \div a)\beta$. The claim follows since $(a'.x') \div a = a'.(x' \div a)$ and $a' \notin (x' \div a)$. □

Fact 40 (Inversion). $a \notin x \to C(a.x)\alpha \to \exists \beta. \ \alpha = \beta.\beta \wedge Cx\beta.$

Proof. Follows with Lemma 39.

Fact 41 (Functionality). $Cx\alpha \to Cx\beta \to \alpha = \beta.$

Proof. By induction on $Cx\alpha$. The case for the first rule is straightforward, and the case for the second rule follows with Fact 40 and the inductive hypothesis.

Fact 42 (Invariance). $x \sim y \to Cx\alpha \to Cy\alpha.$

Proof. By induction on $x \sim y$ using Fact 40.

Fact 43 (Canonicity). *Equipotent ordinals are equal.*

Proof. Let α and β be equipotent ordinals. Then $C\alpha\alpha$ and $C\beta\beta$ by Fact 37. Thus $C\beta\alpha$ by Fact 42, and $\alpha = \beta$ by Fact 41.

We define Γ as $\Gamma x := \pi_1(Tx)$ where T is the function established by Fact 36.

Fact 44. *Γ is an idempotent function such that $Cx(\Gamma x)$, $x \sim \Gamma x$, and Γx is an ordinal for every set x.*

Proof. $Cx(\Gamma x)$ holds by definition for all x. Thus $x \sim \Gamma x$ and Γx is an ordinal by Fact 35.

To show the idempotence of Γ, we fix some x. We have $C(\Gamma x)(\Gamma(\Gamma x))$. Since Γx is an ordinal, we also have $C(\Gamma x)(\Gamma x)$ by Fact 37. Hence $\Gamma(\Gamma x) = \Gamma x$ by Fact 41.

Fact 45 (Coincidence). $x \sim y \leftrightarrow \Gamma x = \Gamma y$.

Proof. Let $x \sim y$. We have $Cx(\Gamma x)$ and $Cy(\Gamma y)$ by Fact 44. Hence $Cy(\Gamma x)$ by Fact 42. Thus $\Gamma x = \Gamma y$ by Fact 41.

Let $\Gamma x = \Gamma y$. Then $x \sim y$ by Facts 38 and 44.   □

Corollary 46 (Equipotence). *Equipotence is a decidable equivalence relation.*

Fact 47 (Fixed Point). *A set x is an ordinal if and only if $\Gamma x = x$.*

Proof. Let α be an ordinal. Then $C\alpha\alpha$ by Fact 37 and $C\alpha(\Gamma\alpha)$ by Fact 44. Thus $\Gamma\alpha = \alpha$ by Fact 41. The other direction follows by Fact 44.

Corollary 48. *It is decidable whether a set is an ordinal.*

8 Binary Trees

We now strengthen the type theory by adding an inductive type $\mathbf{T}$ of binary trees:

$$\mathbf{T} \ := \ 0 \mid \mathbf{T}.\mathbf{T}$$

We will construct an HF structure in the strengthened type theory.

The letters s, t, and u will range over binary trees. We write $s.t.u$ for $s.(t.u)$.

Fact 49. $\mathbf{T}$ *is a discrete type.*

Proof. We obtain a decision function $\forall st.\ s = t \ \triangledown \ s \neq t$ by induction on s using the strong induction principle for trees.

We assume an HF structure X and define a function $S : \mathbf{T} \to X$ mapping trees to sets:

$$S0 \ := \ \emptyset$$
$$S(s.t) \ := \ Ss.St$$

We may see trees as expressions describing sets and S as a function evaluating expressions to sets.

Fact 50. *S has a right inverse. That is, there is a function $I : X \to \mathbf{T}$ such that $S(Ix) = x$ for every set x. Consequently, we have $x = y \leftrightarrow Ix = Iy$ for all sets x and y.*

Proof. With the strong induction principle of X we obtain a certifying function $F : \forall x \bar{\exists} s.\ Ss = x$. We define $Ix := \pi_1(Fx)$.

Lemma 51 (Transfer of Induction Principle). *Let X be a structure that is an HF structures except that it does not come with an induction principle. Let $S : \mathbf{T} \to X$ and $I : X \to \mathbf{T}$ be functions such that $S(Ix) = x$, $S0 = \emptyset$, and $S(s.t) = Ss.St$ for all sets x and all trees s and t. Then X can be extended to an HF structure.*

Proof. Let $p : X \to \mathsf{Prop}$. The strong induction principle for X and p can be obtained from the strong induction principle for $\mathbf{T}$ and $\lambda s.p(Ss)$.

9   Tree Model

We now construct an HF structure as a quotient of the tree type under an equivalence generated by cancellation and swapping. We define this equivalence as an inductive predicate $s \approx t$ and call it *tree equivalence*:

$$\frac{}{s.s.t \approx s.t} \qquad \frac{}{s.t.u \approx t.s.u} \qquad \frac{s \approx s' \quad t \approx t'}{s.t \approx s'.t'}$$

$$\frac{}{s \approx s} \qquad \frac{s \approx t}{t \approx s} \qquad \frac{s \approx t \quad t \approx u}{s \approx u}$$

Tree equivalence satisfies the cancellation and swapping law by definition. It also satisfies the discrimination law.

Fact 52. $s \approx t \to (s = 0 \leftrightarrow t = 0)$.

Proof. By induction on $s \approx t$. We prove $s = 0 \to t = 0$ and $t = 0 \to s = 0$ together so that we can accommodate the symmetry rule.

Fact 53 (Discrimination). $s.t \not\approx 0$.

Proof. Follows with Fact 52.

We define $s \in t := (s.t \approx t)$. Proving that tree equivalence satisfies the membership law takes a little effort. We need an inductive auxiliary predicate $s \dot\in t$ providing a restricted form of membership:

$$\frac{}{s \dot\in s.t} \qquad \frac{s \dot\in u}{s \dot\in t.u}$$

Fact 54. $u \mathbin{\dot\in} s.t \leftrightarrow u = s \lor u \mathbin{\dot\in} t$.

We also need an auxiliary inclusion predicate $s \preceq t := \forall u \mathbin{\dot\in} s \, \exists v \mathbin{\dot\in} t.\ u \approx v$.

Lemma 55. $s \approx t \to s \preceq t \land t \preceq s$.

Proof. By induction on $s \approx t$ using Fact 54. We prove $s \preceq t$ and $t \preceq s$ together so that we can accommodate the symmetry rule.

Lemma 56. $s \mathbin{\dot\in} t \to s \in t$.

Proof. By induction on $s \mathbin{\dot\in} t$.

Fact 57 (Membership). $u \in s.t \to u \approx s \lor u \in t$.

Proof. Let $u \in s.t$. Then $u.s.t \approx s.t$. Thus $u.s.t \preceq s.t$ by Lemma 55. Since $u \mathbin{\dot\in} u.s.t$, we have $u \approx v \mathbin{\dot\in} s.t$ for some v. Thus either $u \approx v = s$ or $u \approx v \mathbin{\dot\in} t$ by Fact 54. The claim follows with Lemma 56.

The quotient of the tree type for tree equivalence will be obtained with a normalizer for tree equivalence using Fact 2. The normalizer will be an idempotent function $\sigma : \mathbf{T} \to \mathbf{T}$ such that $s \approx t \leftrightarrow \sigma s = \sigma t$. Given that tree equivalence is generated by cancellation and swapping, we can obtain a normalizer for tree equivalence as a sorting function for some linear ordering on trees. There is a natural linear ordering on trees based on the idea of lexical ordering:

$$\frac{}{0 < s.t} \qquad \frac{s < s'}{s.t < s'.t'} \qquad \frac{t < t'}{s.t < s.t'}$$

We speak of the *lexical tree ordering*.

Fact 58. *The lexical tree ordering is irreflexive and transitive.*

Proof. Follows with induction on $s < t$.

Fact 59 (Trichotomy). $s < t \mathbin{\bar\lor} s = t \mathbin{\bar\lor} t < s$.

Proof. By nested induction on s and t.

We shall obtain the normalizer by duplicate-eliminating insertion sort. We define a function $\alpha : \mathbf{T} \to \mathbf{T} \to \mathbf{T}$ for order-observing and duplicate-avoiding *insertion* based on the case analysis provided by Fact 59:

$$\alpha s 0 := s.0$$
$$\alpha s(t.u) := \text{case } s < t \Rightarrow s.t.u \mid s = t \Rightarrow t.u \mid t < s \Rightarrow t.\alpha s u$$

Fact 60. $\alpha s t \approx s.t$.

Proof. By induction on t using Fact 59.

We finally define a duplicate-eliminating *sorting* function $\sigma : \mathbf{T} \to \mathbf{T}$:

$$\sigma 0 := 0$$
$$\sigma(s.t) := \alpha(\sigma s)(\sigma t)$$

Fact 61. $\sigma s \approx s$.

Proof. By induction on s using Fact 60.

Next we show that that σ normalizes equivalent trees to identical trees. The key insight behind this result is the fact that insertion respects the cancellation and swap law with respect to equality.

Fact 62. $\alpha s(\alpha s t) = \alpha s t$ *and* $\alpha s(\alpha t u) = \alpha t(\alpha s u)$.

Proof. The first claim follows by induction on t and the second claim follows by induction on u. Both proofs do case analysis according to Fact 59 and eliminate inconsistent cases with Fact 58. There are many cases to consider. The following facts are useful for the case analysis:

- $\alpha s 0 = s.0$ and $\alpha s(s.t) = s.t$.
- If $s < t$, then $\alpha s(t.u) = s.t.u$.
- If $t < s$, then $\alpha s(t.u) = t.\alpha s u$. $\square$

Fact 63. $s \approx t \ \to \ \sigma s = \sigma t$.

Proof. By induction on $s \approx t$ using Fact 62.

Fact 64.

1. $s \approx t \leftrightarrow \sigma s = \sigma t$.
2. σ *is idempotent; that is,* $\sigma(\sigma s) = \sigma s$.
3. *Tree equivalence is decidable.*

Proof. The claims follow with Facts 61 and 63.

Theorem 65 (Model Existence). *There exist an HF structure X and two functions $S : \mathbf{T} \to X$ and $I : X \to \mathbf{T}$ such that:*

1. $S(Ix) = x$ *and* $I(Ss) \approx s$.
2. $Ss = St \leftrightarrow s \approx t$ *and* $Ix = Iy \leftrightarrow x = y$.
3. $S(s.t) = Ss.St$ *and* $I(x.y) \approx Ix.Iy$.
4. $S0 = \emptyset$ *and* $I\emptyset = 0$.

Proof. By Facts 2, 49, and 64 we have a discrete type X and functions $S : \mathbf{T} \to X$ and $I : X \to \mathbf{T}$ such that $S(Ix) = x$ and $I(Ss) = \sigma s$ for all x and s. We define $\emptyset := S0$ and $x.y := S(Ix.Iy)$. The claims (1)–(4) follow with Fact 64. By Lemma 51 it suffices to show that the definitions of $\emptyset$ and adjunction satisfy the cancellation, swap, discrimination, and membership law.

We show the discrimination law. Let $x.y = \emptyset$. Then $S(Ix.Iy) = S0$ by definition. Thus $Ix.Iy \approx 0$ by (2). Contradiction by Fact 53.

We show the swap law. We have $Ix.Iy.Iz \approx Iy.Ix.Iz$ by definition of tree equivalence. Thus $Ix.I(y.z) \approx Iy.I(x.z)$ by (3) and $S(Ix.I(y.z)) = S(Iy.I(x.z))$ by (2). Hence $x.y.z = y.x.z$ by the definition of adjunction.

The cancellation law follows analogously.

We show the membership law. Let $x.y.z = y.z$. By the definition of adjunction and (2) we have $Ix.I(y.z) \approx Iy.Iz$. By (3) we have $Ix.Iy.Iz \approx Iy.Iz$. Hence either $Ix \approx Iy$ or $Ix.Iz \approx Iz$ by Fact 57. Thus either $x = y$ or $x.z = z$ by (2), (1), and the definition of adjunction.   □

We can now transfer results for HF structures to tree equivalence.

Corollary 66 (Extensionality of Tree Equivalence).
$(\forall u.\ u \in s \leftrightarrow u \in t) \rightarrow s \approx t$.

Proof. Let X, S, and I be the objects provided by Theorem 65. Then $s \in t \leftrightarrow Ss \in St$ for all s and t with (2) and (3). Suppose $\forall u.\ u \in s \leftrightarrow u \in t$. Then $\forall x.\ Ix \in s \leftrightarrow Ix \in t$. Thus $\forall x.\ x \in Ss \leftrightarrow x \in St$. Hence $Ss = St$ since X is extensional (Fact 9). Thus $s \approx t$.

10   Conclusion

We have studied finite set theory in constructive type theory. In contrast to a general set theory, finite set theory has a unique model that can be constructed in constructive type theory. We have presented a categorial axiomatization of finite set theory providing for a constructive development of the theory, including the usual set operations, finite ordinals, and cardinality.

We have constructed a model of the axiomatization as a quotient of an inductive type of binary trees. The tree model gives us a natural realization of the type of finite sets in constructive type theory. The operations and results obtained on top of the axiomatization apply to the tree model and all other models. Seen from the perspective of programming, the axiomatization provides an abstraction layer.

We have been careful in spelling out the type theoretic resources needed for the development. For the study of the axiomatization, we work in a type theory with dependent function and pair types, with sum types, and with an impredicative universe of propositions. For the model construction, we add an inductive type of binary trees and a single proof proposition ⊤.

We see finite set theory as a constructive subtheory of general set theory. We believe that the study of finite sets in constructive type theory is instructive for students and also prepares them well for the study of general set theory.

There are many possibilities for future work: Prove that the axiomatization of HF sets is minimal; Find a recursor constructing functions on HF sets in the style of primitive recursion (step functions will have to be provided with admissibility proofs); Study the Peano axiomatization of numbers with strong induction and show that it enables the construction of a model of HF (following Ackermann [1]); Establish categorial axiomatizations for flat finite sets over a base type, for finite multisets, and for finite sets also including non-wellfounded sets; Develop a Coq library supporting the construction of the mentioned inductive quotient types.

The accompanying Coq development follows the presentation of the paper. We wrote a tactic supporting membership-based reasoning in HF structures. With this tactic the proofs of the abstract results turn out to be pleasantly compact. Unexpectedly, some of the proofs for tree sorting took effort because there are so many cases to consider (Lemma 55 and Fact 62). We arrived at compact proofs by devising special-purpose tactics.

Acknowledgement. Denis Müller contributed to the study of tree equivalence during his Bachelor's thesis project on finitary sets.

References

1. Ackermann, W.: Die Widerspruchsfreiheit der allgemeinen Mengenlehre. Math. Ann. **114**(1), 305–315 (1937)
2. Givant, S., Tarski, A.: Peano arithmetic, the Zermelo-like theory of sets with finite ranks. Not. Am. Math. Soc. **77T–E51**, A-437 (1977)
3. Kirby, L.: Finitary set theory. Notre Dame J. Formal Logic **50**(3), 227–244 (2009)
4. Paulson, L.C.: A formalisation of finite automata using hereditarily finite sets. In: Felty, A.P., Middeldorp, A. (eds.) CADE-25. LNCS, vol. 9195, pp. 231–245. Springer International Publishing, Switzerland (2015)
5. Paulson, L.C.: A mechanised proof of Gödel's incompleteness theorems using Nominal Isabelle. J. Autom. Reasoning **55**(1), 1–37 (2015)
6. Previale, F.: Induction and foundation in the theory of hereditarily finite sets. Arch. Math. Logic **33**(3), 213–241 (1994)
7. Świerczkowski, S.: Finite sets and Gödel's incompleteness theorems. Dissertationes Mathematicae, vol. 422. Polish Academy of Sciences, Institute of Mathematics (2003)
8. Takahashi, M.: A foundation of finite mathematics. Publ. Res. Inst. Math. Sci. Kyoto Univ. **12**(3), 577–708 (1977)

Algebraic Numbers in Isabelle/HOL

René Thiemann[(✉)] and Akihisa Yamada

University of Innsbruck, Innsbruck, Austria
{rene.thiemann,akihisa.yamada}@uibk.ac.at

Abstract. We formalize algebraic numbers in Isabelle/HOL, based on existing libraries for matrices and Sturm's theorem. Our development serves as a verified implementation for real and complex numbers, and it admits to compute roots and completely factor real and complex polynomials, provided that all coefficients are rational numbers. Moreover, we provide two implementations to display algebraic numbers, an injective and expensive one, and a faster but approximative version.

To this end, we mechanize several results on resultants, which also required us to prove that polynomials over a unique factorization domain form again a unique factorization domain. We moreover formalize algorithms for factorization of integer polynomials: Newton interpolation, factorization over the integers, and Kronecker's factorization algorithm, as well as a factorization oracle via Berlekamp's algorithm with the Hensel lifting.

1 Introduction

Algebraic numbers, i.e., the numbers that are expressed as roots of non-zero rational (equivalently, integer) polynomials, are an attractive subset of the real or complex numbers. They are closed under arithmetic operations, the arithmetic operations are precisely computable, and comparisons are decidable. As a consequence, algebraic numbers are an important utility in computer algebra systems.

Our original interest in algebraic numbers stems from a certification problem about automatically generated complexity proofs, where we have to compute the Jordan normal form of a matrix in $\mathbb{Q}^{n \times n}$ [16]. To this end, all complex roots of the characteristic polynomial have to be determined.

Example 1. Consider a matrix A whose characteristic polynomial is $f(x) = 1 + 2x + 3x^4$. The complex roots of f are exactly expressed via the real roots of $g = -1 - 12x^2 + 144x^6$ and $h = 7 - 216x^2 - 336x^4 - 1248x^6 + 1152x^8 + 6912x^{12}$:

> root #1 of g + (root #2 of h)i   root #1 of g + (root #3 of h)i
> root #2 of g + (root #1 of h)i   root #2 of g + (root #4 of h)i

Here, real roots are indexed according to the standard order. As the norms of all of these roots are strictly less than 1 (the norms are precisely root #3 and #4 of the polynomial $i = 1 - 3x^4 - 12x^6 - 9x^8 + 27x^{12}$), we can conclude that A^n tends to 0 for increasing n.

© Springer International Publishing Switzerland 2016
J.C. Blanchette and S. Merz (Eds.): ITP 2016, LNCS 9807, pp. 391–408, 2016.
DOI: 10.1007/978-3-319-43144-4_24

In this paper, we provide a fully verified and efficient implementation of algebraic numbers in Isabelle/HOL [14].

- The first problem in computation with algebraic numbers is to obtain a non-zero polynomial which represents a desired algebraic number as its root. To this end, we formalize the theory of *resultants*, and thus provide a verified computation of non-zero polynomials with desired roots (Sect. 2).
- A direct computation of resultants as determinant is infeasible in practice. Hence, we formalize a method based on a Euclid-like algorithm in combination with *polynomial remainder sequences* [1,5] (Sect. 3).
- Polynomials computed via resultants are often not optimal for representing an algebraic number, and lead to exponential growth of degrees during arithmetic operations. To avoid this problem, we formalize polynomial factorization algorithms, including an efficient oracle via Berlekamp's algorithm and the Hensel lifting, and an expensive but certified version of Kronecker's algorithm. To this end, we also formalize algorithms for prime factorization and polynomial interpolation, as well as Gauss' lemma. (Sect. 4)
- An algebraic number a is basically represented by a triple (f, l, r) of rational polynomial f and $l, r \in \mathbb{Q}$ such that a is the unique root of f within the interval $[l, r]$. To compute such an interval, we generalize the existing formalization of Sturm's method [6] to work over the rationals, and precompute the *Sturm sequence* to avoid recomputation. We also take special care for arithmetic operations involving a rational number, and finally provide a *quotient type* for algebraic numbers, which works modulo different representations of the same algebraic numbers. (Sect. 5)
- We also integrate complex algebraic numbers. Our algorithms cover complex root computation, as well as a factorization for rational polynomials over $\mathbb{R}$ or $\mathbb{C}$. (Sect. 6)
- Finally, we develop algorithms for displaying algebraic numbers. A challenge in precisely representing algebraic numbers is to ensure the uniqueness of string representation, independent from the internal representation. Here, the certified factorization algorithm plays a crucial role. (Sect. 7)

For the Coq proof assistant, the Mathematical Components library[1] contains various formalized results around algebraic numbers, e.g., quantifier elimination procedures for real closed fields [4]. In particular, the executable formalization of algebraic numbers for Coq is given by Cohen [2]. He employed Bézout's theorem to derive desired properties of resultants. In contrast, we followed proofs by Mishra [13] and formalized various facts on resultants. We further mechanize an algorithm to compute resultants, as well as the polynomial factorization algorithms. Our work is orthogonal to the more recent work which completely avoids resultants [3].

For Isabelle, Li and Paulson [11] independently implemented algebraic numbers. They however did not formalize resultants; instead, they employed an

---

[1] See http://math-comp.github.io/math-comp.

external tool as an oracle to provide polynomials that represent desired algebraic numbers, and provided a method to validate that the polynomials from the oracle are suitable.[2] Although we also use untrusted oracles for polynomial factorization, the difference is crucial. First, finding polynomials is indispensable for the computation of algebraic numbers, and hence their implementation is not ensured to always succeed. On the other hand, factorization is optional, and is employed only for efficiency. Second, in addition to an external oracle interface, we also provide an internal one, so that no external tools are required. Finally, due to our optimization efforts, we can execute their examples [11, Fig. 3] in 0.03 s on our machine, where they reported 4.16 s.[3]

The whole formalization has been made available in the archive of formal proofs for Isabelle 2016 (http://afp.sourceforge.net), cf. entries Algebraic Numbers, Polynomial Factorization, and Polynomial Interpolation.

2 Resultants

In order to define arithmetic operations over algebraic numbers, the first task is the following: Given non-zero polynomials that have the input numbers as roots, compute a non-zero polynomial that has the output number as a root.

Consider an algebraic number a represented as a root of $f(x) = \sum_{i=0}^{m} f_i x^i$. To represent the unary minus $-a$, clearly *poly-uminus*, defined as the polynomial $f(-x)$, does the job. For the multiplicative inverse $\frac{1}{a}$, it is also not difficult to show that *poly-inverse*, defined as $\sum_{i=0}^{m} f_i x^{m-i}$, has $\frac{1}{a}$ as a root.

For addition and multiplication, given another polynomial $g(x) = \sum_{i=0}^{n} g_i x^i$ representing an algebraic number b, we must compose non-zero polynomials *poly-add f g* and *poly-mult f g* that have $a + b$ and $a \cdot b$ as a root, resp.

For this purpose the resultant is a well-known solution. The resultant of the polynomials f and g above is defined as $\mathrm{Res}(f, g) = \det(S_{f,g})$, where $S_{f,g}$ is the *Sylvester matrix* (blank parts are filled with zeros):

$$S_{f,g} = \begin{bmatrix} f_m & f_{m-1} & \cdots & f_0 & & & \\ & \ddots & \ddots & & \ddots & & \\ & & f_m & f_{m-1} & \cdots & f_0 \\ g_n & g_{n-1} & \cdots & g_0 & & & \\ & \ddots & \ddots & & \ddots & & \\ & & g_n & g_{n-1} & \cdots & g_0 \end{bmatrix}$$

In the remainder of this section, we consider addition – multiplication is treated similarly. The desired result is informally stated as follows, where *poly-add f g* is defined as the resultant of the two bivariate polynomials $f(x - y)$ and $g(y)$, where the resultant is a univariate polynomial over x.

---

[2] Here one cannot just evaluate the polynomial on the algebraic point and test the result is 0; we are defining the basic arithmetic operations needed for this evaluation.

[3] However, we use a faster computer with 3.5 GHz instead of 2.66 GHz.

Lemma 2. *Let f and g be non-zero univariate polynomials with roots a and b, respectively. Then poly-add f g is a non-zero polynomial having $a + b$ as a root.*

The lemma contains two claims: *poly-add* f g has $a+b$ as a root, and *poly-add* f $g \neq 0$. In the next sections we prove each of the claims.

2.1   Resultant Has Desired Roots

For non-constant polynomials f and g over a commutative ring R, we can compute polynomials p and q such that

$$\text{Res}(f, g) = p(x) \cdot f(x) + q(x) \cdot g(x) \tag{1}$$

To formally prove the result, we first define a function *mk-poly* that operates on the Sylvester matrix. For each j-th column except for the last one, *mk-poly* adds the j-th column multiplied by x^{m+n-j} to the last column. Each addition preserves determinants, and we obtain the following equation:

$$\text{Res}(f, g) = \det(\textit{mk-poly } S_{f,g}) = \det \begin{bmatrix} f_m & \cdots & f_1 & f_0 & & & f(x) \cdot x^{n-1} \\ & \ddots & & \ddots & \ddots & & \vdots \\ & & f_m & \cdots & f_1 & f_0 & f(x) \cdot x \\ & & & f_m & \cdots & f_1 & f(x) \\ g_n & \cdots & g_1 & g_0 & & & g(x) \cdot x^{n-1} \\ & \ddots & & \ddots & \ddots & & \vdots \\ & & g_n & \cdots & g_1 & g_0 & g(x) \cdot x \\ & & & g_n & \cdots & g_1 & g(x) \end{bmatrix} \tag{2}$$

Now we apply the *Laplace expansion*, which we formalize as follows.

lemma *assumes* $A \in \textit{carrier}_m$ n n (* **meaning** $A \in R^{n \times n}$ *) *and* $j < n$
    shows $\det A = (\sum i < n.\ A_{(i,\ j)} * \textit{cofactor } A\ i\ j)$

Here, *cofactor* $A\ i\ j$ is defined as $(-1)^{i+j} \cdot \det(B)$, where B is the *minor matrix* of A obtained by removing the i-th row and j-th column. Thus we can remove the last column of the matrix A in (2), by choosing $j = m + n - 1$. Note that then every *cofactor* $A\ i\ j$ is a constant. We obtain p and q in (1) as follows:

$$\text{Res}(f, g) = \left(\sum_{i=0}^{n-1} \textit{cofactor } A\ i\ j \cdot x^i \right) \cdot f(x) + \left(\sum_{i=0}^{m-1} \textit{cofactor } A\ (n+i)\ j \cdot x^i \right) \cdot g(x)$$

Lemma 3. *assumes* degree $f > 0$ *and* degree $g > 0$
    shows $\exists p\ q.$ degree $p <$ degree $g \wedge$ degree $q <$ degree $f \wedge$
        $[: \textit{resultant } f\ g :] = p * f + q * g$

Here, $[: c :]$ is Isabelle's notation for the constant polynomial c. The lemma implies that, if f and g are polynomials of positive degree with a common root a, then $\text{Res}(f, g) = p(a) \cdot f(a) + q(a) \cdot g(a) = 0$. The result is lifted to the bivariate case: for any a and b, $f(a, b) = g(a, b) = 0$ implies $\text{Res}(f, g)(a) = 0$.

lemma *assumes* degree $f > 0 \vee$ degree $g > 0$ *and* poly2 f a b $= 0$
  and poly2 g a b $= 0$
  shows poly (resultant f g) a $= 0$

Here, *poly* is Isabelle's notation for the evaluation of univariate polynomials, and *poly2* is our notation for bivariate polynomial evaluation.

Now for univariate non-zero polynomials f and g with respective roots a and b, the bivariate polynomials $f(x - y)$ and $g(y)$ have a common root at $x = a + b$ and $y = b$. Hence, the univariate polynomial *poly-add* $fg = \text{Res}(f(x - y), g(y))$ indeed has $a + b$ as a root.

lemma *assumes* $g \neq 0$ *and* poly f a $= 0$ *and* poly g b $= 0$
  shows poly (poly-add f g) $(a + b)$ $= 0$

2.2 Resultant Is Non-Zero

Now we consider the second claim: *poly-add* f g is a non-zero polynomial. Note that it would otherwise have any number as a root. Somewhat surprisingly, formalizing this claim is more involving than the first one.

We first strengthen Lemma 3, so that p and q are non-zero polynomials. Here, we require an integral domain *idom*, i.e., there exist no zero divisors.

lemma *assumes* degree $f > 0$ *and* degree $g > 0$
  shows $\exists$ p q. degree $p <$ degree $g \wedge$ degree $q <$ degree $f \wedge$
    $[:$ resultant f g $:] = p * f + q * g \wedge p \neq 0 \wedge q \neq 0$

We further strengthen this result, so that $\text{Res}(f, g) = 0$ implies f and g share a common factor. This requires polynomials over a *unique factorization domain* (UFD), which is available as a locale *factorial-monoid* in HOL/Algebra, but not as a class. We define the class *ufd* by translating the locale as follows:

class *ufd* $= idom +$
  assumes factorial-monoid $($ carrier $= UNIV - \{0\}$, mult $= op *$, one $= 1$ $)$

We also show that polynomials over a UFD form a UFD, a non-trivial proof.

instance *poly* :: (ufd) ufd

Note also that the result is instantly lifted to any multivariate polynomials; if α is of sort *ufd*, then so is α *poly*, and thus so is α *poly poly*, and so on.

Now we obtain the following result, where $coprime_I$ generalizes the predicate *coprime* (originally defined only on the class *gcd*) over *idom* as follows:

definition $coprime_I\ f\ g \equiv \forall h.\ h\ dvd\ f \longrightarrow h\ dvd\ g \longrightarrow h\ dvd\ 1$

Lemma 4. *assumes* $degree\ f > 0 \lor degree\ g > 0$ ***and*** $resultant\ f\ g = 0$
    shows $\neg\ coprime_I\ f\ g$

Now we reason $\mathrm{Res}(f(x - y), g(y)) \neq 0$ by contradiction. If $\mathrm{Res}(f(x - y), g(y)) = 0$, then Lemma 4 implies that $f(x - y)$ and $g(y)$ have a common proper factor. This cannot be the case for complex polynomials: Let $f = f_1 \cdots f_m$ and $g = g_1 \cdots g_n$ be a complete factorization of the univariate polynomials f and g. Then the bivariate polynomials $f(x - y)$ and $g(y)$ are factored as follows:

$$f(x - y) = f_1(x - y) \cdots f_m(x - y) \qquad g(y) = g_1(y) \cdots g_n(y) \qquad (3)$$

Moreover, this factorization is irreducible and unique (up to permutation and scalar multiplication). Since there is a common factor among $f(x - y)$ and $g(y)$, we must have $f_i(x - y) = g_j(y)$ for some $i \leq m$ and $j \leq n$. By fixing y, e.g., to 0, we conclude $f_i(x) = g_j(0)$ is a constant. This contradicts the assumption that f_i is a proper factor of f. We conclude the following result:

lemma *assumes* $f \neq 0$ ***and*** $g \neq 0$ ***and*** $poly\ f\ x = 0$ ***and*** $poly\ g\ y = 0$
    shows $poly\text{-}add\ f\ g \neq 0$

In order to ensure the existence of the complete factorization (3), our original formalization employs the fundamental theorem of algebra, and thus the above lemma is initially restricted to complex polynomials. Only afterwards the lemma is translated to rational polynomials via a homomorphism lemma for *poly-add*. In the development version of the AFP (May 2016), however, we have generalized the lemma to arbitrary field polynomials.

3   Euclid-Like Computation of Resultants

Resultants can be computed by first building the Sylvester matrix and then computing its determinant by transformation into row echelon form. A better way to compute resultants has been developed by Brown via subresultants [1], and a Coq formalization of subresultants exists [12]. We leave it as future work to formalize this algorithm in Isabelle. Instead, we compute resultants using ideas from Collins' primitive PRS (polynomial remainder sequences) algorithm [5].

3.1   The Algorithm and Its Correctness

The algorithm computes resultants $\mathrm{Res}(f, g)$ in the manner of Euclid's algorithm. It repeatedly performs the polynomial division on the two input polynomials and replaces one input of larger degree by the remainder of the division.

We formalize the correctness of this algorithm as follows. Here we assume the coefficients of polynomials are in an integral domain which additionally has a division function such that $(a \cdot b)/b = a$ for all $b \neq 0$. Below we abbreviate $m = degree\ f$, $n = degree\ g$, $k = degree\ r$, and $c = leading\text{-}coeff\ g$.

Lemma 5 (Computation of Resultants)

1. *resultant f g = (− 1)^{n*m} * resultant g f*
2. **assumes** *d ≠ 0* **shows** *resultant (d · f) g = d^n * resultant f g*
3. **assumes** *f = g * q + r and n ≤ m and k < n*
   shows *resultant f g = (− 1)$^{n*(m−k)}$ * c$^{m−k}$ * resultant r g*

Lemma 5(1) allows swapping arguments, which is useful for a concise definition of the Euclid-like algorithm. It is proven as follows: We perform a number of row swappings on the Sylvester matrix $S_{f,g}$ to obtain $S_{g,f}$. Each swap will change the sign of the resultant. In Isabelle, we exactly describe how the transformed matrix looks like after each row-swapping operation.

Lemma 5(2) admits computing Res(f, g) via Res($d·f, g$). As we will see, this is crucial for applying the algorithm on non-field polynomials including bivariate polynomials, which we are dealing with. To prove the result in Isabelle, we repeatedly multiply the rows in $S_{f,g}$ by d, and obtain $S_{d·f,g}$.

The most important step to the algorithm is Lemma 5(3), which admits replacing f by the remainder r of smaller degree. A paper proof again applies a sequence of elementary row transformations to convert $S_{qg+r,g}$ into $S_{r,g}$. We formalize these transformation by a single matrix multiplication, and then derive the property in a straightforward, but tedious way.

To use Lemma 5(3), we must compute a quotient q and a remainder r such that $f = gq + r$. For field polynomials one can just perform polynomial long division to get the corresponding q and r. For non-field polynomials, we formalize the polynomial *pseudodivision*, whose key property is formalized as follows:

lemma assumes *g ≠ 0 and pseudo-divmod f g = (q, r)*
  shows *c$^{1+m−n}$ · f = g * q + r ∧ (r = 0 ∨ k < n)*

Now we compute Res(f, g) as follows: Ensure $m ≥ n$ using Lemma 5(1), and obtain r via pseudodivision. We have Res(f, g) = Res($c^{1+m−n}f, g$)/$c^{(1+m−n)·n}$ by Lemma 5(2), and Res($c^{1+m−n}f, g$) is simplified to Res(g, r) by Lemma 5(3), where the sum of the degrees of the input polynomials are strictly decreased.

The correctness of this reduction is formalized as follows:

lemma assumes *pseudo-divmod f g = (q, r) and m ≥ n and n > k*
  shows *resultant f g = (− 1)^{n*m} * resultant g r / c$^{(1+m−n)*n+k−m}$*

We repeat this reduction until the degree n of g gets to zero, and then use the following formula to finish the computation.

lemma assumes *n = 0* **shows** *resultant f g = c^m*

3.2 Polynomial Division in Isabelle's Class Hierarchy

When formalizing the algorithms in Isabelle (version 2016), we encountered a problem in the class mechanism. There is already the division for field polynomials formalized, and based on this the instance declaration "**instantiation** *poly* :: (*field*) *ring-div*", meaning that α *poly* is in class *ring-div* if and only if α is a field. Afterwards, one cannot have a more general instantiation, such as non-field polynomials to be in class *idom-divide* (integral domains with partial divisions).

As a workaround, we made a copy of *idom-divide* with a different name, so that it does not conflict with the current class instantiation.

class *idom-div* = *idom* + **fixes** *exact-div* :: $\alpha \Rightarrow \alpha \Rightarrow \alpha$
   assumes $b \neq 0 \implies$ *exact-div* $(a * b)$ $b = a$

For polynomials over α :: *idom-div*, we implement the polynomial long division. This is then used as *exact-div* for α *poly* and we provide the following instantiation (which also provides division for multivariate polynomials):[4]

instantiation *poly* :: (*idom-div*) *idom-div*

We further formalize pseudodivision which actually does not even invoke a single division and is thus applicable on polynomials over integral domains.

3.3   Performance Issues

The performance of the algorithm in Sect. 3.1 is not yet satisfactory, due to the repeated multiplication with c^{1+m-n}, a well-known phenomenon of pseudodivision. To avoid this problem, in every iteration of the algorithm we divide g by its *content*, i.e., the GCD of its coefficients, similar to Collins' primitive PRS algorithm. At this point a formalization of the subresultant algorithm will be benefitial as it avoids the cost of content computation.

We further optimize our algorithm by switching from $\mathbb{Q}$ to $\mathbb{Z}$. When invoking *poly-add*, etc., over polynomials whose coefficients are integers (but of type *rat*), we ensure that the intermediate polynomials have integer coefficients. Thus we perform the whole computation in type *int*, and also switch the GCD algorithm from the one for rational polynomials to the one for integer polynomials.

This has a significant side-effect: In Isabelle, the GCD on rational polynomials is already defined and it has to be normalized so that the leading coefficient of the GCD is 1. Thus, the GCD of the *rational* polynomials $1000(x + 1)x$ and $2000(x + 2)(x + 1)$ is just $x + 1$. In contrast, we formalized[5] Collins' primitive PRS algorithm for GCD computation for *integer* polynomials, where the GCD of the above example is $1000(x + 1)$. Hence, dividing by the GCD will eliminate large constants when working on $\mathbb{Z}$, but not when working on $\mathbb{Q}$.

Finally, we provide experimental data in Table 1 in order to compare the various resultant computation algorithms. In each experiment the complex roots

---

[4] We contributed our formalization to the development version of Isabelle (May 2016). There one will find the general "**instantiation** *poly* :: (*idom-divide*) *idom-divide*".

[5] As for the division algorithm, we have not been able to work with Isabelle's existing type class for GCDs, as the GCD on polynomials is only available for fields.

Table 1. Identifying the complex roots of $1 + 2x + 3x^4$ as in Example 1.

algorithm to compute resultants	overall time
(a) algorithm of Sect. 3.1	>24 h
(b) (a) + GCD before pseudodivision	30 m 32 s
(c) (b) with GCD for integer polynomials	34 s

for f of the leading Example 1 are identified. Here, the intermediate computation invokes several times the resultant algorithm on bivariate polynomials of degree 12. Note that in experiment (c) – which applies our final resultant implementation – only 17 % of the time is spent for the resultant computation, i.e., below 6 s.

This and all upcoming experiments have been performed using extracted Haskell code which has been compiled with ghc -O2, and has been executed on a 3.5 GHz 6-Core Intel Xeon E5 with 32 GB of RAM running Mac OS X.

4   Factorization of Rational Polynomials

Iterated resultant computations will lead to exponential growth in the degree of the polynomials. Hence, after computing a resultant to get a polynomial f representing an algebraic number a, it is a good idea to factor $f = f_1^{e_1} \cdots f_k^{e_k}$ and pick the only relevant factor f_i that has a as a root.

Table 2. Computation time/degree of representing polynomials for $\sum_{i=1}^{n} \sqrt{i}$.

factorization	$n = 6$	$n = 7$	$n = 8$	$n = 9$	$n = 10$
none	0.16 s/64	2.78 s/128	2 m 11 s/256	22 m 19 s/512	12 h 19 m/1024
square-free	0.17 s/64	2.86 s/128	2 m 14 s/256	15 m 31 s/384	9 h 31 m/768
complete	0.03 s/8	0.14 s/16	0.35 s/16	0.35 s/16	0.59 s/16

The benefit of factorization is shown in Table 2, where $\sum_{i=1}^{n} \sqrt{i}$ is computed for various n, and the computation time t and the degree d of the representing polynomial is reported as t/d. The table reveals that factorization becomes beneficial as soon as at it can simplify the polynomial.

We provide two approaches for the factorization of rational polynomials. First, we formalize Kronecker's algorithm. The algorithm serves as a verified and complete factorization, although it is not efficient. Second, we also employ factorization oracles, an untrusted code that takes a rational polynomial and gives a list of factors (and the leading coefficient). Validating factorization is easy: the product of the factors should be the input polynomial. On the other hand, completeness is not guaranteed, i.e., the factors are not necessarily irreducible.

4.1   Verified Kronecker's Factorization

We formalize Kronecker's factorization algorithm for integer polynomials. We also formalize Gauss' lemma, which essentially states that factorization over $\mathbb{Q}$ is the same as factorization over $\mathbb{Z}$; thus the algorithm works on rational polynomials. The basic idea of Kronecker's algorithm is to construct a finite set of lists of sample points, and for each list of sample points, one performs polynomial interpolation to obtain a potential factor f and checks if f divides the input polynomial. Formally proving the soundness of this algorithm is not challenging; however, many basic ingredients were not available in Isabelle.

For instance, in order to construct the set of lists of sample points, one has to compute all divisors of an integer $n \neq 0$. If not to be done naively, this basically demands a prime factorization of $|n|$, for which we did not find any useful existing algorithm that has been formalized in Isabelle.

Therefore, we formalize algorithm A of Knuth [9, Sect. 4.5.4] where the list of trial divisors currently excludes all multiples of 2, 3, and 5. Here, the candidate generation works via a function *next-candidates* that takes a lower bound n as input and returns a pair (m, xs) such that xs includes all primes in the interval $[n, m)$, provided that $n = 0$ or $n \mod 30 = 11$. In the following definition, *primes-1000* is a precomputed list consisting of all primes up to 1000.

definition *next-candidates* $n = ($**if** $n = 0$ **then** $(1001,$ *primes-1000*$)$
    else $(n + 30, [n, n+2, n+6, n+8, n+12, n+18, n+20, n+26]))$

Similarly, we did not find formalized results on polynomial interpolation. Here, we integrate both Lagrange and Newton interpolation where the latter is more efficient. Furthermore, we formalize a variant of the Newton interpolation specialized for integer polynomials, which will abort early and conclude that no integer interpolation polynomial exists, namely as soon as the first division of two integers in the interpolation computation yields a non-zero remainder.

Finally, we integrate a divisibility test for integer polynomials, since polynomial divisibility test is by default available only for fields. The algorithm enjoys the same early abortion property as the Newton interpolation for integers.

4.2   Factorization Oracles

We provide two different factorization oracles: a small Haskell program that communicates with Mathematica, and an implementation within Isabelle/HOL. The latter can be used within an Isabelle session (*by eval*, etc.) as well as in generated Haskell or ML code.

They both use the same wrapper which converts the factorization over $\mathbb{Q}$ to a factorization over $\mathbb{Z}$, where the latter factorization can assume a square-free and content-free integer polynomial, represented as a coefficient list. The oracle is integrated as an unspecified constant:

consts *factorization-oracle-int-poly* :: *int list* $\Rightarrow$ *int list list*

The internal oracle implements Berlekamp's factorization algorithm in combination with Hensel lifting [9, Sect. 4.6.2]. Berlekamp's algorithm involves matrices and polynomials over finite fields ($\mathbb{Z}$ modulo some prime p). Here, we reuse certified code for polynomials and matrices whenever conveniently possible; however, the finite fields cannot be represented as a *type* in Isabelle/HOL since the prime p depends on the input polynomial to be factored. As a consequence, we could not use the standard polynomial library of Isabelle *directly*. Instead, we invoke the code generator to obtain the various certified algorithms on polynomials as ML-code, then manually replace the field operations by the finite field operations, and finally define these algorithms as new functions within Isabelle. Eventually, we had a view on all code equations for polynomials, and detected potential optimizations in the algorithm for polynomial long division.[6]

The same problem happens for the matrix operations; however, since the matrix theory is our formalization, we just modified it. We adjusted some of the relevant algorithms so that they no longer rely upon the type class *field*, but instead take the field operations as parameters. Then in the oracle we *directly* apply these generalized matrix algorithms, passing the field operations for finite fields as parameters.

Table 3. Comparing factorization algorithms

	Berlekamp-Hensel	Mathematica	Kronecker
factorization of h, degree 12	0.0 s	0.3 s	0.6 s
factorization of j, degree 27	0.0 s	0.3 s	>24 h
evaluation of $\sum_{i=1}^{5} \sqrt[3]{i}$	17.8 s	9.1 s	–
evaluation of $\sum_{i=1}^{6} \sqrt[3]{i}$	63.9 s	57.7 s	–

We conclude this section with experimental data where we compare the different factorizations in Table 3. Here, polynomial h is taken from Example 1 and j is the unique minimal monic polynomial representing $\sum_{i=1}^{5} \sqrt[3]{i}$, which looks like $-64437024420 + 122730984540x + \ldots + x^{27}$.

The 0.3 s of Mathematica is explained by its start-up time. We can clearly see that Kronecker's algorithm is no match against the oracles, which is why we did not even try Kronecker's algorithm in computing the sums of cubic roots examples – these experiments involve factorizations of polynomials of degree 81. At least on these examples, our internal factorization oracle seems to be not too bad, in comparison with Mathematica (version 10.2.0).

5   Real Algebraic Numbers

At this point, we have fully formalized algorithms which, given algebraic numbers a and b represented as roots of rational polynomials f and g, resp., computes a

---

[6] These optimizations became part of the development version of Isabelle (May 2016).

rational polynomial h having c as a root, where c is any of $a + b$, $a \cdot b$, $-a$, $\frac{1}{a}$, and $\sqrt[n]{a}$. To uniquely represent an algebraic number, however, we must also provide an interval $[l, r]$ in which c is the only root of h.

For $c = -a$ and $c = \frac{1}{a}$, bounds can be immediately given from the bound $[l, r]$ for a: take $[-r, -l]$ and $[\frac{1}{r}, \frac{1}{l}]$, resp. For the other arithmetic operations, we formalized various bisection algorithms.

5.1 Separation of Roots

Our main method to separate roots via bisection is based on a root-counting function ri_f for polynomial f, such that ri_f l r is the number of roots of f in the interval $[l, r]$. Internally, ri_f is defined directly for linear polynomials, and is based on Sturm's method for nonlinear polynomials.

First, we extend the existing formalization of Sturm's method by Eberl [6], which takes a *real* polynomial and *real* bounds, so that it can be applied on *rational* polynomials with *rational* bounds; nevertheless, the number of *real* roots must be determined. This extension is crucial as we later implement the real numbers by the real algebraic numbers via *data refinement* [7]; at this point we must not yet use real number arithmetics. The correctness of this extension is shown mainly by proving that all algorithms utilized in Sturm's method can be homomorphically extended. For instance, for Sturm sequences we formalize the following result:

lemma *sturm (real-of-rat-poly f)* = *map real-of-rat-poly (sturm-rat f)*

For efficiency, we adapt Sturm's method for our specific purpose. Sturm's method works in two phases: the first phase computes a Sturm sequence, and the second one computes the number of roots by counting the number of sign changes on this sequence for both the upper and the lower bounds of the interval. The first phase depends only on the input polynomial, but not on the interval bounds. Therefore, for each polynomial f we precompute the Sturm sequence once, so that when a new interval is queried, only the second phase of Sturm's method has to be evaluated. This can be seen in the following code equation:

definition *count-roots-interval-rat f* =
   (**let** *fs = sturm-squarefree-rat f* (∗ precompute ∗)
   in ...(λ *l r. sign-changes-rat fs l* − *sign-changes-rat fs r* + ...) ...)

For this optimization, besides the essential (f, l, r) our internal representation additionally stores a function $ri :: \mathbb{Q} \to \mathbb{Q} \to \mathbb{N}$ which internally stores the precomputed Sturm sequence for f.

With the help of the root-counting functions, it is easy to compute a required interval. For instance, consider the addition of a and b, each represented by (f, l_a, r_a) and (g, l_b, r_b), and we already have a polynomial h which has $a + b$ as one of its roots. If ri_h $(l_a + l_b)$ $(r_a + r_b) = 1$, then we are done. Otherwise, we repeat bisecting the intervals $[l_a, r_a]$ and $[l_b, r_b]$ with the help of ri_f and ri_g. Similar bisections are performed for multiplication and n-th roots.

For further efficiency, we formalize the bisection algorithms as partial functions [10]. This is motivated by the fact that many of these algorithms terminate only on valid inputs, and runtime checks to ensure termination would be an overhead. In order to conveniently prove the correctness of the algorithms, we define some well-founded relations for inductive proofs, which are reused for various bisection algorithms. For instance, we define a relation based on a decrease in the size of the intervals by at least δ, where δ is the minimal distance of two distinct roots of some polynomial.

Finally, we tighten the intervals more than what is required to identify the root. This is motivated as follows. Assume that the interval $[2, 10000]$ identifies a real root $a \approx 3134.2$ of a polynomial f. Now, consider computing the floor $\lfloor a \rfloor$, which requires us to bisect the interval until we arrive at $[3134.003, 3134.308]$. It would be nice if we could update the bounds for a to the new tighter interval at this point. Unfortunately, we are not aware of how this can be done in a purely functional language. Hence, every time we invoke $\lfloor a \rfloor$ or other operations which depends on a, we have to redo the bisection from the initial interval. Therefore, it is beneficial to compute sufficiently tight intervals whenever constructing algebraic numbers. Currently we limit the maximal size of the intervals by $\frac{1}{8}$.

5.2   Comparisons of Algebraic Numbers

Having defined all arithmetic operations, we also provide support for comparisons of real algebraic numbers, as well as membership test in $\mathbb{Q}$, etc. For membership in $\mathbb{Q}$, we formalize the rational root test which we then integrate into a bisection algorithm. Comparison is, in theory, easy: just compute $x - y$ and determine its sign, which is trivial, since we have the invariant that the signs of the interval bounds coincide. This naive approach however requires an expensive resultant computation. Hence we pursue the following alternative approach: To compare two algebraic numbers a and b, represented by (f, l_a, r_a) and (g, l_b, r_b),

– we first decide[7] $a = b$ by testing whether $gcd\ f\ g$ has a root in $[l_a, r_a] \cap [l_b, r_b]$. The latter property can be determined using Sturm's method; and
– if $a \neq b$, then bisect the intervals $[l_a, r_a]$ and $[l_b, r_b]$ until they become disjoint. Afterwards we compare the intervals to decide $a < b$ or $a > b$.

Note that the recursive bisection in the second step is terminating only if it is invoked with $a \neq b$. At this point, Isabelle's **partial-function** command becomes essential. Note also that specifying the algorithm via **function** prohibits code generation.

If we had a proof that the internal polynomials are irreducible, then the first step could be done more efficiently, since then $f \neq g$ implies $a \neq b$. We leave it for future work to formalize more efficient factorization algorithms.

---

[7] We thank one of the anonymous reviewers for pointing us to this equality test.

5.3　Types for Real Algebraic Numbers

As the internal representation of algebraic numbers, besides the essential (f, l, r) and already mentioned ri, we store another additional information: a flag ty of type *poly-type*, indicating whether f is known to be monic and irreducible (*Monic-Irreducible*) or whether this is unknown (*Arbitrary-Poly*). We initially choose *Arbitrary-Poly* as ty for non-linear polynomials, and *Monic-Irreducible* for linear polynomials after normalizing the leading coefficient. If we have a complete factorization, we may set the polynomial type to *Monic-Irreducible*; however, this would require the invocation of the slow certified factorization algorithm.

In the formalization we create a corresponding type abbreviation for the internal representation (an option type where *None* encodes the number 0), then define an invariant *rai-cond* which should be satisfied, and finally enforce this invariant in the type *real-alg-intern*. For the specification of algorithms on type *real-alg-intern*, the lifting and transfer package has been essential [8].

type-synonym *rai-intern* = (*poly-type* × *root-info* × *rat poly* × *rat* × *rat*)*option*
definition *rai-cond tuple* = (***case*** *tuple* ***of*** *Some* (*ty,ri,f,l,r*) ⇒
　　$f \neq 0 \land$ *unique-root f l r* $\land$ *sgn l = sgn r* $\land$ *sgn r* $\neq 0 \land$ ... | *None* ⇒ *True*)
typedef *real-alg-intern* = *Collect rai-cond*

Then, all arithmetic operations have been defined on type *real-alg-intern*.

In order to implement the real numbers via real algebraic numbers, we did one further optimization, namely integrate dedicated support for the rational numbers. The motivation is that most operations can be implemented more efficiently, if one or both arguments are rational numbers. For instance, for addition of a rational number with a real algebraic number, we provide a function *add-rat-rai* :: *rat* ⇒ *real-alg-intern* ⇒ *real-alg-intern* which does neither require a resultant computation, nor a factorization.

Therefore, we create a new datatype *real-alg-dt*, which has two constructors: one for the rational numbers, and one for the real algebraic numbers whose representing polynomial has degree at least two. This invariant on the degree is then ensured in a new type *real-alg-dtc*, and the final type for algebraic numbers is defined as a quotient type *real-alg* on top of *real-alg-dtc*, which works modulo different representations of the same real algebraic numbers. Here, *real-of-radtc* is the function that delivers the real number which is represented by a real algebraic number of type *real-alg-dtc*.[8]

quotient-type　*real-alg* = *real-alg-dtc* / $\lambda\, x\, y.$ *real-of-radtc x = real-of-radtc y*

Now we provide the following code equations to implement the real numbers via real algebraic numbers by data refinement, where *real-of* :: *real-alg* ⇒ *real* is converted into a constructor in the generated code.

---

[8] Note that the quotient type can be in principle defined also directly on top of *real-alg-dt*, such that the quotient and invariant construction is done in one step, but then code generator will fail in Isabelle 2016.

lemma *plus-real-alg[code]:* $(real\text{-}of\ x) + (real\text{-}of\ y) = real\text{-}of\ (x + y)$
(* similar code lemmas for =, <, -, *, /, floor, etc. *)

Note that in the lemma *plus-real-alg*, the left-hand side of the equality is addition for type *real*, whereas the right is addition of type *real-alg*.

We further prove that real algebraic numbers form an Archimedean field.

instantiation *real-alg :: floor-ceiling* (* includes Archimedean field *)

Finally, we provide a function *real-roots-of-rat-poly :: rat poly $\Rightarrow$ real list* which computes all real roots of a non-zero rational polynomial. It first factors the polynomial, and then for each factor it either uses a closed form to determine the roots, or computes intervals that uniquely identify each root of the factor and returns the corresponding real algebraic numbers. Below, *rpoly* denotes the evaluation of a rational polynomial at a real or complex point.

lemma *assumes* $f \neq 0$
    shows *set* $(real\text{-}roots\text{-}of\text{-}rat\text{-}poly\ f) = \{a :: real.\ rpoly\ f\ a = 0\}$

6 Complex Algebraic Numbers

All of the results on resultants have been developed in a generic way, i.e., they are available for both real and complex algebraic numbers. Hence, in principle one can pursue a similar approach as in Sect. 5 to integrate complex algebraic numbers, one just has to replace Sturm's method by a similar method to separate complex roots, e.g., by using results of Kronecker [15, Sect. 1.4.4].

Since we are not aware of any formalization of such a method, instead we just stick to Isabelle's implementation of complex numbers, i.e., pairs of real numbers representing the real and imaginary part. Note that this is also possible in the algebraic setting: a complex number is algebraic if and only if both the real and the imaginary part are algebraic.

With this representation, all of the following operations become executable on the complex numbers for free: $+$, $-$, $*$, $/$, $\sqrt{\cdot}$, $=$, and complex conjugate. These operations are already implemented via operations on the real numbers, and those are internally computed by real algebraic numbers via data refinement.

The only operation that is not immediate is a counterpart of *real-roots-of-rat-poly* – a method to determine all complex roots of a rational polynomial f. Here, the algorithm proceeds as follows, excluding optimizations.

– Consider a complex root $a+bi$ of f for $a, b \in \mathbb{R}$. Since $a = \frac{1}{2}((a+bi)+(a-bi))$, a is a root of the rational polynomial $g = poly\text{-}mult\text{-}rat\ \frac{1}{2}\ (poly\text{-}add\ f\ f)$. Here, the first f in *poly-add f f* represents $a + bi$ and the second f represents $a - bi$; complex conjugate numbers share the same representing polynomials. Similarly, since $b = \frac{1}{2i}((a + bi) - (a - bi))$, b is a root of $h = poly\text{-}mult\ [:1,0,4:]\ (poly\text{-}add\ f\ (poly\text{-}uminus\ f))$, where $[:1,0,4:]$ is the polynomial $1 + 4x^2$ with root $\frac{1}{2i}$.

– Let C be the set of all numbers $a + bi$ such that $a \in$ *real-roots-of-rat-poly* g and $b \in$ *real-roots-of-rat-poly* h. Then C contains at least all roots of f. Return $\{c \in C. \ f(c) = 0\}$ as the final result.

The actual formalization of *complex-roots-of-rat-poly* contains several special measures to improve the efficiency, e.g., factorizations are performed in between, explicit formulas are used, etc. The soundness result looks as in the real case.

lemma *assumes* $f \neq 0$
    shows set (*complex-roots-of-rat-poly* f) = $\{a :: complex. \ rpoly \ f \ a = 0\}$

The most time-consuming task in *complex-roots-of-rat-poly* is actually the computation of $\{c \in C. \ f(c) = 0\}$ from C. For instance, when testing $f(c) = 0$ in Example 1, multiplications like $b \cdot b$ occur. These result in factorization problems for polynomials of degree 144.

With the help of the complex roots algorithm and the fundamental theorem of algebra, we further develop two algorithms that factor polynomials with rational coefficients over $\mathbb{C}$ and $\mathbb{R}$, resp. Factorization over $\mathbb{C}$ is easy, since then every factor corresponds to a root. Hence, the algorithm and the proof mainly take care of the multiplicities of the roots and factors. Also for the real polynomials, we first determine the complex roots. Afterwards, we extract all real roots and group each pair of complex conjugate roots. Here, the main work is to prove that for each complex root c, its multiplicity is the same as the multiplicity of the complex conjugate of c.

7 Displaying Algebraic Numbers

We provide two approaches to display real algebraic numbers.

The first one displays the approximative value of an algebraic number a. Essentially, the rational number $\frac{\lfloor 1000a \rfloor}{1000}$ is computed and displayed as string. For instance, the first root of polynomial g in Example 1 is displayed as "~ -0.569".

The second approach displays a number represented by (ty, ri, f, l, r) exactly as the string "root #n of f", provided that $ty = $ *Monic-Irreducible* and that n is the number of roots of f in the interval $(-\infty, r]$. In order to determine the value of n, we just apply Sturm's method. In case $ty \neq $ *Monic-Irreducible*, at this point we invoke the expensive certified factorization.

Note that displaying a number must be a function of type *real-alg* $\Rightarrow$ *string*, i.e., the resulting string must be independent of the representative. Clearly, this is the case for the first approach. For the second approach we need a uniqueness result, namely that every algebraic number a is uniquely represented by a monic and irreducible polynomial. To this end, we first formalize the result, that the GCD of two rational polynomials stays the same if we embed $\mathbb{Q}$ into $\mathbb{R}$ or $\mathbb{C}$.

lemma *map-poly of-rat* (*gcd f g*) = *gcd* (*map-poly of-rat f*) (*map-poly of-rat g*)

Using this lemma, we provide the desired uniqueness result.

lemma *assumes* algebraic a *shows* $\exists! f.$ alg-poly a f $\land$ monic f $\land$ irreducible f

Our formalization of this statement works along the following line. Assume f and g are two different monic and irreducible rational polynomials with a common real or complex root a. That is, f and g have a common factor $x - a$ as a real or complex polynomial and hence, the GCD of f and g (over $\mathbb{R}$ or $\mathbb{C}$) is a non-constant polynomial. On the other hand, the GCD of f and g over $\mathbb{Q}$ must be a constant: it cannot be a proper factor of f or g since the polynomials are irreducible over $\mathbb{Q}$, and it cannot be f or g itself, since this contradicts monicity and $f \neq g$.

8   Conclusion

We integrated support for real and complex algebraic numbers in Isabelle/HOL. Although all arithmetic operations are supported, there remain some open tasks.

A formalization of an equivalent to Sturm's method for the complex numbers would admit to represent the roots in Example 1 just as root $\#(1,2,3,4)$ of f, without the need for high-degree polynomials for the real and imaginary part.

A certified efficient factorization algorithm would also be welcome: then the implementation of comparisons of algebraic numbers could be simplified and it would allow to display more algebraic numbers precisely within reasonable time.

Finally, it would be useful to algorithmically prove that the complex algebraic numbers are algebraically closed, so that one is not restricted to rational coefficients in the factorization algorithms over $\mathbb{R}$ and $\mathbb{C}$.

Acknowledgments. We thank the anonymous reviewers for their helpful comments. The early abortion in our divisibility test for integer polynomials is due to Sebastiaan Joosten. This research was supported by the Austrian Science Fund (FWF) project Y757.

References

1. Brown, W.S.: The subresultant PRS algorithm. ACM Trans. Math. Softw. **4**(3), 237–249 (1978)
2. Cohen, C.: Construction of real algebraic numbers in CoQ. In: Beringer, L., Felty, A. (eds.) ITP 2012. LNCS, vol. 7406, pp. 67–82. Springer, Heidelberg (2012)
3. Cohen, C., Djalal, B.: Formalization of a Newton series representation of polynomials. In: CPP 2016, pp. 100–109. ACM (2016)
4. Cohen, C., Mahboubi, A.: Formal proofs in real algebraic geometry: from ordered fields to quantifier elimination. Log. Methods Comput. Sci. **8**(1:02), 1–40 (2012)
5. Collins, G.E.: Subresultants and reduced polynomial remainder sequences. J. ACM **14**, 128–142 (1967)
6. Eberl, M.: A decision procedure for univariate real polynomials in Isabelle/HOL. In: CPP 2015, pp. 75–83. ACM (2015)
7. Haftmann, F., Krauss, A., Kunčar, O., Nipkow, T.: Data refinement in Isabelle/HOL. In: Blazy, S., Paulin-Mohring, C., Pichardie, D. (eds.) ITP 2013. LNCS, vol. 7998, pp. 100–115. Springer, Heidelberg (2013)

8. Huffman, B., Kunčar, O.: Lifting and transfer: a modular design for quotients in Isabelle/HOL. In: Gonthier, G., Norrish, M. (eds.) CPP 2013. LNCS, vol. 8307, pp. 131–146. Springer, Heidelberg (2013)
9. Knuth, D.E.: The Art of Computer Programming. Seminumerical Algorithms, vol. 2, 2nd edn. Addison-Wesley, Boston (1981)
10. Krauss, A.: Recursive definitions of monadic functions. In: PAR 2010. EPTCS, vol. 43, pp. 1–13 (2010)
11. Li, W., Paulson, L.C.: A modular, efficient formalisation of real algebraic numbers. In: CPP 2016, pp. 66–75. ACM (2016)
12. Mahboubi, A.: Proving formally the implementation of an efficient gcd algorithm for polynomials. In: Furbach, U., Shankar, N. (eds.) IJCAR 2006. LNCS (LNAI), vol. 4130, pp. 438–452. Springer, Heidelberg (2006)
13. Mishra, B.: Algorithmic Algebra. Texts and Monographs in Computer Science. Springer, Heidelberg (1993)
14. Nipkow, T., Paulson, L.C., Wenzel, M. (eds.): Isabelle/HOL–A Proof Assistant for Higher-Order Logic. LNCS, vol. 2283. Springer, Heidelberg (2002)
15. Prasolov, V.V.: Polynomials. Springer, Heidelberg (2004)
16. Thiemann, R., Yamada, A.: Formalizing Jordan normal forms in Isabelle/HOL. In: CPP 2016, pp. 88–99. ACM (2016)

Modular Dependent Induction in Coq, Mendler-Style

Paolo Torrini[(⊠)]

Department of Computer Science, KU Leuven, Leuven, Belgium
ptorrx@gmail.com

Abstract. Modular datatypes can be given a direct encoding in Coq using Church-style encodings, but this makes inductive reasoning generally problematic. We show how Mendler-style induction can be used to supplement existing techniques in modular inductive reasoning, and we present a novel technique to apply Mendler-style induction in presence of dependent induction. This results in type-based, conventional-looking proofs that take better advantage of existing Coq tactics and depend less pervasively on general semantic conditions, reducing the need for boilerplate.

1 Introduction

Structural abstraction and modularity are essential to cost-effective software development and verification. However, there is a tension between the modularity of an artifact, which allows for its extensibility, and its datatype structuring. In functional programming, this tension shows in what is known as the *expression problem* [27]: recursive datatypes in their conventional form are associated with fixed sets of shapes, and thus not amenable to extension. The expression problem is even more acute when we consider languages that insist on totality of definitions, as in the case of theorem provers based on type theory, such as Coq [5] and Agda [6]. In such languages, recursive datatypes can also provide the structural characterisation of inductive proofs, which thus suffer from analogous limitations with respect to extensibility and reuse.

In functional programming, an answer to the expression problem has been given in terms of the initial algebra semantics of inductive datatypes [13, 26], with the notion of datatype *à la carte*, or *modular datatype* (MDT), introduced in Haskell by Swierstra [20]. The definition of an MDT consists of two parts: the signature functor, i.e. a non-recursive datatype that can be treated as a module and composed by coproduct, and its recursive closure. In Haskell, the fixpoint closure of the signature functor can be implemented generically. However, this relies on the datatype definition of fixpoint operators that are not strictly positive, hence on a representation that does not work for Coq [5].

One way to overcome this problem is to rely on the container semantics of recursive datatypes [1]. Following this approach, Keuchel *et al.* [15] implemented in Coq an encoding of MDT in the container universe. The container representation of a datatype is a comparatively indirect one. Functors are represented as

J.C. Blanchette and S. Merz (Eds.): ITP 2016, LNCS 9807, pp. 409–424, 2016.
DOI: 10.1007/978-3-319-43144-4_25

container extensions, and fixpoints as W types. Moreover, dealing with recursive relations requires shifting to the universe of indexed containers. A drawback of this approach is that the encoded objects, even the non-recursive ones, are syntactically different from the encoding in which the proofs are carried out. This makes it practically difficult not to distinguish between a specification language and the Coq encoding. Indeed, this is the line consistently pursued by Keuchel *et al.* [16] in their language specification tool.

An alternative way to tackle the positivity problem is to rely on impredicative, higher-order definitions, in the style of Church encodings, thus getting around the problem of actually constructing the fixed points predicatively [19]. This approach has been used by Delaware *et al.* in their Meta-Theory à la Carte (MTC) formalisation of MDT [11], further extended in [9] and implemented in Coq [10]. The main advantage of the impredicative approach is that it provides a direct encoding. Functors are simply encoded as non-recursive datatypes, and dealing with recursive relations does not involve the cost of changing universe. The main problem with Church encodings is that together with the predicative characterisation, we lose the structural induction principle associated with the conventional datatype [19]. This is a consequence of the eliminative character of the impredicative encoding: each object carries its own *fold* with respect to an algebra, as application to it.

The MTC approach builds on top of the algebraic characterisation of inductive principles in terms of the universal properties associated with the initial semantics of the corresponding datatypes, as presented by Hutton [14]. Hutton's technique, known as *universality*, is an algebraic, induction-free proof method for inductive theorems. It permits to dispense with the structural induction principle, relying on the strength of the equational characterisation of the *fold* of an algebra as the unique mediating map from the initial algebra. Delaware *et al.* [11] use universality to prove a fundamental result, that we call the Σ *induction principle*: given a property to prove by induction on a datatype, a proof that the property holds in general can be obtained from an algebra of appropriate type, here called Σ-*proof algebra*. Unlike structural induction, this principle is parametric in the datatype. However, Σ induction can only be applied under two critical conditions. The first one is that the uniqueness of *fold* is provable. The second one is that the Σ-proof algebra is *well-formed*, according to a criterion that addresses the proof term, rather than just the type.

Proving the uniqueness of *fold*, which amounts to proving full initiality of the inductive type, is generally problematic in the impredicative approach. MTC gets around this problem resorting to Σ types, by packing objects together with the associated uniqueness proofs. This approach makes it possible to localise dependency on initiality, but at the cost of making the proof formalism rather cumbersome. It also makes it hard to dispense with functional extensionality. Pervasive reliance on universality is not necessarily a drawback, as indeed one of the original motivations for the universality approach lies in proof search [14]. However, when structural induction is available, the semantic property that is more often needed is one that we call *isorecursiveness*, weaker than initiality, stating that a fixpoint and its unfolding are isomorphic. Moreover, the MTC well-formedness criterion for proof algebras is property-specific. This specificity can be overcome relying on a computational requirement (a weak induction principle) that addresses directly the proof term, breaking the type directedness of proofs.

Is it possible to obtain inductive proofs that are effectively type-directed, making dependency on universality less pervasive? The answer we can give is affirmative, and it relies on a different flavour of initial semantics, originally due to Mendler [18] and Geuvers [12], more recently investigated by various authors [3,4,17,24]. Intuitively, in the Mendler-style approach each inductive type carries its own induction principle, as much as each object carries its own *fold*. In an impredicative encoding *à-la*-Mendler, each inductive type comes with an induction principle which is syntactically built into the encoding. It is then possible to carry out inductive proofs using *Mendler induction*, along the lines of conventional inductive ones. The use of Mendler induction in connection with modular datatypes has been first proposed by Torrini *et al.* [23] with accompanying implementation [21]. However, unlike MTC-style induction, Mendler induction is considerably more restrictive than structural induction. Crucially, the Mendler-style approach cannot handle *dependent induction* – i.e. the cases in which the goals depends on the inductive argument. This can be a big limitation in practice.

Here and in the companion code [22] we tackle this problem, observing that Mendler induction is equivalent to structural induction, provided the goal to be proved does not depend on the inductive argument, and dependency can be eliminated by introducing an additional premise, which we call *predicatisation* of the inductive argument. In particular, under this assumption, it becomes possible to prove isorecursiveness, thus making it possible to dispense with pervasive appeals to initiality. Predicatisation, defined as an indexed Mendler fixpoint object, can be chosen according to the shape of the inductive argument type. Reasoning on MDT by Mendler induction generally extends beyond cases in which predicatisation can be discharged by using basic approaches. Discharging the predicatisation hypothesis generally requires dependent induction, and thus Σ induction, relying on an integration of the Mendler approach with MTC.

2   Initial Semantics and Impredicative Encoding

Modular datatypes rely on the categorical representation of a datatype as the fixed point of the corresponding signature functor [13,26]. In the category of sets and total functions, essentially coinciding with Set in Coq, $F : \mathsf{Set} \to \mathsf{Set}$ is a functor (more precisely, a covariant endofunctor) whenever there exists a corresponding *functor map* that satisfies identity and composition properties (i.e. the *functor laws*). This can be formalised in Coq by instantiating a class Functor [10] where

$$\mathsf{fmap}\ \{A\ B : \mathsf{Set}\} : (A \to B) \to F\ A \to F\ B \tag{1}$$

satisfies the functor laws. Semantically, an algebra ϕ determined by a functor F (F-*algebra*) is a pair $\langle C, f \rangle$ where C is a set (the *carrier*) and $f : F\ C \to C$ a function (the *structure map*). F-algebras together with their homomorphisms form a category. In this category, it is possible to interpret the inductive datatype associated with F as the initial object, denoted $\mu := \langle \mathsf{Fix}\ F, \mathsf{in}_F \rangle$. The initiality of μ boils down to the existence and uniqueness of fold $f :\ \mathsf{Fix}\ F \to C$. These correspond, respectively, to the two directions of a logic equivalence known as the *universal property of fold* [11,14], of which the uniqueness condition (in fact, the critical one) is the following.

$$\mathsf{cfold\_un}\ F\ =_{df}\ \forall w\ h,\ (h \circ \mathsf{in}_F\ =\ f \circ (\mathsf{fmap}\ h))\ \to\ (h\ w\ =\ \mathsf{fold}\ f\ w)\quad(2)$$

Initiality also implies that in_F (the *in-map*) denotes an isomorphism, i.e. it has an inverse $\mathsf{out}_F :\ \mathsf{Fix}\ F \to F\ (\mathsf{Fix}\ F)$ (the *out-map*) such that the following hold (we call these *isorecursive equations*).

$$\begin{aligned} \forall x : \mathsf{Fix}\ F,\ x\ &=\ \mathsf{in}_F\ (\mathsf{out}_F\ x) \\ \forall x : F\ (\mathsf{Fix}\ F),\ x\ &=\ \mathsf{out}_F\ (\mathsf{in}_F\ x) \end{aligned}\quad(3)$$

An alternative initial semantics can be obtained using Mendler algebras, and it is equivalent to the conventional one [24]. A Mendler F-algebra is a pair $\langle C, f \rangle$ where $C :$ Set is the carrier, and $f :\ \forall A :\ \mathsf{Set},\ (A \to C) \to (F\ A \to C)$ is a parametric map from morphisms to morphisms satisfying the following (for F covariant [25]) *strong Mendler algebra condition*.

$$\forall m :\ A \to C,\ f\ A\ m\ =\ (f\ C\ \mathsf{id}) \circ (\mathsf{fmap}\ m)\quad(4)$$

Mendler F-algebras form a category where a morphism between Mendler algebras $\langle C_1, f_1 \rangle$ and $\langle C_2, f_2 \rangle$ is a morphism $h :\ C_1 \to C_2$ that satisfies $h \circ f_1\ C_1\ \mathsf{id}_{C_1} = f_2\ C_1\ h$. Assuming the conventional initial F-algebra μ exists, it can be proved that $\langle \mathsf{Fix}\ F, \lambda A\ m.\ \mathsf{in}_F \circ (\mathsf{fmap}\ m) \rangle$ is the initial Mendler F-algebra. The initiality condition plays an analogous role as in the conventional case. The uniqueness of Mendler-style *fold* can be expressed as follows.

$$\mathsf{mfold\_un}\ F\ =_{df}\ \forall w\ h,\ (h \circ \mathsf{in}_F\ =\ f\ (\mathsf{Fix}\ F)\ h)\ \to\ (h\ w\ =\ \mathsf{fold}\ f\ w)\quad(5)$$

Initial algebra semantics is not restricted to datatypes in Set. An analogous treatment can be given to relations, relying on *indexed algebras*. A relation can be represented as a predicate, i.e. a function from the type of its tupled arguments to Prop. Thinking for simplicity of Prop as a category, given a type K (i.e. $K :$ Type) corresponding to a small category, relations of type $K \to$ Prop can be based on the category of diagrams of type K in Prop where an endofunctor $R : (K \to \mathsf{Prop}) \to (K \to \mathsf{Prop})$, called *indexed functor* here, is associated with an *indexed functor map* that preserves identities and composition, as specified in Coq by a type class Functor[l] where

$$\mathsf{fmap}^{\mathsf{l}}\ \{A\ B :\ K \to \mathsf{Prop}\} :\ (\forall w,\ A\ w \to B\ w) \to \forall w,\ R\ A\ w \to R\ B\ w\quad(6)$$

satisfies the functor laws. The inductively defined relation associated with R can be interpreted as the initial object in the category of K-indexed R-algebras (either of the conventional or the Mendler variety), defined by indexing with K the carriers as well as the associated maps.

Coq is based on the calculus of inductive constructions (CIC) [5] which extends the calculus of constructions (CC) [8] with inductive and coinductive definitions. CC allows for definitions that are impredicative, in the sense of referring in their bodies to collections that are being defined. The solution to the positivity problem adopted in MTC [11] goes back to Pfenning *et al.* [19] in relying on a Church-style encoding of fixpoint operators, thus requiring impredicative definitions. In general, Coq relies on the universe Prop of propositions for impredicative definitions, whereas it uses the universe Set of sets for the predicative hierarchy. However, we need sets in order to take advantage of Σ types. For this reason, we resort to using Coq's `impredicative-set` option, as MTC does.

From the point of view of a type theoretic representation, the type of a conventional algebra (or *Church algebra*) for a functor F and a set C can be identified with the type of its structure map.

$$\mathsf{Alg}^C \ F \ C \ =_{df} \ F \ C \rightarrow C \tag{7}$$

If the initiality property of fixed points is weakened to an existence property, a fixpoint operator can be regarded as a function that maps an algebra to its carrier. An abstract definition of the type-level fixpoint operator Fix^C : (Set $\rightarrow$ Set) $\rightarrow$ Set can then be given, as elimination rule for F-algebras, impredicatively with respect to Set.

$$\mathsf{Fix}^C \ F \ =_{df} \ \forall A : \mathsf{Set}, \ \mathsf{Alg}^C \ F \ A \rightarrow A \tag{8}$$

The map $\mathsf{fold}^C \ F \ C : \mathsf{Alg}^C \ F \ C \rightarrow \mathsf{Fix}^C \ F \rightarrow C$, corresponding to the elimination of a fixpoint value, can be defined as the application of that value.

$$\mathsf{fold}^C \ F \ C \ f \ x \ =_{df} \ x \ C \ f \tag{9}$$

Relying on the functoriality of F, the in-map $\mathsf{in}^C \ F : F(\mathsf{Fix} \ F) \rightarrow \mathsf{Fix} \ F$ and the out-map $\mathsf{out}^C \ F : \mathsf{Fix} \ F \rightarrow F \ (\mathsf{Fix} \ F)$ can be defined as functions.

$$\mathsf{in}^C \ F \ =_{df} \ \lambda x \ A \ f. \ f \ (\mathsf{fmap} \ F \ (\mathsf{fold}^C \ F \ A \ f) \ x) \tag{10}$$

$$\mathsf{out}^C \ F \ =_{df} \ \mathsf{fold}^C \ F \ (F \ (\mathsf{Fix} \ F)) \ (\mathsf{fmap} \ F \ (\mathsf{in}^C \ F)) \tag{11}$$

Also Mendler algebras can be characterised impredicatively by the type of their structure maps. A fixpoint operator can be defined as in the conventional case [11, 18].

$$\mathsf{Alg}^M \ F \ C \ =_{df} \ \forall A : \mathsf{Set}, \ (A \rightarrow C) \rightarrow (F \ A \rightarrow C) \tag{12}$$

$$\mathsf{Fix}^M \ F \ =_{df} \ \forall C : \mathsf{Set}, \ \mathsf{Alg}^M \ F \ C \rightarrow C \tag{13}$$

Unlike the conventional case, the type of a Mendler algebra can be read as the specification of an iteration step, where the bound type variable A represents the type of the recursive calls. The corresponding fold operator

$$\mathsf{fold}^{\mathsf{M}} \ F \ C \ f \ x \ =_{df} \ x \ C \ f \tag{14}$$

indeed has type

$$\mathsf{fold}^{\mathsf{M}} \ F \ C : \ (\forall A : \mathsf{Set}, \ (A \to C) \to (F \ A \to C)) \to (\mathsf{Fix}^{\mathsf{M}} \ F) \to C \tag{15}$$

which can represent an induction principle, under the assumption that the argument to the induction hypothesis is only used therein without further analysis [3,18]. In-maps and out-maps can be defined as follows.

$$
\begin{aligned}
\mathsf{in}^{\mathsf{M}} \ F \ (x : F(\mathsf{Fix}^{\mathsf{M}} \ F)) : \mathsf{Fix}^{\mathsf{M}} \ F \ =_{df} \\
\lambda A \ (f : \mathsf{Alg}^{\mathsf{M}} \ F \ A). \ f \ (\mathsf{Fix}^{\mathsf{M}} \ F) \ (\mathsf{fold}^{\mathsf{M}} \ F \ A \ f) \ x
\end{aligned}
\tag{16}
$$

$$
\begin{aligned}
\mathsf{out}^{\mathsf{M}} \ F \ (x : \mathsf{Fix}^{\mathsf{M}} \ F) : F \ (\mathsf{Fix}^{\mathsf{M}} \ F) \ =_{df} \ x \ (F \ (\mathsf{Fix}^{\mathsf{M}} \ F)) \\
(\lambda A \ (r : A \to F \ (\mathsf{Fix}^{\mathsf{M}} \ F)) \ (a : F \ A). \ \mathsf{fmap} \ F \ (\lambda y : A. \ \mathsf{in}^{\mathsf{M}} \ F \ (r \ y)) \ a)
\end{aligned}
\tag{17}
$$

Whenever $R : (K \to \mathsf{Prop}) \to (K \to \mathsf{Prop})$ is an indexed functor, an indexed R-algebra can be characterised as a K-indexed map, given an indexed carrier $D : K \to \mathsf{Prop}$.

$$\mathsf{Alg}^{\mathsf{Cl}} \ K \ R \ D \ =_{df} \ \forall w : K, \ R \ D \ w \to D \ w \tag{18}$$

Analogously, a K-indexed Mendler R-algebra can be characterised as a function between indexed morphisms.

$$
\begin{aligned}
\mathsf{Alg}^{\mathsf{MI}} \ K \ R \ D \ =_{df} \\
\forall A : K \to \mathsf{Prop}, \ (\forall w : K, \ A \ w \to D \ w) \to \forall w : K, \ R \ A \ w \to D \ w
\end{aligned}
\tag{19}
$$

The corresponding fixpoint operator, with type $((K \to \mathsf{Prop}) \to K \to \mathsf{Prop}) \to K \to \mathsf{Prop}$, and the structuring operators (mediating map, in-map and out-map) can be defined as follows (we show the Mendler variant, the conventional one is analogous).

$$\mathsf{Fix}^{\mathsf{MI}} \ K \ R \ (w : K) \ =_{df} \ \forall A : K \to \mathsf{Prop}, \ \mathsf{Alg}^{\mathsf{MI}} \ K \ R \ A \to A \ w \tag{20}$$

$$\mathsf{fold}^{\mathsf{MI}} \ K \ R \ D \ (f : \mathsf{Alg}^{\mathsf{MI}} \ K \ R \ D) \ (w : K) \ (x : \mathsf{Fix}^{\mathsf{MI}} \ K \ R \ w) \ =_{df} \ x \ D \ f \tag{21}$$

$$
\begin{aligned}
\mathsf{in}^{\mathsf{MI}} \ K \ R \ (w : K) \ (x : R \ (\mathsf{Fix}^{\mathsf{MI}} \ K \ R) \ w) : \mathsf{Fix}^{\mathsf{MI}} \ K \ R \ w \ =_{df} \\
\lambda A \ (f : \mathsf{Alg}^{\mathsf{MI}} \ K \ R \ A). \ f \ (\mathsf{Fix}^{\mathsf{MI}} \ K \ R) \ (\mathsf{fold}^{\mathsf{MI}} \ K \ R \ A \ f) \ w \ x
\end{aligned}
\tag{22}
$$

$$
\begin{aligned}
\mathsf{out}^{\mathsf{MI}} \ K \ R \ (w : K) \ (x : \mathsf{Fix}^{\mathsf{MI}} \ K \ R \ w) : R \ (\mathsf{Fix}^{\mathsf{MI}} \ K \ R) \ w \ =_{df} \\
x \ (R \ (\mathsf{Fix}^{\mathsf{MI}} \ K \ R)) \ (\lambda \ A \ (r : \forall v, \ A \ v \to R \ (\mathsf{Fix}^{\mathsf{MI}} \ K \ R) \ v) \\
(w : K) \ (a : R \ A \ w). \ \mathsf{fmap}^{\mathsf{I}} \ R \ (\lambda y : A \ w. \ \mathsf{in}^{\mathsf{MI}} \ K \ R \ w \ (r \ w \ y)) \ a)
\end{aligned}
\tag{23}
$$

The strong Mendler algebra property, i.e. (4), is not enforced by the impredicative definitions. In the Coq formalisation we introduce a type class (StrongMendlerAlgebra) which requires the satisfaction of (4). This property can be easily discharged, anyway, given the semantic equivalence between Mendler and Church approach, and the possibility to transform Mendler fixpoints into Church ones and vice-versa (see [22]). More critically, in either variant of the impredicative encoding (either Mendler or conventional, and similarly for indexed functors) the definitions of fold do not guarantee full initiality – they only enforce a weaker condition, called quasi-initiality by Wadler [26]. The impredicative definitions do not even suffice to enforce the isorecursive equations. In order to obtain initiality, what is missing is the uniqueness side of the universal properties of fold [11,14], i.e. in Set, (2) for the conventional variant, and (5) for the Mendler one.

The universality of fold is a very strong property which ensures semantic soundness and is required by the Hutton's approach [14], on which MTC relies. Proving universality in the impredicative encoding is still an open challenge to our knowledge. MTC gets around the problem, by packing each fixpoint object together with a proof of uniqueness of its fold, using Σ types. Essentially, this is done by using a lifted form of in-map, that for functor F has type $\mathsf{Alg^C}\ F\ (\Sigma w,\ \mathsf{cfold\_unique}\ F\ w)$, to define an enhanced form of smart constructors [10]. This complicates significantly the syntax, and involves a major part of the MTC development. Our implementation [22] does not follow the MTC workaround. We delegate full initiality to type classes, using FoldUnivProp (with cfold_un) for Church-style fixpoints and MFoldUnivProp (with mfold_un) for Mendler ones, without actually discharging the proof obligations. On the other hand, we distinguish between full initiality and isorecursiveness, which we delegate to a distinct type class, IsoRecursive, associated with the Mendler variant of the isorecursive equations (3).

3   Modular Datatypes: An Example

Our driving example, a simple language $\mathcal{A}$ of arithmetics with natural numbers and sums, is similar to those already used in the relevant literature [11,20]. The syntactic categories of types and values are non-recursively defined. Values are lifted natural numbers. Types include naturals and a bottom type, the latter as a basic exception type (here included for expository purpose – ensuring case analysis does not suffice to prove subject reduction).

$$\mathsf{Inductive\ Val}\ :=\ \mathsf{val}\ (n : \mathsf{nat}) \tag{24}$$

$$\mathsf{Inductive\ Typ}\ :=\ \mathsf{Nat}\ |\ \mathsf{Bot} \tag{25}$$

Expressions include natural numbers, sums, and a catch-all exception. The corresponding signature can be easily shown to instantiate the Functor class.

$$\mathsf{Inductive\ Exp_S}\ (C : \mathsf{Set})\ :=\ \mathsf{lit}\ (v : \mathsf{nat})\ |\ \mathsf{sum}\ (e_1\ e_2 : \ C)\ |\ \mathsf{err} \tag{26}$$

The recursive type of expressions can be obtained as fixpoint of the functor.

$$\text{Definition Exp} := \text{Fix Exp}_S \tag{27}$$

Here and further on we write Fix for fixpoints in Set, dropping the superscript (and so for in, etc., and analogously for relations) when we want to emphasise that we can either use conventional or Mendler fixpoints, bearing in mind that it is always possible to convert between them [22].

The modularity of datatypes à la carte [20] relies on the fact that coproduct preserves functoriality: given functors F_1, F_2, their coproduct $F_1 \oplus F_2$ also defines a functor. We can then take advantage of compositionality as in the non-recursive case. Initial algebra semantics also ensures that recursive functions defined on the structure of an inductive type, represented as *fold* of an algebra, enjoy modularity relying on the equivalence between $(A \to C) \wedge (B \to C)$ and $(A \oplus B) \to C$.

In our example, expression evaluation can be defined recursively, representing partiality with option types, starting from the following, auxiliary non-recursive definitions.

$$
\begin{aligned}
&\text{Definition vsum } (v_1 \; v_2 : \text{ option Val}) : \text{ option Val} := \text{ match } v_1 \text{ with} \\
&\quad | \text{ Some (val } n_1) \; \Rightarrow \text{ match } v_2 \text{ with} \\
&\quad\quad | \text{ Some (val } n_2) \; \Rightarrow \text{ Some (val } (n_1 + n_2)) \\
&\quad\quad | \_ \Rightarrow \text{ None} \\
&\quad | \_ \Rightarrow \text{ None} \\
&\text{Definition exp } (v : \text{ option Val}) : \text{ exp } := \text{ match } v \text{ with} \\
&\quad | \text{ Some (val } n) \; \Rightarrow \text{ lit } n \\
&\quad | \_ \Rightarrow \text{ err}
\end{aligned}
\tag{28}
$$

The evaluation function can be structurally characterised as an algebra. The following definition gives us a Mendler algebra, which unlike a conventional one allows for control over evaluation, as discussed in [11] (indeed, Mendler algebras are used in MTC only for this reason).

$$
\begin{aligned}
&\text{Definition eval}_A \; (A : \text{Set}) \; (r : \; A \to \text{option Val}) \; (e : \text{Exp}_S \; A) : \\
&\text{option Val} := \text{ match } e \text{ with lit } n \; \Rightarrow \text{ Some (val } n) \\
&\quad\quad\quad | \text{ sum } e_1 \; e_2 \; \Rightarrow \text{ vsum } (r \; e_1) \; (r \; e_2) \\
&\quad\quad\quad | \text{ err } \Rightarrow \text{ None}
\end{aligned}
\tag{29}
$$

Finally, the following defines our evaluation function as *fold* of the algebra.

$$\text{Definition eval} : \text{Exp} \to \text{option Val} := \text{fold}^M \text{ eval}_A \tag{30}$$

We can define the typing relation inductively, using the indexed functor Typing_S as signature.

$$
\begin{aligned}
&\text{Inductive Typing}_S \; (T : (\text{Exp} * \text{Typ}) \to \text{Prop}) : (\text{Exp} * \text{Typ}) \to \text{Prop} := \\
&\quad | \text{ LitTyp} : \forall n : \text{nat}, \text{Typing}_S \; T \; (\text{in (lit } n), \; \text{Nat}) \\
&\quad | \text{ SumTyp} : \forall \; (e_1 \; e_2 : \text{Exp}), \\
&\quad\quad T \; (e_1, \; \text{Nat}) \to T \; (e_2, \; \text{Nat}) \to \text{Typing}_S \; T \; (\text{in (sum } e_1 \; e_2), \; \text{Nat}) \\
&\quad | \text{ ErrTyp} : \text{Typing}_S \; T \; (\text{in err}, \; \text{Bot})
\end{aligned}
\tag{31}
$$

Once the appropriate definition of fmap^l is provided, the proof that Typing$_S$ is an instance of Functor^l can be discharged almost automatically. We can then give a modular definition of Typing

$$\text{Definition Typing}: (\text{Exp} * \text{Typ}) \rightarrow \text{Prop} := \text{Fix}^{MI} (\text{Exp} * \text{Typ}) \text{ Typing}_S \quad (32)$$

and use it to define the following notion of type preservation

$$\begin{aligned}
&\text{Definition type\_preservation } (e : \text{Exp}) : \text{ Prop} := \\
&\quad \forall t : \text{Typ}, \text{ Typing}_X (e, t) \rightarrow \text{Typing}_X ((\text{exp} \circ \text{eval}) \ e, \ t)
\end{aligned} \quad (33)$$

relying on Typing$_X$ =$_{df}$ Typing$_S$ Typing.

4   Inductive Proofs with MTC Induction

The impredicative encoding makes it easy to represent modular datatypes in Coq, but leaves us with the problem of how to reason inductively about them. There is no conventional induction principle that can be applied to a term of type Fix^C F, as this type is syntactically a higher-order definition rather than an inductive datatype – and similarly for Mendler-style encodings.

The simplest case of an inductive proof, schematically, is one in which given a type T as the representation of an inductive datatype and a goal G : Prop depending on a context Γ, we can produce a valid sequent

$$\Gamma, \ w : T \vdash g : \ G \quad (34)$$

reasoning semantically by structural induction on T. This is only a special case of the more general one, in which the goal can depend on the inductive argument – as in the following, where we let $P : T \rightarrow$ Prop be a predicate on T, with $T =_{df}$ Fix^C F.

$$\Gamma, w : T \vdash g : P \ w \quad (35)$$

The following can be proved relying on the isorecursiveness of T.

$$\forall v : T, \ \exists w : F \ T, \ P \ v \ = \ P \ (\text{in}^C \ F \ w) \quad (36)$$

Rewriting (35) with (36), we obtain

$$\Gamma, w : F \ T \vdash g' : P \ (\text{in}^C \ F \ w) \quad (37)$$

Here it is possible to apply induction on w, since $F \ T$ is syntactically an inductive datatype: however, what we actually get is case analysis.

The solution adopted by MTC [10,11] relies on the following fundamental result, which we call Σ induction.

$$\begin{aligned}
&\text{Lemma SigmaInduction } \{F : \text{ Set} \rightarrow \text{Set}\} \ \ \{H_F : \text{ Functor } F\} \\
&\quad \{H_{UP} : \ \forall x : \text{Fix}^C \ F, \ \text{FoldUnivProp } x\} \\
&\quad \{P : \text{ Fix}^C \ F \rightarrow \text{Prop}\} \ \ (p\_alg : \text{ Alg}^C \ F \ (\Sigma x, \ P \ x)) \\
&\quad \{H_{WF} : \text{ WellFormedProofAlgebra } p\_alg\} \ : \\
&\qquad \forall x : \text{Fix}^C \ F, \ P \ x
\end{aligned} \quad (38)$$

An analogous result can be proved for Mendler algebras – i.e. for Σ-proof Mendler algebras of type e.g. $\mathsf{Alg^M}\ F\ (\Sigma x,\ P\ x)$ [22]. The essential insight of Σ induction consists of reducing our task to one of proving

$$\Gamma \vdash p\_alg : \mathsf{Alg^C}\ F\ (\Sigma x.\ P\ x) \tag{39}$$

Under the required conditions this suffices to obtain a proof of our original goal. One of the main assumptions is the initiality for F (H_{UP}). As mentioned, MTC can dispense with making explicitly this assumption by packing fixpoint objects with their universal properties. The other critical assumption is the well formedness of p_alg as a Σ-proof algebra (H_{WF}), which boils down to the following:

$$\forall e : F(\Sigma x,\ P\ x), \mathsf{proj1\_sig}\ (\mathsf{p\_alg}\ e) = \\ \mathsf{in}\ (\mathsf{fmap}\ (\Sigma x,\ P\ x)\ (\mathsf{Fix}\ F)\ (\mathsf{proj1\_sig}\ (\mathsf{Fix}\ F)\ P)\ e) \tag{40}$$

Crucially, this condition applies to algebras (i.e. objects) rather than types. Indeed, this is a very strong condition for the object, given the type: it says that the algebra maps each element of the datatype to a dependent sum that has the same element as witness – so, under the Σ cover, the behaviour is actually that of a dependent product. MTC provides a general technique to construct well-formed Σ-proof algebras for a functor. This technique consists of providing a weak structural induction principle for the datatype (also called *poor man's induction*), and on building Σ-proof algebras based on that principle (39). The weak induction trick deflates the problem of proving well-formedness for each object, but it breaks type directedness, introducing a considerable detour: an inductive principle is used to filter terms that are fed to an induction-free proof method – such as Σ induction is.

In our example, the weak induction principle for Exp is the following.

Definition ExpWeakInduction $(P :\ \mathsf{Exp} \to \mathsf{Prop})$
$(H_{lit} :\ \forall n,\ P\ (\mathsf{in^C}\ (\mathsf{lit}\ n)))$
$(H_{sum} :\ \forall\ (e_1\ e_2 :\ \mathsf{Exp})\ (I_1 :\ P\ e_1)\ (I_2 :\ P\ e_2),\ P\ (\mathsf{in^C}(\mathsf{sum}\ e_1\ e_2)))$
$(H_{err} :\ P\ (\mathsf{in^C}\ \mathsf{err}))$
$(e :\ \mathsf{Exp}\ (\Sigma x,\ P\ x)) :\ \Sigma x,\ P\ x :=\ \mathsf{match}\ e\ \mathsf{with}$ $\qquad\qquad$ (41)
$\quad |\ \mathsf{lit}\ n\ \Rightarrow\ \mathsf{exist}\ P\ (\mathsf{in^C}\ (\mathsf{lit}\ n))\ (H_{lit}\ n)$
$\quad |\ \mathsf{sum}\ e_1\ e_2\ \Rightarrow\ \mathsf{exist}\ P\ (\mathsf{in^C}(\mathsf{sum}\ e_1\ e_2))$
$\qquad\quad (H_{sum}\ (\mathsf{proj1\_sig}\ a_1)\ (\mathsf{proj1\_sig}\ a_2)\ (\mathsf{proj2\_sig}\ a_1)\ (\mathsf{proj2\_sig}\ a_2))$
$\quad |\ \mathsf{err}\ \Rightarrow\ \mathsf{exist}\ P\ (\mathsf{in^C}\ \mathsf{err})\ H_{err}$

Using ExpWeakInduction as a refinement step, it is possible to prove a special form of (36), following the main lines of a conventional inductive proof. As expected, the proof thus obtained can be easily shown to be a well-formed Σ-proof algebra. Therefore, under the initiality assumption for Exp, Σ induction can be applied to obtain a modular version of subject reduction (isorecursiveness of Typ is also used) [22].

5 Mendler-Style Induction

Is it possible to obtain modular inductive proofs that are properly type directed? Reconsidering the schematic example in Sect. 4, let us focus on the non-dependent case (34): we already know how to get to

$$\Gamma, w : F \ (\mathsf{Fix}^C \ F) \vdash g' : G \tag{42}$$

using isorecursiveness, and we know that the problem at this point is the lack of an induction hypothesis. Supplying such hypothesis explicitly would give us a generic representation of the step lemma in our inductive proof. This is essentially the idea behind Mendler induction: under the assumption that the argument passed to the induction hypothesis is used only there, without further case analysis, and that therefore we make no use of its type structure, its type can be represented by a fresh type variable [3,18]. Under these restriction, which in fact rules out dependent induction, we can provide the following representation of the step lemma in our inductive proof.

$$\Gamma, \quad A : \mathsf{Type}, \quad h_0 : A \to G, \quad h_1 : F \ A \vdash p : G \tag{43}$$

Given $f =_{df} \lambda A \ h_0 \ h_1. \ p$, the above can be rewritten as a Mendler algebra.

$$\Gamma \vdash f : \mathsf{Alg}^M \ F \ G \tag{44}$$

For the use we are making of it, this is indeed a *Mendler proof algebra*. The original goal (34) can now be obtained by folding, without need of further adjustments.

$$\Gamma \vdash \mathsf{fold}^M \ F \ G \ f \ : \ \forall w : T, \ G \tag{45}$$

In order to prove (43), case analysis (as provided in Coq e.g. by *inversion* and *destruct* tactics [5]) can be applied to h_1, allowing us to reason on the structure of $F \ A$. This actually results in doing induction on that structure, as the induction hypothesis h_0 is already there – granted by the type. In contrast with Σ-proof algebras, Mendler algebras provide us with a proper induction scheme – indeed one that has the same content as non-dependent, structural induction. In relationship with modularity, such algebras can be regarded as proof modules, that can be composed together in the usual sense of case analysis on coproducts. This modularity is entirely type directed and does not involve significant boilerplate, unlike modularity in the MTC approach, which involves not only composing Σ-proof algebra, but also their well formedness proofs.

Although the isorecursive equations are used already in (42), the application of Mendler induction requires neither full initiality, unlike Σ induction, nor even the strong Mendler algebra property. In contrast with the MTC approach, the Mendler-style one maintains a distinction between syntactical reasoning by structural induction, which is directed solely by types, and semantic soundness.

If we consider specifications written in a relational style, as it is particularly natural in structural operational semantics [7], non-dependent induction is expressive enough to cover interesting cases [23]. However, in a broader mathematical perspective, non-dependent induction is rather restrictive. Indeed, it does not suffice to prove a simple type preservation property based on a functional characterisation of evaluation, such as in our example. Can we extend the applicability of this schematic approach beyond non-dependent induction?

We address this problem in terms of a technique which we call *predicatisation*, used to reduce dependent induction, transforming proofs that require dependent structural induction into proofs that rely on non-dependent structural induction with an additional premise, called *predicatisation hypothesis*. Predicatisation makes it possible to switch the inductive argument of the original proof to an inductive predicate over the original argument, thus lifting the proof dependency on that argument to a type-level dependency. In our schematic example (35) this is possible, when there exists a T-indexed functor $R : (T \to \mathsf{Prop}) \to T \to \mathsf{Prop}$ (called *characterising functor*) and a predicate S on T such that $S =_{df} \mathsf{Fix}^{\mathsf{MI}} \, T \, R$ (called *characterising predicate*), for which the following is provable by structural induction on h

$$\Gamma, \; w : T, \; h : S \, w \vdash l : P \, w \tag{46}$$

and moreover the predicatisation hypothesis can be discharged, i.e.

$$\Gamma, \; w : T \vdash S \, w \tag{47}$$

Proving (46) by Mendler induction involves proving the following

$$\Gamma, \; A : T \to \mathsf{Prop}, \; h_0 : \forall v : T, \; A \, v \to P \, v, \; w : T, \; h_1 : R \, A \, w \vdash p : P \, w \tag{48}$$

Given $f =_{df} \lambda A \, h_0 \, w \, h_1. \, p$, this is equivalent to

$$\Gamma \vdash f : \mathsf{Alg}^{\mathsf{MI}} \, T \, R \, P \tag{49}$$

i.e. a T-indexed Mendler algebra. As in the non-dependent example, the original goal can be simply obtained by folding.

$$\Gamma \vdash \mathsf{fold}^{\mathsf{MI}} \, T \, R \, P \, f : \; \forall w : T, \; S \, w \to P \, w \tag{50}$$

The inductive structure of the new proof is determined by S rather than T. In order to obtain a modular proof that can be easily related to the conventional one, it is desirable that S and T have the same shape, and so the associated functors, R and F.

Our Coq formalisation of predicatisation relies on the following type class.

```
Class Predicatisable (F : Set → Set) {H : Functor F} := {
    char_pred_sig : (FixM F → Prop) → (FixM F → Prop);
    char_pred_sig_functor : FunctorI (FixM F) char_pred_sig;
    char_pred : FixM F → Prop := FixMI (FixM F) char_pred_sig;
    total_pred : ∀w : FixM F, char_pred w }
```
(51)

Notice that the characterising predicate char_pred needs to be an indexed Mendler fixpoint, rather than a conventional one, to support its use as inductive argument in Mendler-style induction.

The following lemma provides the key for the induction switch which makes dependent induction reduction possible.

$$\text{Lemma induct\_switch } (F : \text{Set} \to \text{Set}) \quad \{H_F : \text{Functor } F\}$$
$$\{H_P : \text{Predicatisable } F\} \quad (P : \text{Fix } F \to \text{Prop}) : \tag{52}$$
$$(\forall w : \text{Fix } F, \ P \ w) \ = \ (\forall w : \text{Fix } F, \ \text{char\_pred } F \ w \to P \ w)$$

It is then possible to prove the lemma that essentially constitutes our top-level tactic to apply Mendler induction to dependent cases over sets.

$$\text{Lemma Mendler\_induct } (F : \text{Set} \to \text{Set}) \quad \{H_F : \text{Functor } F\}$$
$$\{H_P : \text{Predicatisable } F\} \quad (P : \text{Fix } F \to \text{Prop}) : \tag{53}$$
$$\text{Alg}^{\text{MI}} \ (\text{Fix } F) \ (\text{char\_pred\_sig } F) \ P \to \forall w : \ \text{Fix } F, \ P \ w$$

The H_P premise makes it possible to discharge the predicatisation hypothesis for Fix F. The critical part in discharging this instantiation is proving that the characteristic predicate is total (total_pred), as this may require dependent induction, and therefore Σ induction. For this reason, it is useful to define the characterising predicate so that we can build a well-formed Σ-proof algebra Alg F (Σx, char_pred x). Unlike in MTC, here Σ induction is needed for the proof of a single property for each functor. The natural choice for the characteristic predicate functor is a T-indexed functor that has the same shape as F. A well-formed Σ-proof algebra can then be written down either in a direct way or interactively.

In our concrete example, the following Exp-indexed functor is used to define the characterising predicate IsExp for Exp.

$$\text{Inductive IsExp}_\text{S} \ (T : \text{Exp} \to \text{Prop}) : \ \text{Exp} \to \text{Prop} \ :=$$
$$\mid \text{litIsExp} : \ \forall \ n : \text{nat}, \ \text{IsExp}_\text{S} \ T \ (\text{in } (\text{lit } n))$$
$$\mid \text{sumIsExp} : \ \forall \ e_1 \ e_2, \ T \ e_1 \to T \ e_2 \to \text{IsExp}_\text{S} \ T \ (\text{in } (\text{sum } e_1 \ e_2)) \tag{54}$$
$$\mid \text{errIsExp} : \ \text{IsExp}_\text{S} \ T \ (\text{in err})$$

$$\text{Definition IsExp} \ := \ \text{Fix}^{\text{MI}} \ \text{Exp IsExp}_\text{S} \tag{55}$$

The following Σ-proof algebra can be easily proved to be well-formed.

$$\text{Definition isExpP}_\text{A} \ (t : \ \text{Exp}_\text{S} \ (\Sigma x, \ \text{IsExp } x)) :$$
$$\Sigma x, \ \text{IsExp } x \ := \ \text{match } t \text{ with}$$
$$\mid \text{lit } n \ \Rightarrow \ \text{exist } (\text{in}^{\text{MI}} \ \_ \ (\text{litIsExp } \_ \ n))$$
$$\mid \text{sum } e_1 \ e_2 \ \Rightarrow \ \text{exist } (\text{in}^{\text{MI}} \ \_ \ (\text{sumIsExp } \_ \tag{56}$$
$$(\text{proj1\_sig } e_1) \ (\text{proj1\_sig } e_2) \ (\text{proj2\_sig } e_1) \ (\text{proj2\_sig } e_2)))$$
$$\mid \text{err } \ \Rightarrow \ \text{exist } (\text{in}^{\text{MI}} \ \_ \ (\text{errIsExp } \_))$$

Under the predicatisation hypothesis, one can prove that the recursive function isExpP $=_{df}$ fold$^\text{C}$ isExpP$_\text{A}$ defines an injection from Exp to Σx, IsExp x, i.e. that proj1_sig ∘ isExpP = id. In fact, this is a property closely associated with well formedness, that could indeed be included in the specification of predicatisation.

The totality requirement (total_pred) can be discharged by Σ induction, under the initiality assumption for Exp. Notice that while char_pred has to be an indexed Mendler algebra (as it is used as inductive argument of Mendler induction in the main proof – see below), here we find it more convenient to use the Church fixpoint of $\mathsf{Exp_S}$, constructing the Σ-proof algebra as a conventional algebra, and doing it directly (rather than interactively). Alternatively, we could have constructed a well-formed Σ-proof Mendler algebra and used the Mendler version of Σ induction [22]. Since we have to build a single Σ-proof algebra for each datatype, no issue arise about generalising such constructions.

The inductive reasoning that is really specific to the subject reduction proof can take place independently of universality, though depending on isorecursive equations. Using Mendler induction, isorecursiveness is provable for $\mathsf{Typ_S}$ and $\mathsf{Exp_S}$, under the corresponding predicatisation hypotheses. The subject reduction proof consists of the construction of a Exp-indexed Mendler algebra for $\mathsf{IsExp_S}$ (the characterising functor of Exp) with type_preservation (the property to be proved) as indexed carrier.

Lemma SubRedMAlg $\{H_1$: IsoRecursive $\mathsf{Typ_S}\}$ $\{H_2$: IsoRecursive $\mathsf{Exp_S}\}$:
$\mathsf{Alg^{MI}}$ Exp IsExp$_S$ type_preservation   (57)

We can use the Coq *inversion* tactic to decompose the goal and reason by case analysis, relying on the fact that the induction hypothesis are built into each case. Although we are applying induction on IsExp rather than on Exp, the structure of the proof is essentially unchanged with respect to the conventional proof. The type preservation lemma can be finally proved almost immediately, using Mendler_induct and SubRedMAlg. The proof can be either carried out under the predicatisation hypothesis for Exp_S

Lemma subject_reduction$_1$ $\{H_1$: IsoRecursive $\mathsf{Typ_S}\}$
$\{H_2$: Predicatisable $\mathsf{Exp_S}\}$: $\forall e$: Exp, type_preservation e (58)

or else discharging predicatisation

Lemma subject_reduction$_2$ $\{H_1$: IsoRecursive $\mathsf{Typ_S}\}$
$\{H_2$: $\forall x$: Exp, FoldUnivProp $x\}$: $\forall e$: Exp, type_preservation e (59)

under conditions that match exactly those of the MTC-style proof discussed in Sect. 4 [22].

6   Conclusion

In general, the applicability of Mendler induction is wider than the provability of the corresponding predicatisation hypothesis by Σ induction. For example, in [23] we discussed the purely relational specification based on small step semantics of a language including binders among other features, hence requiring inductive datatypes with negative occurrences. A similar language could be specified by an evaluation function, as in our current example. This would require dependent induction, which could be reduced to non-dependent one using predicatisation. However, in the case of binders we could not discharge the predicatisation hypothesis using the Σ algebra technique as we have just discussed it. This seems to match the fact that in MTC binders require a more sophisticated treatment [11]. This is a limitation of the predicatisation technique, rather than one of Mendler induction. It is comparatively straightforward to extend Mendler-style reasoning to more complex cases. In [23] we discussed basic support for mutual induction, which we have improved in [22] relying on indexed functors on sets.

We have presented a novel modular induction technique, based on Mendler induction and datatype predicatisation, that improves over existing approaches in terms of a broader class of proofs which can be addressed in a style closely resembling the conventional, non-modular one. Different techniques have been considered to get around the problem of non-positive type definitions, including type-based termination [2], and more expressive forms of Mendler induction [4]. Such techniques could be useful in tackling current limitations. On the other hand, a more precise specification of predicatisation would involve relating constructor shapes. This seems intuitively easy to express in a container semantics, as well as in terms of subtyping constraints. Together with the implementation of smart constructor and boilerplate support for datatype à la carte, it is going to be matter for further work. A different question is, whether there are conditions that make initiality provable in the impredicative encoding, or whether the MTC approach of packing proofs with fixpoint objects is the best which can be achieved.

Acknowledgments. We thank Fredrik Nordvall Forsberg, Tom Schrijvers, Steven Keuchel and the anonymous reviewers for important feedback and discussion. This research was supported by EU funding (Horizon 2020, grant 640954) to KU Leuven for the GRACEFUL project

References

1. Abbott, M., Altenkirch, T., Ghani, N.: Containers constructing strictly positive types. Theoret. Comput. Sci. **342**(1), 3–27 (2005)
2. Abel, A.: Type-based termination of generic programs. Sci. Comput. Program. **74**(8), 550–567 (2009)
3. Abel, A., Matthes, R., Uustalu, T.: Iteration and coiteration schemes for higher-order and nested datatypes. Theoret. Comput. Sci. **333**(1–2), 3–66 (2005)
4. Ahn, K.Y., Sheard, T.: A hierarchy of Mendler style recursion combinators: taming inductive datatypes with negative occurrences. In: Proceedings of ICFP 2011, pp. 234–246. ACM (2011)
5. Bertot, Y., Casteran, P.: Interactive Theorem Proving and Program Development – Coq'Art: The Calculus of Inductive Constructions. Springer, Heidelberg (2004)
6. Bove, A., Dybjer, P., Norell, U.: A brief overview of Agda – A functional language with dependent types. In: Berghofer, S., Nipkow, T., Urban, C., Wenzel, M. (eds.) TPHOLs 2009. LNCS, vol. 5674, pp. 73–78. Springer, Heidelberg (2009)

7. Churchill, M., Mosses, P.D., Sculthorpe, N., Torrini, P.: Reusable components of semantic specifications. In: Chiba, S., Tanter, É., Ernst, E., Hirschfeld, R. (eds.) Transactions on AOSD XII. LNCS, vol. 8989, pp. 132–179. Springer, Heidelberg (2015)

8. Coquand, T., Huet, G.: The calculus of constructions. Inf. Comput. **76**, 95–120 (1988)

9. Delaware, B., Keuchel, S., Schrijvers, T., Oliveira, B.C.d.S.: Modular monadic meta-theory. In: ICFP 2013, pp. 319–330. ACM (2013)

10. Delaware, B., Keuchel, S., Schrijvers, T., Oliveira, B.C.d.S.: MTC/3MT-Coq development (2013). http://people.csail.mit.edu/bendy/3MT/

11. Delaware, B., Oliveira, B.C.d.S., Schrijvers, T.: Meta-theory à la carte. In: Proceedings of POPL 2013, pp. 207–218 (2013)

12. Geuvers, H.: Inductive and coinductive types with iteration and recursion. In: Types for Proofs and Programs, pp. 193–217 (1992)

13. Hagino, T.: A typed lambda calculus with categorical type constructors. In: Pitt, D.H., Poignê, A., Rydeheard, D.E. (eds.) Category Theory and Computer Science. LNCS, vol. 283, pp. 140–157. Springer, Heidelberg (1987)

14. Hutton, G.: A tutorial on the universality and expressiveness of fold. J. Funct. Program. **9**(4), 355–372 (1999)

15. Keuchel, S., Schrijvers, T.: Generic datatypes à la carte. In: 9th ACM SIGPLAN Workshop on Generic Programming (WGP), pp. 1–11 (2013)

16. Keuchel, S., Weirich, S., Schrijvers, T.: Needle and Knot: binder boilerplate tied up. In: Thiemann, P. (ed.) ESOP 2016. LNCS, pp. 419–445. Springer, Heidelberg (2016)

17. Matthes, R.: Map fusion for nested datatypes in intensional type theory. Sci. Comput. Program. **76**(3), 204–224 (2011)

18. Mendler, N.P.: Inductive types and type constraints in the second-order lambda calculus. Ann. Pure Appl. Logic **51**(1–2), 159–172 (1991)

19. Pfenning, F., Paulin-Mohring, C.: Inductively Defined types in the calculus of constructions. In: Main, M., Melton, A., Mislove, M., Schmidt, D. (eds.) Mathematical Foundations of Programming Semantic. LNCS, vol. 442, pp. 209–228. Springer, Heidelberg (1989)

20. Swierstra, W.: Data types à la carte. J. Funct. Program. **18**(4), 423–436 (2008)

21. Torrini, P.: Language specification and type preservation proofs in Coq-companion code (2015). http://cs.swan.ac.uk/cspt/MDTC

22. Torrini, P.: Modular induction in Coq – companion code (2016). https://bitbucket.org/ptorrx/modind

23. Torrini, P., Schrijvers, T.: Reasoning about modular datatypes with Mendler induction. In: Matthes, R., Mio, M. (eds.) Proceedings of FICS 2015. EPTCS, pp. 143–157 (2015)

24. Uustalu, T., Vene, V.: Mendler-style inductive types, categorically. Nord. J. Comput. **6**(3), 343 (1999)

25. Uustalu, T., Vene, V.: Coding recursion a la Mendler (extended abstract). Technical report, Department of Computer Science, Utrecht University (2000)

26. Wadler, P.: Recursive types for free! (1990). http://homepages.inf.ed.ac.uk/wadler/papers/free-rectypes/free-rectypes.txt

27. Wadler, P.: The expression problem (1998). http://homepages.inf.ed.ac.uk/wadler/papers/expression/expression.txt

Formalized Timed Automata

Simon Wimmer$^{(\boxtimes)}$

Institut für Informatik, Technische Universität München, Munich, Germany
wimmers@in.tum.de

Abstract. Timed automata are a widely used formalism for modeling real-time systems, which is employed in a class of successful model checkers such as UPPAAL. These tools can be understood as trust-multipliers: we trust their correctness to deduce trust in the safety of systems checked by these tools. However, mistakes have previously been made. This particularly regards an approximation operation, which is used by model-checking algorithms to obtain a finite search space. The use of this operation left a soundness problem in the tools employing it, which was only discovered years after the first model checkers were devised. This work aims to provide certainty to our knowledge of the basic theory via formalization in Isabelle/HOL: we define the main concepts, formalize the classic decidability result for the language emptiness problem, prove correctness of the basic forward analysis operations, and finally outline how both streams of work can be combined to show that forward analysis with the common approximation operation correctly decides emptiness for the class of diagonal-free timed automata.

1 Introduction

The foundations of the theory of timed automata are presented in the seminal work of Alur and Dill [1,2]. They introduced the formalism as a model for systems with real-time constraints and showed how to decide the language emptiness problem via the so-called *region* construction. Unfortunately, the number of regions explored by this algorithm is exponential in the size of the automaton under consideration. Moreover, Alur and Dill also showed that the language emptiness problem for timed automata is PSPACE-hard. Still, the formalism is employed in practical model checking [12,13,19] by means of algorithms based on *Difference Bound Matrices* (DBMs). These algorithms (with some more elaborate optimizations) can cope with many interesting real-life model checking problems. The search space examined by the DBM algorithms is potentially infinite. Therefore an approximation is used to obtain a finite search space. The basic idea is to represent every state (called *zone*) by the smallest set of regions which contains the state.

It took nearly a decade after this operation was initially devised, until Patricia Bouyer discovered [5] that the common algorithmic realization of this operation diverges from its intended result: the computed result is always a convex union of

---

Supported by DFG project NI 491/16-1.

J.C. Blanchette and S. Merz (Eds.): ITP 2016, LNCS 9807, pp. 425–440, 2016.
DOI: 10.1007/978-3-319-43144-4_26

regions, whereas the smallest set of regions containing a zone can be non-convex. This left a soundness problem, which fortunately vanishes for the restricted class of so-called diagonal-free timed automata [6] (Sect. 2.1 precisely characterizes this class). While not as expressive as the full formalism of timed automata, this class is sufficient for modeling most of the problems of practical interest, which explains why the problem was not discovered for many years.

This work aims to solidify the theoretical grounds on which real-time model checking with *diagonal-free* timed automata stands, by formalizing the basic theory and algorithms in Isabelle/HOL, and then going the full length to prove Bouyer's correctness result. Section 2 will present the formalization of the basic notions for diagonal-free timed automata. Then Sect. 3 will show how we formalized DBMs and obtained soundness and completeness results for their basic algorithms. This includes a formalization of the Floyd-Warshall algorithm. Afterwards (Sect. 4) we define the notion of regions and prove that they are suitable for deciding the emptiness problem on timed automata. Finally, in Sect. 5, a refined version of these regions will be used to precisely formalize the approximation operation. To tie the ends of our formalization together, this characterization of approximation will be connected with its algorithmic version. This enables us to reuse the decidability result on the first region construction to prove that DBM-based algorithms together with approximation can decide the language emptiness problem for diagonal-free timed automata. For lack of space, many of our definitions and proofs are shortened or stated informally. We refer the reader to the entry in the Archive of Formal Proofs [15] for the full version (over 18500 lines).

1.1 History and Related Work

As mentioned, the basic theory was devised by Alur and Dill [1,2]. The use of DBMs was also proposed by Dill [10] and brought to practical model checking by Yi et al. [18]. Bouyer's developments of our main correctness results are spread over two papers. The first one presents a generalization of timed automata to *updatable* timed automata and revisits the basic decidability results for this class [7]. The second one [6] connects these results with DBMs to prove that the combination of DBM-based forward analysis operations and approximation decides the language emptiness problem.

We are aware of one previous proof-assistant formalization of timed automata using PVS [16,17]. This work has the basic decidability result using regions and claims to make some attempt to extend the formalization towards DBMs. Another line of work [11,14] aims at modeling the class of *p-automata* [3] (which is undecidable in the general case) in Coq and proving properties of concrete p-automata within Coq. A similar approach was pursued with the help of Isabelle/HOL in the *CClair* project [8]. In contrast, the most important contributions of our work are the full formalization of the relevant DBM algorithms, and particularly the rather intricate developments towards the correctness proof for the approximation operation – both of which pertain to practical real-time model checking.

Unless otherwise stated, our formalizations of the basic notions and DBMs are based on a popular tutorial by Bengtsson and Yi [4],while the developments for the region constructions and the final correctness result follow Bouyer's precise work.

2 Diagonal-Free Timed Automata in Isabelle/HOL

2.1 Syntactic Definition

Compared to standard finite automata, timed automata introduce a notion of clocks. We will fix a type $'c$ for the space of clocks, type $'t$ for time, and a type $'s$ for locations. While most of our formalizations only require $'t$ to belong to a custom type class for totally ordered dense abelian groups, we worked on the concrete type *real* for the region construction for simplicity. Figure 1 depicts an example of a diagonal-free timed automaton.

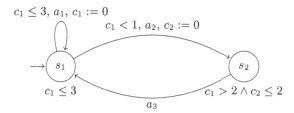

Fig. 1. Example of a diagonal-free timed automaton with two clocks.

Locations and transitions are guarded with *clock constraints*, which have to be fulfilled to stay in a location or to transition between them. The variants of these constraints are modeled by

datatype $('c, 't)$ *cconstraint* $=$
AND $(('c, 't)$ *cconstraint*$)$ $(('c, 't)$ *cconstraint*$)$ $|$
LT $'c$ $'t$ $|$ *LE* $'c$ $'t$ $|$ *EQ* $'c$ $'t$ $|$ *GT* $'c$ $'t$ $|$ *GE* $'c$ $'t$

where the atomic constraints in the second line represent the constraint $c \sim d$ for $\sim = <, \leq, =, >, \geq$, respectively. The sole difference to the full class of timed automata is that those would also allow constraints of the form $c_1 - c_2 \sim d$. We define a timed automaton A as a pair $(\mathcal{T}, \mathcal{I})$ where $\mathcal{I} :: 's \Rightarrow ('c, 't)$ *cconstraint* is an assignment of clock invariants to locations; $\mathcal{T}$ is a set of transitions written as $A \vdash l \longrightarrow^{g,a,r} l'$ where

- $l :: 's$ and $l' :: 's$ are start and successor location,
- $g :: ('c, 't)$ *cconstraint* is the guard of the transition,
- $a :: 'a$ is an action label,
- and $r :: 'c$ *list* is a list of clocks that will be reset to zero when the transition is taken.

Standard definitions of timed automata would include a fixed set of locations with a designated start location and a set of end locations. The language emptiness problem usually asks if any number of legal transitions can be taken to reach an end location from the start location. Thus we can confine ourselves to study reachability and implicitly assume the set of locations to be given by the transitions of the automaton. Note that although the definition of clock constraints allows constants from the whole time space, we will later crucially restrict them to the natural numbers in order to obtain decidability.

2.2 Operational Semantics

We want to define an operational semantics for timed automata via an inductive relation. States of timed automata are pairs of a location and a *clock valuation* of type $'c \Rightarrow 't$ assigning time values to clocks. Time lapse is modeled by shifting a clock valuation u by a constant value d: $u \oplus d = (\lambda x.\ u\ x + d)$. Finally, we connect clock valuations and constraints by writing, for instance, $u \vdash AND\ (LT\ c_1\ 1)\ (EQ\ c_2\ 2)$ if $u\ c_1 < 1$ and $u\ c_2 = 2$. The precise definition is standard.

Using these definitions, the operational semantics can be defined as a relation between pairs of locations and clock valuations. More specifically, we define *action steps*

$$\frac{A \vdash l \xrightarrow{g,a,r} l' \land u \vdash g \land u' \vdash \textit{inv-of } A\ l' \land u' = [r{\rightarrow}0]u}{A \vdash \langle l,\ u \rangle \rightarrow_a \langle l',\ u' \rangle}$$

and *delay steps* via $\dfrac{u \vdash \textit{inv-of } A\ l \land u \oplus d \vdash \textit{inv-of } A\ l \land 0 \le d}{A \vdash \langle l,\ u \rangle \rightarrow^d \langle l,\ u \oplus d \rangle}$. Here *inv-of*

$(\mathcal{T}, \mathcal{I}) = \mathcal{I}$ and the notation $[r \rightarrow 0]u$ means that we update the clocks in r to 0 in u. We write $A \vdash \langle l,\ u \rangle \rightarrow \langle l', u' \rangle$ if either $A \vdash \langle l,\ u \rangle \rightarrow_a \langle l',\ u' \rangle$ or $A \vdash \langle l,\ u \rangle \rightarrow^d \langle l',\ u' \rangle$.

2.3 Zone Semantics

The first conceptual step to get from this abstract operational semantics towards concrete algorithms on DBMs is to consider *zones*. Informally, the concept is simple; a zone is the set of clock valuations fulfilling a clock constraint: $('c,\ 't)\ zone \equiv ('c \Rightarrow 't)\ set$. This allows us to abstract from a concrete state $\langle l,\ u \rangle$ to a pair of location and zone $\langle l,\ Z \rangle$. We need the following operations on zones:

$$Z^{\uparrow} = \{u \oplus d \mid u \in Z \land 0 \le d\} \text{ and } Z_{r \,\rightarrow\, 0} = \{[r{\rightarrow}0]u \mid u \in Z\}.$$

Naturally, we define a zone-based semantics by means of another inductive relation:

$$A \vdash \langle l, Z \rangle \rightsquigarrow \langle l, (Z \cap \{u \mid u \vdash \mathit{inv\text{-}of}\ A\ l\})^\uparrow \cap \{u \mid u \vdash \mathit{inv\text{-}of}\ A\ l\}\rangle$$

$$\frac{A \vdash l \longrightarrow^{g,a,r} l'}{A \vdash \langle l, Z \rangle \rightsquigarrow \langle l', (Z \cap \{u \mid u \vdash g\})_r \to {}_0 \cap \{u \mid u \vdash \mathit{inv\text{-}of}\ A\ l'\}\rangle}$$

With the help of two easy inductive arguments one can show soundness and completeness of this semantics w.r.t. the original semantics (where $*$ is the *Kleene star* operator):

(Sound) $A \vdash \langle l, Z \rangle \rightsquigarrow^* \langle l', Z' \rangle \land u' \in Z' \implies \exists u \in Z.\ A \vdash \langle l, u \rangle \to^* \langle l', u' \rangle$

(Complete) $A \vdash \langle l, u \rangle \to^* \langle l', u' \rangle \land u \in Z$
$\implies \exists Z'.\ A \vdash \langle l, Z \rangle \rightsquigarrow^* \langle l', Z' \rangle \land u' \in Z'$

This is an example of where proof assistants really shine. Not only are our Isabelle proofs shorter to write down than for example the proof given in [18] – we have also found that the less general version given there (i.e. where $Z = \{u\}$) yields an induction hypothesis that is not strong enough in the completeness proof. This slight lapse is hard to detect in a human-written proof.

3 Difference Bound Matrices

3.1 Fundamentals

Difference Bound Matrices constrain differences of clocks (or more precisely, the difference of values assigned to individual clocks by a valuation). The possible constraints are given by:

$$\textbf{datatype}\ 't\ \mathit{DBMEntry} = Le\ 't \mid Lt\ 't \mid \infty$$

This yields a simple definition of DBMs: $'t\ \mathit{DBM} \equiv \mathit{nat} \Rightarrow \mathit{nat} \Rightarrow 't\ \mathit{DBMEntry}$. To relate clocks with rows and columns of a DBM, we use a numbering $v ::'c \Rightarrow \mathit{nat}$ for clocks. DBMs will regularly be accompanied by a natural number n, which designates the number of clocks constrained by the matrix. Although this definition complicates our formalization at times, we hope that it allows us to easily obtain executable code for DBMs while retaining a flexible "interface" for applications. To be able to represent the full set of clock constraints with DBMs, we add an imaginary clock $\mathbf{0}$, which shall be assigned to 0 in every valuation. Zero column and row will always be reserved for $\mathbf{0}$ (i.e. $\forall c.\ v\ c > 0$). If necessary, we assume that v is an injection or surjection for indices less or equal to n. Informally, the zone $[M]_{v,n}$ represented by a DBM M is defined as

$$\{u \mid \forall c_1, c_2, d.\ v\ c_1, v\ c_2 \leq n \longrightarrow$$
$$(M\ (v\ c_1)\ (v\ c_2) = Lt\ d \longrightarrow u\ c_1 - u\ c_2 < d)$$
$$\land\ (M\ (v\ c_1)\ (v\ c_2) = Le\ d \longrightarrow u\ c_1 - u\ c_2 \leq d)\}$$

assuming that $v\ \mathbf{0} = 0$.

Example 1.

$$
\begin{array}{ccc}
\mathbf{0} & c_1 & c_2 \\
\end{array}
\quad
\begin{array}{ccc}
\mathbf{0} & c_1 & c_2 \\
\end{array}
\quad
\begin{array}{ccc}
\mathbf{0} & c_1 & c_2 \\
\end{array}
$$

$$
\begin{matrix}
\mathbf{0} \\
c_1 \\
c_2
\end{matrix}
\begin{pmatrix}
\infty & Lt\ (-3) & Le\ 0 \\
\infty & \infty & \infty \\
Le\ 4 & \infty & \infty
\end{pmatrix}
\quad
\begin{matrix}
\mathbf{0} \\
c_1 \\
c_2
\end{matrix}
\begin{pmatrix}
Le\ 0 & Lt\ (-3) & Le\ 0 \\
\infty & Le\ 0 & \infty \\
Le\ 4 & Lt\ 1 & Le\ 0
\end{pmatrix}
\quad
\begin{matrix}
\mathbf{0} \\
c_1 \\
c_2
\end{matrix}
\begin{pmatrix}
\infty & Le\ 0 & Le\ 0 \\
\infty & \infty & Lt\ (-3) \\
\infty & Le\ 3 & Le\ 0
\end{pmatrix}
$$

The left two DBMs both represent the zone described by the constraint $c_1 > 3 \land c_2 \leq 4$, while the DBM on the right represents the empty zone.[1]

To simplify the subsequent discussion, we will set $'c = nat, v = id$ and assume that the set of clocks of the automaton in question is $\{1..n\}$. We define an ordering relation $\prec$ on $'t\ DBMEntry$ by means of

$$
\frac{a < b}{Le\ a \prec Le\ b} \quad \frac{a < b}{Le\ a \prec Lt\ b} \quad \frac{a < b}{Lt\ a \prec Lt\ b} \quad \frac{a \leq b}{Lt\ a \prec Le\ b} \quad \frac{}{Lt\ \_ \prec \infty} \quad \frac{}{Le\ \_ \prec \infty}
$$

and extend it to $\preceq$ in the obvious way. Observe that $\prec$ and $\preceq$ are total orders. Additionally, we get the following important ordering property of DBMs (by nearly automatic proof):

Lemma 1. $\forall\ i\ j.\ i \leq n \longrightarrow j \leq n \longrightarrow M\ i\ j \preceq M'\ i\ j \implies [M]_{v,n} \subseteq [M']_{v,n}$

We can interpret DBMs as a graph with clocks as vertices and difference constraints as edges between them. To give a concrete meaning to this interpretation, we first define addition on DBM entries: $a \boxplus \infty = \infty; \infty \boxplus b = \infty$; and $(\sim_1 x) \boxplus (\sim_2 y) = \sim' (x + y)$ where $\sim' = Le$ if $\sim_1 = \sim_2 = Le$ and $\sim' = Lt$ if otherwise. Now the length of a path (of DBM indices representing clocks) defined by[2]

$$len\ M\ s\ t\ [] = M\ s\ t \text{ and } len\ M\ s\ t\ (w \cdot ws) = M\ s\ w \boxplus len\ M\ w\ t\ ws$$

gives the key to reasoning about this interpretation: for any $u \in [M]_{v,n}$ and i, j, xs with $set\ (i \cdot j \cdot xs) \subseteq \{0..n\}$,[3] we get $Lt\ (u\ i - u\ j) \prec len\ M\ i\ j\ xs$ via induction on xs. Setting $i = j$, we can immediately conclude that DBMs with negative cycles are always empty. In the following we will make use of a predicate expressing that a DBM does not contain any negative cycles which only consist of vertices less or equal to k for some k:

$cycle\text{-}free\text{-}up\text{-}to\ M\ k\ n \equiv$
$\forall\ i\ xs.\ i \leq n \land set\ xs \subseteq \{0..k\} \longrightarrow Le\ 0 \preceq len\ M\ i\ i\ xs$

We write $cycle\text{-}free\ M\ n$ if $cycle\text{-}free\text{-}up\text{-}to\ M\ n\ n$.

---

[1] We assume a default clock numbering, mapping c_i to index i, for our examples.
[2] $[]$ denotes the empty list and $x \cdot xs$ is a list constructed from head x and tail xs.
[3] $set\ xs$ is the set of elements contained in xs.

3.2 Operations

We define the necessary operations on DBMs to obtain a basic forward analysis algorithm for reachability.

Floyd-Warshall algorithm. From Example 1 we can see that to be able to tell if two DBMs represent the same zone, we first need to put them into some *canonical* form. Formally, this canonical form is characterized by the following property:

$$canonical\ M\ n \equiv \forall i\ j\ k.\ i \le n \wedge j \le n \wedge k \le n \longrightarrow M\ i\ k \preceq M\ i\ j \boxplus M\ j\ k$$

The key property of non-empty canonical DBMs is that we can find a valuation $u \in [M]_{v,n}$ with $u\ i - u\ j = d$ for any d between $-M\ j\ i$ and $M\ i\ j$, or equivalently:

Lemma 2. *Assume* $Le\ d \preceq M\ i\ j$, $Le\ (-d) \preceq M\ j\ i$ *for* M *with cycle-free* $M\ n$, *canonical* $M\ n$, *and* $i, j \le n$ *with* $i \ne j$. *We define* $M\,'$ *by setting* $M\,'\ i\ j = Le\ d$ *and* $M\,'\ j\ i = Le\ (-d)$ *and* $M\,'\ i'\ j' = M\ i'\ j'$ *for all* (i',j') *where* $(i',j') \ne (i,j)$, (j,i). *Then* $[M\,']_{v,n} \subseteq [M]_{v,n}$ *and cycle-free* $M\,'\ n$.

Proof. From Lemma 1, we get $[M\,']_{v,n} \subseteq [M]_{v,n}$. It remains to show that $M\,'$ does not contain a negative cycle. Suppose there is one. Then we can also find a *smallest* negative cycle, which, without loss of generality, is of the form $len\ M\,'\ i\ i\ (j \cdot xs) \prec Le\ 0$ for some xs where $i, j \notin set\ xs$. This proof step is rather intricate in Isabelle. We use a function that explicitly computes smallest negative cycles. An inductive argument yields a result that allows us to rotate cycles. Now, we get $Le\ d \boxplus len\ M\,'\ j\ i\ xs \prec Le\ 0$. We have $xs \ne []$ as this would directly give us the contradiction $Le\ d \boxplus Le\ (-d) \prec Le\ 0$. This means that $Le\ d \boxplus len\ M\ j\ i\ xs \prec Le\ 0$ (by induction on xs), and because M is canonical, $M\ j\ i \prec Le\ (-d)$, which is a contradiction to our assumption. □

An important consequence is that any canonical DBM without a negative diagonal has at least one valuation, which we can construct by repeatedly applying the theorem. Observe that this also implies that a DBM in canonical form is empty iff there is a negative entry on its diagonal.

The canonical form can be computed by the Floyd-Warshall algorithm for the all-pairs shortest paths problem. A simple HOL formulation of the algorithm is

$$fw\text{-}upd\ M\ k\ i\ j \equiv M(i := (M\ i)(j := min\ (M\ i\ j)\ (M\ i\ k \boxplus M\ k\ j)))$$

$$fw\ M\ n\ 0\ 0\ 0 = fw\text{-}upd\ M\ 0\ 0\ 0$$
$$fw\ M\ n\ (Suc\ k)\ 0\ 0 = fw\text{-}upd\ (fw\ M\ n\ k\ n\ n)\ (Suc\ k)\ 0\ 0$$
$$fw\ M\ n\ k\ (Suc\ i)\ 0 = fw\text{-}upd\ (fw\ M\ n\ k\ i\ n)\ k\ (Suc\ i)\ 0$$
$$fw\ M\ n\ k\ i\ (Suc\ j) = fw\text{-}upd\ (fw\ M\ n\ k\ i\ j)\ k\ i\ (Suc\ j)$$

where $f(a := b) \equiv \lambda x.\ if\ x = a\ then\ b\ else\ f\ x$. We abbreviate $fw\ M\ n\ n\ n\ n$ as $FW\ M\ n$. To prove that this algorithm computes the tightest difference constraint for all pairs of clocks, we claim:

Theorem 1
cycle-free-up-to M k n $\wedge$ i' $\leq$ i $\wedge$ j' $\leq$ j $\wedge$ i $\leq$ n $\wedge$ j $\leq$ n $\wedge$ k $\leq$ n $\Longrightarrow$
Min {len M i' j' xs | set xs $\subseteq$ {0..k} $\wedge$ i' $\notin$ set xs $\wedge$ j' $\notin$ set xs $\wedge$ distinct xs}
= fw M n k i j i' j'

The proof is a nested induction, which follows the program structure and uses a standard argument. The theorem implies that *FW* computes a canonical form:

Corollary 1. *cycle-free M n $\Longrightarrow$ canonical (FW M n) n*

The Floyd-Warshall algorithm also *detects* negative cycles by computing a negative diagonal entry. The key observation is that a matrix of this kind either has a negative diagonal entry to start with, or there is a maximal $k < n$ with *cycle-free-up-to M k n*. The latter means that the algorithm computes a negative diagonal entry in iteration $k + 1$. In either case the negative diagonal entry will be preserved by monotonicity of the algorithm. This yields an emptiness check for DBMs.

Intersection. The intersection of two DBMs is trivial to compute. It is simply the point-wise minimum: *And A B $\equiv$ $\lambda i\,j$. min (A i j) (B i j)*. The operation is correct in the following sense: $[A]_{v,n} \cap [B]_{v,n} = [And\ A\ B]_{v,n}$. The $\subseteq$-direction can directly be proved by Isabelle's simplifier, while $\supseteq$ requires a rather lengthy proof by cases.

Reset. We need an operator *reset* such that $u\ c = d$ for all $u \in [reset\ M\ n\ c\ d]_{v,n}$. Thus we define *(reset M n c d) c 0 = Le d* and *(reset M n c d) 0 c = Le (−d)*. By doing so, all difference constraints involving c are invalidated. Therefore we set the corresponding DBM entries to ∞. However, this alone does not yield a correct operation. Consider clocks c_1, c_2 and c_3 and a DBM represented by the clock constraint $c_1 \geq c_2 + 1 \wedge c_1 \leq c_3$. By setting c_1 to 0, we will lose all constraints on c_2 and c_3. This means that the resulting zone will contain a valuation u with $u\ c_1 = u\ c_2 = u\ c_3 = 0$. There is clearly no way to set c_1 back to a different value such the resulting valuation would satisfy the original constraint. The way to resolve this issue is to encode the information we had about c_2 and c_3 in the original constraint (or DBM) also in the new DBM. This is, we derive $c_2 - c_3 \leq -1$. Concretely, we calculate *(reset M n i d) j k = min (M j i + M i k) (M j k)* for all $j, k \leq n$. Note that this computation does nothing if *M* is already in canonical from, allowing a simpler implementation.

For a list of clocks *cs* and a list of time stamps *ts*($|cs| = |ts|$), *set-clocks cs ts u* is the valuation for which *(set-clocks cs ts u)* $cs_i = ts_i$ and the value of $u\ c$ is unchanged for all other clocks $c \notin set\ cs$. We lift *reset* to reset many clocks at once by simply folding it over the list of clocks. We proved correctness of the lifted operation *(reset')*:

(Sound) $(\forall c \in set\ cs.\ 0 < c \wedge c \leq n) \wedge u \in [reset'\ M\ n\ cs\ v\ d]_{v,n}$
$\Longrightarrow \exists ts.\ set\text{-}clocks\ cs\ ts\ u \in [M]_{v,n}$
(Complete) $(\forall c \in set\ cs.\ 0 < c \wedge c \leq n) \wedge u \in [M]_{v,n}$
$\Longrightarrow [cs{\to}d]u \in [reset'\ M\ n\ cs\ v\ d]_{v,n}$

The proofs for these results are among the most complex ones in the whole formalization. The reason is that manual case analyses have to be combined with (linear) arithmetic reasoning, which is hard to automate in Isabelle.

Delay. We need an operation to compute time lapse, i.e. $([M]_{v,n})^{\uparrow}$. For canonical DBMs, this simply amounts to setting $M\ i\ 0 = \infty$ for all $i \leq n$. In the general case, intuitively we can lose information about the difference of two clocks that was recorded between the upper bound of one of them and the lower bound of the other. Accounting for this, we arrive at the following general operation:

up $M \equiv$
$\lambda i\ j.\ \textit{if } 0 < i \textit{ then if } j = 0 \textit{ then } \infty \textit{ else min } (M\ i\ 0 \boxplus M\ 0\ j)\ (M\ i\ j)\textit{ else } M\ i\ j$

Correctness can be obtained similarly to the reset operation.

Abstraction. It is easy to turn an atomic clock constraint into a DBM that represents the same zone. For instance, the zone $\{u \mid u \vdash EQ\ c\ d\}$ is represented by a DBM M where $M\ c\ 0 = Le\ d$ and $M\ 0\ c = Le\ (-d)$, and all other entries are unbounded. Using the already defined intersection operation for constructor *AND*, a function *abstr*, which records entries in this manner while working recursively through a constraint, turns constraints into a DBM-equivalent. Again, we proved correctness (where *collect-clks cc* is the set of all clocks appearing in constraint *cc*):

$$\forall c \in \textit{collect-clks } cc.\ 0 < c \wedge c \leq n \implies [\textit{abstr } cc\ (\lambda i\ j.\ \infty)\ v]_{v,n} = \{u \mid u \vdash cc\}$$

3.3   DBM Operational Semantics

In the last section we have elaborated the adequacy of our DBM-equivalents for all zone operations, allowing us to compute the zone semantics with the help of DBMs. Indeed we can define a new operational semantics based on DBMs:

$$\frac{M_i = \textit{abstr } (\textit{inv-of } A\ l)\ (\lambda i\ j.\ \infty)\ v}{A \vdash \langle l,\ M \rangle \rightsquigarrow_{v,n} \langle l,\ \textit{And } (\textit{up } (\textit{And } M\ M_i))\ M_i \rangle}$$

$$\frac{A \vdash l \xrightarrow{g,a,r} l' \wedge M_i = \textit{abstr } (\textit{inv-of } A\ l')\ (\lambda i\ j.\ \infty)\ v}{A \vdash \langle l,\ M \rangle \rightsquigarrow_{v,n} \langle l',\ \textit{And } (\textit{reset}' (\textit{And } M\ (\textit{abstr } g\ (\lambda i\ j.\ \infty)\ v))\ n\ r\ v\ 0)\ M_i \rangle}$$

Using the correctness results for the DBM operations, it is straightforward to show that this semantics is equivalent to the zone semantics:

$$A \vdash \langle l,\ [M]_{v,n} \rangle \rightsquigarrow^* \langle l',\ Z \rangle$$
$$\longleftrightarrow \exists M'.\ A \vdash \langle l,\ M \rangle \rightsquigarrow^*_{v,n} \langle l',\ M' \rangle \wedge Z = [M']_{v,n}$$

However, we are not done yet: while we can practically compute the semantics of timed automata, the search space could still be infinite. The rest of the paper is concerned with overcoming this problem.

4   From Classic Decidability to a Correct Approximation

4.1   Regions

In their seminal paper, Alur and Dill showed decidability of the emptiness problem for timed automata by giving an adequate finite partitioning of the set of valuations into what they call *regions*. In this section, we will present our formalization of this result and then show how to apply it to obtain a *finite* operational semantics of zones. We use Bouyer's definition of regions as, for one it is more formal and thus easier to formalize, and secondly we will have to use a modified version of it later on.

From now on we will work in a parametric theory (called *locale* in Isabelle), which fixes X as the set of clocks of the automaton. Moreover, a *clock ceiling* k will define an upper bound $k\ c$ for the "relevant" range of any clock $c \in X$ – this ought to correspond to the *maximal* constant appearing for c in any constraint of the timed automaton, e.g., $k\ c_1 = 3$ and $k\ c_2 = 2$ for the automaton of Fig. 1. This is, if $\sim c\ m$ is a constraint of the automaton, we postulate that $m \leq k\ c$, $c \in X$, and that m is a natural number.

A single clock value will always fall into one of three types of intervals from

$$\textbf{datatype}\ intv = Const\ nat \mid Intv\ nat \mid Greater\ nat$$

where the set of values they contain is given by the following rules:

$$\frac{u\ x = d}{intv\text{-}elem\ x\ u\ (Const\ d)} \qquad \frac{d < u\ x \wedge u\ x < d + 1}{intv\text{-}elem\ x\ u\ (Intv\ d)}$$

$$\frac{d < u\ x}{intv\text{-}elem\ x\ u\ (Greater\ d)}$$

Let $I :: {}'c \Rightarrow intv$ be assigning intervals to clocks and r be a finite total preorder over $X_0 \equiv \{x \in X \mid \exists d.\ I\ x = Intv\ d\}$. Then we define the corresponding region $region\ X\ I\ r$ as the set for which[4]

$u \in region\ X\ I\ r$ iff $\forall x{\in}X.\ 0 \leq u\ x \wedge intv\text{-}elem\ x\ u\ (I\ x)$
and $\forall x{\in}X_0.\ \forall y{\in}X_0.\ (x, y) \in r \longleftrightarrow frac\ (u\ x) \leq frac\ (u\ y)$

We will fix a set of regions $\mathcal{R}_\alpha \equiv \{region\ X\ I\ r \mid valid\text{-}region\ X\ k\ I\ r\}$ where *valid-region* $X\ k\ I\ r$ holds if X is finite, r is a total preorder on X_0, and $d \leq k\ x$ if $I\ x = Const\ d$, $d < k\ x$ if $I\ x = Intv\ d$, and $k\ x = d$ if $I\ x = Greater\ d$ for all $x \in X$. Observe that this definition remedies the potential overlap of intervals that the definition of *intv-elem* would admit.

It is clear from Fig. 1, and relatively straightforward to prove in Isabelle/HOL, that $\mathcal{R}_\alpha$ is a finite partitioning of

$$V \equiv \{u \mid \forall x{\in}X.\ 0 \leq u\ x\}\ ,$$

---

[4] *frac r* denotes the fractional part of any real number r.

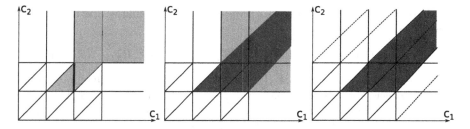

Fig. 2. (1) A region and its time successors in $\mathcal{R}_\alpha$, (2) the α-closure of a zone, and (3) the β-approximation of a zone for $X = \{c_1, c_2\}$ with $k\, c_1 = 3$ and $k\, c_2 = 2$.

the set of all positive valuations. What is not so obvious (and not mentioned by Bouyer) but a useful property to work with, is that any valid region is also non-empty. The crux of this proof is to observe that X_0 can be ordered in equivalence classes according to r such that a valuation u can be chosen for which $frac\,(u\,x) \leq frac\,(u\,y)$ iff $(x,\,y) \in r$. This ordering property of finite total preorders is non-trivial to formalize and makes this step rather technical.

4.2   Decidability with Regions

How are regions and timed automata connected? We will present three key properties that connect regions to time lapse, clock resets, and clock constraints, respectively, allowing us to implement timed automata with the help of regions. Let $[u]_{\mathcal{R}_\alpha} \in \mathcal{R}_\alpha$ be the unique region containing u. We call $[u \oplus t]_{\mathcal{R}_\alpha}$ a *time successor* of $[u]_{\mathcal{R}_\alpha}$ for $t \geq 0$ and denote by $Succ\ \mathcal{R}_\alpha\ R$ the set of all such time successors of all $u \in R$ (cf. Fig. 2.1). Now the three key properties are in order of decreasing difficulty:

(Set of regions) $R \in \mathcal{R}_\alpha \wedge u \in R \wedge R' \in Succ\ \mathcal{R}_\alpha\ R$
$\implies \exists t{\geq}0.\ [u \oplus t]_{\mathcal{R}_\alpha} = R'$
(Compatibility with resets) $R \in \mathcal{R}_\alpha \wedge u \in R \wedge 0 \leq d \wedge d \leq k\, x \wedge x \in X$
$\implies [u(x := d)]_{\mathcal{R}_\alpha} = \{u(x := d) \mid u \in R\}$
(Compatibility with constraints)
$R \in \mathcal{R}_\alpha \wedge \forall (x, m){\in} collect\text{-}clock\text{-}pairs\ cc.\ m \leq k\, x \wedge x \in X \wedge m \in \mathbb{N}$
$\implies R \subseteq \{u \mid u \vdash cc\} \vee \{u \mid u \vdash cc\} \cap R = \emptyset$

Proof. We concentrate on the set of regions property as it has the most interesting formalization. Our proof combines elements of the "classic" result as presented e.g., in [9], and Bouyer's approach. Let $R = region\ X\ I\ r \in \mathcal{R}_\alpha$ for some I, r, let $R' = [v \oplus t]_{\mathcal{R}_\alpha}$, and assume $u, v \in R$ and $t \geq 0$. If $I\,x = Greater\,(k\,x)$ for all $x \in X$ ("upper-right region"), we have $Succ\ \mathcal{R}_\alpha\ R = \{R\} = \{R'\}$ and the proposition is obvious.

Otherwise observe that there exists a single *closest* successor R_{succ} of R (depicted as the thick, dark gray line in Fig. 2.1). We refer to Bouyer for a formal construction of this successor. We can show the characteristic property of this closest successor:

$$\forall u \in R. \; \forall t \geq 0. \; (u \oplus t) \notin R \longrightarrow (\exists t' \leq t. \; (u \oplus t') \in R_{succ} \wedge t' \geq 0)$$

At this point Bouyer states that the proposition follows by "immediate induction". However, regarding formalization, this induction is not quite immediate. For instance, we attempted induction on the set of successors. This necessitates a proof that this set is monotone, which we did not find ourselves able to prove without asserting the very property we were about to prove. Instead, we split the argument in two: one for the case where $t < 1$ and the other for the case where t is an integer. For the first case, consider the "critical" set $C = \{x \in X \mid \exists d. \; I \, x = Intv \, d \wedge d + 1 \leq u \, x + t\}$, the set of clocks for which $u \oplus t$ is shifted beyond R's interval boundaries. Observe that for the closest successor, the critical set is either the same (if $\{x \in X \mid \exists d. \; I \, x = Const \, d\} \neq \emptyset$) or a strict subset (if otherwise). Thus the proposition follows by induction on the cardinality of C. The case where t is an integer follows by direct proof over the structure of regions. Shifting u first by $frac \, t$ and then by $\lfloor t \rfloor$, we arrive at the proposition. $\qquad \square$

This allows us to define a region-based operational semantics for timed automata:

$$\frac{R \in \mathcal{R}_\alpha \wedge R' \in Succ \; \mathcal{R}_\alpha \; R \wedge R \cup R' \subseteq \{u \mid u \vdash inv\text{-}of \; A \; l\}}{A, \mathcal{R}_\alpha \vdash \langle l, R \rangle \rightsquigarrow \langle l, R' \rangle}$$

$$\frac{A \vdash l \longrightarrow^{g,a,r} l' \wedge R \in \mathcal{R}_\alpha}{A, \mathcal{R}_\alpha \vdash \langle l, R \rangle \rightsquigarrow \langle l', \{[r \rightarrow 0] u \mid u \in R \wedge u \vdash g\} \cap \{u \mid u \vdash inv\text{-}of \; A \; l'\}\rangle}$$

From the aforementioned properties, we proved its adequacy w.r.t. to reachability:

$$A, \mathcal{R}_\alpha \vdash \langle l, [u]_{\mathcal{R}_\alpha} \rangle \rightsquigarrow^* \langle l', R' \rangle \wedge R' \neq \emptyset$$
$$\longleftrightarrow \exists u'. \; A \vdash \langle l, u \rangle \rightarrow^* \langle l', u' \rangle \wedge [u']_{\mathcal{R}_\alpha} = R'$$

Note that it is quite natural that this property is weaker compared to previous ones: (sets of) regions only approximate zones and thus can contain valuations that were never reachable in the concrete semantics.

4.3   Approximating Zone Semantics with Regions

From the pure decidability result on regions, we now move back towards zones by *approximating* zones with the smallest set of regions that covers them. Formally we define the α-*closure* of a zone Z: $Closure_\alpha \; Z = \bigcup \{R \in \mathcal{R} \mid R \cap Z \neq \emptyset\}$. Observe that this set need not be convex (cf. Fig. 2.2). We use the α-closure to define an operational semantics on zones that approximates a zone with its α-closure at the end of each step:

$$A \vdash \langle l, Z \rangle \rightsquigarrow \langle l', Z' \rangle \implies A \vdash \langle l, Z \rangle \rightsquigarrow_\alpha \langle l', Closure_\alpha \; Z' \rangle$$

Bouyer would now go and prove from the region properties that the α-closure can be "pushed through" each step:

$$Z \subseteq V \wedge A \vdash \langle l, \; Closure_\alpha \; Z \rangle \rightsquigarrow \langle l', \; Z' \rangle$$
$$\Longrightarrow \exists Z''. \; A \vdash \langle l, \; Z \rangle \rightsquigarrow_\alpha \langle l', \; Z'' \rangle \wedge Z' \subseteq Z''$$

However, we did not find this property strong enough to prove soundness of $\rightsquigarrow_\alpha{}^*$:

$$A \vdash \langle l, \; Z \rangle \rightsquigarrow_\alpha{}^* \langle l', \; Z' \rangle \wedge Z \subseteq V$$
$$\Longrightarrow \exists Z''. \; A \vdash \langle l, \; Z \rangle \rightsquigarrow^* \langle l', \; Z'' \rangle \wedge Closure_\alpha \; Z' \subseteq Closure_\alpha \; Z'' \wedge Z'' \subseteq Z'$$

Note that this property is really what one wants to have since $Closure_\alpha \; Z = \emptyset$ iff $Z = \emptyset$ (assuming that $Z \subseteq V$). We conceived that instead it is sufficient to prove monotonicity of the α-closure w.r.t. to steps in the zone semantics:

$$A \vdash \langle l, \; Z \rangle \rightsquigarrow \langle l', \; Z' \rangle \wedge Closure_\alpha \; Z = Closure_\alpha \; W \wedge W \subseteq Z \wedge Z \subseteq V$$
$$\Longrightarrow \exists W'. \; A \vdash \langle l, \; W \rangle \rightsquigarrow \langle l', \; W' \rangle \wedge Closure_\alpha \; Z' = Closure_\alpha \; W' \wedge W' \subseteq Z'$$

Combining this with the fact that α-closure is an involution, we proved soundness by induction over $\rightsquigarrow_\alpha{}^*$. Completeness follows easily from monotonicity of $\rightsquigarrow^*$:

$$A \vdash \langle l, \; Z \rangle \rightsquigarrow^* \langle l', \; Z' \rangle \wedge Z \subseteq V \wedge Z' \neq \emptyset$$
$$\Longrightarrow \exists Z''. \; A \vdash \langle l, \; Z \rangle \rightsquigarrow_\alpha{}^* \langle l', \; Z'' \rangle \wedge Z' \subseteq Z''$$

While these results are nice from a theoretical standpoint, it is not easier to compute the α-closure than to directly implement timed automata with the region construction presented in the last section. Therefore, the next section will present Bouyer's main insight – that these results can be used to show the correctness of an easily computable approximation operation.

5   Normalization

Consider Fig. 2.3. In addition to $\mathcal{R}_\alpha$ (solid lines), the figure shows a refinement to what we will call $\mathcal{R}_\beta$ (dashed lines). Observe that the smallest set of regions covering the zone painted in dark gray (i.e. its β-closure) is *convex*, whereas its α-closure is not (cf. Fig. 2.2). The idea is to use this β-closure to obtain an effectively computable convex approximation for zones represented by DBMs – DBMs always represent a convex zone and are always covered by a convex β-closure – while inheriting the correctness result from the α-closure as we only refine things.

5.1 $\beta - approximation$

Due to a lack of space, we do not present our construction of $\mathcal{R}_\beta$ and only say that it is can be adopted from $\mathcal{R}_\alpha$ with some modifications. Note that we do not need to transfer the (rather intricate) properties connecting $\mathcal{R}_\alpha$ with transitions of timed automata since we will infer correctness directly from the original construction.

We now want to formalize the notion of a convex approximation of zones with regions from $\mathcal{R}_\beta$. We capture the notion of convexity directly with DBMs. From Example 1, we can see that the types of regions in $\mathcal{R}_\beta$ also induce a specific format for our DBMs: for a DBM entry $M\ i\ j$, we do not need constants outside of $[-\ k\ i;\ k\ j]$ because this is precisely the range to which our regions bound the corresponding values (analogously for constraints involving 0). Thus we use the following notion of *normalized* DBMs:

normalized $M \equiv$
$(\forall i\ j.\ 0 < i \wedge i \leq n \wedge 0 < j \wedge j \leq n \wedge M\ i\ j \neq \infty \longrightarrow$
$\qquad Lt\ (-\ k\ j) \preceq M\ i\ j \wedge M\ i\ j \preceq Le\ (k\ i)) \wedge$
$(\forall i{\leq}n.\ 0 < i \longrightarrow (M\ i\ 0 \preceq Le\ (k\ i) \vee M\ i\ 0 = \infty) \wedge Lt\ (-\ k\ i) \preceq M\ 0\ i)$

Furthermore, all constraints only need to use integer constants, which we denote by *dbm-int M*. Building from these ideas, we define for any zone Z:

$$Approx_\beta\ Z \equiv \bigcap\ \{[M]_{v,n}\ |$$
$$\exists U \subseteq \mathcal{R}_\beta.\ [M]_{v,n} = \bigcup U \wedge Z \subseteq [M]_{v,n} \wedge \textit{dbm-int } M\ n \wedge \textit{normalized } M\}$$

5.2   Connecting $Approx_\beta$ and $Closure_\alpha$

We already argued that is possible to inherit correctness from $Closure_\alpha$ because we only refine regions. Precisely, Bouyer proposed that for any convex zone Z (i.e. $Z = [M]_{v,n}$ for some DBM M), we have $Approx_\beta\ Z \subseteq Closure_\alpha\ Z$, or equivalently:

Theorem 2. $R \in \mathcal{R}_\alpha \wedge Z \subseteq V \wedge R \cap Z = \emptyset \wedge Z = [M]_{v,n} \wedge \textit{dbm-int } M\ n \implies R \cap Approx_\beta\ Z = \emptyset$

The formalization of Bouyer's proof for this proposition is one of the most complicated parts of our development. As the prose proof is already sufficiently complicated, we abstain from presenting our formalization of this result.

Analogously to $\leadsto_\alpha$, we define an approximating semantics $\leadsto_\beta$ using $Approx_\beta$. The main fact we can derive from the Theorem 2 is that $\leadsto_\alpha$ is an approximation of $\leadsto_\beta$:

Lemma 3
$A \vdash \langle l, [M]_{v,n} \rangle \leadsto_\beta \langle l', Z' \rangle \wedge \textit{dbm-int } M\ n \wedge [M]_{v,n} \subseteq W \wedge W \subseteq V \implies \exists W'.\ A \vdash \langle l, W \rangle \leadsto_\alpha \langle l', W' \rangle \wedge Z' \subseteq W'$

Using this result and some additional work, we could infer soundness and completeness of $\leadsto_\beta{}^*$ from the corresponding results for $\leadsto_\alpha{}^*$.

5.3 Computing $Approx_\beta$

So far, we have shown how to obtain a correct approximation operation from $\mathcal{R}_\beta$, which only produces convex sets. The huge gain from that is that this approximation can also be easily computed by *normalizing* DBMs:

norm M k n ≡
$\lambda i\ j.$ *let ub* = *if 0* < *i then k i else 0*; *lb* = *if 0* < *j then* − *k j else 0*
   in if i ≤ *n* ∧ *j* ≤ *n then norm-lower* (*norm-upper* (*M i j*) *ub*) *lb*
      else M i j
norm-upper e t = (*if Le t* ≺ *e then* ∞ *else e*)
norm-lower e t = (*if e* ≺ *Lt t then Lt t else e*)

Lemma 4. *canonical M n* ∧ $[M]_{v,n} \subseteq V$ ∧ *dbm-int M n* $\Longrightarrow$
$Approx_\beta\ ([M]_{v,n}) = [norm\ M\ k\ n]_{v,n}$

Again, we abstain from providing a full presentation of our formalization and only mention that the main ideas are: (1) to observe that normalized integral DBMs can always be represented by an equivalent subset of $\mathcal{R}_\beta$, and (2) that *norm M k n* computes a minimal normalized DBM.

5.4 A Final Semantics

We have assembled all the ingredients to define a semantics for timed automata which captures the essence of what DBM-based model checkers compute:

$$A \vdash \langle l,\ D \rangle \leadsto_{v,n} \langle l',\ D' \rangle \Longrightarrow A \vdash \langle l,\ D \rangle \leadsto_\mathcal{N} \langle l',\ norm\ (FW\ D'\ n)\ k\ n \rangle$$

Combining the fact that β-approximation is computable and the correctness properties of $\leadsto_\beta{}^*and\leadsto^*$, we have achieved our main result: a timed automaton can reach a certain location l' iff we can compute a valid run (using the DBM operations and normalization) that ends in l'.

Theorem 3. $Z = [M]_{v,n}$ ∧ $Z \subseteq V$ ∧ *dbm-int M n* $\Longrightarrow$
$(\exists u{\in}Z.\ \exists u'.\ A \vdash \langle l,\ u \rangle \to^* \langle l',\ u' \rangle)$
$\longleftrightarrow (\exists M'.\ A \vdash \langle l,\ M \rangle \leadsto_\mathcal{N}{}^* \langle l',\ M' \rangle \wedge [M']_{v,n} \neq \emptyset)$

6 Conclusion

We have presented a formalization that, beginning with basic definitions and classic results, closes the loop to show correcntess of the basic DBM-based algorithms that are used in forward analysis of timed automata. However, we have not yet harvested potential practical fruits of this development. A self-evident goal is to obtain an executable version for the algorithms above. By combination with a verified version of e.g., depth-first search, this could already yield a verified tool for deciding language emptiness of timed automata, which could in turn be extended to a fully verified model checker. In another direction of development, the author has already started to reuse the presented formalization to formalize first results about decidability of probabilistic timed automata.

Acknowledgement. I would like to thank Tobias Nipkow and the anonymous reviewers for their helpful comments on earlier versions of this paper.

References

1. Alur, R., Dill, D.L.: Automata for modeling real-time systems. In: Paterson, M. (ed.) ICALP 1990. LNCS, vol. 443, pp. 322–335. Springer, Heidelberg (1990)
2. Alur, R., Dill, D.L.: A theory of timed automata. Theor. Comput. Sci. **126**, 183–235 (1994)
3. Alur, R., Henzinger, T.A., Vardi, M.Y.: Parametric real-time reasoning. In: Proceedings of the Twenty-Fifth Annual ACM Symposium on Theory of Computing, pp. 592–601 (1993)
4. Bengtsson, J.E., Yi, W.: Timed automata: semantics, algorithms and tools. In: Desel, J., Reisig, W., Rozenberg, G. (eds.) Lectures on Concurrency and Petri Nets. LNCS, vol. 3098, pp. 87–124. Springer, Heidelberg (2004)
5. Bouyer, P.: Untameable timed automata! In: Alt, H., Habib, M. (eds.) STACS 2003. LNCS, vol. 2607, pp. 620–631. Springer, Heidelberg (2003)
6. Bouyer, P.: Forward analysis of updatable timed automata. Form. Methods Syst. Des. **24**(3), 281–320 (2004)
7. Bouyer, P., Dufourd, C., Fleury, E., Petit, A.: Are timed automata updatable? In: Emerson, E.A., Sistla, A.P. (eds.) CAV 2000. LNCS, vol. 1855, pp. 464–479. Springer, Heidelberg (2000)
8. Castéran, P., Rouillard, D.: Towards a generic tool for reasoning about labeled transition systems. In: TPHOLs 2001: Supplemental Proceedings (2001). http://www.informatics.ed.ac.uk/publications/report/0046.html
9. Clarke, E.M., Grumberg, O., Peled, D.A.: Model Checking. MIT Press, Cambridge (2001)
10. Dill, D.L.: Timing assumptions and verification of finite-state concurrent systems. In: Sifakis, J. (ed.) CAV 1989. LNCS, vol. 407, pp. 197–212. Springer, Heidelberg (1990)
11. Garnacho, M., Bodeveix, J.P., Filali-Amine, M.: A mechanized semantic framework for real-time systems. In: Braberman, V., Fribourg, L. (eds.) FORMATS 2013. LNCS, vol. 8053, pp. 106–120. Springer, Heidelberg (2013)
12. Henzinger, T.A., Ho, P.-H., Wong-toi, H.: Hytech: a model checker for hybrid systems. Softw. Tools Technol. Transf. **1**(1), 460–463 (1997)
13. Larsen, G.K., Pettersson, P., Yi, W.: Uppaal in a nutshell. Softw. Tools Technol. Transf. **1**(1), 134–152 (1997)
14. Paulin-Mohring, C.: Modelisation of timed automata in Coq. In: Kobayashi, N., Babu, C.S. (eds.) TACS 2001. LNCS, vol. 2215, pp. 298–315. Springer, Heidelberg (2001)
15. Wimmer, S.: Timed automata. Archive of Formal Proofs, March 2016. http://isa-afp.org/entries/Timed_Automata.shtml, Formal proof development
16. Xu, Q., Miao, H.: Formal verification framework for safety of real-time system based on timed automata model in PVS. In: Proceedings of the IASTED International Conference on Software Engineering, pp. 107–112 (2006)
17. Xu, Q., Miao, H.: Manipulating clocks in timed automata using PVS. In: Proceedings of SNPD 2009, pp. 555–560 (2009)
18. Yi, W., Pettersson, P., Daniels, M.: Automatic verification of real-time communicating systems by constraint-solving. In: Proceedings of Formal Description Techniques VII, pp. 243–258 (1994)
19. Yovine, S.: KRONOS: a verification tool for real-time systems. Softw. Tools Technol. Transf. **1**(1), 123–133 (1997)

AUTO2, A Saturation-Based Heuristic Prover for Higher-Order Logic

Bohua Zhan[⊠]

Massachusetts Institute of Technology, Cambridge, USA
bzhan@mit.edu

Abstract. We introduce a new theorem prover for classical higher-order logic named `auto2`. The prover is designed to make use of human-specified heuristics when searching for proofs. The core algorithm is a best-first search through the space of propositions derivable from the initial assumptions, where new propositions are added by user-defined functions called proof steps. We implemented the prover in Isabelle/HOL, and applied it to several formalization projects in mathematics and computer science, demonstrating the high level of automation it can provide in a variety of possible proof tasks.

1 Introduction

The use of automation is a very important part of interactive theorem proving. As the theories to be formalized become deeper and more complex, having a good automatic tool becomes increasingly indispensable. Such tools free users from the tedious task of specifying low level arguments, allowing them to focus instead on the high level outline of the proof.

There is a large variety of existing automatic proof tools. We will be content to list some of the representative ones. Some tools emulate human reasoning by attempting, at any stage of the proof, to apply a move that humans are also likely to make. These include the `grind` tactic in PVS [15], and the "waterfall" algorithm in ACL2 [10]. A large class of automatic provers are classical first-order logic solvers, based on methods such as tableau, satisfiability-modulo-theories (SMT), and superposition calculus. Sledgehammer in Isabelle [3] is a representative example of the integration of such solvers into proof assistants. Finally, most native tools in Isabelle and Coq are based on tactics, and their compositions to realize a search procedure. Examples for these include the `auto` tactic in Isabelle and Coq. The `blast` tactic in Isabelle [13] can also be placed in this category, although it has some characteristics of classical first-order solvers.

All these automatic tools have greatly improved the experience of formalization using proof assistants. However, it is clear that much work still needs to be done. Ideally, formalizing a proof on the computer should be very much like writing a proof in a textbook, with automatic provers taking the place of human readers in filling in any "routine" intermediate steps that are left out in the proof. Hence, one reasonable goal for the near future would be to develop an automatic prover that is strongly enough to consistently fill in such intermediate steps.

© Springer International Publishing Switzerland 2016
J.C. Blanchette and S. Merz (Eds.): ITP 2016, LNCS 9807, pp. 441–456, 2016.
DOI: 10.1007/978-3-319-43144-4_27

In this paper, we describe an alternative approach toward automation in proof assistants. It is designed to combine various desirable features of existing approaches. On the one hand it is able to work with human-like heuristics, classical higher-order logic, and simple type theory. On the other hand it has a robost, saturation-based search mechanism. We discuss these features and their motivations in Sect. 2.

As a first approximation, the algorithm in our approach consists of a best-first search through the space of propositions derivable from the initial assumptions, looking for a contradiction (any task is first converted into contradiction form). New propositions are generated by *proof steps*: user provided functions that match one or two existing propositions, and produce new propositions that logically follow from the matched ones. The order in which new propositions are added is dictated by a scoring function, as in a best-first search framework. There are several elaborations to this basic picture, in order to support case analysis, rewriting, skolemization, and induction. The algorithm will be described in detail, along with a simple example, in Sect. 3.

We implemented our approach in Isabelle/HOL, and used it to develop several theories in mathematics and computer science. In these case studies, we aim to use auto2 to prove all major theorems, either on its own or using a proof outline at a level of detail comparable to that of human exposition. We believe this aim is largely achieved in all the case studies. As a result, the level of automation provided by auto2 in our examples compares favorably with, and in some cases greatly exceeds that of existing tools provided in Isabelle. We give some examples from the case studies in Sect. 4.

The implementation, as well as the case studies, are available at https:// github.com/bzhan/auto2. We choose the name auto2 for two reasons: first, we intend it to be a general purpose prover capable of serving as the main automatic tool of a system, as auto in Isabelle and Coq had been. Second, it relates to one of the main features of the algorithm, which is that any proof step matches at most two items in the state.

In Sect. 5, we compare our approach with other major approaches toward automation, as well as list some related work. We conclude in Sect. 6, and discuss possible improvements and future directions of research.

2   Objectives

In this section, we list the main features our approach is designed to have, and the motivations behind these features.

Use of human-like heuristics: The prover should make use of heuristics that humans employ when searching for proofs. Roughly speaking, such heuristics come in two levels. At the lower level, there are heuristics about when to apply a single theorem. For example, a theorem of the form $A \implies B \implies C$ can be applied in three ways: deriving C from A and B, deriving $\neg A$ from B and $\neg C$, and deriving $\neg B$ from A and $\neg C$. Some of these directions may be more fruitful than others, and humans often instinctively apply the theorem in some of the directions but not in others. At the higher level, there are heuristics concerning induction, algebraic manipulations, procedures for solving certain problems, and so on. Both levels of heuristics are essential for humans to work with any sufficiently deep theory. Hence we believe it is important for the automatic prover to be able to take these into account.

Extensibility: The system should be extensible in the sense that users can easily add new heuristics. At the same time, such additions should not jeopardize the soundness of the prover. This can be guaranteed by making sure that every step taken by the user-added heuristics is verified, following the LCF framework.

Use of higher-order logic and types: The prover should be able to work with higher-order logic, and any type information (in the Isabelle sense) that is present. In particular, we want to avoid translations to and from untyped first-order logic that are characteristic of the use of classical first-order solvers. Avoiding these has several benefits: many heuristics that humans use are best stated in higher-order logic. Also, the statement to be proved is kept short and close to what humans work with, which facilitates printing an informative trace when a proof fails.

Saturation-based search mechanism: Most heuristics are fallible in the sense that they are not appropriate in every situation, and can lead to dead ends when applied in the wrong situations. Moreover, when several mutually-exclusive heuristics are applicable, we would like to consider all of them in turn. Some kind of search is necessary to deal with both of these problems. We follow a saturation-based search strategy in order to obtain the following desirable property: all steps taken by the prover are both permanent and "non-committal". That is, the result of any step is available for use throughout the remainder of the search, but there is never a requirement for it to be used, to allow for the possibility that the step is not appropriate for the proof at hand. The choice of E-matching over simplification to deal with equality reasoning is also chosen with this property in mind.

Having listed the principles motivating our approach, we also want to clarify what are not our main concerns. First, our focus is on proof tasks that occur naturally as intermediate steps during proofs of theorems in mathematics and computer science. We do not intend the prover to be competitive against more specialized algorithms when faced with large tasks that would also be difficult for humans. Second, the prover is not fully automated in the sense that it requires no human intervention – the user still needs to provide heuristics to the prover, including how to use each of the previously proved theorems. Finally, we do not intend to make the prover complete. For more difficult theorems, it expects hints in the form of intermediate steps.

3   Description of the System

In this section, we describe the `auto2` prover in detail, followed by a simple example, and a discussion of how the system is used in practice. We begin with a high-level description of the algorithm, leaving the details to the following subsections.

The algorithm follows a saturation-based strategy, maintaining and successively adding to a list of *items*. We will call this list the *main list* in the remainder of this section. For a first pass, we can think of items as propositions that follow from the initial assumptions, and possibly additional assumptions. Later on (Sect. 3.2) we will see that it can also contain other kinds of information, in addition to or instead of a proposition. Each item is placed in a *box*, which specifies what additional assumptions the item depends on. We discuss boxes in more detail in Sect. 3.1.

New items that may be added to the list are created by *proof steps*, which are user-provided functions that accept as input one or two existing items, and derive a list of new items from the inputs. With a few exceptions (Sect. 3.3), the new items must logically follow from the input items. One common kind of proof steps matches the input items to the one or two assumptions of a theorem, and when there is a match, return the conclusion of the theorem. However, as proof steps are arbitrary functions, they can have more complex behavior.

Reasoning with equalities is achieved by matching up to equivalence (E-matching) using a *rewritetable*. The rewrite table is a data structure that maintains the list of currently known equalities (not containing schematic variables). It provides a matching function that, given a pattern p and a term t, returns all matches of t against p, up to rewriting t using the known equalities. The rewrite table automatically uses transitivity of equality, as well as the congruence property (that is, $a_1 = b_1, \ldots, a_n = b_n$ implies $f(a_1, \ldots, a_n) = f(b_1, \ldots, b_n)$). See [11] for a modern introduction to E-matching. In our implementation, E-matching is essentially a first-order process (we only make use of equalities between terms not in function position), but we also allow matching of certain higher-order patterns, and extend it in other ways (Sect. 3.4). Matching using the rewrite table is used as the first step of nearly all proof steps.

New items produced by proof steps are collected into *updates*, and each update is assigned a score, which indicates its priority in the best-first search. All new updates are first inserted into a priority queue. At each iteration of the algorithm, the update with the lowest score is pulled from the queue. The items contained in the update are then added to the main list and processed one-by-one. Scoring is discussed in Sect. 3.5.

With these in mind, we can give a first sketch of the main loop of the algorithm. We assume that the statement to be proved is written in contradiction form (that is, $[A_1, \ldots, A_n] \implies C$ is written as $[A_1, \ldots, A_n, \neg C] \implies$ False), so the goal is to derive a contradiction from a list of assumptions $A_1, \ldots, A_n$.

- The algorithm begins by inserting a single update to the priority queue, containing the propositions $A_1, \ldots, A_n$.
- At each iteration, the update with the lowest score is pulled from the priority queue. Items within the update are added one-by-one to the main list.
- Upon adding a non-equality item, all proof steps taking one input item are invoked on the item. All proof steps taking two input items are invoked on all pairs of items consisting of the new item and another item in the main list. All updates produced are added to the priority queue.

- Upon adding an equality item (without schematic variables), the equality is added to the rewrite table. Then the procedure in the previous step is redone with the new rewrite table on all items containing up to equivalence either side of the equality (this is called *incremental matching*). All new updates (those that depend on the new equality) are added to the priority queue.
- The loop continues until a contradiction (depending only on the initial assumptions) are derived by some proof step, or if there are no more updates in the queue, or if some timeout condition is reached.

In the current implementation, we use the following timeout condition: the loop stops after pulling N updates from the priority queue, where N is set to 2000 (in particular, all invocations of `auto2` in the given examples involve less than 2000 steps).

3.1 Box Lattice

Boxes are used to keep track of what assumptions each item depends on. Each *primitive* or *composite* box represents a list of assumptions. They are defined recursively as follows: a composite box is a set of primitive boxes, representing the union of their assumptions. The primitive boxes are indexed by integers starting at 0. Each primitive box inherits from a composite box consisting of primitive boxes with smaller index, and contains an additional list of assumptions. It represents the result of adding those assumptions to the parent box. The primitive box 0 (inheriting from {}) contains the list of assumptions in the statement to be proved. Other primitive boxes usually inherit, directly or indirectly, from {0}. The primitive boxes also keep track of introduced variables. From now on we will simply call a composite box as a *box*.

If a contradiction is derived in a box (that is, if `False` is derived from the assumptions in that box), the box is called *resolved*, and appropriate propositions (negations of the assumptions) are added to each of its immediate parent boxes. The overall goal of the search is then to resolve the box {0}, which contains exactly the assumptions for the statement to be proved.

There is a natural partial order on the boxes given by inclusion, and a merge operation given by taking unions, making the set of boxes into a semilattice. New primitive boxes are created by proof steps, and are packaged into updates and added to the queue with a score just like new items. Creating a new primitive box effectively starts a case analysis, as we will explain in the example in Sect. 3.6.

3.2 Item Types

In this section we clarify what information may be contained in an item. In general, we think of an item in a box b as any kind of information that is available under the assumptions in b. One important class of items that are not propositions are the term items. A term item t in box b means t appears as a subterm of some proposition (or another kind of item) in b. The term items can be matched by proof steps just like propositions. This allows the following implementation of directed rewrite rules: given a theorem $P = Q$, where any schematic variable appearing in Q also appears in P, we can add a proof step that matches P against any term item t, and produces the equality $P(\sigma) = Q(\sigma)$ for any match with instantiation σ. This realizes the forward rewrite rule from P to Q.

In general, each item consists of the following information: a string called *item type* that specifies how to interpret the item; a term called *tname* that specifies the content of the item; a theorem that justifies the item if necessary, and an integer score which specifies its priority in the best-first search. The most basic item type is PROP for propositions, for which *tname* is the statement of the theorem, and is justified by the theorem itself. Another basic type is TERM for terms items, for which *tname* is the term itself, and requires no justifying theorems.

The additional information contained in items can affect the behavior of proof steps, and by outputting an item with additional information, a proof step can affect how the output is used in the future. This makes it possible to realize higher level controls necessary to implement more complex heuristics. To give a simple example, in the current implementation, disjunctions are stored under two different item types: DISJ and DISJ_ACTIVE. The latter type induces case analysis on the disjunction, while the former does not. By outputting disjunctions in the appropriate type, a proof step can control whether case analysis will be invoked on the result.

3.3   Skolemization and Induction

Usually, when a proof step outputs a proposition, it must derive the justifying theorem for that proposition, using the justifying theorems of the input items. There are two main exceptions to this. First, given an input proposition $\exists x.P(x)$, a proof step can output the proposition $P(x)$, where x is a previously unused constant. This realizes skolemization, which in our framework is just one of the proof steps.

The second example concerns the use of certain induction theorems. For example, induction on natural numbers can be written as:

$$P(0) \implies \forall n.P(n-1) \longrightarrow P(n) \implies P(n).$$

This form of the induction theorem suggests the following method of application: suppose n is an initial variable in a primitive box i, and proposition $n \neq 0$ is known in (the composite) box $\{i\}$. Then we may insert $P(n-1)$ into box $\{i\}$, where P is obtained from the list of assumptions in i containing n. This corresponds to the intuition that once the zero case is proved, one may assume $P(n-1)$ while proving $P(n)$.

In both cases, any contradiction that depends on the new proposition can be transformed into one that does not. In this first case, this involves applying a particular theorem about existence (exE in Isabelle). In the second case, it involves applying the induction theorem.

3.4 Matching

In this section, we provide more details about the matching process. First, the presence of box information introduces additional complexities to E-matching. In the rewrite table, each equality is stored under a box, and each match is associated to a box, indicating which assumptions are necessary for that match. When new items are produced by a proof step, the items are placed in the box that is the merge of boxes containing the input items, and the boxes associated to all matches performed by that proof step.

We also support the following additional features in matching:

– Matching of associative-commutative (AC) functions: the matching makes limited use of properties of AC functions. For example, if $x = y \star z$ is known, where $\cdot \star \cdot$ is AC, then the pattern $y\star?a$ can match the term $p\star x$, with instantiation $?a := p \star z$ (since $y \star (p \star z) = p \star (y \star z) = p \star x$). The exact policy used in AC-matching is rather involved, as it needs to balance efficiency and not missing important matches.
– Matching of higher-order patterns: we support second-order matching, with the following restriction on patterns: it is possible to traverse the pattern in such a way that any schematic variable in function position is applied to distinct bound variables in its first appearance. For example, in the following theorem:

$$\forall (n :: \mathrm{nat}).f(n) \le f(n+1) \implies m \le n \implies f(m) \le f(n),$$

one can match its first assumption and conclusion against two items, since the left side of the inequality in the first assumption can be matched to give a unique instantiation for f. The condition given here is slightly more general than the condition given by Nipkow [12], where all appearances of a schematic variable in function position must be applied to distinct bound variables.
– Schematic variables for numeric constants: one can restrict a schematic variable to match only to numeric constants (in the current implementation, this is achieved by a special name $?\mathrm{NUMC}_i$). For example, one can write proof steps that perform arithmetic operations, by matching terms to patterns such as $?\mathrm{NUMC}_1+?\mathrm{NUMC}_2$.
– Custom matching functions: one can write custom functions for matching a pattern against an item. This is especially important for items of type other than PROP. But it is also useful for the PROPs themselves. For example, if the pattern is $\neg(p < q)$, one can choose to match $q \le p$ instead, and convert any resulting theorem using the equivalence to $\neg(p < q)$.

3.5 Scoring

The scoring function, which ranks future updates, is crucial for the efficiency of the algorithm as it determines which updates will be explored first in the search. It tries to guess which reasoning steps are more likely to be relevant to the proof at hand. In the current implementation, we choose a very simple strategy. Finding a better scoring strategy will certainly be a major focus in the future.

The current scoring strategy is as follows: the score of any update equals the maximum of the scores of the dependent items, plus an increment depending on the content of the update. The increment is bigger (i.e. the update is discouraged) if the terms in the update are longer, or if the update depends on many additional assumptions.

3.6   A Simple Example

We now give a sample run of `auto2` on a simple theorem. Note this example is for illustration only. The actual implementation contains different proof steps, especially for handling disjunctions. Moreover, we ignore scoring and the priority queue, instead adding items directly to the list. We also ignore items that do not contribute to the eventual proof.

The statement to be proved is

$$\text{prime } p \implies p > 2 \implies \text{odd } p.$$

Converting to contradiction form (and noting that odd p is an abbreviation for $\neg$even p), our task is to derive a contradiction from assumptions prime p, $p > 2$, and even p. The steps are:

1. Add primitive box 0, with variable p, and assumptions prime p, $p > 2$, and even p.
2. Add subterms of the propositions, including TERM prime p and TERM even p.
3. The proof step for applying the definition of prime adds equality

$$\text{prime } p = (p > 1 \land \forall m.m \text{ dvd } p \longrightarrow m = 1 \lor m = p)$$

   from TERM prime p. Likewise, the proof step for applying the definition of even adds equality even $p = 2$ dvd p from TERM even p.
4. When the first equality in the previous step is applied, incremental matching is performed on the proposition prime p. It now matches the pattern $?A \land ?B$, so the proof step for splitting conjunctions produces $p > 1$ and $\forall m.m$ dvd $p \longrightarrow m = 1 \lor m = p$.
5. A proof step matches the propositions $\forall m.m$ dvd $p \longrightarrow m = 1 \lor m = p$ and even p (the second item, when rewritten as 2 dvd p, matches the antecedent of the implication), producing $2 = 1 \lor 2 = p$.
6. The proof step for invoking case analysis matches $2 = 1 \lor 2 = p$ with pattern $?A \lor ?B$. It creates primitive box 1, with assumption $2 = 1$ (see Fig. 1).
7. A proof step matches $2 = 1$ (in box $\{1\}$) with pattern $?\text{NUMC}_1 = ?\text{NUMC}_2$. The proof step examines the constants on the two sides, finds they are not equal, and outputs a contradiction. This resolves box $\{1\}$, adding $2 \neq 1$ into box $\{0\}$.
8. A proof step matches $2 = 1 \lor 2 = p$ with $2 \neq 1$, producing $2 = p$.
9. When the equality in the previous step is added, incremental matching is performed on the proposition $p > 2$ (one of the initial assumptions). This proposition matches pattern $?n > ?n$ (when rewritten as $p > p$ or $2 > 2$), giving a contradiction. This resolves box $\{0\}$ and finishes the proof.

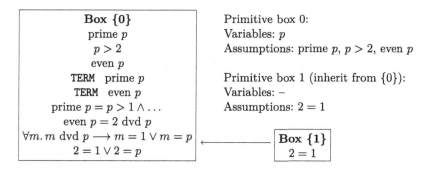

Fig. 1. State of proof after step 6. Arrow indicates inheritance relation on boxes.

3.7 Proof Scripts

For the case studies, we designed our own language of proof scripts for specifying intermediate steps in the proof of a more difficult theorem. The proof scripts are provided as an argument to the auto2 tactic, and are interpreted within the tactic. This requires some straightforward modifications to the main loop and the scoring mechanism, which we will not discuss. The benefit of using an internally interpreted script (instead of Isar) is that the entire state of the proof is maintained between lines of the script, and all previously proved statements are available for use at any given point.

The proof script consists of atomic commands joined together with two connectors: THEN and WITH. Each atomic command specifies an intermediate statement to prove, and what update to add once that statement is proved. The meanings of the two connectors are as follows. The command A THEN B means first process A, and after A is finished, process B. The command A WITH B (with A atomic) means attempt to prove the intermediate statement specified in A, processing B as a part of the attempt.

The simplest atomic commands are OBTAIN and CASE. The command OBTAIN p means attempt to prove p and add it to the list. The command CASE p means attempt to prove that p results in a contradiction, and add $\neg p$ to the list. It is equivalent to OBTAIN $\neg p$.

The command CHOOSE x, $p(x)$ specifies $\exists x.p(x)$ as an intermediate statement. After it is proved, the resulting existence fact is instantiated with variable x (the command fixes variable x so it is not used in other places).

Finally, there are various flavors of induction commands, which specify applications of various kinds of induction theorems. We designed the script system to be extensible: it is possible for the user to add new types of atomic commands.

3.8 Practical Usage

We end this section with a discussion of practical issues concerning the use of the auto2 system.

First, we describe the process of constructing the collection of proof steps. The collection of proof steps specifies exactly what steps of reasoning auto2 may take. With the exception of equality reasoning, which relies on the rewrite table and E-matching, all other forms of reasoning are encoded as proof steps. This includes basic deductions in logic and arithmetic, and the simplification of terms. In particular, auto2 does not invoke any of the other Isabelle commands such as simp and arith, except within the implementation of individual proof steps, for carrying out very specific tasks.

Each proof step is intended to represent a single step of reasoning, and has a clearly-defined behavior. The simplest proof steps apply a single theorem. For example, a theorem of the form $A \implies B \implies C$ can be added for use in either the forward or one of the two backward directions. More complex proof steps are implemented as ML functions. The implementation can make full use of the existing conversion and tactics facility in Isabelle/ML.

In theories developed using auto2, each proof step using theorems in that theory is added right after all required theorems are proved. Once the proof step is added, it is used in all ensuing proofs, both in the current theory and in all descendent theories. For theorems proved in the Isabelle library, "wrapper" theories are created to add proof steps using them. The case studies, for example, use shared wrapper theories for theorems concerning logic, arithmetic, sets, and lists.

There are some circumstances where removing a proof step after using it in a few proofs is acceptable. For example, if a theory introduces constructions, or proves lemmas that are used only within the theory, it is acceptable to remove proof steps related to those constructions and lemmas once they are no longer used. The guiding principle is as follows: by the end of the development of a theory, the collection of proof steps from that theory should form a coherent system of heuristics on how to use the results in that theory. In subsequent theories, auto2 should have a basic competence in using results from that theory, and it should always be possible to specify more involved applications in proof scripts. In particular, the user should never need to add proof steps for using theorems from a previous theory, nor temporarily remove a proof step from a previous theory (to avoid exploding the search space). Realizing this principle means more work is needed when building each theory, to specify the right set of proof steps, but it should pay off in the long run, as it frees the user from having to refer back to the theory in subsequent developments.

Second, we describe the usual interaction loop when proving a theorem or which applying auto2 directly fails. One begins by working out an informal proof of the theorem, listing those steps that appear to require some creativity. One can then try auto2 with these intermediate steps added. If it still does not work, the output trace shows the first intermediate step that auto2 cannot prove, and what steps of reasoning are taken in the attempt to prove that step. If there is some step of reasoning that should be taken automatically but is not, it is an indication that some proof step is missing. The missing proof step should be added, either to a wrapper theory if the relevant theorem is proved in the

Isabelle library, or right after the theorem if it is proved in a theory developed using `auto2`. On the other hand, if one feels the missing step should not be taken automatically, but is a non-obvious step to take in the proof of the current theorem, one should add that step to the proof script instead. The process of adding to the collection of proof steps or to the proof script continues until `auto2` succeeds.

4   Case Studies

In this section, we give some examples from the case studies conducted using `auto2`. We will cover two of the six case studies. Descriptions for the other four (functional data structures, Hoare logic, construction of real numbers, and Arrow's impossibility theorem) can be found in the repository. In writing the case studies, we aim to achieve the following goal: all major theorems are proved using `auto2`, either directly or using proof scripts at a level of detail comparable to human exposition. When a case study parallels an existing Isabelle theory, there may be some differences in the definitions, organization, and method of proof used. The content of the theorems, however, are essentially the same. In the examples below, we will sometimes compare the length of our scripts with the length of Isar scripts for the same theorem in the Isabelle library. We emphasize that this is not intended to be a rigorous comparison, due to the differences just mentioned, and since `auto2` is provided additional information in the form of the set of proof steps, and takes longer to verify the script. The intent is rather to demonstrate the level of automation that can be expected from `auto2`.

Besides the examples given below, we also make a special note of the case study on Arrow's impossibility theorem. The corresponding theory in the Isabelle AFP is one of the seven test theories used in a series of benchmarks on Sledgehammer, starting in [4].

4.1   Elementary Theory of Prime Numbers

The development of the elementary theory of prime numbers is one of the favourites for testing theorem provers [5,14]. We developed this theory starting from the definition of prime numbers, up to the proof of the infinitude of primes and the unique factorization theorem, following `HOL/Number_Theory` in the Isabelle library. For the infinitude of primes, the main lemma is that there always exists a larger prime:

`larger_prime`: $\exists p.\mathsf{prime}\ p \wedge n < p$

`auto2` is able to prove this theorem when provided with the following proof script:

CHOOSE p, prime $p \wedge p$ dvd fact $n + 1$ THEN
CASE $p \leq n$ WITH OBTAIN p dvd fact n

This corresponds to the following proof of `next_prime_bound` in the Isabelle theory `HOL/Number_Theory/Primes` (18 lines).

lemma next_prime_bound: $\exists p.\text{prime } p \wedge n < p \wedge p \leq \text{fact } n + 1$
proof–
   have f1: fact $n + 1 \neq (1 :: \text{nat})$"using fact_ge_1 [of n, where'a=nat] by arith
   from prime_factor_nat [OF f1]
   obtain p where prime p and p dvd fact $n + 1$ by auto
   then have $p \leq$ fact $n + 1$ apply (intro dvd_imp_le) apply auto done
   { assume $p \leq n$
     from prime p have $p \geq 1$
       by (cases p, simp_all)
     with $p \leq n$ have p dvd fact n
       by (intro dvd_fact)
     with p dvd fact $n + 1$ have p dvd fact $n + 1 -$ fact n
       by (rule dvd_diff_nat)
     then have p dvd 1 by simp
     then have $p \leq 1$ by auto
     moreover from prime p have $p > 1$
       using prime_def by blast
     ultimately have False by auto}
   then have $n < p$ by presburger
   with prime p and $p \leq$ fact $n + 1$ show ?thesis by auto
qed

Likewise, we formalized the unique factorization theorem. The uniqueness part of the theorem is as follows (note M and N are multisets, and set M and set N are the sets corresponding to M and N, eliminating duplicates).

factorization_unique_aux:
$\forall p \in \text{set } M.\text{prime } p \implies \forall p \in \text{set } N.\text{prime } p \implies \prod_{i \in M} i \text{ dvd } \prod_{i \in N} i \implies M \subseteq N$

The script needed for the proof is:

CASE $M = \emptyset$ THEN
CHOOSE M', m, $M = M' + \{m\}$ THEN
OBTAIN m dvd $\prod_{i \in N} i$ THEN
CHOOSE n, $n \in N \wedge m$ dvd n THEN
CHOOSE N', $N = N' + \{n\}$ THEN
OBTAIN $m = n$ THEN
OBTAIN $\prod_{i \in M'} i$ dvd $\prod_{i \in N'} i$ THEN
STRONG_INDUCT $(M, [\text{Arbitrary } N])$

This can be compared to the proof of multiset_prime_factorization_unique_aux in the Isabelle theory HOL/Number_Theory/UniqueFactorization (39 lines).

4.2   Verification of Imperative Programs

A much larger project is the verification of imperative programs, building on the Imperative HOL library, which describes imperative programs involving pointers using a Heap Monad [6]. The algorithms and data structures verified are:

- Reverse and quicksort on arrays.
- Reverse, insert, delete, and merge on linked lists.
- Insert and delete on binary search trees.

The proofs are mostly automatic, which is in sharp contrast with the corresponding examples in the Isabelle distribution (in Imperative_HOL/ex). We give one example here. The merge function on two linked lists is defined as:

```
partial_function (heap) merge ::
('a :: {heap, ord}) node ref ⇒'a node ref ⇒'a node ref Heap
where
[code]: merge p q =
  do { np ← !p; nq ← !q;
       if np = Empty then return q
       else if nq = Empty then return p
       else if val np ≤ val nq then
         do { npq ← merge (nxt np) q;
              p := Node (val np) npq;
              return p }
       else
         do { pnq ← merge p (nxt nq);
              q := Node (val nq) pnq;
              return q } }
```

To prove the main properties of the merge function, we used the following two lemmas (commands adding their proof steps are omitted):

```
theorem set_intersection_list: (x ∪ xs) ∩ ys = {} ⇒ xs ∩ ys = {} by auto
```

```
theorem unchanged_outer_union_ref:
  "unchanged_outer h h' (refs_of h p ∪ refs_of h q) ⇒
     r ∉ refs_of h p ⇒
     r ∉ refs_of h q ⇒
     Ref.present h r ⇒
     Ref.get h r = Ref.get h' r"
  by (simp add: unchanged_outer_ref)
```

The statements of the theorems are:

```
theorem merge_unchanged:
  "effect (merge p q) h h' r ⇒ proper_ref h p ⇒ proper_ref h q ⇒
   unchanged_outer h h' (refs_of h p ∪ refs_of h q)"
```

```
theorem merge_local:
  "effect (merge p q) h h' r ⇒ proper_ref h p ⇒ proper_ref h q ⇒
   refs_of h p ∩ refs_of h q = {} ⇒
     proper_ref h' r ∧ refs_of h' r ⊆ refs_of h p ∪ refs_of h q"
```

```
theorem merge_correct:
  "effect (merge p q) h h' r ⇒ proper_ref h p ⇒ proper_ref h q ⇒
   refs_of h p ∩ refs_of h q = {} ⇒
     list_of h' r = merge_list (list_of h p) (list_of h q)"
```

Each of these theorems is proved (in 30–40 s on a laptop) using the same proof script, specifying the induction scheme:

```
DOUBLE_INDUCT (("pl = list_of h p","ql = list_of h q"), Arbitraries ["p","q","h'","r"])
```

In the Isabelle library the proof of the three corresponding theorems, including that of two induction lemmas proved specifically for these theorems, takes 166 lines in total. These theorems also appear to be well beyond the ability of the Sledgehammer tools. It is important to note that this automation is not based on Hoare logic or separation logic (the development here is separate from the case study on Hoare logic), but the proofs here use directly the semantics of commands like in the original examples.

5   Related Work

The author is particularly inspired by the work of Ganesalingam and Gowers [9], which describes a theorem prover that can output proofs in a form extremely similar to human exposition. Our terminology of "box" is taken from there (although the meaning here is slightly different).

There are two ways in which our approach resembles some of the classical first-order solvers. The first is the use of a "blackboard" maintaining a list of propositions, with many "modules" acting on them, as in a Nelson-Oppen architecture [2]. The second is the use of matching up to equivalence (E-matching), which forms a basic part of most SMT solvers. The main differences are explained in the first three items in Sect. 2: our focus on the use of human-like heuristics, and our lack of translation to and from untyped first-order logic.

There have been extensive studies on heuristics that humans use when proving theorems, and their applications to automation. Ganesalingam and Gowers [9] give a nice overview of the history of such efforts. Some of the more recent approaches include the concept of proof plans introduced by Bundy [7,8]. Among proof tools implemented in major proof assistants, the grind tactic [15] and the "waterfall" algorithm in ACL2 [10] both attempt to emulate human reasoning processes. Compared to these studies, we place a bigger emphasis on search, in order to be tolerant to mistaken steps, and to try different heuristics in parallel. We also focus more on heuristics for applying single theorems, although the system is designed with the possibility of higher-level heuristics in mind (in particular with the use of item types).

Finally, tactic-based automation such as auto, simp, and fast in Isabelle also use heuristics in the sense that they apply theorems directionally, and are able to carry out procedures. The main difference with our approach is the search mechanism used. In tactic-based automation, the search is conducted over the space of proof states, which consists of the current goal and a list of subgoals. For blast and other tableau-based methods, the search is over the space of possible tableaux. In our approach, the search is saturation-based, and performed over the space of propositions derivable from the initial assumptions.

A similar "blackboard" approach is used for heuristic theorem proving by Avigad et al. [1], where the focus is on proving real inequalities. The portion of our system concerning inequalities is not as sophisticated as what is implemented there. Instead, our work can be viewed as applying a similar technique to all forms of reasoning.

6 Conclusion

In this paper, we described an approach to automation in interactive theorem proving that can be viewed as a mix of the currently prevailing approaches. While the ideas behind the prover are mostly straightforward, we believe the combination of these ideas is underexplored and, given the examples above, holds significant promise that warrants further exploration.

There are many aspects of auto2 that can be improved in the future. Two immediate points are performance and debugging. The E-matching process is far from optimized, in the sense of [11]. For debugging, the program currently outputs the list of updates applied to the state. One might instead want to view and traverse the dependency graph of updates. One would also like to query the rewrite table at any point in the proof.

There are also many directions of future research. I will just list three main points:

- The scoring function is currently very simple. Except for a few cases, there is currently no attempt at take into account during scoring the proof step used. Instead, one would like to distinguish between proof steps that "clearly should be applied", and those that should be applied "with reluctance". There is also the possibility of using various machine learning techniques to automatically adjust the scoring function for individual proof steps.
- Several aspects of elementary reasoning, such as dealing with associative-commutative functions, and with ordered rings and fields, pose special challenges for computers. While the current implementation is sufficient in these aspects for the examples at hand, more will need to be done to improve in both completeness and efficiency.
- Finally, one would like to improve auto2's ability to reason in other, diverse areas of mathematics and computer science. On the verification of imperative programs front, one would like to know how well auto2 can work with separation logic, or perhaps a framework based on a mix of separation logic and "natural" reasoning used in the given examples will be ideal. On the mathematical front, each field of mathematics offers a distinctive system of heuristics and language features. One would like to expand the collection of proof steps, as well as proof script syntax, to reflect these features.

Acknowledgements. The author thanks Jasmin Blanchette for extensive comments on a first draft of this paper, and also Jeremy Avigad, Adam Chlipala, and Larry Paulson for feedbacks and discussions during the project. The author also thanks the anonymous referees for their comments. The work is done while the author is supported by NSF Award No. 1400713.

References

1. Avigad, J., Lewis, R.Y., Roux, C.: A heuristic prover for real inequalities. In: Klein, G., Gamboa, R. (eds.) ITP 2014. LNCS, vol. 8558, pp. 61–76. Springer, Heidelberg (2014)
2. Barrett, C., Sebastiani, R., Seshia, S.A., Tinelli, C.: Satisability modulo theories. In: Biere, A., et al. (ed.) Handbook of Satisability, pp. 825–885. IOS Press (2008)
3. Blanchette, J.C., Kaliszyk, C., Paulson, L.C., Urban, J.: Hammering towards QED. J. Formaliz. Reason. 9(1), 101–148 (2016)
4. Böhme, S., Nipkow, T.: Sledgehammer: judgement day. In: Giesl, J., Hähnle, R. (eds.) IJCAR 2010. LNCS, vol. 6173, pp. 107–121. Springer, Heidelberg (2010)
5. Boyer, R.S., Moore, J.S.: A Computational Logic. ACM Monograph Series. Academic Press, New York (1979)
6. Bulwahn, L., Krauss, A., Haftmann, F., Erkök, L., Matthews, J.: Imperative functional programming with Isabelle/HOL. TPHOLs 2008. LNCS, vol. 5170, pp. 134–149. Springer, Heidelberg (2008)
7. Bundy, A.: A science of reasoning. In: Lassez, J.-L., Plotkin, G. (eds.) Computational Logic: Essays in Honor of Alan Robinson, pp. 178–198. MIT Press (1991)
8. Bundy, A.: The use of explicit plans to guide inductive proofs. CADE-9. LNCS, vol. 310, pp. 111–120. Springer, Heidelberg (1988)
9. Ganesalingam, M., Gowers, W.T.: A fully automatic problem solver with human-style output. arXiv:1309.4501
10. Kaufmann, M., Manolios, P., Moore, J.S.: Computer-Aided Reasoning: An Approach. Kluwer Academic Publishers, Boston (2000)
11. Moura, L., Bjourner, N.: Efficient E-matching for SMT solvers. In: Pfenning, F. (ed.) CADE 2007. LNCS (LNAI), vol. 4603, pp. 183–198. Springer, Heidelberg (2007)
12. Nipkow, T.: Functional unification of higher-order patterns. In: Vardi, M. (ed) Eighth Annual Symposium on Logic in Computer Science, pp. 64–74. IEEE Computer Society Press (1993)
13. Paulson, L.C.: A generic tableau prover and its integration with Isabelle. J. Univ. Comp. Sci. 5(3), 73–87 (1999)
14. Wenzel, M., Wiedijk, F.: A comparison of the mathematical proof languages Mizar and Isar. J. Autom. Reason. 29, 389–411 (2002)
15. Shankar, N., Owre, S., Rushby, J.M., Stringer-Calvert, D.W.J.: PVS prover guide. Comput. Sci. Lab. (2001). SRI International, Menlo Park

Rough Diamonds

What's in a Theorem Name?

David Aspinall[1] and Cezary Kaliszyk[2(✉)]

[1] LFCS, School of Informatics, University of Edinburgh, Edinburgh, UK
[2] Institut für Informatik, University of Innsbruck, Innsbruck, Austria
cezary.kaliszyk@uibk.ac.at

Abstract. ITPs use names for proved theorems. Good names are either widely known or descriptive, corresponding to a theorem's statement. Good names should be consistent with conventions, and be easy to remember. But thinking of names like this for every intermediate result is a burden: some developers avoid this by using consecutive integers or random hashes instead. We ask: is it possible to relieve the naming burden and automatically suggest sensible theorem names? We present a method to do this. It works by learning associations between existing theorem names in a large library and the names of defined objects and term patterns occurring in their corresponding statements.

1 Introduction

How do we name theorems? Science has a tradition of *historical reference*, attaching names by attribution to their discoverer. HOL Light contains fine examples such as RIEMANN_MAPPING_THEOREM:

$$\forall s.\ \text{open } s \wedge \text{simply\_connected } s \Longleftrightarrow s = \{\} \vee s = (\text{:real}\hat{\ }2) \vee$$
$$\exists f\ g.\ f \text{ holomorphic\_on } s \wedge g \text{ holomorphic\_on ball}(0,\ 1) \wedge$$
$$(\forall z.\ z \text{ IN } s \Longrightarrow f\ z \text{ IN ball}(Cx(\&0),\ \&1) \wedge g(f\ z) = z) \wedge$$
$$(\forall z.\ z \text{ IN ball}(Cx(\&0),\&1) \Longrightarrow g\ z \text{ IN } s \wedge f(g\ z) = z)$$

Mere lemmas are seldom honoured with proper names. Papers and textbooks use localised index numbers instead ("see Lemma 5.7 on p. 312") which is succinct but unhelpful. For practical proof engineering, working with large developments and libraries, we are quickly swamped with intermediate lemmas. Despite search facilities in some ITPs, users still need to name statements, leading to the proliferation of less profound *descriptive names* such as:

$$\text{ADD\_ASSOC}:\ \forall m\ n\ p.\ m+(n+p)=(m+n)+p$$

$$\text{REAL\_MIN\_ASSOC}:\ \forall x\ y\ z.\ \min x\,(\min y\,z)=\min\,(\min x\,y)\,z$$

$$\text{SUC\_GT\_ZERO}:\ \forall x.\ \text{Suc}\,x>0$$

Descriptive names are convenient and mnemonic, based on the statement of a theorem, and following conventions due to the author, library or system. But inventing names requires thought, and remembering them exactly later can be tricky. Unsurprisingly, people complain about the burden of naming, even inventing schemes that generate automatic names like (in Flyspeck [4]):

© Springer International Publishing Switzerland 2016
J.C. Blanchette and S. Merz (Eds.): ITP 2016, LNCS 9807, pp. 459–465, 2016.
DOI: 10.1007/978-3-319-43144-4_28

HOJODCM LEBHIRJ OBDATYB MEEIXJO KBWPBHQ RYIUUVK DIOWAAS

Such (fuzzy) hashes of theorem statements or sequential numbers give no clue of content, and we are back to old-fashioned indices. Some readers may be happy, accepting that names should have "denotation but not connotation" (as Kripke recalled the position of Mill [6]). But like Kripke, we see value in connotation: after all, a name is the handle for a potentially much longer statement.

So we wonder: *is it possible to relieve theorem naming burden by automatically naming theorems following established conventions?* Given that we have large corpora of carefully named theorems, it is natural to try a learning approach.

2   Parts of Names and Theorems

To start, we examine the form of human-generated descriptive theorem names. These are compound with separators: $l_0\_ \ldots \_l_m$, where the l_i are commonly used stem words (labels). Examples like REAL_MIN_ASSOC show a connection between names of constants (MIN), their types (REAL), and the structure of the statement (ASSOC). Using previous work on features for learning-assisted automated reasoning [5], we extract three characterizations of theorem statements:

- **Symbols**: constant and type names (including function names);
- **Subterms**: parts of the statement term where no logical operators appear;
- **Patterns**: subterms, with abstraction over names of defined objects.

Patterns allow us to model certain theorem shapes, such as commutativity, associativity, or distributivity, without the actual constants these properties talk about [2]. For examples like SUC_GT_ZERO, we see that *where* the name parts occur is important. So we also collect:

- **Positions**: for each feature f, a position $p(f)$, normalised so that $0 \le p(f) \le 1$, given by the position of f in the print order of the statement.

The leftmost feature has position 0 and the rightmost 1; if only a single feature is found, it has position $\frac{1}{2}$. Names are treated correspondingly: for $l_0\_ \ldots \_l_m$, the stems are assigned equidistant positions $p(l_0) = 0, \ldots, p(l_m) = 1$.

3   Learning Associations Between Names and Statements

We investigate two schemes for associating theorem statements with their names:

- **Consistent**: builds an association between symbols (e.g., constant and type names) and parts of theorem names;
- **Abstract**: uses patterns to abstract from concrete symbols, building a matching between positions in statements and name parts.

We hypothesise that the first scheme might be the more successful when used *within* a specific development, (hopefully) consistently re-using the same sub-components of relevant names, whereas the second may do better *across* different developments (perhaps ultimately, even across different provers).

To try these schemes out, we implement a k-Nearest Neighbours (kNN) multi-label classifier. A proposed label is computed together with a weighted average of positions. The algorithm first finds a fixed number k of training examples (named theorem statements), which are most similar to a set of features being considered. The stems and positions from the training examples are used to estimate the relevance of stems and proposed positions for the currently evaluated statement. The *nearness* of two statements s_1, s_2 is given by

$$n(s_1, s_2) = \sqrt{\sum_{f \in \overline{f}(s_1) \cap \overline{f}(s_2)} w(f)^2}$$

where $w(f)$ is the IDF (inverse document frequency) weight of a feature f. To efficiently find nearest neighbours, we index the training examples by features, so we can ignore examples that have no features in common with the currently considered statement. Given the set of k training examples nearest to the current statement, we evaluate the relevance of a label as follows:

$$R(l) = \sum_{s_1 \in N, l \in \overline{l}(s_1)} \frac{n(s_1, s_2)}{|\overline{l}(s_1)|}$$

We propose positions for a stem using the weighted average of the positions in the recommendations; weights are the corresponding nearness values.

Table 1. Selected features extracted from the statement of ADD_ASSOC

Feature	Frequency	Position	IDF
$(V_0 + (V_1 + V_2) = (V_0 + V_1) + V_2)$	1	0.37	7.82
$((V_0 + V_1) + V_2)$	1	0.75	7.13
$(V_0 + V_1)$	1	0.84	3.95
+	4	0.72	2.62
num	3	0.21	1.15
=	1	0.43	0.23
∀	3	0.15	0.03

As an example, we examine how the process works with the consistent naming scheme and the ADD_ASSOC statement shown previously. Our algorithm uses the statement in a fully-parenthesised form with types attached to binders:

$$\forall num.(\forall num.(\forall num.((V_0 + (V_1 + V_2)) = ((V_0 + V_1) + V_2))))$$

Table 2. Nearest neighbours found for ADD_ASSOC.

Theorem name	Statement	Nearness
MULT_ASSOC	(V0 * (V1 * V2)) = ((V0 * V1) * V2)	553
ADD_AC_1	((V0 + V1) + V2) = (V0 + (V1 + V2))	264
EXP_MULT	(EXP V0 (V1 * V2)) = (EXP (EXP V0 V1) V2)	247
HREAL_ADD_ASSOC	(V0 $+_H$ (V1 $+_H$ V2)) = ((V0 $+_H$ V1) $+_H$ V2)	246
HREAL_MUL_ASSOC	(V0 $*_H$ (V1 $*_H$ V2)) = ((V0 $*_H$ V1) $*_H$ V2)	246
REAL_ADD_ASSOC	(V0 $+_R$ (V1 $+_R$ V2)) = ((V0 $+_R$ V1) $+_R$ V2)	246
REAL_MUL_ASSOC	(V0 $*_R$ (V1 $*_R$ V2)) = ((V0 $*_R$ V1) $*_R$ V2)	246
REAL_MAX_ASSOC	(MAX$_R$ V0 (MAX$_R$ V1 V2)) = (MAX$_R$ (MAX$_R$ V0 V1) V2)	246
INT_ADD_ASSOC	(V0 $+_Z$ (V1 $+_Z$ V2)) = ((V0 $+_Z$ V1) $+_Z$ V2)	246
REAL_MIN_ASSOC	(MIN$_R$ V0 (MIN$_R$ V1 V2)) = (MIN$_R$ (MIN$_R$ V0 V1) V2)	246

From this statement, features are extracted, computing their frequency, average position and then IDF across the statement. A total of 46 features are extracted; Table 1 shows a selection ordered by rarity (IDF). The highest IDF value is for the feature most specific to the overall statement: it captures associativity of +.

Next, Table 2 shows the nearest neighbours for the features of ADD_ASSOC among the HOL Light named theorems, discounting ADD_ASSOC itself. Most of these are associativity statements. The stem AC_ is commonly used in HOL to denote associative-commutative properties.

Finally, the first predicted stems with their predicted positions are presented in Table 3. With ASSOC and ADD being the first two suggested stems, taking into account their positions, ADD_ASSOC is indeed the top prediction for the theorem name. The following predicitons are reasonable too:

AC_ADD_ASSOC, AC_NUM_ADD_ASSOC, AC_ADD, NUM_ADD_ASSOC.

Table 3. Stems and positions suggested by our algorithm, sorted by relevance measure.

Stem	Positions	Stem	Positions
ADD	[0.18; 0.14; 0.12; 0; 0.12]	MONO	[0.44; 0.49]
ASSOC	[1]	REFL	[1]
AC	[0; 1]	SELECT	[0]
NUM	[0.67; 0.45; 0; 0.70]	SPLITS	[0]
FIXED	[0]	RCANCEL	[1]
EQUAL	[0.25; 0.25]	IITN	[0]
ONE	[0]	GEN	[1]

Unsurprisingly, the consistent scheme performs less well in situations where new defined objects appear in a statement, it can only suggest stems it has seen before. The abstract scheme addresses this. For this, we first gather all symbol names in the training examples and order them by decreasing frequency. Next, for every training example we find the non-logical objects that appear, and replace their occurrences by object placeholders, with the constants numbered in the order of their global frequencies. For example, a name like DIV_EQ_0 becomes C_0_EQ_C_3, and the theorem statement is abstracted similarly.

For the statement of the theorem ADD_ASSOC the first three names predicted by the abstract naming scheme are: +_ASSOC, num_+, and num_+_ASSOC. The use of the stem ADD is only predicted with $k > 20$.

4 Preliminary Evaluation

We perform a standard leave-one-out cross-validation to evaluate how good names predicted on a single dataset are. The predictor is trained on all the examples apart from the current one to evaluate, and is tested on the features of the current one.

For 2298 statements in the HOL Light core library, the results are presented in Table 4. We explored four different options for the algorithm:

- Upper: names are canonicalised to upper case.
- AbsN: we use the abstraction scheme described above for naming.
- AbsT: we use abstraction in statements before training.
- Stem: we use a stemming operation to break down names.

The first option is useful in the libraries of HOL Light and HOL4 where the capitalization is mostly uniform, but may not be desired in proof assistant libraries where this is not the case. For example min and Min are used with different semantics in Isabelle/HOL.

The last option allows us to model the naming convention in HOL Light where a statement is relative to a type, but the type name does not get repeated. For example, a theorem that relates the constants REAL_ABS and REAL_NEG can be given the name REAL_ABS_NEG rather than REAL_ABS_REAL_NEG.

Each row in Table 4 is a combination of options. Results are split in the columns: the number of statements for which the top prediction is the same as the human-given name (First Choice); the number where the human name is in place 2–10 (Later Choice); one of the ten names is correct modulo stem order (Same Stems); the number where the human-used stems are predicted but not combined correctly (All Stems); and the number for which at least part of the prediciton is correct (Stem Overlap). Altogether, the human-generated names are among the top ten predictions by some instance of the algorithm in over 50 % of cases; 40 % for the best performing version based on the abstract scheme. The abstract scheme beats the consistent mechanism even in the same library.

The final column shows the number of cases that fail completely. These include cases with familiar (historical) names such as:

– EXCLUDED_MIDDLE (proposed reasonable name: DISJ_THM)
– INFINITY_AX (proposed reasonable name: ONTO_ONE).

We would not expect to predict these names, unless they are already given in the training data. In other cases, failure might indicate inconsistency: the human names may not always be "correct". We uncovered some cases of inconsistency such as where multiplication was occasionally called MULT rather than MUL, for example.

Table 4. Leave-one-out cross-validation on the HOL Light core dataset.

Setup Options	First Choice	Later Choice	Same Stems	All Stems	Stem Overlap	Fail Fail
-	118	533	180	911	516	40
Upper	134	583	182	901	474	24
AbsN	187	517	222	849	474	49
AbsN+Stem	218	530	208	849	460	33
AbsN+Upper	203	461	250	888	478	18
AbsN+AbsT	172	243	125	919	755	84
AbsN+AbsT+Upper	206	387	211	771	680	43
AbsN+AbsT+Stem	214	455	178	881	532	38
AbsN+Stem+Upper	238	491	299	835	418	17
AbsN+AbsT+Stem+Upper	273	501	291	757	459	17
Combined	336	728	271	632	321	10

5   Conclusions

The initial results are encouraging and suggest foundations for name-recommender systems that might be built into ITP interfaces. Even if the perfect name is not proposed, suggestions may spark an idea for the user. We plan to go further and look at case studies such as renaming Flyspeck, or using naming maps as a bridge between different ITPs. Moreover, more advanced machine learning schemes could be used to distinguish the use of the same symbol for different operators.

Related work. This appears to be the first attempt to mine named theorems and produce a recommender system for naming new theorems in a *meaningful* way. There have been a number of non-meaningful proposals, e.g., Flyspeck's random 8-character identifiers [4]; Mizar's naming scheme of theory names and numbers (examples like WAYBEL34:67 [3]); the use of MD5 recursive statement hashes [8].

Identifier naming has been studied in software engineering. Lawrie et al. [7] investigated name consistency in large developments. Deissenboeck and Pizka [1] build a recommender similar to our consistent scheme (but using different methods). They note that programming style guides say identifiers should be "self-describing" to aid comprehension. Indeed, program obfuscators, intended to *hinder* comprehension, randomize identifiers, producing names like those in Flyspeck.

Acknowledgments. This work has been supported by UK EPSRC (EP/J001058/1) and the Austrian Science Fund FWF (P26201).

References

1. Deissenboeck, F., Pizka, M.: Concise and consistent naming. Softw. Q. J. **14**(3), 261–282 (2006)
2. Gauthier, T., Kaliszyk, C.: Matching concepts across HOL libraries. In: Watt, S.M., Davenport, J.H., Sexton, A.P., Sojka, P., Urban, J. (eds.) CICM 2014. LNCS, vol. 8543, pp. 267–281. Springer, Heidelberg (2014)
3. Grabowski, A., Korniłowicz, A., Naumowicz, A.: Four decades of Mizar. J. Autom. Reason. **55**(3), 191–198 (2015)
4. Hales, T., et al.: A formal proof of the Kepler conjecture. CoRR, 1501.02155 (2015)
5. Kaliszyk, C., Urban, J., Vyskočil, J.: Efficient semantic features for automated reasoning over large theories. In: IJCAI 2015, pp. 3084–3090. AAAI Press (2015)
6. Kripke, S.A.: Naming and necessity. In: Semantics of Natural Language, pp. 253–355. Springer, Netherlands, Dordrecht (1972)
7. Lawrie, D., Morrell, C., Feild, H., Binkley, D.: What's in a name? A study of identifiers. In: ICPC 2006, pp. 3–12. IEEE (2006)
8. Urban, J.: Content-based encoding of mathematical and code libraries. In: MathWikis 2011, CEUR-WS, vol. 767, pp. 49–53 (2011)

Cardinalities of Finite Relations in Coq

Paul Brunet[1], Damien Pous[1], and Insa Stucke[2]($^{(\boxtimes)}$)

[1] Univ Lyon, CNRS, ENS de Lyon, UCB Lyon 1, LIP, Lyon, France
[2] Institut für Informatik, Christian-Albrechts-Universität zu Kiel, Kiel, Germany
ist@informatik.uni-kiel.de

Abstract. We present an extension of a Coq library for relation algebras, where we provide support for cardinals in a point-free way. This makes it possible to reason purely algebraically, which is well-suited for mechanisation. We discuss several applications in the area of graph theory and program verification.

1 Introduction

Binary relations have a rich algebraic structure: rather than considering relations as objects relating points, one can see them as abstract objects that can be combined using various operations (e.g., union, intersection, composition, transposition). Those operations are subject to many laws (e.g., associativity, distributivity). One can thus use equational reasoning to prove results about binary relations, graphs, or programs manipulating such structures. This is the so-called relation-algebraic method [12,14,15].

Lately, the second author developed a library for the Coq proof assistant [9,10], allowing one to formalise proofs using the relation algebraic approach. This library contains powerful automation tactics for some decidable fragments of relation algebra (Kleene algebra and Kleene algebra with tests), normalisation tactics, and tools for rewriting modulo associativity of relational composition.

The third author recently relied on this library to formalise algebraic correctness proofs for several standard algorithms from graph theory: computing vertex colourings [1] and bipartitions [2].

Here we show how to extend this library to deal with cardinals of relations, thus allowing one to reason about quantitative aspects. We study several applications in [3]; in this extended abstract we focus on a basic result about the size of a linear order and an intermediate result from graph theory.

2 Preliminaries

Given two sets X, Y, a *binary relation* is a subset $R \in \mathcal{P}(X \times Y)$. The set X (resp. Y) is called the *domain* (resp. *codomain*) of the relation.

This work was supported by the project ANR 12IS02001 PACE.

D. Pous—This author is supported by the European Research Council (ERC) under the European Union's Horizon 2020 programme (CoVeCe, grant agreement No 678157).

J.C. Blanchette and S. Merz (Eds.): ITP 2016, LNCS 9807, pp. 466–474, 2016.
DOI: 10.1007/978-3-319-43144-4_29

With the usual set-theoretic operations of inclusion ($\subseteq$), union ($\cup$), intersection ($\cap$), complement ($\bar{\cdot}$), the empty relation (O_{XY}) and the universal relation (L_{XY}), binary relations between two sets X and Y form a Boolean lattice. Given three sets X, Y, Z and relations $R \in \mathcal{P}(X \times Y)$ and $S \in \mathcal{P}(Y \times Z)$ we also consider the operations of *composition* ($RS \in \mathcal{P}(X \times Z)$) and *transposition* ($R^{\mathsf{T}} \in \mathcal{P}(Y \times X)$), as well as the *identity relation* ($I_X \triangleq \{(x, x) \mid x \in X\} \in \mathcal{P}(X \times X)$). These operations can be abstracted through the axiomatic notion of *relation algebra*. Binary relations being the standard model of such an algebra, we use the same notations.

Definition 2.1 (Relation Algebra). *A relation algebra is a category whose homsets are Boolean lattices, together with an operation of transposition ($\cdot^{\mathsf{T}}$) such that:*

(P_1) *composition is monotone in its two arguments, distributes over unions and is absorbed by the bottom elements;*

(P_2) *transposition is monotone, involutive ($R^{\mathsf{T}^{\mathsf{T}}} = R$), and reverses compositions: for all morphisms R, S of appropriate types, we have $(RS)^{\mathsf{T}} = S^{\mathsf{T}}R^{\mathsf{T}}$;*

(P_3) *for all morphisms Q, R, S of appropriate types, $QR \subseteq S$ iff $Q^{\mathsf{T}}\bar{S} \subseteq \bar{R}$ iff $\bar{S}R^{\mathsf{T}} \subseteq \bar{Q}$;*

(P_4) *for all morphism $R : X \to Y$, $R \neq O$ iff for all objects X', Y', $LRL = L_{X'Y'}$.*

From properties (P_2), we deduce that transposition commutes with all Boolean connectives, and that $I^{\mathsf{T}} = I$. Equivalences (P_3) are called *Schröder equivalences* in [12]; they correspond to the fact that the structure is residuated [5]. The last property (P_4) is known as *Tarski's rule*; it makes it possible to reason algebraically about non-emptiness.

Important classes of morphisms can be defined algebraically. For instance, we say in the sequel that a morphism $R : X \to Y$ is:

- *injective* if $RR^{\mathsf{T}} \subseteq I$,
- *surjective* if $I \subseteq R^{\mathsf{T}}R$,
- *univalent* if its transpose is injective (i.e., $R^{\mathsf{T}}R \subseteq I$),
- *total* if its transpose is surjective (i.e., $I \subseteq RR^{\mathsf{T}}$),
- a *mapping* if R is total and univalent.

One can easily check that these definitions correspond to the standard definitions in the model of binary relations.

Before introducing cardinals, we need a way to abstract over the singleton sets from the model of binary relations; we use the following definition:

Definition 2.2 (Unit in a Relation Algebra). *A unit in a relation algebra is an object 1 such that $O_{11} \neq L_{11}$ and $I_1 = L_{11}$.*

In other words, there are only two morphisms from a unit to itself. In the model of binary relations, every singleton set is a unit. Using units, we can axiomatise the notion of cardinal in a relation algebra; we mainly follow Kawahara [8]:

Definition 2.3 (Cardinal). *A relation algebra with* cardinal *is a relation algebra with a unit* 1 *and a monotone function* $|\cdot|$ *from morphisms to natural numbers such that for all morphisms* Q, R, S *of appropriate types:*

(C_1) $|\mathsf{O}| = 0$,
(C_2) $|\mathsf{I}_1| = 1$,
(C_3) $|R^{\mathsf{T}}| = |R|$,
(C_4) $|R \cup S| + |R \cap S| = |R| + |S|$,
(C_5) *if* Q *is univalent, then* $|R \cap Q^{\mathsf{T}}S| \leq |QR \cap S|$ *and* $|Q \cap SR^{\mathsf{T}}| \leq |QR \cap S|$.

Note that these requirements for a cardinal rule out infinite binary relations: we have to restrict to binary relations between finite sets, i.e., graphs. Typically, in this model, the cardinal of a relation is the number of pairs it contains. This restriction is harmless in practice: we only work with finite sets when we study, for example, algorithms.

Many natural facts of cardinal can be derived just from conditions (C_1) to (C_4), e.g., monotonicity. The last condition (C_5) is less intuitive; it is called the *Dedekind inequality* in [8]. It allows one to compare cardinalities of morphisms of different types. Kawahara uses it to obtain, e.g., the following result:

Lemma 2.4. *Assume a relation algebra with cardinal. For all morphisms* Q, R, S *of appropriate type, we have:*

1. *If* R *and* S *are univalent, then* $|RS \cap Q| = |R \cap QS^{\mathsf{T}}|$.
2. *If* R *is univalent and* S *is a mapping, then* $|RS| = |R|$.

Leaving cardinals aside, two important classes of morphisms are that of vectors and points, as introduced in [11], for providing a way to model subsets and single elements of sets, respectively:

– *vectors*, denoted with lower case letters v, w in the sequel, are morphisms $v : X \to Y$ such that $v = v\mathsf{L}$. In the standard model, this condition precisely amounts to being of the special shape $V \times Y$ for a subset $V \subseteq X$.
– *points*, denoted with lower case letters p, q in the sequel, are injective and nonempty vectors. In the standard model, this condition precisely amounts to being of the special shape $\{x\} \times Y$ for an element $x \in X$.

In the binary relations model, one can characterise vectors and points from their Boolean-matrix representation of binary relations: a vector is a matrix whose rows are either zero everywhere or one everywhere, and a point is a matrix with a single row of ones and zeros everywhere else. Every morphism with unit as its codomain is a vector; points with unit as their codomain have cardinal one:

Lemma 2.5. *Let* $p : X \to 1$ *be a point in a relation algebra with a cardinal (and unit). We have* $|p| = 1$.

We conclude this preliminary section with the notion of pointed relation algebra. Indeed, in the model of binary relations, the universal relation between X and Y is the least upper bound of all points between X and Y. This property is called the *point axiom* in [4]. Since we restrict to finite relations, we give a finitary presentation of this law.

Definition 2.6. (Pointed Relation Algebra). *A relation algebra is* pointed *if for all* X, Y *there exists a (finite) set* P_{XY} *of points such that* $\mathsf{L}_{XY} = \bigcup_{p \in P_{XY}} p$.

As a consequence, in pointed relation algebras it holds $\mathsf{I}_X = \bigcup_{p \in P_{XX}} pp^\mathsf{T}$. When working in pointed relation algebras with cardinal, we also have results like the following, where we use $|X|$ as a shorthand notation for $|\mathsf{L}_{X1}|$:

Lemma 2.7. *For all objects* X *and* Y *we have* $|\mathsf{L}_{XY}| = |X| \cdot |Y|$ *and* $|\mathsf{I}_X| = |X|$.

Any pointed relation algebra with cardinal is in fact isomorphic to an algebra of relations on finite sets; therefore, the above list of axioms can be seen as a convenient list of facts about binary relations which make it possible to reason algebraically. Still, our modular presentation of the theory makes it possible to work in fragments of it where this representation theorem breaks, i.e., for which other models exist than that of binary relations.

3   Relation Algebra in Coq

The Coq library RelationAlgebra [9,10] provides axiomatisations and tools for various fragments of the calculus of relations: from ordered monoids to Kleene algebra, residuated structures, and Dedekind Categories. It is structured in a modular way: one can easily decide which operations and axioms to include.

In the present case, these are Boolean operations and constants, composition, identities, transposition. We extended the library by a module `relalg` containing definitions and facts about this particular fragment. For instance, this module defines many classes of relations, some of which we already mentioned in Sect. 2. For those properties we use classes in Coq:

`Class is_vector (C: ops) X Y (v: C X Y) := vector: v*top == v.`

Here we assume an ambient relation algebra `C`, `ops` being the corresponding notion, as exported by the RelationAlgebra library. Variables `X,Y` are objects of the category, and `v: C X Y` is a morphism from `X` to `Y`. The symbols `*` and `==` respectively denote composition and equality; `top` is the top morphism of appropriate type: its source and target (`Y` twice) are inferred automatically.

The RelationAlgebra library provides several automation tactics to ease equational reasoning [9,10]. The most important ones are:

- `ra_normalise` for normalising the current goal w.r.t. the simplest laws (mostly about idempotent semirings, units and transposition),
- `ra` for solving goals by normalisation and comparison,
- `lattice` for solving lattice-theoretic goals,
- `mrewrite` for rewriting modulo associativity of categorical composition.

The library also contains a decision procedure for Kleene algebra with tests, which we do not discuss here for lack of space. Those tactics are defined either by reflection, where a decision procedure is certified within Coq (`ra_normalise`, `ra`); by exhaustive proof search (`lattice`); or as ad hoc technical solutions (`mrewrite`, which is a plugin in OCaml that applies appropriate lemmas to reorder parentheses and generalise the considered (in)equation.

A crucial aspect for this work is the interplay between the definitions from this library and Coq's support for setoid rewriting [13], which makes it possible to rewrite using both equations and inequations in a streamlined way, once the monotonicity or anti-monotonicity of all operations has been proved.

This is why we use a class to define the above predicate is_vector: in this case, the tactic `rewrite vector` will look for a subterm of a shape v*top where v is provably a vector using typeclass resolution, and replace it with v. Similar classes are set-up for all notions discussed in the sequel (injective, surjective, univalent, total, mapping, points, and many more).

We also define classes to represent relation algebra with unit, relation algebra with cardinal, and pointed relation algebra. Units are introduced as follows:

```
Class united (C: ops) := {
  unit: ob C;
  top_unit: top' unit unit == 1;
  nonempty_unit:> is_nonempty (top' unit unit) }.
```

The field `unit` is the unit object; the two subsequent fields correspond to the requirements from Definition 2.2. The symbol 1 is our notation for identity morphisms. Assuming units, one can then define cardinals:

```
Class cardinal (C: ops) (U: united C) := {
  card: forall X Y, C X Y → nat;
  card0: forall X Y, @card X Y 0 = 0;
  card1: @card unit unit 1 = 1;
  cardcnv: forall X Y (R: C X Y), card Rᵀ = card R;
  cardcup: forall X Y (R S: C X Y), card (R ∪ S) + card (R ∩ S) = card R + card S;
  cardded: forall X Y Z (R: C X Y) (S: C Y Z) (T: C X Z),
      is_injective R → card (T ∩ (R*S)) ≤ card (Rᵀ * T ∩ S);
  cardded': forall X Y Z (R: C Y X) (S: C Y Z) (T: C Z X),
      is_univalent R → card (R ∩ (S*T)) ≤ card (R * Tᵀ ∩ S) }.
```

The first field is the cardinal operation itself. The remaining ones correspond to the conditions from Definition 2.3.

Next we give two Coq proofs about cardinals, to show the ease with which it is possible to reason about them. The first one correspond to Lemma 2.4(2).

```
Lemma card_unimap X Y Z (R: C X Y) (S: C Y Z):
      is_univalent R → is_mapping S → card (R*S) = card R.
Proof. rewrite ←capxt, card_uniuni, surjective_tx. apply card_weq. ra. Qed.
```

Here, Lemma `uniuni` corresponds to Lemma 2.4(1); `capxt` states that top is a unit for meet; `surjective_tx` that every surjective morphism R satisfies $\mathsf{L}R = \mathsf{L}$; and `card_weq` that cardinals are preserved by equality.

The second illustrative proof is that of Lemma 2.5, which becomes a oneliner:

```
Lemma card_point X (R: C X unit): is_point R → card R = 1.
Proof. rewrite ←cardcnv, ←dot1x. rewrite card_unimap. apply card1. Qed.
```

(Lemma `dot1x` states that I is a left unit for composition.)

4   Applications

We first detail an easy example where we link the cardinality of morphisms representing linear orders to the cardinality of their carrier sets. The second example is based on a graph theoretic result giving a lower bound for the cardinality of an independent set.

4.1   Linear Orders

A morphism $R : X \to X$ is a *partial order* on X if R is *reflexive, antisymmetric* and *transitive* (i.e., $I \subseteq R$, $R \cap R^\mathsf{T} \subseteq I$ and $RR \subseteq R$). If R is additionally *linear* (i.e., $R \cup R^\mathsf{T} = \mathsf{L}$) we call R a *linear order*. Recall that for an object X, $|X|$ is a shorthand for $|\mathsf{L}_{X1}|$. We have

Theorem 4.1. *If $R : X \to X$ is a linear order, then $|R| = \frac{|X|^2 + |X|}{2}$.*

Proof. Since R is antisymmetric we have $R \cap R^\mathsf{T} \subseteq I$. Furthermore, we have $I \subseteq R$ since R is reflexive so that $R \cap R^\mathsf{T} = I$. Now we can calculate as follows:

$$
\begin{aligned}
|X|^2 + |X| &= |\mathsf{L}_{XX}| + |\mathsf{I}_X| && \text{(by Lemma 2.7)} \\
&= |R \cup R^\mathsf{T}| + |\mathsf{I}_X| && \text{(R linear)} \\
&= |R \cup R^\mathsf{T}| + |R \cap R^\mathsf{T}| && \text{(R reflexive and antisymmetric)} \\
&= |R| + |R^\mathsf{T}| && \text{(by (C_4))} \\
&= |R| + |R| && \text{(by (C_3))}
\end{aligned}
$$

$\square$

With the presented tools, this lemma can be proved in Coq in a very same way. First we need to define a notation for the cardinal of an object:

```
Notation card' X := card (top' X unit).
Lemma card_linear_order X (R: C X X): is_order R → is_linear R →
  2*card R = card' X * card' X + card' X.
Proof.
  intros Ho Hli.
  rewrite ←card_top, ←card_one.
  rewrite ←Hli.
  rewrite ←kernel_refl_antisym.
  rewrite capC, cardcup.
  rewrite cardcnv. lia.
Qed.
```

The standard Coq tactic `lia` solves linear integer arithmetic. The lemmas `card_top` and `card_one` correspond to the statements of Lemma 2.7, i.e.,

```
Lemma card_top X Y: card (top' X Y) = card' X * card' Y.
Lemma card_one X: card (one X) = card' X.
```

Lemma `kernel_refl_antisym` states that the kernel of a reflexive and antisymmetric morphism is just the identity.

4.2   Independence Number of a Graph

In this section we prove bounds for the *independence number* of an undirected graph [16]. An *undirected (loopfree) graph* $g = (X, E)$ has a symmetric and irreflexive adjacency relation. It can thus be represented by a morphism $R :$ $X \to X$ that is *symmetric* (i.e., $R^\mathsf{T} \subseteq R$) and *irreflexive* (i.e., $R \cap I = 0$).

An *independent set* (or *stable set*) of g is a set of vertices S such that any two vertices in S are not connected by an edge, i.e., $\{x, y\} \notin E$, for all $x, y \in S$. Independent sets can be modelled abstractly using vectors: a vector $s : X \to 1$ models an independent set of a morphism R if $Rs \subseteq \bar{s}$. Furthermore, we say that an independent set S of g is *maximum* if for every independent set T of g we have $|T| \leq |S|$. The maximum size of an independent set is defined as:

$$\alpha_R \triangleq \max \{|s| \mid s \text{ is an independent set of } R\} \ .$$

One easily obtain the lower bound $\alpha_R \leq \sqrt{|\overline{R}|}$. In fact, we have $|s| \leq \sqrt{|\overline{R}|}$ for every independent set s, which we can prove in two lines using our library.

The upper bound is harder to obtain. We have $\frac{|R|}{k+1} \leq \alpha_R$, where k is the maximum degree of R. Call *maximal* an independent set which cannot be enlarged w.r.t. the preorder $\subseteq$:

```
Definition maximal (v: C X unit) := forall w, v <== w → R * w <== !w → w <== v.
```

As expected, maximum independent sets are maximal:

```
Lemma maximum_maximal (v: C X unit):
  R*v <== !v → card v = independent_number R → maximal v.
```

(Note that the converse is not necessarily true.) Then we prove the following algebraic characterisation of maximal independent sets: while independent sets are characterised by an inequality ($Rv \subseteq \bar{v}$), maximal are characterised by an equality ($Rv = \bar{v}$).

```
Lemma maximal_independent_iff (v: C X unit):
  R*v <== !v → (maximal v ↔ R*v == !v).
```

Finally, obtaining the lower bound for the independence number consists in proving that maximal independent sets, defined algebraically, satisfy this bound:

```
Lemma maximal_lower_bound (v: C X unit):
  R*v == !v → card' X ≤ (maximum_degree R + 1) * card v.
Theorem independent_lower_bound:
  card' X <== (maximum_degree R + 1) * independent_number R.
```

Including the proofs of the three key lemmas, the final theorem is eventually proved in 41 lines of Coq. We consider this a success as this is comparable to what is required for a detailed paper proof.

5   Conclusion

We presented an extension of the Coq RelationAlgebra library [3], that makes it possible to reason algebraically about cardinalities of binary relations. A key feature of the Coq proof assistant for this work is *dependent types*: they allow us to define relation algebras as categories in a straightforward way, so that we can talk about vectors or units as one would do on paper. While our approach to cardinals would certainly work when starting from Kahl's implementation of allegories in Agda [7], it remains unclear to us whether it could be adapted to his formalisation of relation algebra in Isabelle/Isar [6].

References

1. Berghammer, R., Höfner, P., Stucke, I.: Tool-based verification of a relational vertex coloring program. In: Kahl, W., Winter, M., Oliveira, J. (eds.) RAMiCS 2015. LNCS, vol. 9348, pp. 275–292. Springer, Heidelberg (2015). doi:10.1007/978-3-319-24704-5_17
2. Berghammer, R., Stucke, I., Winter, M.: Investigating and computing bipartitions with algebraic means. In: Kahl, W., Winter, M., Oliveira, J. (eds.) RAMiCS 2015. LNCS, vol. 9348, pp. 257–274. Springer, Heidelberg (2015). doi:10.1007/978-3-319-24704-5_16
3. Brunet, P., Pous, D., Stucke, I.: Cardinalities of relations in Coq. Coq Development and full version of this extended abstract (2016). http://media.informatik.uni-kiel.de/cardinal/
4. Furusawa, H.: Algebraic formalisations of fuzzy relations and their representation theorems. Ph.D. thesis, Department of Informatics, Kyushu University (1998)
5. Galatos, N., Jipsen, P., Kowalski, T., Ono, H., Lattices, R.: An Algebraic Glimpse at Substructural Logics. Elsevier, Oxford (2007)
6. Kahl, W.: Calculational relation-algebraic proofs in Isabelle/Isar. In: Berghammer, R., Möller, B., Struth, G. (eds.) RelMiCS 2003. LNCS, vol. 3051, pp. 178–190. Springer, Heidelberg (2004)
7. Kahl, W.: Dependently-typed formalisation of relation-algebraic abstractions. In: de Swart, H. (ed.) RAMICS 2011. LNCS, vol. 6663, pp. 230–247. Springer, Heidelberg (2011)
8. Kawahara, Y.: On the cardinality of relations. In: Schmidt, R.A. (ed.) RelMiCS/AKA 2006. LNCS, vol. 4136, pp. 251–265. Springer, Heidelberg (2006)
9. Pous, D.: Relation Algebra and KAT in Coq. http://perso.ens-lyon.fr/damien.pous/ra/
10. Pous, D.: Kleene algebra with tests and coq tools for while programs. In: Blazy, S., Paulin-Mohring, C., Pichardie, D. (eds.) ITP 2013. LNCS, vol. 7998, pp. 180–196. Springer, Heidelberg (2013)
11. Schmidt, G., Ströhlein, T.: Relation algebras: concept of points and representability. Discrete Math. **54**(1), 83–92 (1985)
12. Schmidt, G., Ströhlein, T.: Relations and Graphs - Discrete Mathematics for Computer Scientists. EATCS Monographs on Theoretical Computer Science. Springer, Berlin (1993)
13. Sozeau, M.: A new look at generalized rewriting in type theory. J. Formalized Reason. **2**(1), 41–62 (2009)

14. Tarski, A.: On the calculus of relations. J. Symbolic Log. **6**(3), 73–89 (1941)
15. Tarski, A., Givant, S.: A Formalization of Set Theory without Variables, vol. 41. Colloquium Publications, AMS, Providence, Rhode Island (1987)
16. Wei, V.: A lower bound for the stability number of a simple graph. Bell Laboratories Technical Memorandum 81-11217-9 (1981)

Formalising Semantics for Expected Running Time of Probabilistic Programs

Johannes Hölzl[(⊠)]

Fakultät für Informatik, TU München, Munich, Germany
hoelzl@in.tum.de

Abstract. We formalise two semantics observing the expected running time of pGCL programs. The first semantics is a denotational semantics providing a direct computation of the running time, similar to the weakest pre-expectation transformer. The second semantics interprets a pGCL program in terms of a Markov decision process (MDPs), i.e. it provides an operational semantics. Finally we show the equivalence of both running time semantics.

We want to use this work to implement a program logic in Isabelle/HOL to verify the expected running time of pGCL programs. We base it on recent work by Kaminski, Katoen, Matheja, and Olmedo. We also formalise the expected running time for a simple symmetric random walk discovering a flaw in the original proof.

1 Introduction

We want to implement expected running time analysis in Isabelle/HOL based on Kaminski *et al.* [9]. They present semantics and proof rules to analyse the expected running time of probabilistic guarded command language (pGCL) programs. pGCL is an interesting programming language as it admits probabilistic and non-deterministic choice, as well as unbounded while loops [12].

Following [9], in Sect. 3 we formalise two running time semantics for pGCL and show their equivalence: a denotational one expressed as expectation transformer of type $(\sigma \Rightarrow \mathsf{ennreal}) \Rightarrow (\sigma \Rightarrow \mathsf{ennreal})$, and a operational one defining a Markov decision process (MDP). This proof follows the equivalence proof of pGCL semantics on the expectation of program variables in [4] derived from the pen-and-paper proof by Gretz *et al.* [3].

Based on these formalisations we analyse the simple symmetric random walk, and show that the expected running time is infinite. We started with the proof provided in [9], but we discovered a flaw in the proof of the lower ω-invariant based on the denotational semantics. Now, our solution combines results from the probability measure of the operational semantics and the fixed point solution from the denotational semantics.

Both proofs are based on our formalisation of Markov chains and MDPs [4]. The formalisation in this paper is on BitBucket[1].

---

[1] https://bitbucket.org/johannes2011/avgrun.

© Springer International Publishing Switzerland 2016
J.C. Blanchette and S. Merz (Eds.): ITP 2016, LNCS 9807, pp. 475–482, 2016.
DOI: 10.1007/978-3-319-43144-4_30

2 Preliminaries

The formulas in this paper are oriented on Isabelle's syntax: type annotations are written $t :: \tau$, type variables can be annotated with type classes $t :: \tau :: tc$ (i.e. t has type τ which is in type class tc), and type constructors are written in post-fix notation: e.g. α set. We write int for integers, ennreal for extended non-negative real numbers $[0, \infty]$, α stream for infinite streams of α, α pmf for probability mass functions (i.e. discrete distributions) on α. The state space is usually the type variable σ. On infinite streams sdrop $n\ \omega$ drops the first n elements from the stream ω: sdrop $0\ \omega = \omega$ and sdrop $(n+1)\ (s\cdot\omega) =$ sdrop $n\ \omega$.

Least Fixed Points. A central tool to define semantics are least fixed points on complete lattices: $\alpha \Rightarrow (\beta :: \text{complete-lattice})$, bool, enat, and ennreal. Least fixed points are defined as lfp $f = \bigsqcap\{u \mid f\ u \leq u\}$. For a monotone function f, we get the equations lfp $f = f\,(\text{lfp}\, f)$. Fixed point theory also gives nice algebraic rules: the rolling rule "rolls" a composed fixed point: $g\,(\text{lfp}\,(\lambda x.\ f\,(g\,x))) = \text{lfp}\,(\lambda x.\ g\,(f\,x))$ for monotone f and g, and the diagonal rule for nested fixed points: lfp $(\lambda x.\ \text{lfp}\,(f\,x)) = \text{lfp}\,(\lambda x.\ f\,x\,x)$, for f monotone in both arguments.

To use least fixed points in measure theory, countable approximations are necessary. This is possible if the function f is sup-continuous: $f\,(\bigsqcup_i C\,i) = \bigsqcup_i f\,(C\,i)$ for all chains C. Then f is monotone and lfp $f = \bigsqcup_i f^i \perp$. For our proofs we also need an induction and a transfer rule[2]:

$$\frac{\text{mono}\,f \qquad \forall x \leq \text{lfp}\,f.\ P\,x \longrightarrow P\,(f\,x) \qquad \forall S.\ (\forall x \in S.\ P\,x) \longrightarrow P\,(\bigsqcup S)}{P\,(\text{lfp}\,f)}$$

$$\frac{\text{sup-continuous}\,f, g, \text{and}\ \alpha \qquad \alpha\,\perp\ \leq \text{lfp}\,g \qquad \alpha \circ f = g \circ \alpha}{\alpha(\text{lfp}\,f) = \text{lfp}\,g}$$

Markov Chains (MCs) and Markov Decision Processes (MDPs). An overview of Isabelle's MC and MDP theory is found in [4,5]. A MC is defined by a transition function $K :: \alpha \Rightarrow \alpha$ pmf, inducing an expectation: $\mathbb{E}_s^K[f]$ is the expectation of f over all traces in K starting in s. A MDP is defined by a transition function $K :: \alpha \Rightarrow \alpha$ pmf set, inducing the maximal expectation: $\hat{\mathbb{E}}_s^K[f]$ is the supremum of all expectation of f over all traces in K starting in s. Both expectations $\mathbb{E}_s^K[f]$ and $\hat{\mathbb{E}}_s^K[f]$ have values in ennreal, which is a complete lattice. Both are sup-continuous on measurable functions (called *monotone convergent* in measure theory), which allows us to apply the transfer rule when f is defined as a least fixed point. Also both expectations support an iteration rule, i.e. we can compute them by first taking a step in K and then continue in the resulting state t:

$$\mathbb{E}_s^K[f] = \int_t \mathbb{E}_t^K[\lambda\omega.\ f(t \cdot \omega)]dK_s \quad \text{and} \quad \hat{\mathbb{E}}_s^K[f] = \bigsqcup_{D \in K_s} \int_t \hat{\mathbb{E}}_t^K[\lambda\omega.\ f(t \cdot \omega)]dD.$$

---

[2] In our formalisation, the transfer rule is stronger: expectation requires measurability, hence we restrict the elements to which we apply α by some predicate P.

$$\sigma \text{ pgcl} = \text{Empty} \quad | \quad \text{Skip} \quad | \quad \text{Halt} \quad | \text{Assign } (\sigma \Rightarrow \sigma \text{ pmf})$$
$$| \text{ Seq } (\sigma \text{ pgcl}) (\sigma \text{ pgcl}) \qquad\qquad | \text{ Par } (\sigma \text{ pgcl}) (\sigma \text{ pgcl})$$
$$| \text{ If } (\sigma \Rightarrow \text{bool}) (\sigma \text{ pgcl}) (\sigma \text{ pgcl}) | \text{While } (\sigma \Rightarrow \text{bool}) (\sigma \text{ pgcl})$$

Fig. 1. pGCL syntax

$$\text{ert} :: \sigma \text{ pgcl} \Rightarrow (\sigma \Rightarrow \text{ennreal}) \Rightarrow (\sigma \Rightarrow \text{ennreal})$$
$$\text{ert Empty} \quad f = f$$
$$\text{ert Skip} \quad\ f = 1 + f$$
$$\text{ert Halt} \quad\ f = 0$$
$$\text{ert (Assign } u)\ \ f = 1 + \lambda x. \int_y f\, y\, d(u\, x)$$
$$\text{ert (Seq } c_1\ c_2)\ f = \text{ert } c_1\ (\text{ert } c_2\ f)$$
$$\text{ert (Par } c_1\ c_2)\ f = \text{ert } c_1\ f \sqcup \text{ert } c_2\ f$$
$$\text{ert (If } g\ c_1\ c_2)\ f = 1 + \lambda x.\ \text{if } g\ x \text{ then ert } c_1\ f\ x \text{ else ert } c_2\ f\ x$$
$$\text{ert (While } g\ c)\ f = \text{lfp } (\lambda W\ x.\ 1 + \text{if } g\ x \text{ then ert } c\ W\ x \text{ else } f\ x)$$

Fig. 2. Expectation transformer semantics for pGCL running times

Where $t \cdot \omega$ is the stream constructor and $\int f dD$ is the integral over the pmf D.

3 Probabilistic Guarded Command Language (pGCL)

The probabilistic guarded command language (pGCL) is a simple programming language allowing probabilistic assignment, non-deterministic choice and arbitrary While-loops. A thorough description of it using the weakest pre-expectation transformer (wp) semantics is found in McIver and Morgan [12]. Gretz *et al.* [3] shows the equivalence of wp with a operational semantics based on MDPs. Hurd et al. [8] and Cock [2] provide a shallow embedding of pGCL in HOL4 and Isabelle/HOL. We follow the definition in Kaminski *et al.* [9].

In Fig. 1 we define a datatype representing pGCL programs over an arbitrary program state of type σ. Empty has not running time. Halt immediately aborts the program. Seq is for sequential composition. Par is for non-deterministic choice, i.e. both commands are executed and then one of the results is chosen. Assign, If, and While have the expected behaviour, and all three commands require one time step. A probabilistic choice is possible with Assign u, where u is a probabilistic state transformer ($\sigma \Rightarrow \sigma$ pmf). The expected running time of Assign u weights each possible running time with the outcome of u. The assignment is deterministic is u is a Dirac distribution, i.e. assigning probability 1 to exactly one value. We need the datatype to have a deep embedding of pGCL programs, which is necessary for the construction of the MDP.

Expected Running Time. The denotational semantics for the running time is given as an expectation transformer, which is similar to the denotational seman-

$K :: (\sigma\ \mathsf{pgcl} \times \sigma) \Rightarrow (\sigma\ \mathsf{pgcl} \times \sigma)\ \mathsf{pmf}\ \mathsf{set}$

$K(\mathsf{Empty},\quad s) = \ll \mathsf{Empty}, s \gg$

$K(\mathsf{Skip},\qquad s) = \ll \mathsf{Empty}, s \gg$

$K(\mathsf{Halt},\qquad s) = \ll \mathsf{Halt}, s \gg$

$K(\mathsf{Assign}\ u,\ s) = \{[\lambda s'. (\mathsf{Empty}, s')]\ (u\ s)\}$

$$K(\mathsf{Seq}\ c_1\ c_2, s) = \left[\lambda(c', s'). \left(\left\{\begin{array}{ll} c_2 & \text{if } c' = \mathsf{Empty} \\ \mathsf{Halt} & \text{if } c' = \mathsf{Halt} \\ \mathsf{Seq}\ c'\ c_2 & \text{otherwise}\end{array}\right\}, s'\right)\right]\ K\ (c_1, s)$$

$K(\mathsf{Par}\ c_1\ c_2,\ s) = \ll c_1, s \gg \cup \ll c_2, s \gg$

$K(\mathsf{If}\ g\ c_1\ c_2,\ s) = \text{if } g\ s \text{ then } \ll c_1, s \gg \text{ else } \ll c_2, s \gg$

$K(\mathsf{While}\ g\ c,\ s) = \text{if } g\ s \text{ then } \ll \mathsf{Seq}\ c\ (\mathsf{While}\ g\ c), s \gg \text{ else } \ll \mathsf{Empty}, s \gg$

$\mathsf{cost} :: (\sigma \Rightarrow \mathsf{ennreal}) \Rightarrow \sigma\ \mathsf{pgcl} \Rightarrow \sigma \Rightarrow \mathsf{ennreal} \Rightarrow \mathsf{ennreal}$

$\mathsf{cost}\ f\ \mathsf{Empty}\quad s\ \_ = f s$ $\mathsf{cost}\ \_\ (\mathsf{Seq}\ \mathsf{Empty}\ \_)\ \_\ x = x$

$\mathsf{cost}\ \_\ \mathsf{Halt}\qquad \_\ \_ = 0$ $\mathsf{cost}\ f\ (\mathsf{Seq}\ c\ \_)\qquad s\ x = \mathsf{cost}\ f\ c\ s\ x$

$\mathsf{cost}\ \_\ (\mathsf{Par}\ \_\ \_)\ s\ x = x$ $\mathsf{cost}\ \_\ \_\qquad\qquad \_\ x = 1 + x$

$\ll c, s \gg$ is the singleton set of the singleton distribution (c, s).

$[f]\mu$ maps f over all elements of μ

Fig. 3. MDP semantics for pGCL running times

tics for the expectation of program variables as weakest pre-expectation transformers. Again we follow the definition in Kaminski *et al.* [9]. In Fig. 2 we define the expectation transformer ert taking a pGCL command c and an expectation f, where f assigns an expected running time to each terminal state of c. This gives a simple recursive definition of the Seq case, for the expected running time of a pGCL program we will set $f = 0$. We proved some validating theorems about expectation transformer ert, i.e. continuity and monotonicity of ert c, closed under constant addition for Halt-free programs, sub-additivitiy, etc.

MDP Semantics. For the operational small-step semantics we introduce a MDP constructed per pGCL program, and compute the expected number of steps until the program terminates. In Fig. 3 we define the MDP by its transition function K and the per-state cost function cost $f\ c\ s\ x$. The per-state cost cost $f\ c\ s\ x$ computes the running time cost associated with the program c at state s. Here the program is seen as a list of statements, hence we walk along a list of Seq and only look at its left-most leaf. If the program is Empty the MDP is stopped and we return $f\ s$ containing further running time cost we want to associated to a finished state s (in most cases this will be 0, but it is essential in the induction case of Theorem 1). When the execution continues we also add x, c.f. the definition of $\mathsf{cost}_{\mathsf{stream}}$.

The transition function K induces now a set of trace spaces, one for each possible resolution of the non-deterministic choices introduced by Par. We write $\hat{\mathbb{E}}^K_{(c,s)}[f]$ for the maximal expectation of $f :: (\sigma\ \mathsf{pgcl} \times \sigma)\ \mathsf{stream} \Rightarrow \mathsf{ennreal}$ when the MDP starts in (c, s). We define the cost of a trace as the sum of cost over all states in the trace:

$$\text{cost}_{\text{stream}} \ f \ ((c, s) \cdot \omega) \overset{\text{lfp}}{=} \text{cost} \ f \ c \ s \ (\text{cost}_{\text{stream}} \ f \ \omega)$$

Finally the maximal expectation of $\text{cost}_{\text{stream}}$ computes ert:

Theorem 1. $\hat{\mathbb{E}}^K_{(c,s)}[\text{cost}_{\text{stream}} \ f] = \text{ert} \ c \ f \ s$

Proof (Induction on c). The interesting cases are Seq and While. For Seq we prove the equation $\hat{\mathbb{E}}^K_{(\text{Seq} \ a \ b,s)}[\text{cost}_{\text{stream}} \ f] = \hat{\mathbb{E}}^K_{(a,s)}[\text{cost}_{\text{stream}} \ (\lambda s. \ \hat{\mathbb{E}}^K_{(b,s)}[\text{cost}_{\text{stream}} \ f])]$, by fixed point induction in both directions. For While we prove

$$\hat{\mathbb{E}}^K_{(\text{While} \ g \ c,s)}[\text{cost}_{\text{stream}} \ f] = \text{lfp} \ (\lambda F \ s. \ 1 + \text{if} \ g \ s \ \text{then} \ \hat{\mathbb{E}}^K_{(c,s)}[\text{cost}_{\text{stream}} \ f] \ \text{else} \ f \ s) \ s$$

by equating it to a completely unrolled version using fixed point induction and then massaging it in the right form using the rolling and diagonal rules.   □

4   Simple Symmetric Random Walk

As an application for the expected running time analysis Kaminski *et al.* [9] chose the simple random walk. As difference to [9] we do not use ω-invariants to prove the infinite running time, but the correspondence of the program with a Markov chain (there is no non-deterministic choice).

The simple symmetric random walk (srw) is a Markov chain on $\mathbb{Z}$, in each step i it goes uniformly to $i + 1$ or $i - 1$ (i.e. in both cases with probability $1/2$). Surprisingly, but well known (and formalised by Hurd [7]), it reaches each point with probability 1. Equally surprising, the expected time for the srw to go from i to $i + 1$ is infinite! Kaminski *et al.* [9] prove this by providing a lower ω-invariant. Unfortunately, this proof has a flaw: in Appendix B.1 of [10] (the extended version of [9]), the equation $1 + [\![x > 0]\!] \cdot 2 + [\![1 < x \le n+1]\!] \cdot \infty + [\![0 < x \le n-1]\!] \cdot \infty = 1 + [\![x > 0]\!] \cdot 2 + [\![0 < x \le n+1]\!] \cdot \infty$ does not hold for $n = 0$ and $x = 1$. The author knows from private communication with Kaminski *et al.* that it still is possible to use a lower ω-invariant. Unfortunately, the necessary invariant gets much more complicated.

After discovering the flaw in the proof, we tried a more traditional proof. The usual approach in random walk theory uses the generating function of the first hitting time. Unfortunately, this would require quite some formalizations in combinatorics, e.g. Stirling numbers and more theorems about generating functions than available in [4]. Finally, we choose an approach similar to [7], i.e. we set up a linear equation system and prove that the only solution is infinity.

Now, srw :: int $\Rightarrow$ int pmf is the transition function for the simple symmetric random walk. The expected time to reach j when started in i is written $H \ i \ j \overset{\text{def}}{=} \mathbb{E}^{\text{srw}}_i[f \ j]$, where $f \ j \ (k \cdot \omega) \overset{\text{lfp}}{=} \text{if} \ j = k \ \text{then} \ 0 \ \text{else} \ 1 + f \ j \ \omega$ is the first hitting time. Now we need to prove the following rules: (I) $H \ j \ i = H \ j \ k + H \ k \ i$ if $i \le j \le k$, (II) $H \ (i + t) \ (j + t) = H \ i \ j$, (III) $H \ i \ j = H \ j \ i$ and (VI) $H \ i \ j = (\text{if} \ i = j \ \text{then} \ 0 \ \text{else} \ 1 + (H \ i \ (j + 1) + H \ i \ (j - 1))/2)$. From these rules we can derive $H \ i \ j = \infty$ for $i \ne j$.

Rule (VI) is derived the expectation transformer semantics. But it is not clear to us how to prove rule (I) by only applying fixed point transformations or induction. Instead we prove (I) in a measure theoretic way:

$$H\ j\ k + H\ k\ i = \mathbb{E}_j^{\mathsf{srw}}[f\ j + H\ k\ i]$$

$$= \sum_n (n + H\ k\ i) \cdot \Pr_j(f\ k = n) \tag{1}$$

$$= \sum_n \mathbb{E}_j^{\mathsf{srw}}[\lambda\omega.\ (n + f\ i\ (\mathsf{sdrop}\ n\ \omega)) \cdot [\![f\ k\ \omega = n]\!]]$$

$$= \sum_n \mathbb{E}_j^{\mathsf{srw}}[f\ i] = H\ j\ i \tag{2}$$

Equation 1 requires that $f\ k$ is finite with probability 1, we do a case distinction: if it is not finite a.e. the result follows from $H\ j\ i \geq H\ j\ k = \infty$. Equation 2 is now simply proved by induction on n. The proofs for Eqs. 1 and 2 essentially operate on each trace ω in our probability space, making them inherently dependent on the trace space.

Theorem 2 (The running time of srw is infinite). $H\ i\ j = \infty\ if\ i \neq j.$

5   Coupon Collector

Another example we formalised is the coupon collector example from [9]. The idea is to compute the expected time until we collect N different coupons from a uniform, independent and infinite source of coupons. The left side of Fig. 4 shows our concrete implementation CC_N, the right side is its refinement (there is no array cp necessary). By fixed point transformations we show that the (refined) inner loop's running time has a Geometric distribution, and hence the expected running time for CC_N is: $\mathsf{ert}\ \mathsf{CC}_N\ 0\ s = 2 + 4N + 2N\sum_{i=1}^{N}\frac{1}{i}$ for $N > 0$.

6   Related Work

The first formalisation of probabilistic programs was by Hurd [7] in hol98, formalising a trace space for a stream of probabilistic bits. Hurd et al. [8] is different approach, formalising the weakest pre-expectation transformer semantics

$$\begin{array}{lll}
\overbrace{x := 0, cp := [F, \ldots, F]}^{N\ \text{times}}, i := 0 & & c := 0, b := F \\
\mathtt{WHILE}\ x < N\ \mathtt{DO} & x \quad \rightarrow \quad c & \mathtt{WHILE}\ c < N\ \mathtt{DO} \\
\quad \mathtt{WHILE}\ cp[i]\ \mathtt{DO}\ i :\sim \mathsf{Unif}\{0, \ldots, N\} & cp[i] \quad \rightarrow \quad b & \quad \mathtt{WHILE}\ b\ \mathtt{DO}\ b :\sim \mathsf{Bern}(x/N) \\
\quad cp[i] := T, x := x + 1 & |cp| = x & \quad b := T, c := c + 1
\end{array}$$

Fig. 4. The Coupon Collector in pGCL and its refinement

of pGCL in HOL4. Both formalisations are not related. Audebaud and Paulin-Mohring [1] use a shallow embedding of a probability monad in Coq. Cock [2] provides a VCG for pGCL in Isabelle/HOL. Hölzl and Nipkow [5,6] formalises MCs and analyses the expected running time of the ZeroConf protocol. On the basis of [5] formalises MDPs and shows the equivalence of the weakest pre-expectation transformer (based on the pen-and-paper proof in [3]).

Unlike Theorem 1, these formalisations either define denotational semantics [1,2,8], or operational semantics [5–7], none of them relate both semantics.

7 Conclusion and Future Work

While formalising the random walk example in [9] we found an essential flaw in the proof in [10]. Our solution seams to indicate, that for the verification of expected running times an ω-invariant approach is not enough. While the expectation transformer gives us a nice verification condition generator (e.g. [2]), the trace space might be required to get additional information i.e. fairness and termination. The equivalence between the expectation transformer semantics and the MDP semantics provides the required bridge between both worlds. Also we might require a probabilistic, relational Hoare logic (maybe based on [11]) to automate tasks like Fig. 4.

References

1. Audebaud, P., Paulin-Mohring, C.: Proofs of randomized algorithms in Coq. Sci. Comput. Prog. **74**(8), 568–589 (2009)
2. Cock, D.: Verifying probabilistic correctness in Isabelle with pGCL. In: SSV 2012. EPTCS, vol. 102, pp. 167–178 (2012)
3. Gretz, F., Katoen, J., McIver, A.: Operational versus weakest pre-expectation semantics for the probabilistic guarded command language. Perform. Eval. **73**, 110–132 (2014)
4. Hölzl, J.: Markov chains and Markov decision processes in Isabelle/HOL. Submitted to JAR in December 2015. http://in.tum.de/~hoelzl/mdptheory
5. Hölzl, J.: Construction and Stochastic Applications of Measure Spaces in Higher-Order Logic. Ph.D. thesis, Technische Universität München (2013)
6. Hölzl, J., Nipkow, T.: Interactive verification of Markov chains: two distributed protocol case studies. In: QFM 2012. EPTCS, vol. 103 (2012)
7. Hurd, J.: Formal Verification of Probabilistic Algorithms. Ph.D. thesis (2002)
8. Hurd, J., McIver, A., Morgan, C.: Probabilistic guarded commands mechanized in HOL. Theoret. Comput. Sci. **346**(1), 96–112 (2005)
9. Kaminski, B.L., Katoen, J.-P., Matheja, C., Olmedo, F.: Weakest precondition reasoning for expected run–times of probabilistic programs. In: Thiemann, P. (ed.) ESOP 2016. LNCS, vol. 9632, pp. 364–389. Springer, Heidelberg (2016). doi:10.1007/978-3-662-49498-1_15
10. Kaminski, B.L., Katoen, J., Matheja, C., Olmedo, F.: Weakest precondition reasoning for expected run-times of probabilistic programs. CoRR abs/1601.01001v1 (Extended version) (2016)

11. Lochbihler, A.: Probabilistic functions and cryptographic oracles in higher order logic. In: Thiemann, P. (ed.) ESOP 2016. LNCS, vol. 9632, pp. 503–531. Springer, Heidelberg (2016). doi:10.1007/978-3-662-49498-1_20
12. McIver, A., Morgan, C.: Abstraction, Refinement and Proof for Probabilistic Systems. Monographs in Computer Science. Springer, New York (2004)

On the Formalization of Fourier Transform in Higher-order Logic

Adnan Rashid$^{(\boxtimes)}$ and Osman Hasan

School of Electrical Engineering and Computer Science (SEECS)
National University of Sciences and Technology (NUST), Islamabad, Pakistan
{adnan.rashid,osman.hasan}@seecs.nust.edu.pk

Abstract. Fourier transform based techniques are widely used for solving differential equations and to perform the frequency response analysis of signals in many safety-critical systems. To perform the formal analysis of these systems, we present a formalization of Fourier transform using higher-order logic. In particular, we use the HOL-Light's differential, integral, transcendental and topological theories of multivariable calculus to formally define Fourier transform and reason about the correctness of its classical properties, such as existence, linearity, frequency shifting, modulation, time reversal and differentiation in time-domain. In order to demonstrate the practical effectiveness of the proposed formalization, we use it to formally verify the frequency response of an automobile suspension system.

Keywords: Higher-order logic · HOL-Light · Fourier transform

1 Introduction

It is customary to use differential equations for capturing the dynamic behavior of engineering and physical systems for their continuous-time analysis [9]. The complexity of the analysis varies with their size, nature of the input signals and the design constraints. Fourier Transform [2] is a transform method, which converts a time varying function to its corresponding ω-domain representation, where ω is its corresponding angular frequency [1]. In this way, the differentiation and integration in time domain analysis are transformed into multiplication and division operators in the frequency domain and thus are easily solved through algebraic manipulation. Moreover, the ω-domain representations of the differential equations can also be used for the frequency response analysis of the corresponding systems.

The first step in the continuous-time system analysis, using Fourier transform, is to model the dynamics of the system using a differential equation. This differential equation is then transformed into its equivalent ω-domain representation by using the Fourier transform. Next, the resulting ω-domain equation is simplified using various Fourier transform properties, such as existence, linearity, frequency shifting, modulation, time reversal and differentiation. The main purpose is to either solve the differential equation to obtain values for the variable

© Springer International Publishing Switzerland 2016
J.C. Blanchette and S. Merz (Eds.): ITP 2016, LNCS 9807, pp. 483–490, 2016.
DOI: 10.1007/978-3-319-43144-4_31

ω or obtain the frequency response of the system corresponding to the given differential equation. Once the frequency response is obtained, it can be used to analyze the dynamics of the system by studying the impact of different frequency components on the intended behaviour of the given system.

Traditionally, the transform methods based analysis has been done using paper-and-pencil, numerical methods and symbolic techniques. However, all of these techniques cannot ascertain accurate analysis due to their inherent limitations, like human-error proneness, numerical errors and discretization errors. Given the wide-spread usage of physical systems in many safety-critical domains, such as medicine and transportation, accurate transform methods based analysis has become a dire need. With the same motivation, higher-order-logic theorem proving has been used for the formalization of Z [7] and Laplace [8] transforms. However, the formalization of Z-transform can only be utilized for discrete-time system analysis. Similarly, Laplace transform based analysis is only limited to causal functions, i.e., the functions that fulfill the condition: $f(x) = 0$ for all $x < 0$. Physical systems are often modeled by the non-causal continuous functions, i.e., the functions with infinite extent. Fourier transform can cater for the analysis involving both continuous and non-causal functions and thus can overcome the above-mentioned limitations of Z and Laplace transforms.

In this paper, we propose to formalize Fourier transform in higher-order logic to leverage upon its benefits for formally analyzing physical continuous-time linear systems. In particular, we formalize the definition of Fourier transform in higher-order logic and use it to verify the classical properties of Fourier transform, such as existence, linearity, frequency shifting, modulation, time reversal and differentiation. These foundations can be built upon to reason about the analytical solutions of differential equations or frequency responses of the physical systems. In order to demonstrate the practical effectiveness of the reported formalization, we present a formal analysis of an automobile suspension system.

2   Formalization of Fourier Transform

Mathematically, the Fourier transform is defined for a function $f : \mathbb{R}^1 \to \mathbb{C}$ as:

$$\mathcal{F}[f(t)] = F(\omega) = \int_{-\infty}^{+\infty} f(t)e^{-j\omega t}dt, \ \omega \ \epsilon \ \mathbb{R} \tag{1}$$

We formalize Eq. 1 in HOL-Light as follows:

Definition 1. Fourier Transform
$\vdash \forall$ w f. fourier f w =
    integral UNIV (λt. cexp (--((ii * Cx w) * Cx (drop t))) * f t)

The function fourier accepts a complex-valued function f : $\mathbb{R}^1 \to \mathbb{R}^2$ and a real number w and returns a complex number that is the Fourier transform of f as represented by Eq. 1. In the above function, we used complex exponential function cexp : $\mathbb{R}^2 \to \mathbb{R}^2$ because the return data-type of the function f is $\mathbb{R}^2$. To multiply w with ii, we first converted w into a complex number $\mathbb{R}^2$ using Cx. Similarly, t has data-type $\mathbb{R}^1$ and to multiply it with ii $*$ Cx w, it is first converted into a real number by using drop and then it is converted to data-type $\mathbb{R}^2$ using Cx. Next, we use the vector function integral to integrate the expression $f(t)e^{-i\omega t}$ over the whole real line since the data-type of this expression is $\mathbb{R}^2$. Since the region of integration of the vector integral function must be a vector space, therefore we represented the interval of the integral by UNIV : $\mathbb{R}^1$ which represents the whole real line.

The Fourier transform of a function f exists, i.e., the integrand of Eq. 1 is integrable, and the integral has some converging limit value, if f is piecewise smooth and is absolutely integrable on the whole real line [1,5]. A function is said to be piecewise smooth on an interval if it is piecewise differentiable on that interval. Similarly, a function f is absolutely integrable on the whole real line if it is absolutely integrable on both the positive and negative real lines. The Fourier existence condition can thus be formalized in HOL-Light as follows:

Definition 2. *Fourier Exists*
⊢ ∀ f g w a b. fourier_exists f =
   (∀ a b. f piecewise_differentiable_on interval [lift a, lift b]) ∧
               f absolutely_integrable_on {x | &0 <= drop x} ∧
               f absolutely_integrable_on {x | drop x <= &0}

In the above function, the first conjunct expresses the piecewise smoothness condition for the function f. In the second conjunct, {x | &0 <= drop x} represents the interval $[0, \infty)$, whereas {x | drop x <= &0} represents the interval $(-\infty, 0]$ in the last conjunct. Both these conjuncts jointly ensure that the function f is absolutely integrable on whole real line.

3   Formal Verification of Fourier Transform Properties

In this section, we use Definitions 1 and 2 to verify some of the classical properties of Fourier transform in HOL-Light. The verification of these properties not only ensures the correctness of our definitions but also plays a vital role in minimizing the user intervention and time consumption in reasoning about Fourier transform based analysis of systems.

The existence of the improper integral of Fourier Transform is a pre-condition for most of the arithmetic manipulations involving the Fourier transforms. This condition is formalized in HOL-Light as follows:

Theorem 1. Integrability of Integrand of Fourier Transform Integral
$\vdash \forall$ f w. fourier_exists f $\Rightarrow$
(λt. cexp (--((ii * Cx w) * Cx (drop t))) * f t) integrable_on UNIV

Table 1. Properties of Fourier Transform

Mathematical Form	Formalized Form
	Linearity

$\mathcal{F}[\alpha f(t) + \beta g(t)] =$ $\alpha F(\omega) + \beta G(\omega)$	$\vdash \forall$ f g w a b.
	   fourier_exists f $\wedge$ fourier_exists g $\Rightarrow$
	   fourier (λt. a * f t + b * g t) w =
	     a * fourier f w + b * fourier g w

	Frequency Shifting

$\mathcal{F}[e^{i\omega_0 t} f(t)] = F(\omega - \omega_0)$	$\vdash \forall$ f w w0.
	   fourier_exists f $\Rightarrow$
	   fourier (λt. cexp ((ii * Cx (w0)) * Cx (drop t)) * f t) w =
	   fourier f (w - w0)

	Modulation

$\mathcal{F}[cos(\omega_0 t) f(t)] =$ $\dfrac{F(\omega - \omega_0) + F(\omega + \omega_0)}{2}$	$\vdash \forall$ f w w0.
	   fourier_exists f $\Rightarrow$
	   fourier (λt. ccos (Cx w0 * Cx (drop t)) * f t) w =
	   (fourier f (w - w0) + fourier f (w + w0)) / Cx (&2)

$\mathcal{F}[sin(\omega_0 t) f(t)] =$ $\dfrac{F(\omega - \omega_0) - F(\omega + \omega_0)}{2i}$	$\vdash \forall$ f w w0.
	   fourier_exists f $\Rightarrow$
	   fourier (λt. csin (Cx w0 * Cx (drop t)) * f t) w =
	   (fourier f (w - w0) - fourier f (w + w0)) / (Cx (&2) * ii)

	Time Reversal
$\mathcal{F}[f(-t)] = F(-\omega)$	$\vdash \forall$ f w. fourier_exists f $\Rightarrow$ fourier (λt. f (--t)) w = fourier f (--w)
	First-order Differentiation

$\mathcal{F}[\frac{d}{dt}f(t)] = i\omega F(\omega)$	$\vdash \forall$ f w.
	   fourier_exists f $\wedge$
	   fourier_exists (λt. vector_derivative f (at t)) $\wedge$
	   ($\forall$t. f differentiable at t) $\wedge$
	   ((λt. f (lift t)) $\rightarrow$ vec 0) at_posinfinity $\wedge$
	   ((λt. f (lift t)) $\rightarrow$ vec 0) at_neginfinity
	   $\Rightarrow$ fourier (λt. vector_derivative f (at t)) w =
	     ii * Cx w * fourier f w

	Higher-order Differentiation

$\mathcal{F}[\frac{d^n}{dt^n}f(t)] = (i\omega)^n F(\omega)$	$\vdash \forall$ f w n.
	   fourier_exists_higher_deriv n f $\wedge$
	   ($\forall$t. differentiable_higher_derivative n f t) $\wedge$
	   ($\forall$p. p < n $\Rightarrow$
	     ((λt. higher_vector_derivative p f (lift t)) $\rightarrow$ vec 0)
	     at_posinfinity) $\wedge$
	   ($\forall$p. p < n $\Rightarrow$
	     ((λt. higher_vector_derivative p f (lift t)) $\rightarrow$ vec 0)
	     at_neginfinity)
	   $\Rightarrow$ fourier (λt. higher_vector_derivative n f t) w =
	     (ii * Cx w) pow n * fourier f w

The proof of above theorem is based on splitting of the region of integration, i.e., the whole real line UNIV : $\mathbb{R}^1$, as a union of positive real line (interval $[0, \infty)$) and negative real line (interval $(-\infty, 0]$). Then, some theorems regarding integration and integrability are used to conclude the proof of Theorem 1.

Next, we verified some of the classical properties of Fourier transform, given in Table 1.

The above-mentioned formalization is done interactively and it took around 4000 lines of code and approximately 600 man-hours. The first author started working with HOL-Light as a novice user and it took him about 200 man-hours to get familiar with its proof styles and procedures. About another 100 man-hours were spent in understanding the Multivariate theories of HOL-Light, which are the foundational theories towards this work. The actual formalization task took about 300 man-hours. The major difficulty faced during the formalization was the unavailability of detailed proofs for the properties of Fourier transform in literature. The available paper-and-pencil based proofs were found to be very abstract and missing the complete reasoning about the steps. The source code of our formalization is available for download [6] and can be utilized for further developments and the analysis of physical systems.

4   Application: Automobile Suspension System

In this section, we provide the verification of the frequency response of an automobile suspension system, depicted in Fig. 1. An automobile suspension system consists of the chassis connected to the wheels through a spring and dashpot (shock absorber). The road surface can be thought of as a superposition of rapid and gradual small-amplitude changes in elevation, which represents the roughness of the surface. These rapid and gradual changes are acting like high and low frequencies, respectively. The automobile suspension system is intended to filter out the rapid variations on the road surface, i.e., to act as a low pass filter. We perform the formal analysis of this system using our proposed formalization of Fourier transform within the sound core of HOL-Light theorem prover.

The behaviour of a automobile suspension system with input $u(t)$ and output $y(t)$ can be expressed by the following differential equation [5]:

$$M\frac{d^2y(t)}{dt^2} + b\frac{dy(t)}{dt} + ky(t) = ku(t) + b\frac{du(t)}{dt}, \tag{2}$$

In the above equation, M is the mass of the chassis, whereas, k is the spring constant and b represents the shock absorber constant, as shown in Fig. 1. All of these are design parameters of the underlying system and can have positive values only.

The corresponding frequency response of the automobile suspension system is given as follows [5]:

$$\frac{Y(\omega)}{U(\omega)} = \frac{\frac{b}{M}(i\omega) + \frac{k}{M}}{(i\omega)^2 + \frac{b}{M}(i\omega) + \frac{k}{M}} \tag{3}$$

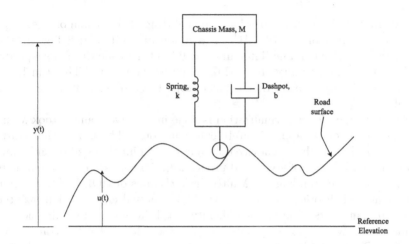

Fig. 1. Automobile Suspension System [5]

We aim to verify this frequency response using Eq. 2, which can be verified as the following theorem in HOL-Light.

Theorem 2. Frequency Response of Automobile Suspension System
⊢ ∀ y u w a. &0 < M ∧ &0 < b ∧ &0 < k ∧
  (∀t. differentiable_higher_derivative 2 y t) ∧
  (∀t. differentiable_higher_derivative 1 u t) ∧
  fourier_exists_higher_deriv 2 y ∧
  fourier_exists_higher_deriv 1 u ∧
  (∀p. p < 2 ⇒
        ((λt. higher_vector_derivative p y (lift t)) → vec 0)
        at_posinfinity) ∧
  (∀p. p < 2 ⇒
        ((λt. higher_vector_derivative p y (lift t)) → vec 0)
        at_neginfinity) ∧
  ((λt. u (lift t)) → vec 0) at_posinfinity ∧
  ((λt. u (lift t)) → vec 0) at_neginfinity ∧
  (∀t. diff_eq_ASS y u a b c) ∧ ∼(fourier u w = Cx (&0)) ∧
  ∼((ii * Cx w) pow 2 + Cx (b / M) * ii * Cx w
                        + Cx (k / M) = Cx (&0))
    ⇒ (fourier y w / fourier u w =
                        (Cx (b / M) * ii * Cx w + Cx (k / M)) /
        ((ii * Cx w) pow 2 + Cx (b / M) * ii * Cx w + Cx (k / M))

The first three assumptions ensure that the variables corresponding to mass of chassis (M), spring constant (k) and shock absorber constant (b) cannot be negative or zero. The next two assumptions ensure that the functions y and u are differentiable up to the second-order and first-order, respectively. The next assumption represents the Fourier transform existence condition upto the second-order derivatives of function y. Similarly, the next assumption ensures that the Fourier transform exists up to the first-order derivative of function u. The next two assumptions represent the condition $\lim\limits_{t\to\pm\infty} y^{(k)}(t) = 0$ for each $k = 0, 1$, i.e., $\lim\limits_{t\to\pm\infty} y^{(1)}(t) = 0$ and $\lim\limits_{t\to\pm\infty} y^{(0)}(t) = \lim\limits_{t\to\pm\infty} y(t) = 0$, where $y^{(k)}$ is the kth derivative of y. The next two assumptions provide the condition $\lim\limits_{t\to\pm\infty} u(t) = 0$. The next assumption represents the formalization of Eq. 2 and the last two assumptions provide some interesting design related relationships, which must hold for constructing a reliable automobile suspension system. Finally, the conclusion of the above theorem represents the frequency response given by Eq. 3. The proof of Theorem 2 is based on Definition 1 and the property of Fourier transform of higher-order derivative of a function, along with some arithmetic reasoning. The proof script for this application consists of approximately 500 lines of HOL-Light code [6] and the proof process took just a couple of hours, which clearly indicates the usefulness of our proposed formalization in conducting the Fourier transform analysis of real-world applications. Given the continuous and non-causal nature of the functions involved in this analysis, the existing Z [7] and Laplace transform [8] formalizations cannot be used for conducting the above-mentioned formal analysis.

5   Conclusions

In this paper, we proposed a formalization of Fourier transform in higher-order logic. We presented the formal definition of Fourier transform and based on it, verified its properties, namely existence, linearity, frequency shifting, modulation, time reversal and differentiation in time-domain. Lastly, in order to demonstrate the practical effectiveness of the proposed formalization, we presented a formal analysis of an automobile suspension system.

The proposed formalization of Fourier transform can be utilized to conduct the formal analysis of many safety-critical systems involving signal processing filters, such as low-pass, high-pass, band-pass and band-stop [5] and in wireless communication systems, such as antenna [2] and signal transmission [3]. Similarly, in optics, it can be used to formally study the behaviour of light, such as intensity and diffraction, in different optical devices [2], which can be very useful for the recently initiated project on the usage of higher-order-logic theorem proving for the formal analysis of optics [4].

References

1. Beerends, R.J., Morsche, H.G., Van den Berg, J.C., Van de Vrie, E.M.: Fourier and Laplace Transforms. Cambridge University Press, Cambridge (2003)
2. Bracewell, R.N.: The Fourier Transform and its Applications. McGraw-Hill, New York (1978)

3. Du, K.L., Swamy, M.N.S.: Wireless Communication Systems: from RF Subsystems to 4G Enabling Technologies. Cambridge University Press, Cambridge (2010)
4. Khan-Afshar, S., Siddique, U., Mahmoud, M.Y., Aravantinos, V., Seddiki, O., Hasan, O., Tahar, S.: Formal analysis of optical systems. Math. Comput. Sci. **8**(1), 39–70 (2014)
5. Oppenheim, A.V., Willsky, A.S., Hamid Nawab, S.: Signals and Systems. Prentice Hall Processing Series, 2nd edn. Prentice Hall, Inc., Englewood Cliffs (1996)
6. Rashid, A.: On the Formalization of Fourier Transform in Higher-order Logic (2016). http://save.seecs.nust.edu.pk/projects/fourier/
7. Siddique, U., Mahmoud, M.Y., Tahar, S.: On the formalization of Z-transform in HOL. In: Klein, G., Gamboa, R. (eds.) ITP 2014. LNCS, vol. 8558, pp. 483–498. Springer, Heidelberg (2014)
8. Taqdees, S.H., Hasan, O.: Formalization of laplace transform using the multivariable calculus theory of HOL-light. In: McMillan, K., Middeldorp, A., Voronkov, A. (eds.) LPAR-19 2013. LNCS, vol. 8312, pp. 744–758. Springer, Heidelberg (2013)
9. Yang, X.S.: Mathematical Modeling with Multidisciplinary Applications. Wiley, Hoboken (2013)

CoqPIE: An IDE Aimed at Improving Proof Development Productivity

Kenneth Roe[✉] and Scott Smith

The Johns Hopkins University, Baltimore, USA
kendroe@hotmail.com

Abstract. In this paper we present CoqPIE(CoqPIE is available for download at http://github.com/kendroe/CoqPIE), a new development environment for Coq which delivers editing functionality centered around common prover usage workflow not found in existing tools. The main contributions of CoqPIE build from having an integrated parser for both Coq source and for prover output. The primary novelty is not the parser but how it is used: CoqPIE includes tools to carry out complex editing functions such as lemma extraction and replay. In proof replay for example both new and old outputs of the proof script are parsed into ASTs. These ASTs allow replay to do updates such as fixing hypothesis references.

1 Introduction

In this paper we present CoqPIE, a new development environment for Coq which delivers editing functionality centered around common prover usage workflow that is not found in existing tools. The design of CoqPIE was driven by the author's frustrating with a few of the existing proof development workflows. First, when a proof gets to be more than about 300 steps, the time it takes for `coqtop` to process a single tactic slows; this makes browsing quite tedious. Second, when developing a large proof with many lemmas, proving a lemma often reveals an error in the lemma itself. This change then propagates and requires the statements of other lemmas to be changed. Since many of these lemmas have likely already been proven, they need to be replayed (likely with proof script editing), a tedious process.

Improving the above and similar workflows is the primary goal of the design of CoqPIE, which we now describe.

2 An Overview of CoqPIE

The diagram in Fig. 1 shows the CoqPIE UI with a sample proof derivation open. There are three views shown. On the left is a tree view of the entire project similar to the tree view found in modern IDEs. The top level of the tree view shows the files in the project; opening a file node displays a list of all the Coq declarations in that file. Opening a theorem declaration in turn shows

© Springer International Publishing Switzerland 2016
J.C. Blanchette and S. Merz (Eds.): ITP 2016, LNCS 9807, pp. 491–499, 2016.
DOI: 10.1007/978-3-319-43144-4_32

the steps used to prove that theorem, with steps arranged in a tree based on subgoal relationships.

The middle view displays the source file based on the selection made in the tree view on the left. This view functions in a manner similar to the source file view in CoqIDE or Proof General. As with those tools, shading is used to indicate the portion of the file already processed by coqtop. Unlike Proof General and CoqIDE, the CoqPIE process management system automatically recompiles dependent source files.

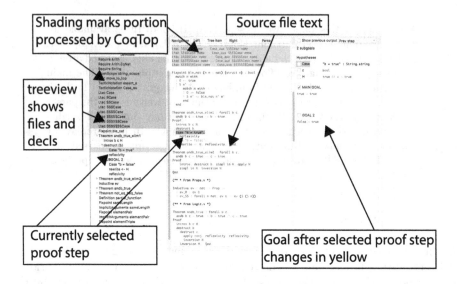

Fig. 1. The main CoqPIE window

The window on the right is similar to the Coq state window in CoqIDE or Proof General: it shows the current goal and hypotheses. However, instead of showing the state at the current processing point of Coq, it shows the state just *after* the selected definition or proof step from the tree view at the left. This is possible because CoqPIE runs the entire project and saves all output from coqtop before editing can commence. With this initial pass it is possible to very quickly browse theorems and to see the state after each step. This full proof tree state is also maintained during editing: as the user edits a source file and reruns coqtop to verify the updates, the cached outputs are updated. Differences from the state just before the most recent tactic was executed are highlighted in yellow. One can also view differences between hypotheses and the goal or differences between old and new versions of a state (useful for the replay assist described later), via the combo box just above the window on the right which allows selection of which differences to show.

Since CoqPIE keeps intermediate proof state around it can be more intelligent about whether definitions and lemmas are up-to-date: definitions with out-of-date Coq output information are color coded so the user knows they need to be replayed.

Parsing. Coq has an internal CoqAst data structure, but it is not easily accessible with the current API. So, for the current implementation of CoqPIE, we chose to create our own parser. This choice has a number of ramifications. First, the Coq language is quite large and complex; we are only able to parse the commonly-used subset. Second, Coq has a `Notation` construct that can add new syntax to the language. We currently do not have the capability to handle this construct. Longer-term we hope to see a CoqAst API exposed which we will directly be able to use. If a definition or proof step cannot be parsed, then CoqPIE inserts a bad declaration or bad step AST node in the proof tree. The end point is determined by looking for a period.

Dependency management. CoqPIE maintains dependencies between definitions and theorems. When a theorem or definition is changed, all dependent theorems and definitions are highlighted in the project treeview. Creating an exact algorithm for tracking dependencies is very difficult [6] due to the complexities of Coq's higher-order semantics. Many other issues arise in doing dependency analysis, see [22], including opaque vs transparent proof dependencies. An opaque transparency is a dependency that can be identified by the proof statement alone. Transparent dependencies occur when a tactic in the proof script depends on another theorem. These can sometimes be hard to identify as theorems may be chosen automatically by tactics such as `auto`. Our current approach is to use an incomplete dependency tracking algorithm: CoqPIE bases dependency relationships only on identifiers that explicitly appear in a proof or definition.

Lemma extraction. It is often useful to extract one of the goals of a theorem as a lemma in order to break a large proof into more manageable pieces. Coq can process two theorems of 100 steps each much faster than one theorem of 200 steps. CoqPIE provides a command that automates this extraction. The extraction is done in the following steps:

1. The statement of the new theorem is constructed by taking the goal as the consequent. Each hypothesis becomes an antecedent. If the hypothesis appears to be a variable, then it is encoded as part of a `forall` construct. Otherwise it is encoded as an antecedent of the form *hyp* ->.
2. The steps used to prove the goal are extracted and become the script for the theorem. One can find the end of the sequence of steps used to prove the current goal at the goal state of each subsequent step. The first step after the current step for which the number of goals is one less than that of the current goal is the last step that needs to be extracted with the theorem.
3. In front of the script from the previous step an `intros` statement is added to introduce all of the generated antecedents.

4. The steps to prove the goal are commented out in the main theorem.
5. An `apply` of the newly generated theorem plus an `apply` for each hypothesis is generated in place of those steps that have been commented out.
6. Finally, if there are existential variables in the goal (such as ?508), the lemma extraction tactic tries to figure out how to fill in this variable. The trick here is to realize that this variable is likely filled in by the steps that prove this goal in the parent theorem. The heuristic is to compare the subgoals after these steps have executed in the main goal to the corresponding subgoals from before they were executed.

This tactic is only a heuristic, and there are several cases in which it will fail. For example, a `Focus` in the middle will break the algorithm for finding the end of the steps for the lemma.

Replay assist. When the statement of a theorem changes, most of the old proof script may still be correct, but at each step minor changes may need to be made. One common example is that hypothesis names may have changed. For example, `apply H` may need to become `apply H0`. To improve the workflow we have implemented a replay assistant which automatically will replay proof and apply heuristics to patch the proof back together. Replay assist saves both the `coqtop` output from before the theorem changed and the output of the new theorem up to the point where a patch may need to be made. One can then compare the two texts and see that `H` has been renamed `H0`, and patch the proof script accordingly.

The replay assistant provides a semi-automated assistant to help with the task of proof patching. There is a "Replay" button that advances `coqtop` past one proof step in a manner similar to "Right." However, steps will be edited if necessary. So, `unfold noFind in H` will be changed to `unfold noFind in H0` if the hypothesis was renamed, and then `coqtop` will advance. There also is a "Show previous output" button to show the old output that can be used to see the old goal state. This is useful if hand editing is necessary. Goal information is attached as annotations to the text of the proof steps. Hence if steps are inserted, then the goals will automatically retain its connection to the original steps.

The current replay algorithm only makes updates to hypothesis labels, but we are planning to extend the functionality in the near future. To update hypothesis labels, CoqPIE finds the renaming by looking at both the old and new result from the previous step and choosing the hypothesis from the new state that is the closest match to the one from the old state. Matches are scored by doing a top down comparison of the two AST trees and counting the number of nodes that match.

Coq users will often explicitly name hypotheses that keep changing position during proof development in order to make direct replay more reliable; while this approach improves the odds of a successful replay, the CoqPIE replay tool allows users to skip this step. In addition, we aim to extend CoqPIE replay to support other changes including detecting when a new subgoal has been added, commenting out a subgoal that has been removed, and reordering proof steps.

Admittedly it will never be possible to patch back every single proof, but it should be possible to eliminate many of the tedious steps users must take when patching a proof.

3   Experience with Implementation

The current implementation has all of the functionality described in this paper. The first author has been using the tool exclusively for proof editing in a multi-file project containing around 10000 lines of Coq code. The tool has also been used to read in a couple of other large derivations including a microprocessor verification example [26][1] and the first few chapters of Software Foundations [20]. We needed to make some very minor edits to get Software Foundations to compile.

There is an up-front cost of using CoqPIE: the full project needs to be run and intermediate goals parsed and cached. The table in Fig. 2 shows times for processing some projects from scratch. The times are taken from runs on a 2011 MacBook Pro with a 2.7 Ghz Intel i5 core and 8G of memory. Since this only needs to re-run if the state of the tool becomes inconsistent, it should be an infrequent event.

Project	Compile time	CoqPIE initialization time	Memory usage (Python process+ largest Coq process)
Model.v	0:03	0:46	35M+163M
DPLL	1:36	9:08	94M+581M
Microprocessor	3:14	4:19:29	12M+825M
Software Foundations	0:06	4:01	47M+187M

Fig. 2. Times and memory usage of CoqPIE on different test cases.

Initialization times for CoqPIE are a few times slower than what is needed to compile the project. While for our current projects the initialization time is tolerable, as shown in the table, for larger projects it will be problematic and we will need to do background updating as is done in PIDE.

Future implementation plans. There are a number of areas where improvement is needed before CoqPIE is ready for widespread adoption. We are looking into integration with PIDEtop. The `coqtop` parser may be integrated directly into CoqPIE if we can get some cooperation from the Coq development team. We plan to add additional heuristics to replay as we work with more complex theorems. We also anticipate adding other high level heuristics beyond replay.

---

[1] A couple of type checking errors showed up in CoqPIE but not when compiling outside of CoqPIE. We are still working to find the source of these errors.

4   Related Work

In addition to CoqIDE and Proof General, there are several other Coq IDE development efforts. PIDE/jedit [8,27,28] introduces asynchronous communication between the IDE and the theorem prover to improve the user experience. The idea is that as text is being edited in a proof script, the theorem prover is continuously running in the background verifying the new text and all dependencies. Concurrency is used to speed up theorem proving tasks. The tool saves all output and adds markups to the text in appropriate places. Our system currently does not run the prover as a background task or do automatic updating.

CoqPIE provides a goal state window that highlights differences and allows the showing/hiding of individual hypotheses, whereas PIDE/jedit simply stores the text of the theorem prover's output. We do parsing of the output both for the above functionality and replay. CoqPIE also replaces proof scripts with admit for proofs on which the user is not working. This gains much of the same performance advantage as concurrency.

The IDE supplied with Coq 8.5 also introduces concurrency and dependency analysis to speed up processing of files. We aim to add support for concurrency in CoqPIE in the future.

Coqoon [15] is an effort to integrate Coq into Eclipse. It provides a tree view to show all files and declarations in the Coq input, similar to our tree view. Parsing is less developed than what exists in CoqPIE: Cocoon provides a simple lexer for tokens and determines the dividing point between definitions by finding periods. CoqPIE on the other hand provides full AST generation along with links between the nodes and positions in the text. There is no concept of storing both the old an new versions of goals in Coqoon and hence no framework for the style of replay assist provided by CoqPIE. Since there are no ASTs, refactoring operations such as lemma extraction are not possible in Coqoon. Finally, there is no difference highlighting since that feature is also dependent on having a full AST. Coqoon is built on top of PIDE and so it allows for asynchronous recompilation of proofs. The PIDE protocol also allows Coqoon to have cached output at each step. The CoqPIE initialization process is not needed; instead, theorem proving is a background task and annotations are collected as they become available.

There also are efforts to build Coq IDEs at MIT and UCSD [3,4]. Both are web-based. However, these tools are primarily intended for teaching.

Proviola [25] is a tool that compiles Coq source code and captures the output at each step. The tool then generates a Javascript-based web page that can display the outputs as the user hovers over each tactic in a proof. Our tool in addition to caching output also parses the output so it can be used by editing macros. CoqPIE also provides algorithms for updating the cache when the source code is edited and the Coq process is rerun.

Pcoq [10] is an earlier UI for Coq. It features a window showing the proof script, another window showing the Coq output and a third window showing a list of potential theorems that can be applied at the current step. The first two windows are similar to what exists in Proof General and Coq IDE. The third window is unique to Pcoq and would be a useful feature to add to CoqPIE. CtCoq [9,12] builds on Pcoq. It provides the same basic windows as Pcoq, and also parses Coq syntax. It is integrated directly with the CoqAst data structure. Unlike CoqPIE, this AST parsing is used to create a tree-oriented editing paradigm. UI-based point/click/drag and drop commands are used for constructing proofs in place of entering commands. In comparison, our system uses the ASTs to implement many heuristic operations such as replay assist and lemma extraction.

Company Coq [21] is an extension to Proof General that adds many useful features, including shortcut text entry, completion, and reference to Coq documentation. These features would also be useful to add to CoqPIE but they are not our primary focus. Company Coq also includes a lemma extraction feature. However, its implementation does not use an actual AST and hence is less developed.

Proof script transformations have been discussed in [18]. The method involves creating a few correctness preserving transformations. Since the transformations must be formally verified it limits the scope of what tasks can be performed. The refactoring operations in CoqPIE are heuristic in nature so correctness all falls back on Coq.

5   Conclusion

We have presented CoqPIE, a novel Coq editing framework. A key feature of CoqPIE is use of an integrated parser that links AST nodes to source text, which then allows us to create several different forms of intelligent editing functionality, including proof refactoring, showing differences between terms to help guide proof development, and maintaining dependencies so that out-of-date information is clearly highlighted. The current implementation develops a few refactoring tools, but we have only scratched the surface of what refactoring tools can be built over the CoqPIE foundation.

Acknowledgements. The authors would like to thank Gregory Malecha, Valentin Robert and Jesper Bengston for their feedback.

References

1. Coq 8.5 beta release. https://coq.inria.fr/news/123.html. Accessed 20 Mar 2015
2. Coqoon home page. https://itu.dk/research/tomeso/coqoon/. Accessed 19 Mar 2015
3. MIT proofs page. http://proofs.csail.mit.edu/. Accessed 19 Mar 2015
4. Peacoq home page. http://goto.ucsd.edu/peacoq/. Accessed 19 Mar 2015
5. Proof general. http://proofgeneral.inf.ed.ac.uk/. Accessed 20 Mar 2015
6. Alama, J., Mamane, L., Urban, J.: Dependencies in formal mathematics: applications and extraction for Coq and Mizar. In: AISC/MKM/Calculemus, pp. 1–16 (2012)

7. Ayache, N.: Combining the Coq proof assistant with first-order decision procedures (2006)
8. Barras, B., Tankink, C., Tassi, E.: Asynchronous processing of Coq documents: from the kernel up to the user interface. In: Urban, C., Zhang, X. (eds.) ITP 2015. LNCS, vol. 9236, pp. 51–66. Springer International Publishing, Switzerland (2015)
9. Bertot, J., Bertot, Y.: CtCoq: a system presentation. In: McRobbie, M.A., Slaney, J.K. (eds.) CADE 1996. LNCS, vol. 1104, pp. 231–234. Springer, Heidelberg (1996)
10. Bertot, Y.: Pcoq: a graphical user-interface for Coq. https://www-sop.inria.fr/lemme/pcoq/
11. Bertot, Y.: The CtCoq system: design and architecture. Formal Aspect Comput. 11(3), 225–243 (1999)
12. Bertot, Y., Kahn, G., Théry, L.: Proof by pointing. In: Hagiya, M., Mitchell, J.C. (eds.) TACS 1994. LNCS, vol. 789, pp. 141–160. Springer, Heidelberg (1994)
13. Boite, O.: Proof reuse with extended inductive types. In: Slind, K., Bunker, A., Gopalakrishnan, G.C. (eds.) TPHOLs 2004. LNCS, vol. 3223, pp. 50–65. Springer, Heidelberg (2004)
14. Chlipala, A., Malecha, G., Morrisett, G., Shinnar, A., Wisnesky, R.: Effective interactive proofs for higher-order imperative programs. In: 14th ICFP (2009)
15. Faithfull, A., Bengtson, J., Tassi, E., Tankink, C.: Coqoon: an IDE for interactive proof development in Coq. In: Chechik, M., Raskin, J.-F. (eds.) TACAS 2016. LNCS, vol. 9636, pp. 316–331. Springer, Heidelberg (2016). doi:10.1007/978-3-662-49674-9_18
16. Gonthier, G., Mahboubi, A., Tassi, E.: A small scale reflection extension for the Coq system. Research Report RR-6455, Inria Saclay Ile de France (2014)
17. Hasker, R.: The replay of program derivations. Ph.D. thesis, University of Illinois at Urbana-Champaign (1995)
18. Whiteside, I., Aspinall, D., Dixon, L., Grov, G.: Towards formal proof script refactoring. In: Davenport, J.H., Farmer, W.M., Urban, J., Rabe, F. (eds.) MKM 2011 and Calculemus 2011. LNCS, vol. 6824, pp. 260–275. Springer, Heidelberg (2011)
19. Malecha, G., Chlipala, A., Braibant, T.: Compositional computational reflection. In: Klein, G., Gamboa, R. (eds.) ITP 2014. LNCS, vol. 8558, pp. 374–389. Springer, Heidelberg (2014)
20. Pierce, B.C., Casinghino, C., Gaboardi, M., Greenberg, M., Hritcu, C., Sjoberg, V., Yorgey, B.: Software foundations. https://www.cis.upenn.edu/~bcpierce/sf/current/index.html
21. Pit-Claudel, C., Courtieu, P.: Company-Coq: taking proof general one step closer to a real IDE. In: Coq PL (2016)
22. Pons, O., Bertot, Y., Rideau, L.: Notions of dependency in proof assistants. In: UITP (1998)
23. Tankink, C.: PIDE for asynchronous interation with Coq. http://arxiv.org/pdf/1410.8221.pdf
24. Tankink, C.: Proof in context - web editing with rich modeless contextual feedback. In: 10th International Workshop on User Interfaces for Theorem Provers, pp. 42–56 (2012)
25. Tankink, C., Geuvers, H., McKinna, J., Wiedijk, F.: Proviola: a tool for proof re-animation. In: 9th International Conference on Mathematical Knowledge Management (2010)
26. Vijayaraghavan, M., Chlipala, A., Arvind, Dave, N.: Modular deductive verification of multiprocessor hardware designs. In: Kroening, D., Păsăreanu, C.S. (eds.) CAV 2015. LNCS, vol. 9207, pp. 109–127. Springer, Heidelberg (2015)

27. Wenzel, M.: Asynchronous user interaction and tool integration in Isabelle/PIDE. In: Klein, G., Gamboa, R. (eds.) ITP 2014. LNCS, vol. 8558, pp. 515–530. Springer, Heidelberg (2014)
28. Wenzel, M.: Isabelle/jedit (2014). http://isabelle.in.tum.de/dist/doc/jedit.pdf. Accessed 19 Mar 2015

Author Index

Printed in the United States
By Bookmasters